Computational Fluid Dynamics and Energy Modelling in Buildings

Computational Fluid Dynamics and Energy Modelling in Buildings

Fundamentals and Applications

Parham A. Mirzaei

WILEY Blackwell

Registered Office(s)
John Wiley & Sons, Inc., 111 River Street, Hoboken, NJ 07030, USA
John Wiley & Sons Ltd, The Atrium, Southern Gate, Chichester, West Sussex, PO19 8SQ, UK

Editorial Office
The Atrium, Southern Gate, Chichester, West Sussex, PO19 8SQ, UK

For details of our global editorial offices, customer services, and more information about Wiley products visit us at www.wiley.com.

Wiley also publishes its books in a variety of electronic formats and by print-on-demand. Some content that appears in standard print versions of this book may not be available in other formats.

Library of Congress Cataloging-in-Publication Data
Names: Mirzaei, Parham A., author.
Title: Computational fluid dynamics and energy modelling in buildings : fundamentals and applications / Parham A Mirzaei.
Description: Hoboken, NJ : Wiley-Blackwell, 2022. | Includes index.
Identifiers: LCCN 2022026337 (print) | LCCN 2022026338 (ebook) | ISBN 9781119743514 (paperback) | ISBN 9781119743521 (adobe pdf) | ISBN 9781119743538 (epub) | ISBN 9781119815099 (obook)
Subjects: LCSH: Buildings–Environmental engineering. | Computational fluid dynamics.
Classification: LCC TH6021 .M56 2022 (print) | LCC TH6021 (ebook) | DDC 696–dc23/eng/20220722
LC record available at https://lccn.loc.gov/2022026337
LC ebook record available at https://lccn.loc.gov/2022026338

Cover Image:
Cover Design by

Set in 9.5/12.5pt STIXTwoText by Straive, Pondicherry, India
Printed and bound by CPI Group (UK) Ltd, Croydon, CR0 4YY

C9781119743514_121022

Contents

Preface

Emerging of Building Engineering

The age of first buildings dates to the era of early human species to three million years ago. The adaption to environment forced our hunter-gatherer ancestors to eventually dwell in permanent settlements after the invention of agriculture about 12 000 years ago. The buildings since those times evolved not only to shelter them from their predators but also to provide them with thermally comfortable environments, offering habitable spaces. Various building designs and technologies have been developed around the world since prehistoric era, inherently nurtured from their local resources, inspired from their surrounding nature, and adapted to their regional climate. Even with the scarcity of resources in different time periods in diverse worldwide regions, building design technologies were consistently flourished to respond to various personal needs of their dwellers.

It worth saying that the science employed in building houses was behind of early residents' common sense to understand particular concepts such as air quality or energy efficiency subjects, which only being understood in the modern time. From the delighted scent of foods to unpleasant smell of molds were found to be managed by inhabitants' instinct though the physics behind these concepts was just being explored after classic comprehension of fluid mechanics and particulate physics. As another example, energy shortage of fossil fuels was only being perceived a few decades ago, which enforced governments to admit buildings as a major consumer and urged them to seek energy efficiency actions. Surprisingly, global warming itself is being admitted only in the recent decades, yet the denial discourse, even at this time, has voices as strong as the observed scientific facts. This struggle evidently impacted the necessary global actions in development of net-zero energy buildings as one of the countermeasure strategies against the global warming. All these examples and many more per se acknowledge the fact that buildings were around for millennium, although the accompanying science was only founded just few decades ago as a newborn discipline of building science. In other words, building science was historically undermined by scientist, authorities, and public due to the flawed assumption of dealing with simplified systems as well as their inability to scrutinize the butterfly effect, initiating from numerous buildings in a global scale, impacting vital issues such as energy crisis, disease transmissions and development, and environmental damages. The modern time, nevertheless, bluntly mirrored these misbelieves to the face of the new world.

After industrialization, modern universities have stablished engineering schools and after about two centuries many standards disciplines remained intact, including Mechanical Engineering, Electrical Engineering, Civil Engineering, Industrial Engineering, Chemical Engineering, Manufacturing Engineering, Aerospace Engineering, Metallurgical Engineering, Biomedical Engineering, etc. In a worldwide spectrum, the name of such disciplines might be slightly different, but their content and course structures barely alter from a country to another and from a university to another. This has not happened by coincidence, but on the essence of decades of knowledge being developed and thus related books being authored by scientists, scholars, and educators. For example, Mechanical Engineering in its contemporary term was offered as a separate course initially back in the nineteenth century, and from those days, tons of fluid mechanics book have been written across the world. One can now access hundreds of books that correspond to fluid mechanics and select a suitable one in accordance with an institution's demand and culture.

Nonetheless, and as it was explained earlier, recognition of buildings as a bundle of complex systems connecting many players from randomly behaving bio-occupants to multipart heating, cooling, and ventilation systems occurred only in the recent years, while the role of near future fossil fuel depletion and global warming cannot be denied on paving the road to reach such a global agreement. The result of revealing undermined science of buildings has led to many countries to launch a new engineering field just starting from few decades ago. The newborn engineering claimed, sometimes borrowed, and eventually assembled its diverse and scattered knowledge from other engineering programs and yet is recognized with various names in different regions of the world such as Architectural Engineering, Building Engineering, and Architectural Environmental Engineering. Here, for the sake of simplicity, the author refers to all of them as Building Engineering in the following paragraphs.

Necessity of Fundamental Books for Building Engineering

The author had the privilege to work and collaborate with many Building Engineering programs across the world. Through about past two decades, the author could clearly observe that the curriculum of such programs in different countries is yet to be frequently revisited and modified to become more effective and agile in response to the demands of the local and global industry and academia. This means that Building Engineering still might need years to become a fully self-sufficient engineering program and declare its independency from other engineering disciplines. Nevertheless, this is merely plausible if the related fundamentals and textbooks are not simultaneously planned and authored to outline a clear boundary for Building Engineering programs.

In this aspect and aligned with the scope of this book, the author can frankly refer to a systematic lack of fundamental books related to numerical modelling of heat and mass transfer in the realm of building science. While great books are articulated in the field, those manuscripts are mainly suitable for graduate students, explaining the applications, tools, and knowledge around the topic. Nevertheless, they can be barely referred to undergraduate students due to the fact that the background level of students in Building Engineering disciplines may significantly vary in different countries in terms of mathematical level as some have allocated more focus on the architectural aspects of buildings while others have stressed more weight on the engineering aspects. On the other hand, books from other engineering programs like Mechanical Engineering have different approaches towards numerical solutions of heat and mass transfer phenomena. For example, Mechanical Engineering students are well equipped with mathematics and thermo-fluid modules across their first two to three years of study to comprehend computational fluid dynamics (CFD) towards their final years of undergraduate or possibly graduate studies. However, Building Engineering students are more focused on other aspects of a particular system, which is a building, rather than being solemnly educated with the advanced mathematical background necessary to follow a CFD module. The result on many occasions is an agonizing struggle for Building Engineering students to follow CFD books tailored for Mechanical and Aerospace Engineering students. The same argument can be valid for heat transfer and fluid mechanics books. Even one further step backward would be lack of books for emerging, but on demand, topics such as building energy simulation. While a diverse range of commercial, in-house, and governmental funded tools are around and used in many simulation modules, a fundamental book to explain working mechanism of these tools, specifically written for Building Engineering undergraduate students can be barely addressed. Despite the fact that there is a lack of such textbooks about numerical approaches in Building Engineering, it would be fair to highlight that some limited branches of heat and mass transfer approaches in Building Engineering have generated suitable textbooks such as design of HVAC systems in buildings.

Structure of This Book

The aim of 'Computational Fluid Dynamics and Energy Modelling in Buildings – Fundamentals and Applications' is to respond to the identified and explained lack in the latter section and thus to provide the fundamental knowledge of common numerical methods in understanding heat and mass transfer in buildings. While the book comprehensively elaborates on the dynamic energy simulation in building and the way airflow is simplified in the building energy simulation models, it also offers background knowledge of an alternative detailed and advanced technique of finite volume method to model heat and mass flow in buildings.

The key aspect of 'Computational Fluid Dynamics and Energy Modelling in Buildings – Fundamentals and Applications' is that it is tailored for audiences without extensive past experiences on numerical methods achieved from many years of the author's experience in delivering related courses to a diverse range of graduate and undergraduate students with various backgrounds. Hence, undergraduate or graduate students in Building Engineering, Architecture, Urban Planning, Geography, Architectural Engineering, and other engineering fields, along with building performance and simulation professionals, can read this book to gain additional clarity on the topics of building energy simulation and CFD. This book comprises three seasons each containing different chapters.

In the first season of 'Computational Fluid Dynamics and Energy Modelling in Buildings – Fundamentals and Applications', which has seven chapters, the author intends to review the fundamentals of fluid mechanics, thermodynamics, and heat transfer in three separate chapters with a specific focus on the related knowledge to building physics. Examples are Bernoulli equation in fluid mechanics in Chapter 2, first laws of thermodynamics in Chapter 4, and convection in internal flows in Chapter 6. Hence, the readers with even a minimum level of familiarity with thermo-fluid subjects can learn the very essential topics while being directed to the associated resources if they are seeking for further explanations over those topics. This background knowledge sets the scene for readers to discover the extent in which these fundamentals are applied in buildings with the explanations of commonly employed simplifications and assumptions in three following chapters (Chapters 3, 5, and 7). Again, some examples are the application of Bernoulli equation in airflow modelling within buildings that is related to fluid mechanics, conventional processes in air-conditioning systems of buildings that is associated with thermodynamics, and heat loss in piping system of buildings that corresponds to heat transfer chapters.

In the second season, which has two chapters (Chapters 8 and 9), the author elaborates on the implementation of the fundamentals explained in the first season to model energy flow in buildings. This season, therefore, explains the basis of all the commercial and educational building energy simulation tools. In this sense, an innovative, illustrative nodal network concept is introduced in this book to help readers to easily comprehend the basics of conservation laws in buildings. The application of numerical techniques to form dynamic simulation tools is further presented in this season. In general, these understandings help readers to identify and justify their choices when working with building energy simulation tools rather than being a default user.

Detailed airflow information in buildings cannot be obtained in the building energy simulation techniques. Therefore, the third season, which has three chapters (Chapters 10, 11, and 12), is focused to introduce CFD as a detailed and powerful airflow modelling technique in buildings. While the related challenges and considerations are discussed, this season starts with an overview of the fundamentals of the finite volume method to solve the governing equation of fluid introduced in the first season. The last chapter of this season (Chapter 12) is specifically allocated to solve various practical problems of airflow within and around buildings using a commercial, but popular, tool.

How to Read This Book

The author suggests different ways to read this book. For those readers with a related background in thermo-fluids from disciplines such as Mechanical Engineering, they might consider to directly start the book from the second season while they can use the first season as a quick review text on the fluid mechanics, thermodynamics, and heat transfer. Inversely, for those readers with less background in these subject areas, for example from Architectural programs, the author recommends that they carefully read the book from the beginning to equip themselves with the necessary fundamental knowledge that is used in the latter seasons.

Furthermore, undergraduate students, building engineers, architects, etc., keen on learning building energy simulation but might not be interested in learning advanced numerical modelling of fluid flows, can comprehensively study the dynamic building energy simulation by only reading the second season.

The season three is also designed to offer fundamentals of CFD and necessary procedures for pre-processing, solving, and post-processing of CFD problems particularly in buildings. This season is suitable for undergraduate and graduate students in addition to engineers who are interested in learning CFD from a less mathematically oriented textbook. This does not imply that all the necessary knowledge and elements in a CFD simulation are not covered and addressed in this season. Inversely, this season offers a practical and step-by-step procedure to first form a CFD problem and then to solve and analyse it. This season not only is a suitable reference book for Building Engineering disciplines, but it can be a robust textbook for other disciplines such as Mechanical Engineering courses.

Acknowledgement and Dedication

I am truly indebted to my former supervisors, mentors, colleagues, and friends in the field of building science who helped me to learn and improve in the past two decades.

I would like to specifically express my gratitude to my former and current PhD and MSc students and researchers who helped me with the solution of some of the examples in various chapters, including Ruijun Zhang, Hamid Motamedi, Mohammadreza Shirzadi, Marzieh Fallahi, and Yangyu Gan.

I would like to dedicate this book to my kindhearted wife Dr Sanam Akhavannasab who supported me with her seamless love and encouragement to complete this book, to my beautiful daughter Sophie who fueled my life with joy and energy needed to carry out such an interminable project, and to my late farther Prof. Hassan A. Mirzaei and my mother Mansoureh Vilataj who were my first inspiration to pursue science and to achieve my academic dreams.

March 2022   Dr Parham A. Mirzaei

1

An Overview of Heat and Mass Transport in Buildings

1.1 Introduction

Heat and mass transports are essential contributors of all phenomena on the planet earth and are key parts of many implications from large-scale events such as continental hurricanes to small-scale ones such as diffusion in cells' membranes known as osmosis. Life with any means from unicellular organism to complex mammals from its formation and then further to its evolution is highly indebted to the fundamental laws of transports in water. Humankind could master the planet only by controlling the heat and mass transport from learning to set campfires to properly utilize water for farming and agriculture. In the modern era, the industrial revolution happened when human race could tame the power of steam and formulate how to effectively transfer heat to liquid water and to extract energy from its vapour. There are countless evidences apart from these few examples, indicating that the transport phenomena in almost any organic or non-organic system have to be explicated if that system is to be well-understood.

Buildings are complex systems, being primarily our shelter throughout history to protect our essential body functionality against change in weather conditions while they were sanctuary places to our other societal demands such as security against predators, reproduction of next generations, storing belongings, and food, etc. Hence, like any other systems, buildings are edifices highly interlinked with heat and mass transport as it can be seen in Figure 1.1. In other words, from a building physics standpoint, heat and mass transports are directly associated with the level of energy demand, comfort, and air quality in buildings. Fundamentals of heat and mass transport in buildings, in addition to their governing equations, will be explained in Part 1 (Chapters 2–7). Here, Figure 1.1 briefly addresses three main means of heat transfer mechanisms, including radiation, convection, and conduction, to be actively present in the building physics. As we elaborate in more details, each mechanism plays its role to transport heat and mass between buildings and their surrounding environments.

As it can be seen in Figure 1.1, buildings exchange heat with the environment via convection mechanism through their skin and surfaces, and via mass transport of moist air throughout buildings known as natural ventilation by means of wind-driven and buoyancy-driven flows. The latter transport phenomenon also is responsible for inducing fresh air into the buildings. Moisture transports in our spaces are usually impacted by various

Computational Fluid Dynamics and Energy Modelling in Buildings: Fundamentals and Applications, First Edition. Parham A. Mirzaei.
© 2023 John Wiley & Sons Ltd. Published 2023 by John Wiley & Sons Ltd.

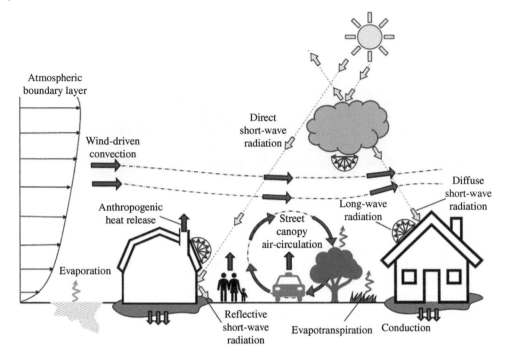

Figure 1.1 Heat and mass transport around and within buildings.

means of evaporation, evapotranspiration, and water exchanges in addition to inhalation and exhalation from human respiratory events. As it will be comprehensively discussed in the following chapters, human comfort is defined in direct relation with the room temperature and the level of moisture in addition to the local air velocity. In general, buildings and infrastructure in built environment alter the wind-driven or buoyancy-driven convective exchanges compared to the scenario where the land is inhabited with wildlife landscape.

The same narrative is valid for the complex radiative exchanges, including short- and long-wave radiation, amongst all artificial surfaces of a built environment in comparison with initial plain green landscapes either as grasslands, forests, or jungles. Surface materials and buildings geometry, hence, are essential factors in alteration of this energy contributor as they will be discussed in the proceeding chapters. Eventually, anthropogenic heat release from inhabitants, vehicles, infrastructures, and more importantly numerous buildings is another key player in alteration of energy and mass balance of a built environment that cannot and shall not be disregarded.

1.2 Heat and Mass Transport in Traditional Buildings

Before recent awareness related to depletion of natural resources and fossil fuels in addition to their health-related issues and the negative footprints on the global warming, our insight from heat and mass transport in traditional buildings was limited to ensuring comfortable spaces for living. In other words, concepts such as energy efficiency and air quality were barely taken into account until the recent years. Therefore, in this section, we focus on some of the application of heat and mass transport in traditional buildings and in the next section, we introduce some of the new and recent considerations related to the transport phenomena.

We lived in buildings for thousands of years, and we used our common sense to transform them into comfortable spaces in many strands. Thermal comfort, as we scrutinize in more details in Chapter 5, is essential for our bodies' psychological activities although a part of it is related to psychological aspects, which enabled us to further adapt to various range of climates from harsh arctic and equatorial to moderate ones. Yet, over centuries inhabitants spread in different landscapes discovered techniques and technologies to adapt their spaces more towards their comfort zones.

Examples are many, ageing back to cave dwelling of humankind. Not necessary to go very far beyond, we can bring a famous example of **hypocaust** being used in Ancient Greek and Roman empire [1]. An old portrait and ruins of a hypocaust can be seen in Figure 1.2. This system served as a central heating system, which circulated warmed air produced in a firehouse toward the conditioned rooms, halls, walls, or baths through cavities built under the floors or walls. While such systems were transformed to more contemporary formats during centuries, the central heating was again back to the life in the past decades as a viable technology in energy saving of buildings. The modern equivalent of the floor pathways system in distributing of heat is also known as the underfloor distribution system, a widely adapted technique in the recent energy efficient buildings.

Despite invention of many traditional heating systems, people dwelling in hot climates similarly benefited from different technologies to better adapt to their built environment. Badgir or windcatcher dating back to Achaemenid dynasty of Persian empire is a famous evidence of passive ventilation systems, benefiting from wind in hot and mainly dry climates (see Figure 1.3). As we explore about the pressure difference in Chapter 3 and evaporative cooling in Chapter 5, as the essence of windcatchers, cold water extracted from wells was integrated on some occasions with windcatchers to supply fresh and cold air for spaces in many regions. Windcatchers as attractive architectural elements are studied and integrated in many modern buildings across the world, admitting the fact that not hundreds but thousands of years of humankind experience in utilization of passive and natural ventilation strategies should not and shall not be neglected.

Windows are also crucial elements in natural ventilation as it will be explored in Chapters 2 and 3 when Bernoulli equation and discharge coefficients are introduced. These concepts without being formulated in the modern formats are being applied to the various designs of windows in different climates. An attractive example can be seen in Figure 1.4 where air circulation occurs in the building throughout numerous windows with the cross-ventilation and side-ventilation strategies as the fundamentals behind will be explored in the following chapters.

Figure 1.5 demonstrates another example of an architectural element called Cumba in Ottoman dynasty [2]. Aside from aesthetic features of such elements, they were adapted to promote more fresh air circulation in such spaces, normally used as gathering and socializing halls. As we discuss this in Chapter 3, while this was understood by experience

(a)

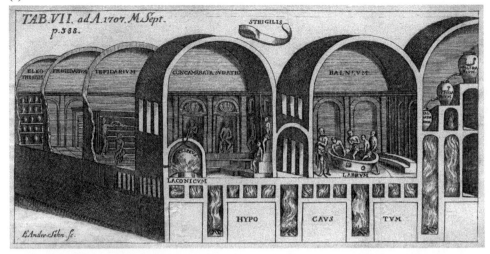

(b)

Figure 1.2 (a) A hypocaust portrait. (b) An underfloor heating system of a hypocaust in a Roman bath. *Source:* (a) Mary Evans Library/Adobe Stock. (b) RnDmS/Adobe Stock.

through centuries, such alterations in the aerodynamics of buildings can potentially modify pressure distribution around them and thus cause stronger air movement within them. Again, a recent trend in inclusion of natural ventilation in modern buildings is to benefit from aerodynamic feature of their architectural design.

Figure 1.3 Windcatchers (Badgir) used as a natural cooling system in Iranian traditional architecture, City of Yazd. *Source:* Dario Bajurin/Adobe Stock.

Figure 1.4 The Palace of Winds built in 1799, Rajasthan, India. *Source:* grafixme/Adobe Stock.

For thousands of years, humankind learned how to aesthetically adorn dwellings with plants, floras, trees, and vegetations by attaching specialized spaces such as courtyard, patios, and gardens to residences. For example, Figure 1.6 demonstrates a 200-year-old courtyard as a very common gathering area in the traditional housing of central Iran. Furthermore,

Figure 1.5 A traditional Ottoman-era Anatolian with architectural element of Cumba, Sivrihisar, Turkey. *Source:* daphnusia/Adobe Stock.

Figure 1.6 A 200-year-old traditional courtyard house, Isfahan, Iran. *Source:* efesenko/Adobe Stock.

Figure 1.7 exhibits another traditional architectural element in the Roman era, known as peristyle, which is a porch surrounding the perimeter of courtyard by row of columns.

The freshness and positive feeling of being exposed to such vibrant areas was appreciated and frequently referenced in history books, literature pieces, and artworks of old

Figure 1.7 Ancient Roman Peristylium in the House of Golden Cupids, Pompeii, Italy. *Source:* BlackMac/Adobe Stock.

and recent civilizations. Nonetheless, it was just in the recent times that the impact of vegetation on the respiratory health of dwellers was known as modern concepts of air quality. Heat and mass transfer again are the key elements in the diffusion, advection, and dilution of contaminants such as CO_2 or other artificially released small particles (such as PM2.5 and PM10) and volatile organic compounds (VOC). Without understanding the physics behind moisture transport or heat exchanges between an occupant and its surrounding environment as discussed through Chapters 2–7, it is almost impossible to evaluate air quality of a space. The recent solutions also suggest greenery elements such as green roofs and green facades as an energy saving strategy, amongst many others. Again, looking back through our achievements since antiquity can offer easy and available solutions to control moisture and excess heat in our spaces without being tremendously dependent on fossil fuels to condition them.

In the case on dwellings with capacity of more occupants such as apartments and high-rise buildings, again we can find integrated solutions by investigating our past history. Traditional Hakka Tulou homes as seen in Figure 1.8 are examples of dwellings with multiple occupants, protecting them from natural disasters and enemy threats, yet capable of providing comfortable spaces using local materials. While sustainability elements can be evidently seen in such structures, these shelters are well designed to enhance natural ventilation and to provide effective shading for enhancing the thermal comfort. Again, the future chapters will help us to comprehend heat and mass transport mechanism in any dwelling similar to the traditional and modern building examples. As we explain in the proceeding sections, understanding of underlying physics by defining suitable assumptions and simplifications is the essential first step to model a building.

(a)

(b)

Figure 1.8 (a) Traditional Hakka Tulou homes with (b) their typical interior spaces, Yongding, China. *Source:* Top: san724/Adobe Stock, Bottom: xiaoliangge/Adobe Stock.

1.3 Heat and Mass Transports in Modern Buildings

Mechanical ventilation systems such as combi-boilers, chillers, evaporative cooling systems, etc. still are dominant systems conditioning our spaces, around the world. These systems are working based on heat and mass transfer, which can be explored from both thermodynamical aspects as explained in Chapters 4 and 5 or a heat transfer point of view as described in Chapters 6 and 7. The role of fluid dynamics in the movement of working fluid is also further examined in Chapters 2 and 3. If we expand our control volume of the investigated system to a room of a building, then we can briefly explore how heat and mass are key elements of their thermal system conditions. This later will enable us to translate the underlying physics of transport to the modelling concept, which itself can be also simulated with numerical approach. Example of such room is shown in Figure 1.9 where some of the common cooling strategies are demonstrated to condition this room. In all these systems, heat is removed from the room and occupants by the convection through movement of air with different techniques. To justify the correct choice of cooling and ventilation system for this room, aside from experimental works that are expensive and unrepeatable for some scenarios, one can use modelling technique.

Moreover, in the following section, several technologies are introduced to achieve the goal of energy efficiency in buildings. Concepts such as net-zero energy buildings

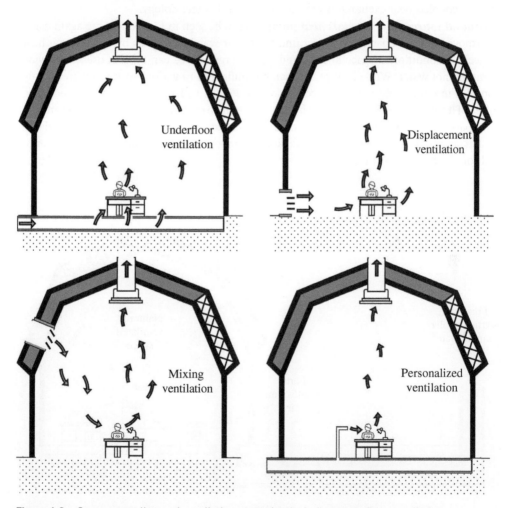

Figure 1.9 Common cooling and ventilation strategies, including underfloor ventilation, displacement ventilation, mixing ventilation, and personalized ventilation.

(NZEB), green buildings, sustainable buildings, etc. are developed based on integration of multiple renewable energy sources into the buildings. Nevertheless, all these technologies and concepts are relied on the heat and mass transport between them and building envelopes. Here, we introduce some of these technologies and techniques.

Ground soil has a tremendous thermal mass or heat capacity with almost a constant temperature during a season below a certain height from its surface. This thermal characteristic sets soil as a thermal storage sink or source and thus many technologies are introduced to benefit from the soil thermal capacity. As shown in Figure 1.10, **ground-coupled heat exchangers** or **earth-to-air heat exchangers** are a natural ventilation technique, composed of a ducting system, which brings the fresh warm air and transfers heat to the ground, and thus become colder during summertime. The inverse is also possible in the wintertime, when the fresh and colder air can extract heat from the soil and become warmer for the space heating purposes. Thus, the convective heat transfer between the moist air and ground-coupled heat exchangers are the essence of this technology.

Ground source (geothermal) heat pumps as can be seen in Figure 1.11 are again using enormous storage capacity of the ground to cool or warm buildings by transferring heat between a conditioned space and ground. The energy source can be also used to provide domestic hot water. While heat pumps can be configured in various forms and are mainly complex in design and maintenance, the one in Figure 1.11 represents a simplified schematic of it that utilizes a primary and a secondary loop to exchange heat between the ground and space. Again, convective heat transfers between exchangers and air/soil are essential to understand their performances. One can add the importance of conductive heat transfer in soil for geothermal heat pumps.

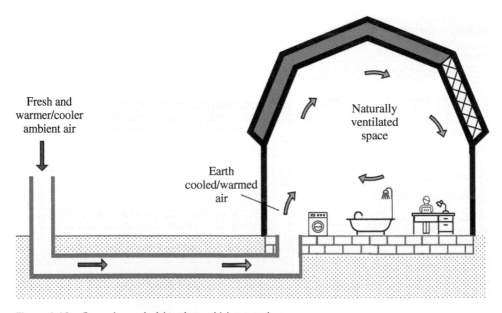

Figure 1.10 Ground-coupled (earth-to-air) heat exchanger.

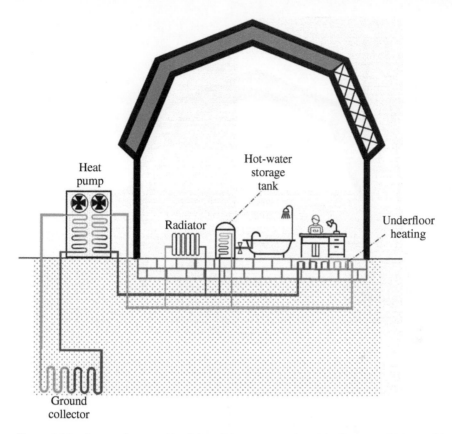

Figure 1.11 A ground source (geothermal) heat pump schematic diagram, which provides space heating/cooling and domestic hot water.

There are a series of technologies, which benefit from abundant solar energy impacting the earth surface on a daily basis, to facilitate space heating in buildings. Figure 1.12 demonstrates a **solar chimney** (**thermal chimney**) that is designed to warm up air within its vertical shaft and thus impose a buoyant force on air to form an updraft, which then promotes a natural flow to suck fresher and cooler air towards the buildings. The convection and short-wave radiation are essential factors in solar chimney mechanism. As illustrated in Figure 1.13, in another similar passive design, known as **Trombe wall** (**mass wall**) [3], the solar energy absorbed using a dark colour painted wall with a high thermal mass is used to provide natural ventilation and space heating. Trombe walls are normally faced towards the equator and are covered with high-insulated glazing systems. The heated air by the thermal wall is circulated in the space. While convection and short-wave radiation are the heat transfer mechanisms in this passive strategy, long-wave radiation similarly plays a pivotal role in the exchange of energy amongst heated wall and the space as explained in Chapters 6 and 7.

If the wind is abundant in a region, it can be directly driven to a building to provide fresh air. While air movement can enhance thermal comfort as will be discussed in the future

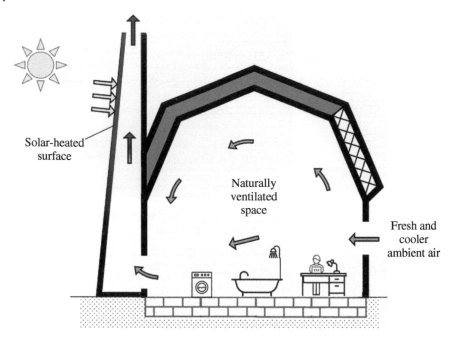

Figure 1.12 Solar chimney.

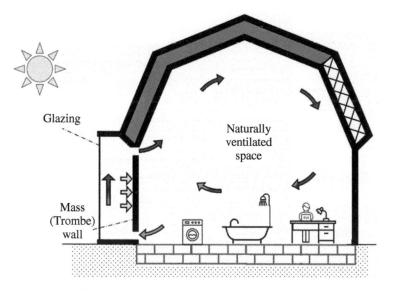

Figure 1.13 Trombe wall.

chapters, it is feasible to benefit from directing the air over a cold water or through an underground duct to also reduce its temperature. As it was stated in the previous section, the pressure difference around buildings helps windcatchers (see also Figure 1.14) to flush the fresh air throughout a building [4]. If this flushing strategy only occurs due to a pair of windows

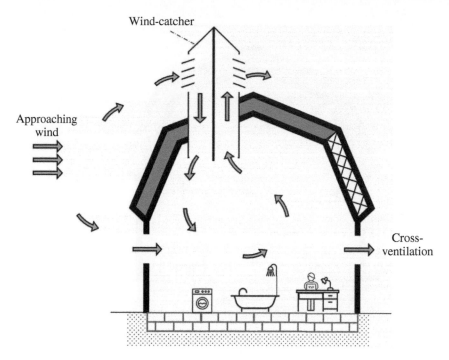

Figure 1.14 Windcatchers and cross-ventilation in a building.

positioned in windward and leeward positions, the caused natural ventilation is called **cross-ventilation**. The role of moist air advection or mass transfer is the essential element in the described passive ventilation systems.

Photovoltaic elements are broadly integrated to buildings as roof- and façade-mounted modules. They can effectively transform solar energy to electricity, especially in the altitudes with a higher solar gain. Nonetheless, the heat and mass transport become essential around these units when these systems are combined with air thermal collectors to form building integrated photovoltaics/thermal air collector as depicted in Figure 1.15. The heated dark-coloured photovoltaics warm up the air beneath the cavities of these modules and then a fan pushes the air throughout the space for the heating purposes. In such systems, short-wave radiative and convective fluxes can be assumed as the main means of heat transfer.

Thermal collectors have a broad range and format from air to water operating ones and can be integrated with various other technologies such as BIPV/Ts [5]. Note that we are only demonstrating a plain diagram of them in this section. For example, a hot water thermal collector is shown in Figure 1.16, which operates with water to absorb solar radiation and then to transfer it to a storage tank to be used as domestic hot water. These systems are popular in regions with a high level of solar gain. From a heat and mass point of view, the convection and conduction in the operating fluid (here water) in addition to the solar radiation as the energy source are the main contributors in these systems.

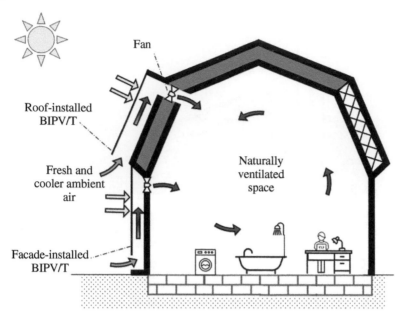

Figure 1.15 Building integrated photovoltaics/thermal air collector (BIPV/T).

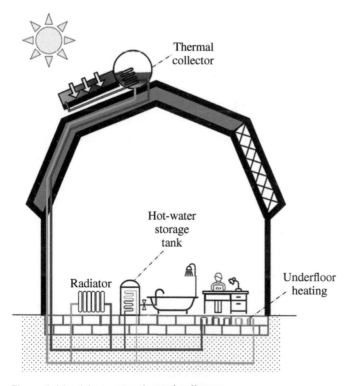

Figure 1.16 A hot-water thermal collector.

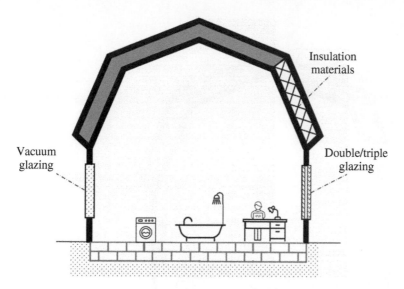

Figure 1.17 Insulation materials and insulating glasses.

Figure 1.17 presents utilization of advanced materials such as insulations and windows such as double/triple glazing systems to minimize heat loss through the skin of buildings. As we explain thermal gaps in Chapters 6 and 7, minimizing conduction through buildings' surface materials can considerably result in energy saving. While conduction through solid materials and elements such as windows is the essential mechanism in the heat transfer of a building and its surrounding environment, the role of convective exchanges between solid surfaces and moist air around them is equally of paramount of importance.

Figure 1.18 demonstrates the usage of green surface technologies in energy efficiency in buildings. As it will be explained in Chapter 6, a significant part of solar radiation can be absorbed on the surfaces. The utilization of green roofs and green facades [6], thus, can reduce this absorption, specifically in hot climates. Figure 1.18 also depicts a ***phase change material*** (PCM) [7] unit that is embedded as a wallboard into the conditioned space. PCMs are thermal storage units that are recently integrated to materials such as mortar or heating, ventilation, and air conditioning (HVAC) elements – such as storage tanks – to reduce the peak temperature and to increase minimum temperature of conditioned spaces. PCMs are both liquid and solid at a same time that impose both conduction and convection to be equally important factors in their functionalities.

1.4 Modelling of Heat and Mass Transport in Buildings

Modelling is redefinition, representation, or abstraction of the reality with some level of simplifications. In the building physics, modelling is defined as the physical representation or mathematical description of a system or phenomenon. Both approaches are helpful to realize how a building is operating to further be designed/redesigned, constructed/

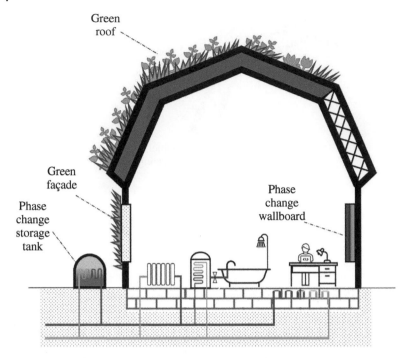

Figure 1.18 Thermal storage and greening strategies.

retrofitted, or controlled. In terms of modelling of heat and mass transport in buildings, same goals are expected in this aspect in building envelopes. Nonetheless, the purpose of design/redesign, construction/retrofit, or control is related to the objective of the modelling.

1.4.1 Modelling Objectives

In general, modelling objectives encompass energy efficiency, cost saving, air quality enhancement, environmental preservation, thermal comfort improvement, heat-related morbidities reduction, or a combination of the latter objectives (see Figure 1.19).

The modelling purpose in terms of assumptions on underlying physics and the level of simplifications in the mathematical expressions is therefore aligned with the objectives, which can be defined from various perspectives in different communities, including building engineers, architects, urban climatologist, meteorologists, and geographers. For example, when improvement of energy efficiency of a building is defined as the main objective of the modelling by a building engineer, then a high level of details in the heat transport between the building surfaces and the surrounding environment is a pivotal parameter to be considered in the calculations. On the contrary, the details of airflow regime formed in the building's room are not within the interest of the modelling. Nonetheless, in the design of a clean room for a high-tech industry, the primary objective is to provide a high level of air quality, and thus the level of details on the airflow movement within the room is crucial in the modelling process. In this case, however, the energy demand of the building is

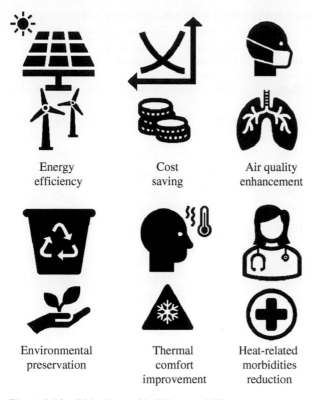

Energy
efficiency

Cost
saving

Air quality
enhancement

Environmental
preservation

Thermal
comfort
improvement

Heat-related
morbidities
reduction

Figure 1.19 Objectives of building modelling.

an unimportant issue. Therefore, the objective of an investigation always reveals the direction of the transport modelling in a specific subject in terms of the level of the details and accuracy required. For example, assume all the passive and active technologies in the previous section. While we expect all means of heat transfer mechanisms to contribute to all these technologies, modelling direction enforces us to always apply assumptions to simplify our models. Table 1.1 presents the importance of heat and mass transport elements in each introduced technology.

1.5 Modelling Approaches

Once the importance of each contributing parameter is recognized, the modelling procedure can be designed. In general, there are two main approaches of observational and mathematical modelling.

1.5.1 Observational Methods

Observational (experimental) methods are about construction of a physical or mocked prototype of a building system, or an entire building envelope. Then, a single or multiple

Table 1.1 Importance of heat and mass transport elements in some building integrated technologies.

Technology	Convection	Conduction	Long-wave radiation	Short-wave radiation	Moisture transport
Ground-coupled heat exchanger	High	High	Low	Low	Medium
Geothermal heat pump	High	High	Low	Low	Low
Solar chimney	High	High	Low	High	Low
Trombe wall	High	High	High	High	Low
Windcatcher	High	Low	Low	Low	High
BIPV/T	High	High	Low	High	Low
Water thermal collector	High	High	Low	High	Low
Insulation materials	Medium	High	Low	Low	Low
Double/triple glazing	Medium	High	Low	Medium	Low
Green-roof and green-façade	Medium	High	High	High	High
Phase change materials	High	High	Low	Low	Low

sensors are employed to monitor the required parameters (e.g., temperature, humidity, solar radiative flux, and convective flux). Figure 1.20 presents some of the common sensors used in this aspect. Investigation of the observational methods are not within the focus of this book though we can, in general, address these approaches to be accurate in the outcome. However, they are mainly expensive and time-consuming in the setup. Moreover, limited information only can be extracted from the observational methods, depending on the accuracy and practicality of a utilized sensor. Hence, we focus on mathematical models in the next section.

1.5.2 Mathematical Methods

As it will be explained in the following chapters of this book, extensive efforts have been conducted to provide mathematical expressions for heat and mass transport phenomena. These expressions are mainly in the form of partial differential equations such as Euler equations or Navier–Stokes equations, representing the fluid behaviour. Nonetheless, simplified models are also widely used to present transport phenomena in buildings. The famous example is Bernoulli equation which will be discussed in Chapter 2. Once mathematical models are presented, we have analytical and numerical approaches to solve these expressions. Analytical methods are bounded to only specific problems as the equations become complex in most of the real-word scenarios. This is further discussed in Part 2 and Part 3 of this book (Chapters 8–12) when numerical methods or simulation methods

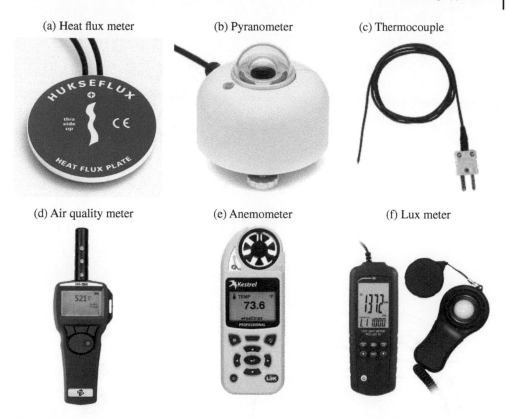

Figure 1.20 Various sensors utilized in building modelling, including (a) heat flux meter, (b) pyranometer (c) thermocouple, (d) air quality meter, (e) anemometer, and (f) lux meter. *Source:* (a, b) Hukseflux, (c) MADGTECH, (d) Testmeter, (e) r-p-r LTD, and (f) PCE.

are introduced as flexible and powerful options to provide solutions for almost all the defined problems. We scrutinize some of these numerical techniques in detail in the future chapters. Here, we only explain the necessity to understand simplification of models associated with the spatial and temporal scales of a problem due to the plausible extensive computational costs of a simulation.

1.5.2.1 Spatial Scale of Modelling

The objective of the modelling of transport phenomena in the building science is closely a parameter of the scale of interest. As shown in Figure 1.21, and discussed earlier, the physics around a building includes a combination of complex and diverse phenomena, interacting in different scales from human body to city size.

The objectives, hence, find meaning in a specified scale. For example, presented passive and active technologies in the previous section are mainly related to a building-scale spectrum. For example, the interaction between multiple buildings can be of interest in urban or neighbourhood scale (e.g., urban heat island study [8, 9]), and thus requires certain attention to model geostrophic winds and breezes coming from seas or mountains. In such scale, however, a transport related to

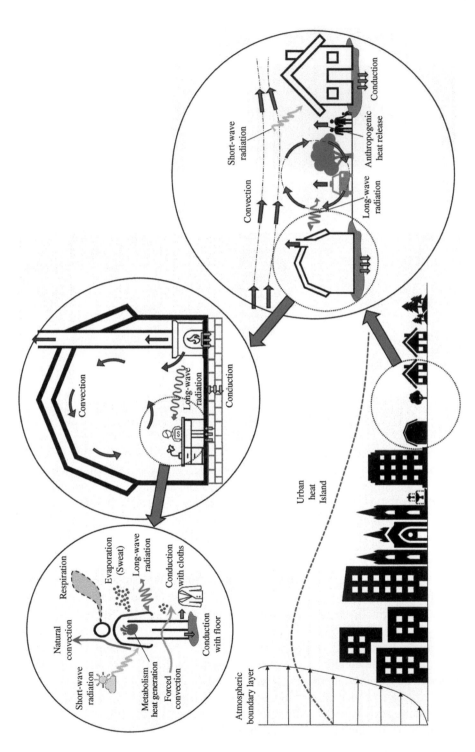

Figure 1.21 Various scales in building and urban sciences.

the heat exchange between human bodies and the environment can be neglected to immensely simplify the problem. Inversely, if a human-scale model for comfort studies is the focus of a research, then it is most likely to exclude the exact effect of the geostrophic winds and breezes on the human body, especially when the occupant is located in an enclosed building. Therefore, it is very crucial as a first step to simplify physics and scale of the investigated subject. This implies that a suitable modelling scale can be only identified in compliance with the defined objective in order to minimize the complexities and computational costs of the study. Note that the simplification is imperative to reflect upon models and to manage experimental and computational costs. In general, the diverse transport phenomena in a built environment are embedded as sub-models into the modelling. We can categorize these sub-models to be effective in four scales as presented in Table 1.2, including human scale, building scale, microclimate scale (neighbourhood scale), and urban scale.

As shown in Figure 1.21, human-scale models are mainly dealing with the energy balance over a control volume defined as a human body. Therefore, clothing, respiration, body metabolism, etc. are crucial factors in which the role of transport phenomena is important, occurring as all the means of heat transfer mechanism.

Models in the building-scale group are mainly limited to an isolated building envelope where the influence of neighbouring buildings on the transport equation is neglected. Although outdoor parameters such as temperature, solar radiation, long-wave radiation, and moisture are external inputs into such models, these models are simplistic in representing the mutual impact of a building with its surrounding environment. As we explore this in Chapters 8 and 9, these models are mainly following building energy simulation concept if a numerical technique is selected for the modelling.

The transport phenomena in the microclimate-scale studies include the interaction of a building with its surrounding environment (see Figure 1.21). In principle, solar radiation, soil conduction, and surface convection from the buildings' surfaces are highly influential

Table 1.2 Importance of sub-models effective in transport phenomena in different scales of studies.

Scale	Size	Long-wave	Wind	Cloud	Solar	Tree and vegetation	Soil	Pond	Human metabolism
		Sub-models effective in transport phenomena							
Human (1 person)	1 m–4 m	M	M	L	M	L	L	L	H
Building (1 room–1 building)	1 m–10 m	M	L	M	L	L	L	L	M
Microclimate (1 building–neighbourhood)	10 m–100 m	M	H	M	M	M	M	M	L
Urban (neighbourhood–City)	100 m–Few km	H	H	H	H	H	H	H	L

L, low; M, medium; H, high.

in such models and the airflow patterns around and within buildings should be well resolved via experimental approaches or solving transport equations using computational fluid dynamics (CFD) or urban canopy models (UCMs). As CFD technique will be explained in Chapters 10–12, the weakness of the CFD models is mainly associated with their limited domain size (few hundred meters) due to the extensive computational cost while UCMs are weak in the detailed presentation of airflow around the buildings for example in thermal comfort associated studies.

Investigation of the urban-scale subjects are broadly adopted in urban climatology and meteorology fields. The impact of urban-scale policies to mitigate the urban heat island, urban ventilation, pollution dispersion management, and greening depend on solving the transport phenomena in this scale. The developed models are based on the governing equations of fluid dynamics whilst equally important models such as radiation, cloud cover, and soil are integrated into the calculations. As the major limitation of these models, only very coarse cells (control volumes) can be employed, implying a weak resolution on the surface layer to observe interactions between buildings and their environment.

1.5.2.2 Temporal Scale of Modelling

Similar to the spatial scale, the temporal scale can define the objective, detail level, and study approach of a study. For example, if a subject is related to the natural, mechanical, or mixed ventilation, then such mechanisms are assumed to perform for a long period of time and in an almost similar efficiency. Therefore, a steady-state modelling of the airflow field would be a suitable strategy. Such strategy is also practically valid for the thermal comfort studies. On the other hand, let us assume that a short-term smoke dispersion generated by cooking facilities within a room is likely to be modelled. Then, obviously, a transient model is necessary to capture the temporal snapshots of the smoke dispersion in the room. The dynamic modelling of the annual/seasonal building energy simulation with suitable time steps is another practical example. There are also scenarios in which a semi-transient strategy is the best option to model the transport behaviour. This strategy applies to the conditions when sequential snapshots are selected for steady-state modelling of the transport equation. A well-known example is coupling of steady-state CFD with dynamic building energy simulation. Temporal strategy to model transport phenomena in buildings should be therefore chosen based on the defined objective, available time, and computational resources.

References

1 Forbes, R.J. (1965). *Studies in Ancient Technology*, vol. 22. Leiden: E.J. Brill.

2 Eldem, S.H. (1986). *Turkish houses, ottoman period: Türkiye Anıt*. Turizm Değerlerini Koruma Vakfı: Çevre.

3 Hu, Z., He, W., Ji, J., and Zhang, S. (2017). A review on the application of Trombe wall system in buildings. *Renewable and Sustainable Energy Reviews* 70: 976–987.

4 Saadatian, O., Haw, L.C., Sopian, K., and Sulaiman, M.Y. (2012). Review of windcatcher technologies. *Renewable and Sustainable Energy Reviews* 16 (3): 1477–1495.

5 Norton, B. (2014). Harnessing solar heat.

6 Sutton, R.K. (2015). Green roof ecosystems.

7 Fleischer, A.S. (2015). *Thermal Energy Storage Using Phase Change Materials: Fundamentals and Applications*. Springer International Publishing.

8 Mirzaei, P.A. (2015). Recent challenges in modeling of urban heat island. *Sustainable Cities and Society*. 19: 200–206.

9 Mirzaei, P.A. and Haghighat, F. (2010). Approaches to study urban heat island–abilities and limitations. *Building and Environment*. 45 (10): 2192–2201.

2

An Overview on Fundamentals of Fluid Mechanics in Buildings

2 An Overview of Fluid

2.1 Definition of Fluid

From a molecular perspective, materials with a dense molecular structure and large inter-molecular cohesive forces are known to exist in the solid state while in conditions when the distance between structures of the molecule is higher and therefore the intermolecular forces are relatively weaker, the material is known to be in the liquid state (see Figure 2.1). The weaker internal forces grant liquid molecules more freedom to take the shape of their containers, although to a certain extent, and to not be easily compressed. Once the distance is further increased, mainly due to the external conditions, and molecules found more freedom to move under negligible intermolecular forces, then the material is in its gaseous state. Consequently, gases can be deformed and take the shape of their containers similar to liquids. In general, both liquids and gases are known as fluids and their behaviour and characteristics will be further investigated in this chapter.

2.1.1 System of Units

In the SI system, the **primary quantities** and their units are defined as length (L), mass (M), time (T), and temperature (Θ). These primary units can be used to define **secondary quantities**. Table 2.1 shows a range of common secondary quantities widely used in engineering problems.

Example 2.1
Find the dimension of the secondary quantity of 'pressure'.

Solution

Pressure is defined as the perpendicular force over the unit of area:

$$p = \frac{\text{Force}}{\text{Area}} = \frac{F}{A} = \frac{ma}{A}$$

$$\Rightarrow p = \frac{\text{MLT}^{-2}}{L^2} = \text{ML}^{-1}\text{T}^{-2}$$

Computational Fluid Dynamics and Energy Modelling in Buildings: Fundamentals and Applications,
First Edition. Parham A. Mirzaei.
© 2023 John Wiley & Sons Ltd. Published 2023 by John Wiley & Sons Ltd.

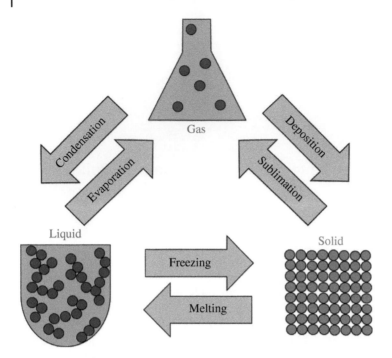

Figure 2.1 Different states of materials.

2.2 Properties of Fluid

2.2.1 Density

Mass per unit of volume is called density and is shown with the Greek symbol of ρ 'rho'. As it can be seen in Table 2.1, the unit of density is kg/m^3 or ML^{-3} in the SI system. In general, the density of liquid has a minimal variation in different pressures and temperatures. The example is water density, which is about 998 kg/m^3 in 293 K that can barely change during the seasonal temperature variations in relation with many problems related to the built environment. Nonetheless, the irregular change of water density in sub-zero temperature (solid state) should be considered in the related calculations, especially in the piping design of hot and cold water systems. Unlike liquids, gases density is highly impacted by pressure and temperature of an environment. This relation is discussed in more details in the following sections.

2.2.2 Specific Weight

The Greek symbol of 'gamma' (γ) denotes the specific density and stands for the weight of a material over the unit of volume:

$$\gamma = \rho g \tag{2.1}$$

where, g is the gravity acceleration ($g = 9.807$ m/s^2) and thus the specific density has the unit of N/m^3 or ML^{-2} T^{-2} in the SI system.

Table 2.1 Common secondary quantities in building physics.

	FLT system	MLT system		FLT system	MLT sSystem
Acceleration	LT^{-2}	LT^{-2}	Power	FLT^{-1}	ML^2T^{-3}
Angle	$F^0L^0T^0$	$M^0L^0T^0$	Pressure	FL^{-2}	$ML^{-1}T^{-2}$
Angular acceleration	T^{-2}	T^{-2}	Specific heat	$L^2T^{-2}\theta^{-1}$	$L^2T^{-2}\theta^{-1}$
Angular velocity	T^{-1}	T^{-1}	Specific weight	FL^{-3}	$ML^{-2}T^{-2}$
Area	L^2	L^2	Strain	$F^0L^0T^0$	$M^0L^0T^0$
Density	$FL^{-4}T^2$	ML^{-3}	Stress	FL^{-2}	$ML^{-1}T^{-2}$
Energy	FL	ML^2T^{-2}	Surface tension	FL^{-1}	MT^{-2}
Force	F	MLT^{-2}	Temperature	θ	θ
Frequency	T^{-1}	T^{-1}	Time	T	T
Heat	FL	ML^2T^{-2}	Torque	FL	ML^2T^{-2}
Length	L	L	Velocity	LT^{-1}	LT^{-1}
Mass	$FL^{-1}T^2$	M	Viscosity (dynamic)	$FL^{-2}T$	$ML^{-1}T^{-1}$
Modulus of elasticity	FL^{-2}	$ML^{-1}T^{-2}$	Viscosity (kinematic)	L^2T^{-1}	L^2T^{-1}
Moment of a force	FL	ML^2T^{-2}	Volume	L^3	L^3
Moment of inertia (area)	L^4	L^4	Work	FL	ML^2T^{-2}
Moment of inertia (mass)	FLT^2	ML^2			
Momentum	FT	MLT^{-1}			

2.2.3 Viscosity

Viscosity is a property of materials. To understand its definition, we explain a hypothetical experiment. Hence, we assume that a fluid material exists between two very long parallel plates. The upper plate is pulled with a constant force of F to consequently impose this force over the fluid material while the lower plate is fixed. As illustrated in Figure 2.2, if the material is attached to the upper plate, it consequently moves from point B to point $\acute{B}$ under the employed force F and therefore creates a displacement of $B\acute{B} = \delta a$. We also notice that the

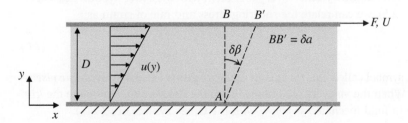

Figure 2.2 Material deformation between a fixed and a pulled plate.

displacement of δa causes a small angular displacement of $\delta\beta$ at point A, which is known as the **shearing strain**. We can also write the velocity of each vertical point in accordance with δa via the below equation:

$$u = \frac{Uy}{D}$$
$$\Rightarrow \frac{du}{dy} = \frac{U}{D} \tag{2.2}$$

Note that $\delta\beta$ has a very small magnitude, so we have:

$$\tan \delta\beta \approx \delta\beta = \frac{\delta a}{D}$$
$$\Rightarrow \delta\beta = \frac{U\delta t}{D} \tag{2.3}$$

Now, if we define the **shearing strain rate** ($\dot{\gamma}$) to represent the angular displacement in time, then combining Eqs. (2.2) and (2.3) gives:

$$\dot{\gamma} = \lim_{\delta t \to 0} \frac{\delta\beta}{\delta t}$$
$$\Rightarrow \dot{\gamma} = \frac{U}{D} = \frac{du}{dy} \tag{2.4}$$

Now, we should note that if the constant force of F is imposed on the upper plate, then this force needs to be in an equilibrium with $F = \tau A$ where τ is the **shear stress** developed on the moving plate interface and A is its effective area. The rate of shearing strain ($\dot{\gamma}$) is proportional to the shearing stress (τ):

$$\tau \propto \dot{\gamma}$$
$$\Rightarrow \tau \propto \frac{du}{dy} \tag{2.5}$$

In fluid materials, it is valid to assume that the particles are attached to the upper plate, thus in this case, they move with the velocity of U. In other words, the condition of **no-slip** is a valid assumption on the surface of the fluid. Due to the imposed shearing stress, a continuous deformation occurs in the layers of a fluid while the upper layer has the velocity of U and the bottom plate, which is in contact with the fixed plate, has the velocity of zero. The layers in between move linearly between zero and U as shown in Figure 2.2 and therefore form a velocity gradient of $\frac{du}{dy}$. The behavior of many common fluids (water, oil, gasoline, air, etc.), known as **Newtonian fluids**, is very consistent with the above hypothetical experiment. Using Eq. (2.5), we can relate the shearing stress and rate of strain as:

$$\tau = \mu \frac{du}{dy} \tag{2.6}$$

where μ (a Greek symbol called 'mu') is known as the **absolute viscosity**, **dynamic viscosity**, or **viscosity**. When the viscosity is combined with the density, we can define the **kinematic viscosity** in fluid mechanic as below with the Greek symbol of ν ('nu'):

$$\nu = \frac{\mu}{\rho} \tag{2.7}$$

Note that the above behaviour, where the shear stress is proportional to the rate of strain, is associated with Newtonian fluids although there are fluids that will not follow this rule known as **non-Newtonian fluids**. Some famous examples are toothpaste, paint, and blood.

2.3 Pressure and State of Fluid

2.3.1 Definition

In a given point of a fluid, the pressure term (p) is defined as the normal force (ΔF_n) acting on the unit area (ΔA):

$$p = \lim_{\Delta A \to 0} \frac{|\Delta F_n|}{\Delta A} = \frac{dF_n}{dA} \tag{2.8}$$

Hence, the unit of the pressure is N/m^2 or $ML^{-1}T^{-2}$ (kg/m s^2) in the SI system as it is calculated in Example 2.1. A unit of N/m^2 is also called one Pascal (Pa) as illustrated in Figure 2.3.

2.3.2 Static Pressure

To develop the equation for the pressure and understand how it is acting on fluid particles, we start to scrutinize the forces acting on a material element (**control volume**) as seen in Figure 2.4. A control volume is a volume independent of mass in space from which matter can pass. There are two types of forces imposing on the control volume, including body and surface forces. Here, the gravity ($\gamma = \rho g$) is accounted as the acting body force. Note that the pressure ($p_{@y} = p$) acts on the centre of the control volume (x, y, z) and therefore to express the pressure of the surfaces ($@ \mp \frac{\delta y}{2}$), we have to use approximations such as the Taylor series. Therefore, we use Taylor series with neglecting higher order terms as the dimension of an infinitesimal control volume is approaching to zero and the higher order products approach to zero as well. We can now define pressure on the surfaces of the control volume as:

$$p_{@y - \frac{\delta y}{2}} = p_{@y} - \frac{\partial p_{@y}}{\partial y} \frac{\delta y}{2} + O(\delta y^2) = p - \frac{\partial p}{\partial y} \frac{\delta y}{2} \tag{2.9}$$

$$p_{@y + \frac{\delta y}{2}} = p_{@y} + \frac{\partial p_{@y}}{\partial y} \frac{\delta y}{2} + O(\delta y^2) = p + \frac{\partial p}{\partial y} \frac{\delta y}{2} \tag{2.10}$$

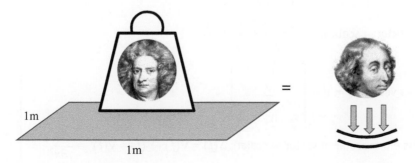

1m

1m

Figure 2.3 One unit of Newton over one unit of area equals one unit of Pascal.

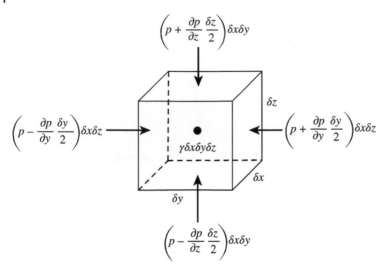

Figure 2.4 Surface and body forces on a material control volume.

Thus, the acting force on the *y*-direction surfaces becomes:

$$\delta F_y = \left(P_{@y-\frac{\delta y}{2}}\right)A_y - \left(P_{@y+\frac{\delta y}{2}}\right)A_y = \left[\left(P - \frac{\partial p}{\partial y}\frac{\delta y}{2}\right) - \left(p + \frac{\partial p}{\partial y}\frac{\delta y}{2}\right)\right]A_y$$

$$\Longrightarrow \delta F_y = -\frac{\partial p}{\partial y}\delta y \delta x \delta z \tag{2.11}$$

Similarly, we can derive surface pressures on the other surfaces as illustrated in Figure 2.4. Thus, we can obtain the resultant surface force (δF_s) on the control volume in the vector form as follows:

$$\delta F_s = -\frac{\partial p}{\partial y}\delta y \delta x \delta z - \frac{\partial p}{\partial x}\delta x \delta y \delta z - \frac{\partial p}{\partial z}\delta z \delta x \delta y = -\nabla p(\delta z \delta x \delta y) \tag{2.12}$$

Reminder:

Gradient (scalar → vector): $\nabla f = \dfrac{\partial f}{\partial x}\hat{i} + \dfrac{\partial f}{\partial y}\hat{j} + \dfrac{\partial f}{\partial z}\hat{k}$

Divergence (vector → scalar): $\nabla \cdot \vec{v} = \dfrac{\partial v}{\partial x} + \dfrac{\partial v}{\partial y} + \dfrac{\partial v}{\partial z}$

Curl (vector → vector): $\nabla \times \vec{v} = \begin{bmatrix} \hat{i} & \hat{j} & \hat{k} \\ \dfrac{\partial}{\partial x} & \dfrac{\partial}{\partial y} & \dfrac{\partial}{\partial z} \\ v_x & v_y & v_z \end{bmatrix}$

Laplacian (*f* is either scalar or vector → scalar): $\Delta(f) = \nabla(f) \cdot \nabla(f) = \nabla(f)^2 = \dfrac{\partial^2(\ )}{\partial x^2} + \dfrac{\partial^2(\ )}{\partial y^2} + \dfrac{\partial^2(\ )}{\partial z^2}$

Using the second law of Newton, we can write the below equation for the control volume mass of δm:

$$\sum \delta F = \delta m \mathbf{a} \tag{2.13}$$

Note that gravity is a body force acting on the control volume in z-direction ($\gamma = \rho g$):

$$\sum \delta F = \delta F_s - F_g = \delta m \mathbf{a}$$
$$\implies -\nabla p(\delta z \delta x \delta y) - \gamma(\delta z \delta x \delta y)\hat{k} = \rho(\delta z \delta x \delta y)\mathbf{a} \tag{2.14}$$
$$\implies -\nabla p - \gamma \hat{k} = \rho \mathbf{a}$$

If the fluid is at rest or static condition, then $\mathbf{a} = \mathbf{0}$ and Eq. (2.14) becomes:

$$-\nabla p - \gamma \hat{k} = 0 \tag{2.15}$$

Thus, Eq. (2.15) implies that the only non-zero term of the pressure term exists in the z-direction:

$$-\frac{dp}{dz} = \gamma \tag{2.16}$$

Now, we can integrate Eq. 2.16 over any two vertical points with pressures p_1 and p_2, and find the pressure difference between them as:

$$\int_{p_1}^{p_2} dp = \int_{z_1}^{z_2} -\gamma dz \tag{2.17}$$

Here, if the fluid is incompressible, then density will not change and thus Eq. (2.17) can be simplified as:

$$p_2 - p_1 = -\gamma(z_2 - z_1) = -\gamma h \tag{2.18}$$

where, h is the height between points 1 and 2, known also as the **pressure head**. This simply implies that the pressure at point 1 can be written as:

$$p_1 = \gamma h + p_2 \tag{2.19}$$

As it can be seen, pressure is a quantity that can be presented relative to other points' pressure. Hence, the **absolute pressure** is defined relative to the zero pressure or **perfect vacuum pressure**. For example, the pressure of point A in Figure 2.5 in relation to the absolute zero pressure. Nevertheless, the **gage pressure** represents a relative pressure to the local atmosphere pressure reference. For example, gage pressure of point A and B are shown in Figure 2.5 relative to the atmosphere pressure as a common reference pressure. A negative gage pressure also can be called as a **vacuum pressure** (here, gage pressure of point B relative to the atmosphere pressure). One atmospheres or **bar** is defined as 101.3 kPa.

2.3.3 Hydrostatic Pressure

All the surfaces submerged in a fluid are exposed to a force acted on them. This force in a static fluid is perpendicular to the object's surface and is linearly developing with the depth

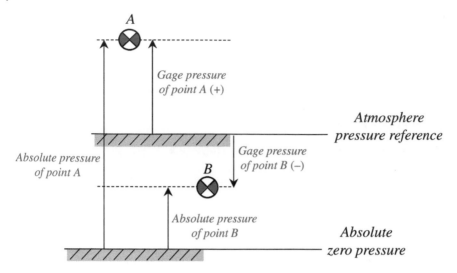

Figure 2.5 Absolute and gauge pressures.

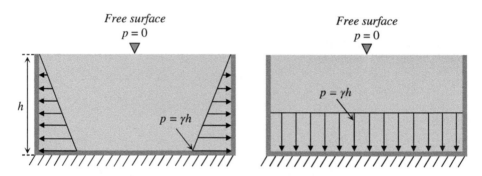

Figure 2.6 Pressure on the bottom and side walls of a fluid container.

of the free surface (gage pressure of zero relative to the atmosphere pressure) in the fluid. For the bottom of the fluid tank or container, this force is constant and uniformly equal to $F_R = pA = \gamma hA$ while it acts as a non-uniform force on the side walls (see Eq. (2.19)) as shown in Figure 2.6.

When an object with a complex shape is submerged in a fluid, the calculation of the resultant force is also more complex though it is a required information for the design purposes in many applications. Consider the complex object of Figure 2.7 in which it has an area of A and is under a force (F_R), acting perpendicularly on its surface. Hence, the acting force can be obtained with the integration of the pressure column of fluid over the surface area of the submerged object:

$$F_R = \int_A \gamma h dA = \int_A \gamma y \sin \theta dA \qquad (2.20)$$

Figure 2.7 Hydrostatic force on a general submerged object.

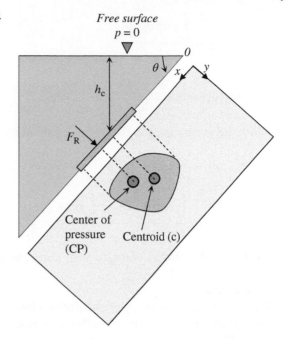

where, θ is the angle of the floated object with the free surface. When this angle θ and γ are constant, we can rewrite Eq. (2.20) as:

$$F_R = \gamma \sin \theta \int_A y dA$$

$$\Longrightarrow F_R = \gamma A y_c \sin \theta = \gamma A h_c$$

(2.21)

Recall that $y_c = \dfrac{\int_A y dA}{A}$ is defined as the y-coordinate of the centroid of the area. Note that h_c is the vertical distance of the area centroid from the fluid surface ($h_c = y_c \sin \theta$). Also, we found, as a general statement, that the acting force is independent of the object angle of θ and has a value of pressure at the centroid ($p = \gamma h_c$).

As it can be seen, the location of the resultant force can have a distance with the centroid of the object. The distributed pressure at all points, thus, generates a moment around the x-axis ($y dF$) as expected for a complex object, which should be equal to the one created by the resultant force (F_R) applied at y_R (the y-coordinate of F_R):

$$M_x = F_R y_R = \int_A y dF$$

(2.22)

Now, we replace F from Eq. (2.20):

$$M_x = \int_A \gamma \sin \theta y^2 dA$$

(2.23)

We know that $I_x = \int_A y^2 dA$ is the **second moment** of the area or **moment of inertia**. Therefore, we can rewrite Eq. (2.23) when θ and γ are constant replacing F_R from Eq. (2.21):

$$y_R = \frac{\int_A \gamma \sin\theta y^2 dA}{F_R}$$

(2.24)

$$\Rightarrow y_R = \frac{\int_A \gamma \sin\theta y^2 dA}{\gamma A y_c \sin\theta} = \frac{\int_A y^2 dA}{A y_c} = \frac{I_x}{A y_c}$$

With the same approach the moment around the x-coordinate can be found as:

$$M_y = F_R x_R = \int_A x dF = \int_A \gamma \sin\theta xy dA$$

(2.25)

$$\Rightarrow x_R = \frac{\int_A \gamma \sin\theta xy dA}{F_R} = \frac{\int_A xy dA}{A y_c} = \frac{I_{xy}}{A y_c}$$

where, I_{xy} is the product of inertia with respect to the x and y areas.

Example 2.2

Water is trapped within a cavity wall during a heavy rain through an unwanted crack while all the drainage pathways and weep holes are blocked by residual of construction materials, which are not properly cleaned (see Figure 2.E2). What is the pressure on walls if the water starts to be filled in the wall up to a height of 2.4 m?

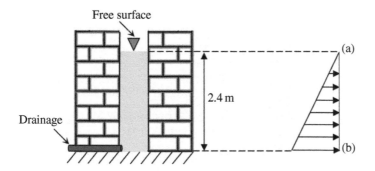

Figure 2.E2 Rain trapped in a wall cavity.

Solution

We assume the cavity to be in a 2D shape and the water density to be 1000 kg/m³. The water surface is connected to the atmosphere pressure through the crack and thus it is assumed to be equal to 1 bar. Therefore, the pressure exerted on the adjacent walls will be equal to the sum of the water column and the surrounding air pressures. The resultant pressure distribution will have a form as depicted in Figure 2.6. So, the pressure at point '*a*' can be calculated as:

$$p_a = 1\,\text{bar} = 10^5\,\text{Pa}$$

And at point 'b':

$$p_b = p_a + p_{\text{water column}} = p_a + \gamma h$$

$$\Longrightarrow p_b = 10^5 + 1000 \times 9.81 \times 2.4 = 123,544 \, \text{Pa}$$

At any arbitrary point between 'a' and 'b', the water column pressure is linearly changing with the height and can be shown as:

$$p(h) = p_a + \rho g h$$

2.3.4 Buoyancy

In the case that a stationary body is completely or partially submerged in a fluid, there is a resultant force acting from the fluid on the body, which is called **buoyancy** force. A balloon, for example, is completely submerged in the air while a fleet is partially submerged in the water. The resultant buoyancy force is a result of the vertical pressure difference through the height of the body with an upward direction as the pressure increases with the depth of a fluid. To investigate the impact of the buoyancy force, consider an immersed body with an arbitrary shape in a fluid as shown in Figure 2.8.

Assume that the body has a volume of V and as shown in the free diagram of Figure 2.8, several forces are acting on its surfaces. F_B is the exerted force by the body to the fluid. It should be noted that all the horizontal forces are equal and eventually cancel out with each other. Thus, one can derive the balance equation of the acting forces on the body as below:

$$F_B = F_2 - F_1 \tag{2.26}$$

We know that F_2 is the pressure force acting upwards to the lower surface area of the body, which is equal to the weight of fluid needed to fill the volume above the lower surface of the body ($\acute{V} = h_1 A$) with any arbitrary shape and the volume of the body (V):

$$F_2 = \gamma\left(V + \acute{V}\right) = \gamma(V + h_1 A) \tag{2.27}$$

Figure 2.8 Buoyancy force acting on a submerged body.

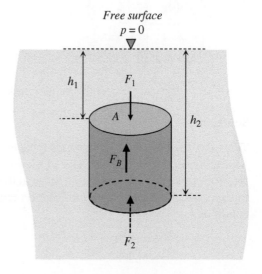

Similarly, F_1 is the acting downwards force on the upper surface area equal to the volume of the fluid above this surface:

$$F_1 = \gamma \acute{\forall} = \gamma h_1 A \tag{2.28}$$

Hence, we can replace Eqs. (2.27) and (2.28) into Eq. (2.26) and find the upwards buoyancy force:

$$F_B = \gamma(\forall + h_1 A) - \gamma h_1 A = \gamma \forall \tag{2.29}$$

It can be concluded that the buoyancy force is equal to the weight of the fluid that needs to fill the volume of the body. This rule is known as **Archimedes' principle**. In the case that a body is floated rather than being in a static immersing condition, the upwards buoyancy force is again equal to the weight of the fluid that is required to fill the floated volume ($\forall_D$):

$$F_B = \gamma \forall_D \tag{2.30}$$

Example 2.3

The water level within a traditional flush tank is controlled by a floating operated valve, which is set to allow water to enter when the water level drops below the required level as shown in Figure 2.E3. The float ball (half of the sphere is submerged) is 0.1 m in diameter, which is mounted on a frictionless hinge. The value should undergo on a 5N force to be completely closed at an angle of $120°$. If the float rod length is 0.3 m, what would be the weight of the spherical valve to maintain the required condition for the flush system.

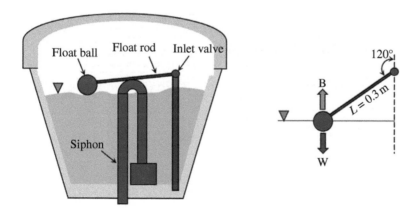

Figure 2.E3 Schematic of a traditional flush tank.

Solution

The valve needs a net of 5N upwards force to remain fully closed. As seen in Figure 2.E3, neglecting the frictions and weight of the rod, weight (W) and buoyancy (B) are the only forces exerted on the float valve. As stated in the question, half of the float ball is always submerged. So, if we write a torque balance equation around the hinge, assuming a static condition:

$$(F_B - W)L\cos(60) = 5L\cos(60)$$
$$\Longrightarrow F_B - W = 5$$

While the exerted force is independent of the rod characteristics, replacing the buoyancy force for half of the volume ($F_B = \rho_w g \dfrac{\forall}{2} = \dfrac{4}{6}\pi\rho_w gr^3$) into the equation gives:

$$\frac{4\pi}{6}\rho_w gr^3 - W = 5$$

Replacing the density of water ($\rho_w = 1000 \dfrac{kg}{m^3}$) and rearranging the above equation:

$$F = 5 - \frac{4\pi}{6} \times 0.05^3 \times 1000 \times 9.81 = 5 - 2.57 = 2.43\,N \Longrightarrow W = 0.2\,kg$$

2.3.5 Vapour Pressure and Boiling

Liquids will evaporate from the surface due to the dominance of momentum over intermolecular forces occurring over some molecules. Assume the evaporation process in a container that the evaporation from the surface of a liquid is equal to gas molecules entering the liquid phase. In other words, the materials in both states are in the equilibrium condition. In this case, the vapour state of the material is called to exist in the saturation state or **vapour pressure** (p_v).

The vapour pressure magnitude of a material is highly related to its temperature. In a boiling condition, for example, the absolute pressure of the liquid reaches the vapour pressure and thus some molecules start to leave the liquid. In this case, the vapour pressure of the material is equal to the pressure exerted on the liquid by the surrounding environmental pressure at vapour pressure. Evidently, the environmental pressure changes in a relation with the elevation and thus we expect the boiling point of liquid to change accordingly as depicted for water in Figure 2.9.

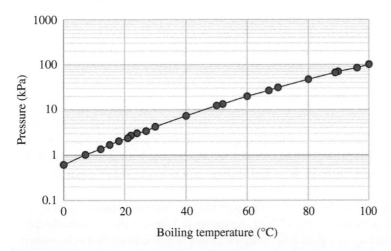

Figure 2.9 Alteration of the boiling point of water in accordance with the atmospheric pressure.

Example 2.4

Atmospheric pressure has a relation with the elevation similar to the shown graph in Figure 2.9. What would be the water boiling point at top of Everest Mountain with 8848 m height?

Solution

A common equation to calculate the atmosphere pressure (p) above sea level (h) is proposed as below:

$$p = 101,325(1 - 2.2557710 - 5\,h)^{5.25588}$$

From this equation, we can calculate the atmosphere pressure at the top of Everest Mountain with 8848 m height to be about 0.314 bar (31.4 kPa). The pressure can be also shown in Figure 2.E4.

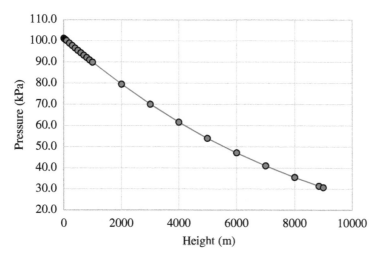

Figure 2.E4 Pressure variation in accordance to elevation.

Refer to Figure 2.9, thus, the associated boiling point for this pressure can be approximately found as 70 °C.

2.3.6 Pressure Measurement Devices

Through the past centuries various measurement apparatus were built to measure the pressure, most of them benefited from the concept of gage pressure or pressure column created by a secondary fluid. As illustrated in Figure 2.10, these devices are known as ***manometers*** and are created in various forms such as ***barometers***, ***piezometer tubes***, ***U-tubes***, and ***inclined-tubes***.

Barometer, for example, is mainly used to measure the atmospheric pressure, which was invented by Evangelista Torricelli around 1644. As shown in Figure 2.10a, it normally consists of a column of mercury in a glass tube, which is closed in one side while it is in contact with the atmosphere pressure at the other end. The column of mercury stays in a certain height to obtain an equilibrium between its weight (plus the vapour pressure of mercury, which is a very small number and can be neglected) and the atmosphere pressure:

$$p_{\text{atmosphere}} = \rho_{\text{mercury}}gh \qquad\qquad (2.31)$$

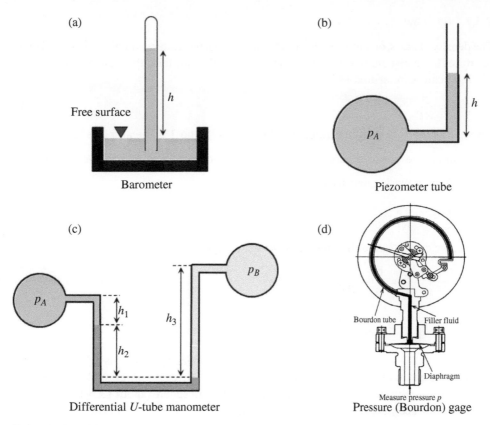

(a)

Free surface

h

Barometer

(b)

p_A

h

Piezometer tube

(c)

p_A

h_1

h_2

h_3

p_B

Differential U-tube manometer

(d)

Bourdon tube

Filler fluid

Diaphragm

Measure pressure p

Pressure (Bourdon) gage

Figure 2.10 Various pressure measurement devices. *Source:* (d) Based on Nakayama [1].

Simple piezometers have a vertical tube, which is attached to the container or pipe that the pressure is about to be measured (see Figure 2.10b). Note that if the end of the tube is open, then it can be set to zero and the concept of the gage pressure can be used:

$$p_A = \rho_{\text{fluid}}gh \tag{2.32}$$

The main disadvantage of piezometer tubes is their weakness in measuring pressures lower than that of the atmosphere (or the fluid pressure in the other end of the tube). Also, normally a long tube is required when the pressure is very high, and thus this is not a practical option in many conditions. As seen in Figure 2.10c, U-tube manometers are normally used when a difference pressure between two containers is desired. We can derive the pressure difference equation between points and therefore reach to an equation that links p_A and p_B together:

$$p_A + \gamma_1 h_1 + \gamma_2 h_2 = p_B + \gamma_3 h_3$$
$$\implies p_A - p_B = \gamma_3 h_3 - \gamma_1 h_1 - \gamma_2 h_2 \tag{2.33}$$

2.3.7 Gas Law

The density of gases, or compressible fluids, can change with alteration of pressure and temperature. If we assume that the elevation is not significant, then the state equation of an ideal gas can be presented as below:

$$\rho = \frac{p}{RT} \tag{2.34}$$

where, R is the gas constant number (8.314 J/K · mol), and T is the absolute temperature.

Assuming the pressure behaviour in a static condition ($-\frac{dp}{dz} = \rho g$), we can rewrite Eq. (2.16) by replacing density from Eq. (2.34):

$$\frac{dp}{dz} = -\frac{pg}{RT} \tag{2.35}$$

Now, we can integrate the pressure between two points:

$$\int_{p_1}^{p_2} \frac{dp}{p} = \int_{z_1}^{z_2} -\frac{g}{RT} dz$$

$$\implies \ln \frac{p_2}{p_1} = -\frac{g}{R} \int_{z_1}^{z_2} \frac{dz}{T} \tag{2.36}$$

Equation (2.36) can lead to various behaviour of the ideal gas from this point due to different assumption of terms' variation in the equation. The common example is when the temperature is constant (T_0), for example within the stratosphere (about 13.7–16.8 km from the earth' surface), Eq. (2.36) can be written as:

$$p_2 = p_1 e^{\left[\frac{g(z_2 - z_1)}{RT_0}\right]} \tag{2.37}$$

In general, the gas law of Eq. (2.34) can be represented in terms of the variation of pressure against volume, mainly used in thermodynamics as further explained in Chapter 4:

$$PV^n = \text{constant} \tag{2.38}$$

where n is the polytropic index.

Table 2.2 represents various relationships between pressure and volume. Note that γ is defined as the adiabatic index (C_p/C_v).

Table 2.2 Different relationships between pressure and volume.

	Process name	Effect
$n = 0$	Isobaric	Constant pressure
$n = 1$	Isothermal	Constant temperature
$1 < n < \gamma$	Quasi-adiabatic	
$n = \gamma$	Adiabatic	No heat transfer
$n = \infty$	Isochoric	Constant volume

Note that other forms of gas models to represent gas behaviours rather than ideal gas law, such as Van der Waals model, result in different equations though their explanations are beyond the scope of this book.

2.3.8 Bernoulli Equation

If the flow is assumed to be *inviscid*, then the viscosity effect would be negligible, meaning that the shear stress and the rate of strain displacement are also negligible. In reality, there is no such fluid, but from a practical point of view, there are conditions that the viscosity is relatively small compared to the other effect such as inertia. If a fluid particle is inviscid, then the flow movement is only under the impact of the gravity (or other significant body forces) and pressure forces. Thus, we can derive Newton's second law for an inviscid fluid flow similar as Eq. (2.14):

$$-\nabla p - \gamma \hat{k} = \rho \mathbf{a} \tag{2.39}$$

Note that although the motion is normally occurring in three dimensions, here for the simplicity purposes, we use a two-dimensional coordinate to describe the motion. Also, we start with the steady flow in which a particle in a certain point passes through same path in the flow called *streamlines* (see Figure 2.11). This implies that the streamlines as the path of a particle consist of tangent lines to its velocity vectors throughout the flow field. It is easier to define the coordinate of the flow based on these streamlines, and hence s (t) is defined as the distance of the particle from the origin and $R(s)$ is the local radius of the curvature at a particular point. Using a differentiation chain rule, therefore, the acceleration of a particle on a streamline coordinate can be written as below:

$$a_s = \frac{dV}{dt} = \frac{\partial V}{\partial s}\frac{ds}{dt} = \frac{\partial V}{\partial s}V \tag{2.40}$$

where, V is the velocity of the particle.

We can also obtain the centrifugal acceleration as:

$$a_n = \frac{V^2}{R} \tag{2.41}$$

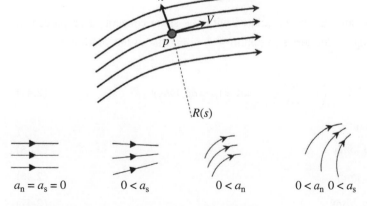

$$a_n = a_s = 0 \qquad 0 < a_s \qquad 0 < a_n \qquad 0 < a_n\ 0 < a_s$$

Figure 2.11 Streamlines and motion of a particle.

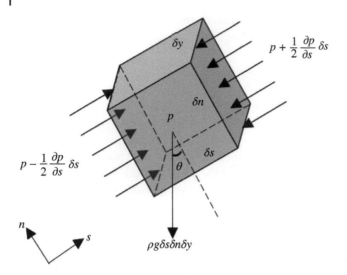

Figure 2.12 Free-body diagram of the fluid particle.

As shown in Figure 2.11, we expect that the acceleration components to be non-zero for various conditions of streamlines when they become closer to each other ($0 < a_s$) and/or have curvatures ($0 < a_n$). In the case of the straight streamline scenarios, $a_n = a_s = 0$ (see Figure 2.11).

Now, to derive the Newton second law on a streamline, we consider the free-body diagram of a fluid particle as shown in Figure 2.12. Consequently, we can find the balance of the forces applied to the fluid particle as below:

$$\sum \delta F_s = \delta m a_s = \rho \delta \forall \frac{\partial V}{\partial s} V \tag{2.42}$$

where $\delta \forall = \delta s \delta n \delta y$ is the infinitesimal volume of the small particle (control volume).

Moreover, the component of the particle weight in the streamline direction is:

$$\delta W_s = -\delta W \sin\theta = -\gamma \delta \forall \sin\theta \tag{2.43}$$

The impact of the pressure can be calculated with the same concept as employed in Eqs. (2.9) and (2.10) using Taylor series to obtain $\frac{\partial p}{\partial s} \frac{\delta s}{2} \approx \delta p_s$ due to the small dimensions of the fluid particle:

$$\delta F_{@s-\frac{\delta s}{2}} = \left(p_{@s-\frac{\delta s}{2}}\right) A_s = \left(p - \frac{\partial p}{\partial s} \frac{\delta s}{2}\right) \delta n \delta y \approx (p - \delta p_s)\delta n \delta y \tag{2.44}$$

Similarly, we can obtain:

$$\delta F_{@s+\frac{\delta s}{2}} \approx (p + \delta p_s)\delta n \delta y \tag{2.45}$$

$$\delta F_{@n-\frac{\delta s}{2}} \approx (p - \delta p_n)\delta s \delta y \tag{2.46}$$

$$\delta F_{@n+\frac{\delta s}{2}} \approx (p + \delta p_n)\delta s \delta y \tag{2.47}$$

Now, we can insert Eqs. (2.43)–(2.47) into Eq. (2.42) and write the net pressure force on the particle in addition to the gravity force in the s-direction:

$$\sum \delta F_s = -\gamma \delta \forall \sin\theta + [(p - \delta p_s)\delta n \delta y - (p + \delta p_s)\delta n \delta y]$$

$$\Longrightarrow \sum \delta F_s = -\gamma \delta \forall \sin\theta - 2\delta p_s \delta n \delta y = -\gamma \delta \forall \sin\theta - 2\frac{\partial p}{\partial s}\frac{\delta s}{2}\delta n \delta y$$

$$\Longrightarrow \sum \delta F_s = -\gamma \delta \forall \sin\theta - \frac{\partial p}{\partial s}\delta \forall \tag{2.48}$$

Similarly, for the n-direction, we can derive the below equation:

$$\sum \delta F_n = \delta W_n - [(p - \delta p_n)\delta s \delta y - (p + \delta p_n)\delta s \delta y] = \gamma \delta \forall \cos\theta - 2\frac{\partial p}{\partial n}\frac{\delta n}{2}\delta s \delta y$$

$$\Longrightarrow \sum \delta F_n = -\gamma \delta \forall \cos\theta - \frac{\partial p}{\partial s}\delta \forall \tag{2.49}$$

Again, recalling the Newton second law, Eq. (2.48) becomes:

$$\rho \delta \forall \frac{\partial V}{\partial s}V = -\gamma \delta \forall \sin\theta - \frac{\partial p}{\partial s}\delta \forall \tag{2.50}$$

With further simplifications and implying that $\sin\theta = \dfrac{dz}{ds}$ along the streamline, the final form of Eq. (2.50) can be written as:

$$\rho V \frac{\partial V}{\partial s} = -\gamma \frac{dz}{ds} - \frac{\partial p}{\partial s} \tag{2.51}$$

Also, in the normal direction, we can follow a same approach and simplify Eq. (2.49) as:

$$\rho \delta \forall \frac{V^2}{R} = -\gamma \delta \forall \cos\theta - \frac{\partial p}{\partial n}\delta \forall \tag{2.52}$$

And, again replacing $\cos\theta = \dfrac{dz}{dn}$, the final form of Eq. (2.52) gives:

$$\rho \frac{V^2}{R} = -\gamma \frac{dz}{dn} - \frac{\partial p}{\partial n} \tag{2.53}$$

With performing integration of two final forms of the equations over the streamlines, we can reach the below equation for the stream-wise direction:

$$dp = \frac{\partial p}{\partial s}ds + \frac{\partial p}{\partial n}dn$$

$$\overset{(dn\,=\,0)}{\Longrightarrow} \frac{\partial p}{\partial s} = \frac{dp}{ds} \tag{2.54}$$

Similarly, $\dfrac{\partial V}{\partial s} = \dfrac{dV}{ds}$ and since we have $\left(\rho V \dfrac{dV}{ds} = \dfrac{\rho}{2}\dfrac{d(V^2)}{ds}\right)$, we can simplify Eq. (2.51) alongside of the streamlines as below:

$$\frac{\rho}{2}\frac{d(V^2)}{ds} = -\gamma \frac{dz}{ds} - \frac{dp}{ds}$$

$$\Longrightarrow \gamma dz + dp + \frac{1}{2}\rho d(V^2) = 0 \tag{2.55}$$

And after integration along a streamline, we can express:

$$\int \frac{dp}{\rho} + \frac{1}{2}V^2 + gz = C \tag{2.56}$$

where C is a constant value. If the density and gravity acceleration are not changing, then the **Bernoulli equation**, named after Daniel Bernoulli (1700–1782), for an inviscid, steady, and incompressible flow alongside of a streamline can be written as below:

$$p + \frac{1}{2}\rho V^2 + \gamma z = C$$

or

$$\frac{p}{\gamma} + \frac{V^2}{2g} + z = C \tag{2.57}$$

In this equation, $\frac{p}{\gamma}$ is called **pressure head**, $\frac{V^2}{2g}$ is called **velocity head,** and z is called **elevation head**.

If we follow a same procedure for the particle movement in a normal direction to a streamline, and after integration across the streamline, we can rewrite Eq. (2.53) as:

$$\int \frac{dp}{\rho} + \int \frac{V^2}{R}dn + gz = \acute{C} \tag{2.58}$$

Note that the variation of these parameter across a streamline is mainly negligible in comparison with the stream-wise variations. So, we can find the final form of Eq. (2.58) for an inviscid, steady, and incompressible flow across the streamline as follows:

$$p + \rho \int \frac{V^2}{R}dn + \gamma z = \acute{C} \tag{2.59}$$

Example 2.5

A cold-water tank is used to supply the cold water of a piping system as illustrated in Figure 2.E5.1. Each floor has a 2.8 m height and the water surface in the tank has a 1.4 m height from the sink in Floor-3. Calculate the pressure behind each sink valves of the cold-water supply pipes at each floor, neglecting the losses. Assume all the valves to be closed.

Solution

It is assumed that all valves in each floor have a same pressure level. The water level height from the bottom of the extraction pipe is denoted by L as illustrated in Figure 2.E5.2. We can, therefore, assume that $L + h \approx \frac{h'}{2}$. Since all valves are closed, the velocity in all pipelines is zero.

Atmospheric pressure is exerted on the water level inside the tank. Thus, the Bernoulli equation can be written between point-1 at the tank surface and points 2, 3, and 4 at different floors as follows:

Between Point-1 and Point-2:

$$P_1 + \rho g(h' + h' + h + L) = P_2 + \rho g(h' + h')$$
$$\Longrightarrow P_1 + \rho g(h + L) = P_2$$

$$\Longrightarrow P_2 = 10^5 + 1000 \times 9.81 \times \frac{2.8}{2} = 113{,}734 \text{ Pa}$$

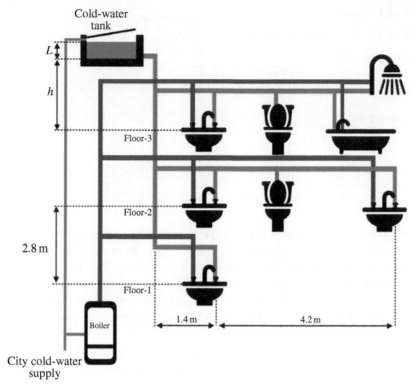

Figure 2.E5.1 A piping system in a multi-story building with a cold-water tank.

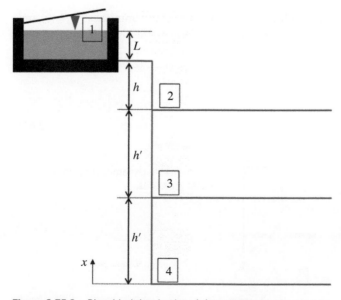

Figure 2.E5.2 Pipes' heights in the piping system.

Between Point-1 and Point-3:

$$P_1 + \rho g(h' + h' + h + L) = P_3 + \rho g(h')$$
$$\Longrightarrow P_1 + \rho g(h + L + h') = P_3$$

$$\Longrightarrow P_3 = 10^5 + 1000 \times 9.81 \times \left(\frac{2.8}{2} + 2.8\right) = 141,202 \text{ Pa}$$

Between Point-1 and Point-4:

$$P_1 + \rho g(h' + h' + h + L) = P_4$$

$$\Longrightarrow P_4 = 10^5 + 1000 \times 9.81 \times \left(\frac{2.8}{2} + 2.8 + 2.8\right) = 168,670 \text{ Pa}$$

2.3.9 Dynamic Pressure

Bernoulli equation can be used to define the concept of dynamic pressure as well as the stagnation pressure. As illustrated in Figure 2.13, when there is an inviscid, steady, and incompressible flow through a conduit, we can investigate the pressure in different points. Point-1 almost represents the static pressure as it will not change alongside of a streamline. Static pressure can be measured by using a piezometer attached to a hole on the surface of the pipe. As we understood before, this pressure is equal to **hydrostatic pressure** of $p_1 = \gamma h_1$. Moreover, we can write Bernoulli equation between point-1 and point-2 as we assumed both are along one streamline and have the same elevation ($z_1 = z_2$). Point-2 represents the end of a small tube inserted into the flow towards the stream:

$$\frac{p_1}{\gamma} + \frac{V_1^2}{2g} + z_1 = \frac{p_2}{\gamma} + \frac{V_2^2}{2g} + z_2 = C$$

$$\Longrightarrow \frac{p_1}{\gamma} + \frac{V_1^2}{2g} = \frac{p_2}{\gamma} + \frac{V_2^2}{2g} \tag{2.60}$$

Note that point-2 is the location that fluid's momentum energy diminishes ($V_2 = 0$) and transforms to potential energy that fills the tube against the gravity up to a height of H. hence, we can rewrite Eq. (2.60) as:

$$\frac{p_1}{\gamma} + \frac{V_1^2}{2g} = \frac{p_2}{\gamma}$$

$$\Longrightarrow p_1 + \frac{\rho V_1^2}{2} = p_2 \tag{2.61}$$

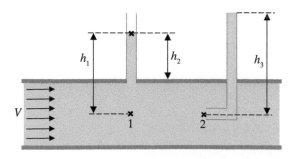

Figure 2.13 Dynamic, stagnation and static pressures.

Here, $\dfrac{\rho V_1^{\,2}}{2}$ is called the **dynamic pressure** term and point-2 is called the **stagnation point** where the moving fluid particles become stationary. The pressure attributed to the stagnation point is the maximum obtainable pressure in a streamline. In general, static pressure (p), hydrostatic pressure (γz), and dynamic pressure ($\dfrac{\rho V^2}{2}$) together are called the **total pressure**. Therefore, another interpretation of the Bernoulli equation is that the total pressure is always constant along a streamline.

Understanding of the dynamic and static pressures is widely used for the velocimetry purposes in different applications. A famous example, as depicted in Figure 2.14a,b, is the Pitot-static tube, which is mainly used to measure the aircrafts' velocity. Between an upstream point and the stagnation point-2, we can develop the Bernoulli equation as:

$$p + \frac{\rho V^2}{2} = p_2 \tag{2.62}$$

Note that the elevation between point-2 and point-3 as well as point-1 and point-4 is neglected in the tube and thus we can write:

$$p_2 = p_3 \tag{2.63}$$

$$p_1 = p_4 = p \tag{2.64}$$

(a)

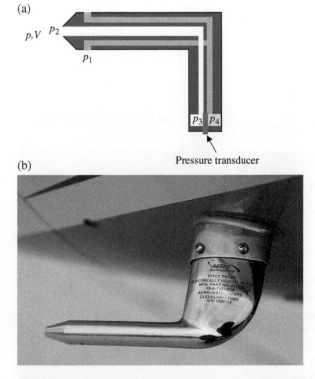

(b)

Figure 2.14 Pitot-static tube (a) the schematic sketch and (b) mounted in an airplane. *Source:* (b) CJ1 Pitot Tube: Manufacturer: Cessna.

Thus, by replacing p_4 from Eq. (2.64) and p_3 from Eqs. (2.63) and (2.62), we can subtract p_4 from p_3 as:

$$p_3 - p_4 = \left[p + \frac{\rho V^2}{2} \right] - p$$

$$\Rightarrow V = \sqrt{2 \left(\frac{p_3 - p_4}{\rho} \right)} \tag{2.65}$$

The $p_3 - p_4$ is calculated with a pressure transducer and hence the velocity (V) can be obtained using Eq. (2.65).

In addition to Pitot-static tube, the Bernoulli equation is also broadly used for flow rate measurement in pipes and open channels. The example is the orifice meter, nozzle meter and Venturi meter as shown in Figure 2.15. Using the concept of the conservation of mass before and after the obstacles ($Q_1 = Q_2$). This means that the low-pressure and high-velocity region before the obstacle transforms to the high-pressure and low-velocity one, respectively. The measurement of the pressures before and after the obstacle can be thus used to calculate the velocity and mass flow rate as follows:

$$Q = Q_1 = Q_2$$
$$\Rightarrow Q = A_1 V_1 = A_2 V_2 \ (A_1 > A_2) \tag{2.66}$$

Applying Bernoulli equation to the points before and after the obstacle and perform simplifications:

$$p_1 + \frac{\rho V_1^2}{2} = p_2 + \frac{\rho V_2^2}{2}$$

$$\Rightarrow V_2^2 - V_1^2 = 2 \left(\frac{p_1 - p_2}{\rho} \right)$$

p_1, V_1, A_1 p_2, V_2, A_2

(1)   (2)

(a)

(b)

(c)

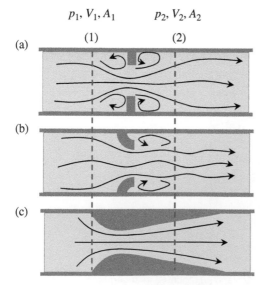

Figure 2.15 (a) Orifice, (b) Nozzle, and (c) Venturi.

$$\Longrightarrow V_2^2 \left(1 - \left(\frac{A_2}{A_1} \right)^2 \right) = 2 \left(\frac{p_1 - p_2}{\rho} \right)$$

$$\Longrightarrow V_2 = \sqrt{ 2 \left(\frac{p_1 - p_2}{\rho} \right) \Big/ \left(1 - \left(\frac{A_2}{A_1} \right)^2 \right) } \tag{2.67}$$

Now, we can replace Eq. (2.67) into Eq. (2.66) as follows:

$$\Longrightarrow Q = A_2 \sqrt{ 2 \left(\frac{p_1 - p_2}{\rho} \right) \Big/ \left(1 - \left(\frac{A_2}{A_1} \right)^2 \right) } \tag{2.68}$$

Note that the elevation is assumed to be equal at point-1 and point-2. With the measurement of the areas and pressures at point-1 and point-2, one can calculate the mass flow rate (Q) using Eq. (2.68). It should be considered, however, in reality, the Bernoulli conditions are not fully satisfied in the conduits, and therefore the discrepancy is normally corrected with an empirical **discharge coefficient**, depending on the geometry of the obstacle (see Chapter 3 for more details).

Example 2.6

What would be the exhausting cold-water velocities from the sinks of Example 2.5 if the pipes' diameters are all ¼ inch. Pipeline's losses are neglected, and other consuming units are supposed to be fully closed when a specific valve is open. The level of water inside the tank is also assumed to be constant as it is continuously fed by the city supply water.

Solution

Sink outflow is discharged to the atmosphere ($P_1 = P_2 = P_3 = P_4 = P_{\text{atm}}$). Referring to Figure 2.E6, the Bernoulli equation can be again written between point-1 on the water surface in the tank and points 2, 3, and 4 at the sinks outflow levels.
Between Point-1 and Point-2:

$$P_1 + \rho g(h' + h' + h + L) = P_2 + \frac{1}{2}\rho V^2 + \rho g(h' + h')$$

$$\Longrightarrow P_1 + \rho g(h + L) = P_2 + \frac{1}{2}\rho V^2$$

$$\Longrightarrow V = \sqrt{2 \times 9.81 \times 1.4} = 5.24 \, \text{m/s}$$

Between Point-1 and Point-3:

$$P_1 + \rho g(h' + h' + h + L) = P_2 + \frac{1}{2}\rho V^2 + \rho g(h')$$

$$\Longrightarrow P_1 + \rho g(h + L + h') = P_2 + \frac{1}{2}\rho V^2$$

$$\Longrightarrow V = \sqrt{2 \times 9.81 \times (1.4 + 2.8)} = 9.08 \, \text{m/s}$$

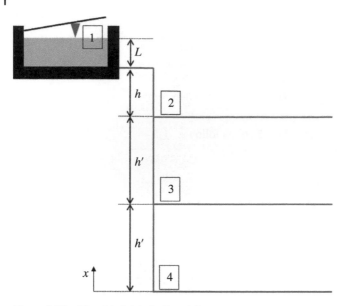

Figure 2.E6 Pipes' heights in the piping system.

Between Point-1 and Point-4:

$$P_1 + \rho g(h' + h' + h + L) = P_2 + \frac{1}{2}\rho V^2$$

$$\Longrightarrow P_1 + \rho g(h + L + h' + h') = P_2 + \frac{1}{2}\rho V^2$$

$$\Longrightarrow V = \sqrt{2 \times 9.81 \times (1.4 + 2.8 + 2.8)} = 11.72 \, \text{m/s}$$

Note that the velocities are independent from the pipe diameters in an inviscid flow.

2.4 Fluid in Motion

2.4.1 Steady and Unsteady Flows

When a fluid is investigated, two methods can be used, including Eulerian and Lagrangian methods. In the **Eulerian method**, the flow characteristics are represented in a spatial and temporal coordinate system. Therefore, each fluid particle has a position vector relative to the coordinate system. Similarly, other field representations of fluid such as velocity and acceleration can be defined relative to the coordinate system. In the **Lagrangian method**, however, each fluid particle has its own system of coordinates, and the fluid properties associated with the particles change as a function of time. In fluid mechanics, it is common to use Eulerian method though the Lagrangian methods does have its own applications.

When fluid characteristics do not vary in a given point in the space, then the flow is assumed to be steady. Steady flows are simpler than unsteady ones, hence, in many engineering applications, it is quite reasonable to simplify unsteady flows to steady ones if the

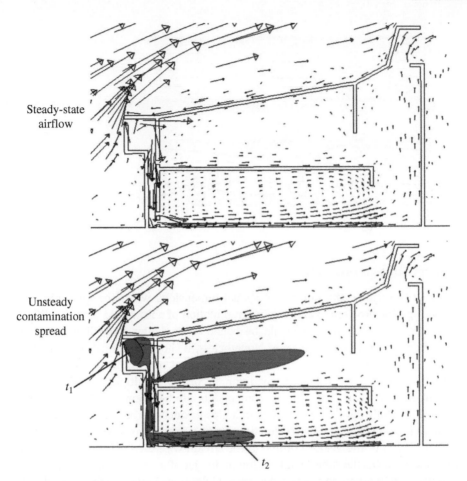

Steady-state
airflow

Unsteady
contamination
spread

t_1

t_2

Figure 2.16 Steady natural ventilation in a room.

compromise dose not significantly impact the results. For example, a jet flow in a room can be assumed to be a steady flow if only few initial seconds after starting the system is neglected. As another example, as shown in Figure 2.16, we can always assume a natural ventilation to be steady in a room as the environmental conditions change in an hourly basis. On the contrary, if a sudden contamination from a vehicle is spread into the building by this natural ventilation strategy, then the concentration decay from t_1 to t_2 is a highly unsteady phenomenon and thus cannot be simplified to a steady condition.

2.4.2 Laminar and Turbulent Flows

Laminar flow is a state of flow in which fluid's layer are moving smoothly aligned with each other. As it can be depicted in Figure 2.17, the arrows or injected drops of ink in a hypothetical experiment in the laminar flow moves parallel to the flow and attached to the layers of fluid without moving upwards/downwards to other layers. In the shown turbulent flow, however, arrow or ink drops are moving with unsteady flow with highly cross-stream mixing between layers of fluid particles. Examples of turbulent flows are plumes or jets in rooms.

(a)

Laminar
flow

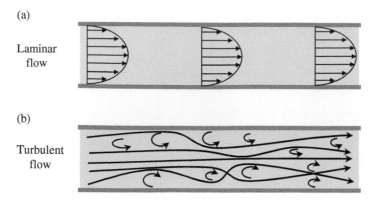

(b)

Turbulent
flow

Figure 2.17 (a) Laminar (b) turbulent flows.

2.4.3 Multiple-Dimensional Flow

It is essential to recognize the number of spatial dimensions when the quantities are described in a specific problem to allude misrepresentation of a flow field. The simplification from a realistic 3D flow field would be beneficial in better modelling and thus simulation of the flow on many occasions. Nonetheless, it should be noted that vigorous justifications should be offered when a 3D flow is reduced to a problem with a lower dimension. Identification of symmetry condition in one or multiple boundaries and negligible value of one or multiple velocity components can be amongst such reasonings.

For example, the fully developed flow field in a pipe or duct can be assumed to be one-dimensional as the impact of gravity acceleration can be neglected as seen in Figure 2.18a. There is no variation of the velocity profile in the stream-wise direction due to the fully developed condition. On the other hand, an impinging jet of a ventilation system from a circular diffuser can be assumed as a two-dimensional flow (see Figure 2.18b). The velocity profile has a symmetric shape around the axis of the diffuser though it highly changes when the jet is developing. Finally, the general flow filed around a building (bluff body) can be reckoned as a three-dimensional velocity profile as presented in Figure 2.18c.

2.5 Governing Equation of Fluids

2.5.1 Reynolds Transport Theorem

Reynolds stress theorem represents a relation between the governing equation defined in a system and control volume of a velocity field. This implies that we can transform the Lagrangian representation of a transport phenomenon to its Eulerian form using Reynolds stress theorem. This approach leads to the understanding of transport phenomena in many fluid problems when either of the forms is necessary coexists.

First, we assume that ϕ is a property of a system and Φ is amount of that property per unit of mass transported by mass of fluid ($\phi = m\Phi$). So, to calculate the total amount of ϕ in a system, we can sum all amount of ϕ in infinitesimal fluid mass of $\rho\delta\forall$ as:

$$\phi_{\text{sys}} = \int_{\text{sys}} \Phi(\rho\delta\forall) \tag{2.69}$$

(a)

(b)

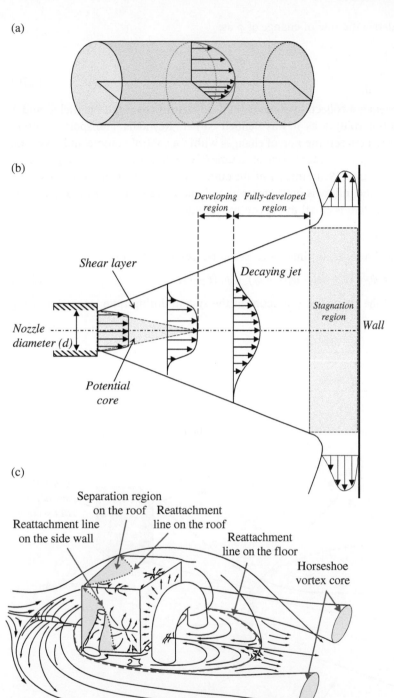

(c)

Figure 2.18 (a) One-dimensional flow in a pipe (b) two-dimensional flow in an impinging jet a (c) three-dimensional flow around a bluff body. *Source:* Based on Hunt et al. [2].

Hence, we can derive the rate of change of ϕ as:

$$\frac{d\phi_{sys}}{dt} = \frac{d\left(\int_{sys} \Phi(\rho\delta\forall)\right)}{dt} \tag{2.70}$$

Note that the system is a collection of matter of fixed identity (here fluid particles) and is different form control volume as it was defined earlier. Reynolds Transport Theorem defines a relationship between the rate of changes within a control volume and a system.

For this purpose, first we consider a control volume (CV) at time t as shown in Figure 2.19. As it can be seen in Figure 2.19, at time $t + \delta t$, the outflow from the control volume is shown with volume II while the inflow is presented with volume I. We know, for time t, that the amount of ϕ in the system is equal to the control volume as below:

$$\phi_{sys}(t) = \phi_{C.V.}(t) \tag{2.71}$$

The system value, however, in time $t + \delta t$ would be (see Figure 2.19):

$$\phi_{sys}(t + \delta t) = \phi_{C.V.}(t + \delta t) - \phi_I(t + \delta t) + \phi_{II}(t + \delta t) \tag{2.72}$$

Now we can find the change of ϕ amount in the system during time interval δt using Eqs. (2.71) and (2.72):

$$\begin{aligned} \frac{\delta\phi_{sys}}{\delta t} &= \frac{\phi_{sys}(t + \delta t) - \phi_{sys}(t)}{\delta t} \\ \Longrightarrow \frac{\delta\phi_{sys}}{\delta t} &= \frac{[\phi_{C.V.}(t + \delta t) - \phi_I(t + \delta t) + \phi_{II}(t + \delta t)] - \phi_{sys}(t)}{\delta t} \end{aligned} \tag{2.73}$$

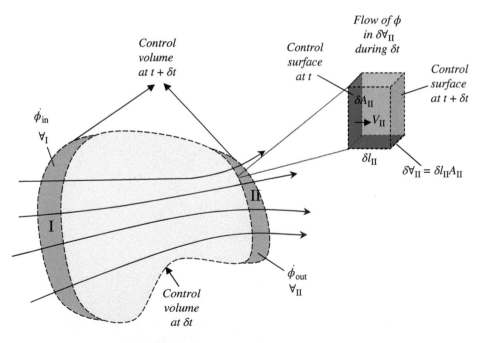

Figure 2.19 System progression in a control volume.

Knowing that $\phi_{sys}(t) = \phi_{C.V.}(t)$, then we further can rewrite Eq. (2.73) as:

$$\frac{\delta\phi_{sys}}{\delta t} = \frac{[\phi_{C.V.}(t + \delta t) - \phi_I(t + \delta t) + \phi_{II}(t + \delta t)] - \phi_{C.V.}(t)}{\delta t}$$

$$\implies \frac{\delta\phi_{sys}}{\delta t} = \frac{\phi_{C.V.}(t + \delta t) - \phi_{C.V.}(t)}{\delta t} - \frac{\phi_I(t + \delta t)}{\delta t} + \frac{\phi_{II}(t + \delta t)}{\delta t} \tag{2.74}$$

When δt approaches to zero ($\delta t \longrightarrow 0$), then the left-hand side of Eq. (2.74) is represented by material derivative notation of $\dfrac{\delta\phi_{sys}}{\delta t} = \dfrac{D\phi_{sys}}{Dt}$, which is a Lagrangian representation of the rate of change of ϕ. The rate of change of ϕ in control volume can be also shown as below equation when $\delta t \longrightarrow 0$ using Eq. (2.70):

$$\lim_{\delta t \to 0} \frac{\phi_{C.V.}(t + \delta t) - \phi_{C.V.}(t)}{\delta t} = \frac{\partial\phi_{C.V.}}{\partial t}$$

$$\implies \frac{\partial\phi_{C.V.}}{\partial t} = \frac{\partial\left(\int_{C.V.} \rho\Phi d\forall\right)}{\partial t} \tag{2.75}$$

The control volume (control surfaces) of Figure 2.19 can be represented by replacing the infinitesimal volumes I and II with infinitesimal areas (A_I and A_{II}) multiplied by the distances $\delta\ell_I$ and $\delta\ell_{II}$ in which the control surfaces are moved during δt:

$$\delta\forall_I = A_I\delta\ell_I \tag{2.76}$$

$$\delta\forall_{II} = A_{II}\delta\ell_{II} \tag{2.77}$$

Equations (2.76) and (2.77) can be further represented in terms of constant fluid velocities of V_I and V_{II} of the moved control surfaces, which are also assumed to be normal to the surfaces (we present a more general form later in this section):

$$\delta\forall_I = A_I V_I \delta t \tag{2.78}$$

$$\delta\forall_{II} = A_{II}V_{II}\delta t \tag{2.79}$$

Knowing that the amount of ϕ in regions I and II can be obtained as:

$$\phi_I(t + \delta t) = (\rho_I\Phi_I)(\delta\forall_I) \tag{2.80}$$

$$\phi_{II}(t + \delta t) = (\rho_{II}\Phi_{II})(\delta\forall_{II}) \tag{2.81}$$

We replace Eqs. (2.78) and (2.79) into Eqs. (2.80) and (2.81), respectively:

$$\phi_I(t + \delta t) = (\rho_I\Phi_I)(A_I V_I \delta t) \tag{2.82}$$

$$\phi_{II}(t + \delta t) = (\rho_{II}\Phi_{II})(A_{II}V_{II}\delta t) \tag{2.83}$$

Therefore, the rate of change in outflow and inflow of ϕ through the control surfaces can be shown as:

$$\dot{\phi}_{in} = \lim_{\delta t \to 0} \frac{\phi_I(t + \delta t)}{\delta t} = \frac{(\rho_I\Phi_I)(A_I V_I \delta t)}{\delta t} = \rho_I\Phi_I A_I V_I \tag{2.84}$$

$$\dot{\phi}_{out} = \lim_{\delta t \to 0} \frac{\phi_{II}(t + \delta t)}{\delta t} = \frac{(\rho_{II}\Phi_{II})(A_{II}V_{II}\delta t)}{\delta t} = \rho_{II}\Phi_{II}A_{II}V_{II} \tag{2.85}$$

This will result in rearrangement of the Reynolds transport theorem (Eq. (2.74)) as follows:

$$\frac{D\phi_{\text{sys}}}{Dt} = \frac{\partial\phi_{C.V.}}{\partial t} - \dot\phi_{\text{out}} - \dot\phi_{\text{in}}$$

$$\Longrightarrow \frac{D\phi_{\text{sys}}}{Dt} = \frac{\partial\phi_{C.V.}}{\partial t} - \rho_{\text{I}}\Phi_{\text{I}}A_{\text{I}}V_{\text{I}} + \rho_{\text{II}}\Phi_{\text{II}}A_{\text{II}}V_{\text{II}} \tag{2.86}$$

The above equation provides a relationship between the left-hand side Lagrangian form of the transport equation, which is presented in the system, and its Eulerian form in the right-hand side, which is evaluated as the change of quantity ϕ in the control volume and its fluxes in the control surfaces. Obviously, a more general form of the Reynolds transport theorem, when there are multiple control surfaces can be written as:

$$\frac{D\phi_{\text{sys}}}{Dt} = \frac{\partial\phi_{C.V.}}{\partial t} - \sum_{\text{Control Surface}}\rho\Phi V \cdot A \tag{2.87}$$

Also, one of the assumptions in the above development of the expression for Reynolds transport theorem was that the velocity is normal to the boundary surfaces. Here, we further assume that the velocity on a boundary surface is not normal to it and has an angle of θ between the velocity vector and the outwards pointing normal vector to the surface ($\boldsymbol{n}$) (see Figure 2.20). Hence, the amount of ϕ carried across the infinitesimal area of δA in the time interval of δt is defined as:

$$\delta\phi = \rho\Phi\delta\forall = \rho\Phi(V\cos\theta\delta t)\delta A$$

$$\Longrightarrow \delta\dot\phi_{\text{out}} = \lim_{\delta t \to 0}\frac{\rho\Phi(V\cos\theta\delta t)\delta A}{\delta t} = \rho\Phi V \cdot \boldsymbol{n}\delta A \tag{2.88}$$

Now, we can integrate the entire outflow over the control surface and thus Eq. (2.88) becomes:

$$\dot\phi_{\text{out}} = \int_{C.S.\text{out}} d\dot\phi_{\text{out}} = \int_{C.S.\text{out}} \rho\Phi V \cdot \boldsymbol{n} dA \tag{2.89}$$

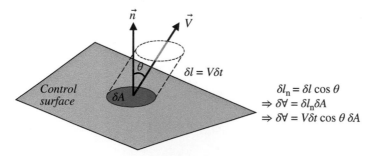

Figure 2.20 Outflow across an infinitesimal control surface area for a non-normal velocity to the boundary surface.

Similarly, we can derive the same equation for $\dot{\phi}_{\text{in}}$ as the inflow rate into the control volume:

$$\dot{\phi}_{\text{in}} = -\int_{C.S.\text{in}} \rho\Phi V \cdot n dA \tag{2.90}$$

Obviously, the net flux of ϕ over the entire control surface becomes:

$$\dot{\phi}_{\text{out}} - \dot{\phi}_{\text{in}} = \int_{C.S.} \rho\Phi V \cdot n dA \tag{2.91}$$

Thus, the Reynolds transport theorem for a non-deforming control volume can be obtained in a general form as:

$$\frac{D\phi_{\text{sys}}}{Dt} = \frac{\partial}{\partial t}\left(\int_{C.V.} \rho\Phi d\forall\right) + \int_{C.S.} \rho\Phi V \cdot n dA \tag{2.92}$$

The flux from boundaries is zero when there is no velocity $V = 0$ (fluid particles are stuck onto the surface) or $\cos\theta$ is zero, implying that fluid particles slide along the surface without entering the boundary surfaces.

When a flow is in steady condition ($\frac{\partial}{\partial t} = 0$), then Eq. (2.92) is simplified to:

$$\frac{D\phi_{\text{sys}}}{Dt} = \int_{C.S.} \rho\Phi V \cdot n dA \tag{2.93}$$

Note that $\frac{D()}{Dt} = \frac{\partial()}{\partial t} + V \cdot \nabla()$, so in the steady case, the whole term of $\frac{D\phi_{\text{sys}}}{Dt}$ is not necessary zero. In an unsteady condition nonetheless, we have $\frac{\partial}{\partial t} \neq 0$, and thus, the unsteady term is non-zero. In case, there is a balance between inflow and outflow fluxes, the $\int_{C.S.}\rho\Phi V \cdot n dA$ term is zero. If the control volume is moving, Eq. (2.92) becomes:

$$\frac{D\phi_{\text{sys}}}{Dt} = \frac{\partial}{\partial t}\left(\int_{C.V.} \rho\Phi d\forall\right) + \int_{C.S.} \rho\Phi W \cdot n dA \tag{2.94}$$

where, W is the relative velocity ($W = V - V_{C.V.}$) and $V_{C.V.}$ is the control volume velocity.

2.5.2 Continuity Equation

One of the first application of Reynolds transport theorem is derivation of the continuity equation. For a non-deforming control volume, we can assume ϕ to be the mass ($\Phi = 1$ and $M_{\text{sys}} = \int_{\text{sys}}\rho d\forall$) and thus we can rewrite Eq. (2.92) to obtain:

$$\frac{DM_{\text{sys}}}{Dt} = \frac{D}{Dt}\int_{\text{sys}} \rho d\forall$$

$$\Longrightarrow \frac{D}{Dt}\int_{\text{sys}} \rho d\forall = \frac{\partial}{\partial t}\left(\int_{C.V.} \rho d\forall\right) + \int_{C.S.} \rho V \cdot n dA \tag{2.95}$$

This equation states that the time rate of change of the mass in a system in an instant of time is equal to summation of the time rate change of the mass in the control volume and the net time rate of mass flow across the control surfaces. Continuity equation is about **conservation of mass** or unchanging contents in a system, so we can simplify Eq. (2.95) as:

$$\frac{DM_{sys}}{Dt} = \frac{D}{Dt} \int_{sys} \rho d\forall = 0$$

$$\Longrightarrow \frac{\partial}{\partial t} \left(\int_{C.V.} \rho d\forall \right) + \int_{C.S.} \rho V \cdot n dA = 0 \tag{2.96}$$

Evidently, in a steady flow, density in addition to other field properties are constant and time rate of change of mass in the control volume is zero ($\frac{\partial}{\partial t} \left(\int_{C.V.} \rho d\forall \right) = 0$). Eq. (2.96) also states that the rates of mass accumulation/depletion in a control volume is equal to **mass flow rates** from control surfaces. Mass flow rate on many occasions is defined as:

$$\dot{m} = \rho Q = \rho A V \tag{2.97}$$

where Q is the volume flow rate (m^3/s). Thus, the mass flow rates across the surface controls can be expressed as:

$$\int_{C.S.} \rho V \cdot n dA = \sum \dot{m}_{out} - \sum \dot{m}_{in} \tag{2.98}$$

Example 2.7
A dehumidifier is extracting water from the moist air in a room. The schematic of a humidifier system is depicted in Figure 2.E7. When the room temperature is 19°C and the relative humidity (RH) is 85%, the well-insulated humidifier extracts 0.16 L/h of water while it has a nominal working airflow of 165 m^3/h. What would be the mass and humidity ratio of the leaving moist air? Note as it is introduced in Chapter 4, the ratio of water vapour mass (m_v) to dry air mass (m_a) is $\omega_1 = 11.730$ g/kg of dry air when RH is 85%.

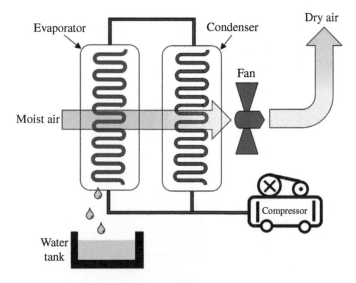

Figure 2.E7 Schematic of a humidifier.

Solution

We assume the control volume as the flowing masses of air and moisture, excluding coils and other elements. We also consider the process as a steady state and thus we can apply Eq. (2.98) to the mass flow rates:

$$\sum \dot{m}_{out} - \sum \dot{m}_{in} = 0$$

$$\Longrightarrow \dot{m}_{a1} + \dot{m}_{v1} = \dot{m}_{a2} + \dot{m}_{v2} + \dot{m}_{l3}$$

Here, point-1, point-2, and point-3 are associated with the inlet, outlet, and water drain points.

We have $\dot{m}_{l3}$ as ($\rho_l = 1000$ kg/m^3):

$$\dot{m}_{l3} = \rho Q_l = 1000 \times 0.00016 = 0.16 \text{ kg/h}$$

For the moist air $\dot{m}_{in} = \dot{m}_{a1} + \dot{m}_{v1}$, the density of the moist air can be calculated approximately from the related tables $\rho_a = 1.200$ kg/m^3 (see Chapter 4):

$$(\dot{m}_{a1} + \dot{m}_{v1}) = \rho_a Q_a = 1.2 \times 165 = 198 \text{ kg/h}$$

Thus, we can also calculate $\dot{m}_{out} = \dot{m}_{a2} + \dot{m}_{v2}$ as:

$$\dot{m}_{out} = 197.5 \text{ kg/h}$$

Also, knowing that $\omega_1 = 11.730$ g/kg:

$$\dot{m}_{a1} + \omega_1 \dot{m}_{a1} = 198 \text{ kg/h}$$

$$\Longrightarrow \dot{m}_{a1} = \dot{m}_{a2} = 195.7 \text{ kg/h}$$

$$\Longrightarrow \dot{m}_{v1} = 2.30 \text{ kg/h}$$

And, for the water:

$$\dot{m}_{v1} = \dot{m}_{v2} + \dot{m}_{l3} = 2.3 \text{ kg/h}$$

$$\Longrightarrow \dot{m}_{v2} = 2.14 \text{ kg/h}$$

This results in: $\omega_2 = \dot{m}_{v2}/\dot{m}_{a2} = 0.0109$ kg/kg of dry air

2.5.3 Momentum Equation

Momentum equation starts with the Newton's second law, stating that the time change rate of the linear momentum in a system is equal to the external forces acting on the system:

$$\frac{D}{Dt} \int_{sys} \rho V d\forall = \sum F_{sys} \tag{2.99}$$

Here, we should note that F_{sys} is identical at an instant of time with forces acting on the content of the control volume as there is no acceleration involved with system as it is moving with a constant velocity, so $F_{sys} = F_{C.~v.}$. Now, for such a system if we consider Eq. (2.92) in a non-deforming control volume, we can replace ϕ with the system momentum

($\int_{sys}\rho V d\forall$) and Φ with the velocity (momentum per unit mass) to achieve the linear momentum equation:

$$\frac{D}{Dt}\int_{sys}\rho V d\forall = \frac{\partial}{\partial t}\left(\int_{C.V.}\rho V d\forall\right) + \int_{C.S.}V\rho V \cdot ndA \tag{2.100}$$

Now, combining Eqs. (2.99) and (2.100), we reach ($F_{sys} = F_{C.V.}$):

$$\sum F_{C.V.} = \frac{\partial}{\partial t}\left(\int_{C.V.}\rho V d\forall\right) + \int_{C.S.}V\rho V \cdot ndA \tag{2.101}$$

2.5.4 Energy Equation

If we apply the Reynolds transfer theorem for the energy in a system with non-deforming control volume when Φ is set to e (thermal energy per unit of mass), we can write:

$$\frac{D}{Dt}\int_{sys}e\rho d\forall = \frac{\partial}{\partial t}\left(\int_{C.V.}e\rho d\forall\right) + \int_{C.S.}e\rho V \cdot ndA \tag{2.102}$$

We can also derive the first law of thermodynamics to state a relation for energy in a system as:

$$\frac{D}{Dt}\int_{sys}e\rho d\forall = \left(\sum\dot{Q}_{in} - \sum\dot{Q}_{out}\right)_{sys} + \left(\sum\dot{W}_{in} - \sum\dot{W}_{out}\right)_{sys} \tag{2.103}$$

This implies that the time rate of increase of energy stored in a system is equal to the net time rate of energy transferred to it with the net heat transfer ($\sum\dot{Q}_{in} - \sum\dot{Q}_{out} = \dot{Q}_{net}$) and the net conducted work ($\sum\dot{W}_{in} - \sum\dot{W}_{out} = \dot{W}_{net}$):

$$\frac{D}{Dt}\int_{sys}e\rho d\forall = \left(\sum\dot{Q}_{in} - \sum\dot{Q}_{out}\right)_{sys} + \left(\sum\dot{W}_{in} - \sum\dot{W}_{out}\right)_{sys}$$
$$\Longrightarrow \frac{D}{Dt}\int_{sys}e\rho d\forall = \left(\dot{Q}_{net} + \dot{W}_{net}\right)_{sys} \tag{2.104}$$

We know from before that we can describe the total energy per unit of mass of particles (e) as the internal energy per unit of mass (u), the kinetic energy per unit of mass ($\frac{V^2}{2}$), and the potential energy per unit of mass (gz):

$$e = u + \frac{V^2}{2} + gz \tag{2.105}$$

Again, we can argue that the net heat transfer and work for the system at an instant time are similarly acting on the control volume, thus:

$$\left(\dot{Q}_{net} + \dot{W}_{net}\right)_{sys} = \left(\dot{Q}_{net} + \dot{W}_{net}\right)_{C.V.}$$
$$\Longrightarrow \frac{\partial}{\partial t}\left(\int_{C.V.}e\rho d\forall\right) + \int_{C.S.}e\rho V \cdot ndA = \left(\dot{Q}_{net} + \dot{W}_{net}\right)_{C.V.} \tag{2.106}$$

So, we can now write the work as the shaft work ($\dot{W}_{shaft}$) plus the normal stress force ($\dot{W}_{ns}$) acting on the control surface as ($\dot{W}_{ns} = \int_{C.S.} \sigma V \cdot ndA$) and combining Eqs. (2.105) and (2.106):

$$\frac{\partial}{\partial t} \left(\int_{C.V.} e\rho d\forall \right) + \int_{C.S.} e\rho V \cdot ndA = \left(\dot{Q}_{net} + \dot{W}_{shaft} \right)_{C.V.} + \int_{C.S.} \sigma V \cdot ndA$$

$$\implies \frac{\partial}{\partial t} \left(\int_{C.V.} e\rho d\forall \right) + \int_{C.S.} \left(u - \frac{\sigma}{\rho} + \frac{V^2}{2} + gz \right) \rho V \cdot ndA = \left(\dot{Q}_{net} + \dot{W}_{shaft} \right)_{C.V.}$$

(2.107)

where we can assume $\sigma = -p$ and, therefore, $h \equiv u + \frac{p}{\rho}$ in Eq. (2.107).

Example 2.8

Assume the dehumidifier of Example 2.7. With a same condition as before, for the well-insulated humidifier extracts 0.16 L/h of water while it has a nominal working airflow of 165 m³/h. If the temperature of the leaving moist air is 21°C, determine the temperature of the drained water.

Solution

The condition is again steady state ($\frac{\partial}{\partial t} \left(\int_{C.V} e\rho d\forall \right) = 0$), and the system is adiabatic ($\dot{Q}_{net} = 0$). We also define the control volume same as before in a sense that it does not include the refrigerant that removes the heat and the compressor, which perform the shaft work ($\dot{W}_{net}$). The mass flow of the moist air is also equal to those existing as water and drier moist air from the control volume as shown in Example 2.7 ($\dot{m}_{a1} + \dot{m}_{v1} = \dot{m}_{a2} + \dot{m}_{v2} + \dot{m}_{l3}$). Now, if the elevation changes are neglected, Eq. (2.107) becomes:

$$\sum \dot{m}_{out} h_{out} - \sum \dot{m}_{in} h_{in} = 0$$

This implies that:

$$\dot{m}_{a1} h_{a1} + \dot{m}_{v1} h_{v1} = \dot{m}_{a2} h_{a2} + \dot{m}_{v2} h_{v2} + \dot{m}_{l3} h_{l3}$$

Enthalpy of the water vapour and liquid can be found from standard tables [3] while air can be assumed as an ideal gas in which the enthalpy can be calculated from $h_a = C_{pa} T_a$ ($C_{pa} = 1.004$ kJ/kg K):

	Mass $\left(\frac{kg}{h}\right)$	Enthalpy (kJ/kg)	Energy (kJ/h)
Air at point-1	195.7	293.3	$\dot{m}_{a1} C_{pa} T_1 = 57,403.7$
Air at point-2	195.7	295.3	$\dot{m}_{a2} C_{pa} T_2 = 57,796.7$
Vapour at point-1	2.30	2536.26	$\dot{m}_{v1} h_{v1} = 5822.3$
Vapour at point-2	2.14	2539.92	$\dot{m}_{v2} h_{v2} = 5424.3$

Hence, for liquid at point-3 ($\dot{m}_{l3} = 0.16$ kg/h):

$$\dot{m}_{l3} h_{l3} = 5 \text{ kJ/h}$$

$$\implies h_{l3} = 31.32 \text{ kJ/kg}$$

There, we can refer to the table of liquid water at saturation pressure and find the temperature to be about $7°$C.

2.6 Differential Form of Fluid Flow

2.6.1 Fluid Element Kinematic

Before we start to derive the governing equation of moving fluid in a differential form, we need to understand the mechanisms in which an infinitesimal fluid element can change in a velocity filed. If this element is assumed to have a rectangular shape as depicted in Figure 2.21, the fluid element's motion from ABCD to reach A'B'C'D' in a short time interval of δt can be then approximated by the superposition of a series of movements and deformations. In specific, we can expect the element to simultaneously undergo linear and angular deformations in addition to translation and rotation in the flow field.

If the fluid element has the velocity of $V = u\hat{i} + v\hat{j} + w\hat{k}$, then the acceleration based on the substantial (material) derivative $\left(\frac{D()}{Dt} = \frac{\partial()}{\partial t} + (V \cdot \nabla)()\right)$ can be expressed as:

$$a = \frac{\partial v}{\partial t} + u\frac{\partial V}{\partial x} + v\frac{\partial V}{\partial y} + w\frac{\partial V}{\partial z} \tag{2.108}$$

With regards to the linear translation of a fluid element as shown in Figure 2.22, we can easily realize that all the nodes in the fluid element will move with a same velocity in the imposed direction, resulting in the relocation of the fluid element. For example, the cuboid element will move by $u\delta t$ in the x-direction and by $v\delta t$ in the y-direction.

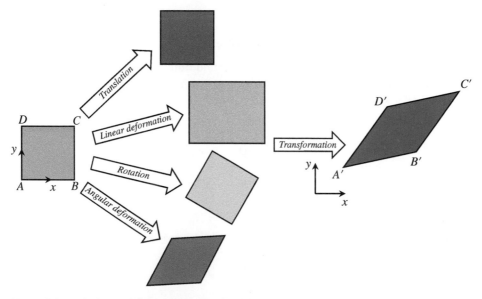

Figure 2.21 Motion and deformation of a fluid element.

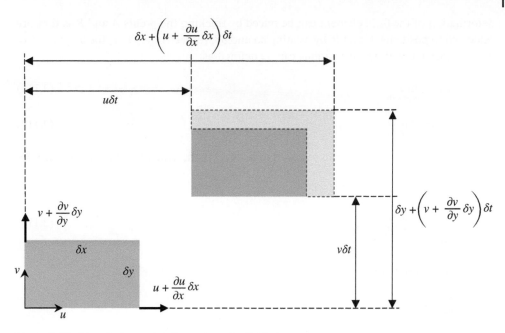

Figure 2.22 Translation and deformation of a fluid element.

If the fluid element is linearly deformed by a single velocity gradient $\frac{\partial u}{\partial x}$, then the stretched fluid element in x-direction will result the points A and C to be stretched and form the new cuboid (see Figure 2.22). The stretched length has the amount of $\frac{\partial u}{\partial x}\delta x\delta t$ during the time interval of δt. A same analogy can be applied to stretch the fluid element with a single velocity gradient $\frac{\partial v}{\partial y}$ to reach a deformation of $\frac{\partial v}{\partial y}\delta y\delta t$ in y-direction.

Rate of the volume change per unit volume in x-direction can be thus calculated as ($\delta\forall = \delta x\delta y\delta z$):

$$\frac{\delta\forall}{\delta t} = \left(\frac{\partial u}{\partial x}\delta x\right)\delta y\delta z$$
$$\Longrightarrow \frac{1}{\delta\forall}\frac{d\forall}{dt} = \frac{\partial u}{\partial x} \tag{2.109}$$

With a similar approach, we can obtain the volume changer per unit volume for y- and z-directions and rewrite the **volumetric dilatation** rate as:

$$\frac{1}{\delta\forall}\frac{d\forall}{dt} = \frac{\partial u}{\partial x} + \frac{\partial v}{\partial y} + \frac{\partial w}{\partial z} = \nabla\cdot\mathbf{V} \tag{2.110}$$

Now, we consider that the fluid element has undergone the angular deformation and rotation. As it can be observed in Figure 2.23, the angular

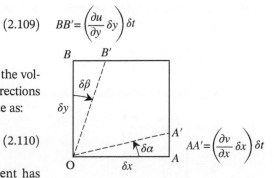

$$BB' = \left(\frac{\partial u}{\partial y}\delta y\right)\delta t$$

$$AA' = \left(\frac{\partial v}{\partial x}\delta x\right)\delta t$$

Figure 2.23 Angular deformation of a fluid element.

deformation of the fluid element can be traced by tracking the points A and B as they are relocated to positions A' and B' by rotating $\delta\alpha$ and $\delta\beta$, respectively. Thus, the angular velocities associated with these rotations can be defined as:

$$\omega_{OA} = \lim_{\delta t \to 0} \frac{\delta\alpha}{\delta t} \tag{2.111}$$

$$\omega_{OB} = \lim_{\delta t \to 0} \frac{\delta\beta}{\delta t} \tag{2.112}$$

The angular deformation (α) is assumed to be very small, which means that $\tan(\delta\alpha) \approx \delta\alpha$. This further results in:

$$\frac{AA'}{OA} = \delta\alpha$$
$$\Longrightarrow \frac{(\partial v/\partial x)\delta t}{\delta x} = \frac{\partial v}{\partial x}\delta t = \delta\alpha \tag{2.113}$$

Also, we know $\tan(\delta\beta) \approx \delta\beta$:

$$\frac{\partial u}{\partial y}\delta t = \delta\beta \tag{2.114}$$

Therefore, we can simplify the angular velocity of OA using Eqs. (2.111) and (2.113):

$$\omega_{OA} = \lim_{\delta t \to 0} \frac{\delta\alpha}{\delta t} = \lim_{\delta t \to 0} \frac{\left(\frac{\partial v}{\partial x}\delta t\right)}{\delta t} = \frac{\partial v}{\partial x} \tag{2.115}$$

With a same approach, we can obtain ω_{OB} from Eqs. (2.112) and (2.114):

$$\omega_{OB} = \lim_{\delta t \to 0} \frac{\delta\beta}{\delta t} = \frac{\partial u}{\partial y} \tag{2.116}$$

Therefore, one can describe the angular deformation to be caused by $\frac{\partial u}{\partial y}$ in x-direction and $\frac{\partial v}{\partial x}$ in y-direction. In a more general form, ω_z can be defined as the rotation about the z-axis, which is the average angular velocities (here, the counterclockwise rotation is assumed to be positive):

$$\omega_z = \frac{1}{2}\left(\frac{\partial v}{\partial x} - \frac{\partial u}{\partial y}\right) \tag{2.117}$$

Similarly, we can derive the below rotations about y- and z-axes:

$$\omega_x = \frac{1}{2}\left(\frac{\partial w}{\partial y} - \frac{\partial v}{\partial z}\right) \tag{2.118}$$

$$\omega_y = \frac{1}{2}\left(\frac{\partial u}{\partial z} - \frac{\partial w}{\partial x}\right) \tag{2.119}$$

Eventually, one can rewrite the rotation vector (ω) as:

$$\omega = \omega_x\hat{i} + \omega_x\hat{j} + \omega_x\hat{k} \tag{2.120}$$

The rotation vector is half of the curl of the velocity vector:

$$\omega = \frac{1}{2}\nabla \times \mathbf{V} \tag{2.121}$$

Hence, the vorticity (ξ) is defined as twice of the rotation vector:

$$\xi = 2\omega = \nabla \times \mathbf{V} \tag{2.122}$$

2.6.2 Differential Form of Continuity Equation

Continuity equation for a non-deforming control volume was expressed by Eq. (2.95):

$$\frac{\partial}{\partial t}\left(\int_{C.V.} \rho d\forall\right) + \int_{C.S.} \rho \mathbf{V} \cdot \mathbf{n} dA = 0 \tag{2.123}$$

For an infinitesimal cubic control volume as seen in Figure 2.24, one can assign all the fluid properties to the centre of the element. Thus, the integral form of the net rate of mass increase can be expressed as:

$$\frac{\partial}{\partial t}\int_{C.V.} \rho d\forall \approx \frac{\partial \rho}{\partial t}\delta x \delta y \delta z \tag{2.124}$$

On the other hand, the net rate of mass inflow through control surfaces can be calculated knowing the mass flow rate per unit of area for different directions. For the x-direction, the mass flow rate entering the centre of the control volume through the control surface of $x - \dfrac{\delta x}{2}$ by using Taylor series and neglecting higher order terms can be derived as:

$$\rho u|_{x-\frac{\delta x}{2}} = \rho u - \frac{\partial(\rho u)}{\partial x}\frac{1}{2}\delta x \tag{2.125}$$

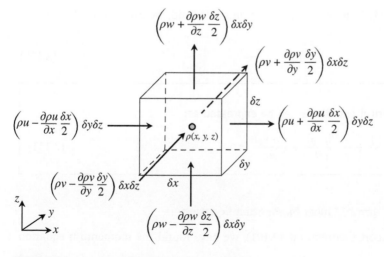

Figure 2.24 A cubic infinitesimal control volume.

Similarly, the mass flow rate exiting the control volume through the control surface of $x = \frac{\delta x}{2}$ can be shown as:

$$\rho u|_{x + \frac{\delta x}{2}} = \rho u + \frac{\partial(\rho u)}{\partial x} \frac{1}{2} \delta x \tag{2.126}$$

If we apply a same approach to the other control surfaces in the y- and z-directions, then we can write the net mass flow rate through the control volume as below:

$$
\left[\left(\rho u - \frac{\partial(\rho u)}{\partial x} \frac{1}{2} \delta x \right) \delta y \delta z - \left(\rho u + \frac{\partial(\rho u)}{\partial x} \frac{1}{2} \delta x \right) \delta y \delta z \right]
$$
$$
+ \left[\left(\rho v - \frac{\partial(\rho v)}{\partial y} \frac{1}{2} \delta y \right) \delta x \delta z - \left(\rho v + \frac{\partial(\rho v)}{\partial y} \frac{1}{2} \delta y \right) \delta x \delta z \right]
$$
$$
+ \left[\left(\rho w - \frac{\partial(\rho w)}{\partial z} \frac{1}{2} \delta z \right) \delta x \delta y - \left(\rho w + \frac{\partial(\rho w)}{\partial z} \frac{1}{2} \delta z \right) \delta x \delta y \right] \tag{2.127}
$$

Now, we can rewrite the net rate of mass flow using Eq. (2.127) as:

$$- \frac{\partial(\rho u)}{\partial x} \delta x \delta y \delta z - \frac{\partial(\rho v)}{\partial y} \delta y \delta x \delta z - \frac{\partial(\rho w)}{\partial z} \delta z \delta x \delta y \tag{2.128}$$

Therefore, as we know that the net rate of mass increase in the control volume is equivalent to the net rate of mass flow through its control surfaces, we can obtain the continuity equation adding Eqs. (2.124) and (2.128) together:

$$\frac{\partial \rho}{\partial t} \delta x \delta y \delta z = - \frac{\partial(\rho u)}{\partial x} \delta x \delta y \delta z - \frac{\partial(\rho v)}{\partial y} \delta y \delta x \delta z - \frac{\partial(\rho w)}{\partial z} \delta z \delta x \delta y$$

$$\implies \frac{\partial \rho}{\partial t} + \frac{\partial(\rho u)}{\partial x} + \frac{\partial(\rho v)}{\partial y} + \frac{\partial(\rho w)}{\partial z} = 0$$

$$\implies \frac{\partial \rho}{\partial t} + \nabla \cdot \rho \mathbf{V} = 0 \tag{2.129}$$

Obviously, in a steady state flow of an incompressible fluid, we can further simplify the continuity equation of Eq. (2.129) to:

$$\nabla \cdot \mathbf{V} = 0 \tag{2.130}$$

Note

Continuity in cylindrical coordinate can be expressed as:

$$\frac{\partial \rho}{\partial t} + \frac{1}{r} \frac{\partial(r \rho v_r)}{\partial r} + \frac{1}{r} \frac{\partial(\rho v_\theta)}{\partial \theta} + \frac{\partial(\rho v_z)}{\partial z} = 0 \tag{2.131}$$

2.6.3 Differential Form of Linear Momentum Equation

From Reynolds transport theorem, Eq. (2.101), we have found the momentum equation as below:

$$\sum F_{CV} = \frac{\partial}{\partial t} \left(\int_{C.V.} \rho \mathbf{V} d\mathbb{V} \right) + \int_{C.S.} \mathbf{V} \rho \mathbf{V} . \mathbf{n} dA \tag{2.132}$$

Figure 2.25 Normal and searing stresses on a control volume.

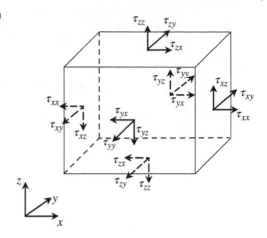

Forces that act on the infinitesimal control volume are in general described as those acting on the surface of the control volume known as ***surface forces*** (F_s), and those distributed through the control volume are known as ***body forces*** (F_b). Examples of surface forces are pressure and viscous forces, and examples for body forces are gravity, centrifuge, Coriolis, and electromagnetic forces. To provide a practical ***double subscript*** notation to identify acting forces and the related stresses of τ_{ij} on a particular control surface, we consider the first subscript 'i' as the direction of the normal vector acting on that control surface, and the second subscript 'j' as the direction of that particular force. For example, τ_{zz} (also shown by σ_{zz}) illustrates the normal stress in Figure 2.25 while τ_{zy} and τ_{zx}, respectively depict shearing stresses in the y- and x-directions on the associated control surface. Here, we can also define the positive value of a stress as its direction is aligned with the positive coordinate directions. Hence, τ_{zz} is assumed to be positive in the top plane. Also, it has a positive value in the bottom plane since it has a negative direction aligned with the negative coordinate direction. This implies that outwards stresses have positive values.

Similar to before, we can express the value of all stresses in the centre of the infinitesimal control volume using Taylor series though neglecting the higher order terms. Thus, in the x-direction we can write both surface forces as:

$$\delta F_{sx} = \left[\left(\sigma_{xx} + \frac{\partial \sigma_{xx}}{\partial x}\frac{1}{2}\delta x \right)\delta y \delta z + \left(\tau_{yx} + \frac{\partial \tau_{yx}}{\partial y}\frac{1}{2}\delta y \right)\delta x \delta z + \left(\tau_{zx} + \frac{\partial \tau_{zx}}{\partial z}\frac{1}{2}\delta z \right)\delta x \delta y \right]$$

$$- \left[\left(\sigma_{xx} - \frac{\partial \sigma_{xx}}{\partial x}\frac{1}{2}\delta x \right)\delta y \delta z + \left(\tau_{yx} - \frac{\partial \tau_{yx}}{\partial y}\frac{1}{2}\delta y \right)\delta x \delta z + \left(\tau_{zx} + \frac{\partial \tau_{zx}}{\partial z}\frac{1}{2}\delta z \right)\delta x \delta y \right]$$

$$\implies \delta F_{sx} = \frac{\partial \sigma_{xx}}{\partial x}\delta x \delta y \delta z + \frac{\partial \tau_{yx}}{\partial y}\delta y \delta x \delta z + \frac{\partial \tau_{zx}}{\partial z}\delta z \delta x \delta y = \left(\frac{\partial \sigma_{xx}}{\partial x} + \frac{\partial \tau_{yx}}{\partial y} + \frac{\partial \tau_{zx}}{\partial z} \right)\delta x \delta y \delta z$$

$$(2.133)$$

Similarly, in y- and z-directions, we can find (Figure 2.26):

$$\delta F_{sy} = \left(\frac{\partial \tau_{xy}}{\partial x} + \frac{\partial \sigma_{yy}}{\partial y} + \frac{\partial \tau_{zy}}{\partial z} \right)\delta x \delta y \delta z \qquad (2.134)$$

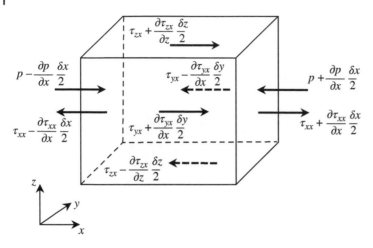

Figure 2.26 Normal and searing stresses.

$$\delta F_{sz} = \left(\frac{\partial \tau_{xz}}{\partial x} + \frac{\partial \tau_{yz}}{\partial y} + \frac{\partial \sigma_{zz}}{\partial z}\right)\delta x \delta y \delta z \tag{2.135}$$

We develop the equation of linear momentum in the x-direction as:

$$\delta F_x = \delta F_{sx} + \delta F_{bx} = \delta m a_x \tag{2.136}$$

Knowing that δF_{bx} is the body force and $\delta m = \rho \delta x \delta y \delta z$, we can apply Eqs. (2.133)–(2.135) into the linear momentum equation of Eq. (2.136):

$$\left(\frac{\partial \sigma_{xx}}{\partial x} + \frac{\partial \tau_{yx}}{\partial y} + \frac{\partial \tau_{zx}}{\partial z}\right)\delta x \delta y \delta z + \rho g_x \delta x \delta y \delta z = \rho \delta x \delta y \delta z \left(\frac{\partial u}{\partial t} + u\frac{\partial u}{\partial x} + v\frac{\partial u}{\partial y} + w\frac{\partial u}{\partial z}\right)$$

$$\Longrightarrow \left(\frac{\partial \sigma_{xx}}{\partial x} + \frac{\partial \tau_{yx}}{\partial y} + \frac{\partial \tau_{zx}}{\partial z}\right) + \rho g_x = \rho\left(\frac{\partial u}{\partial t} + u\frac{\partial u}{\partial x} + v\frac{\partial u}{\partial y} + w\frac{\partial u}{\partial z}\right) \tag{2.137}$$

Note, the body force here is the gravitational force of $\delta m g_x$ in the x-direction and a_x is replaced using Eq. (2.108). For other directions, we can also find:

$$\left(\frac{\partial \tau_{xy}}{\partial x} + \frac{\partial \sigma_{yy}}{\partial y} + \frac{\partial \tau_{zy}}{\partial z}\right) + \rho g_y = \rho\left(\frac{\partial v}{\partial t} + u\frac{\partial v}{\partial x} + v\frac{\partial v}{\partial y} + w\frac{\partial v}{\partial z}\right) \tag{2.138}$$

$$\left(\frac{\partial \tau_{xz}}{\partial x} + \frac{\partial \tau_{yz}}{\partial y} + \frac{\partial \sigma_{zz}}{\partial z}\right) + \rho g_z = \rho\left(\frac{\partial w}{\partial t} + u\frac{\partial w}{\partial x} + v\frac{\partial w}{\partial y} + w\frac{\partial w}{\partial z}\right) \tag{2.139}$$

These equations are general differential equations of motion for both solid and fluid continuums. However, from what we learned before from Newtonian fluids, we can replace the stresses with the related equations. We start with a fluid with negligible viscosity, which enable us to neglect shearing stresses and has an ***inviscid*** (***non-viscous***) fluid. In fluids, we can also define pressure as the negative of normal stresses independent from directions $(-p = \sigma_{xx} = \sigma_{yy} = \sigma_{zz})$.

2.6.4 Euler's Equation of Motion

The derived equation of the motion in the previous section was found by **Leonard Euler** (1707–1783) where he assumed the fluid to be inviscid and thus shearing stresses to be negligible in Eqs. (2.137)– (2.139) ($-p = \sigma_{xx} = \sigma_{yy} = \sigma_{zz}$):

$$\frac{\partial p}{\partial x} + \rho g_x = \rho \delta \left(\frac{\partial u}{\partial t} + u\frac{\partial u}{\partial x} + v\frac{\partial u}{\partial y} + w\frac{\partial u}{\partial z} \right) \tag{2.140}$$

$$\frac{\partial p}{\partial y} + \rho g_y = \rho \left(\frac{\partial v}{\partial t} + u\frac{\partial v}{\partial x} + v\frac{\partial v}{\partial y} + w\frac{\partial v}{\partial z} \right) \tag{2.141}$$

$$\frac{\partial p}{\partial z} + \rho g_z = \rho \left(\frac{\partial w}{\partial t} + u\frac{\partial w}{\partial x} + v\frac{\partial w}{\partial y} + w\frac{\partial w}{\partial z} \right) \tag{2.142}$$

Euler's equations can be further simplified under different assumptions to find simpler sets of equations. Example is the Bernoulli equation, which can be derived across a streamline in Euler flow. If the flow is irrational (**potential flow**), then Bernoulli equation can be developed in any two points of an arbitrary flow. Potential flow is a traditional assumption to address various type of flows around certain body shapes such as circular cylinder or Rankine oval immersed in a uniform flow. Whilst this technique is widely adapted to many engineering problems and has provided useful information in many regions of a flow where the viscosity is negligible, and velocity is high. The major weakness of Euler's equations is related to its accuracy in regions where the flow cannot be assumed to be inviscid, or the flow is decelerating (e.g. boundary layer or wake region).

2.6.5 Navier–Stokes Equations

For incompressible Newtonian fluids, we can relate the normal and shearing stresses to the rate of deformation and further develop Eqs. (2.137)–(2.139). For example, in the x-direction, we can apply the below values for the normal and shearing stresses (for further details see the recommended reading references in the end of this chapter):

$$\sigma_{xx} = -p + 2\mu\frac{\partial u}{\partial x} \tag{2.143}$$

$$\tau_{xy} = \tau_{yx} = \mu \left(\frac{\partial u}{\partial y} + \frac{\partial v}{\partial x} \right) \tag{2.144}$$

Also, in the y-direction:

$$\sigma_{yy} = -p + 2\mu\frac{\partial v}{\partial y} \tag{2.145}$$

$$\tau_{yz} = \tau_{zy} = \mu \left(\frac{\partial v}{\partial z} + \frac{\partial w}{\partial y} \right) \tag{2.146}$$

And, finally in the z-direction, we can write:

$$\sigma_{zz} = -p + 2\mu\frac{\partial w}{\partial z} \tag{2.147}$$

$$\tau_{zx} = \tau_{xz} = \mu\left(\frac{\partial w}{\partial x} + \frac{\partial u}{\partial z}\right) \tag{2.148}$$

Now, we rearrange the general form of linear momentum equations of Eq. (2.137) in the x-direction as below:

$$\rho\left(\frac{\partial u}{\partial t} + u\frac{\partial u}{\partial x} + v\frac{\partial u}{\partial y} + w\frac{\partial u}{\partial z}\right) = \rho g_x + \left(+\frac{\partial\left(-p + 2\mu\frac{\partial u}{\partial x}\right)}{\partial x} + \frac{\partial\mu\left(\frac{\partial u}{\partial y} + \frac{\partial v}{\partial x}\right)}{\partial y} + \frac{\partial\mu\left(\frac{\partial w}{\partial x} + \frac{\partial u}{\partial z}\right)}{\partial z}\right)$$

$$\Longrightarrow \rho\left(\frac{\partial u}{\partial t} + u\frac{\partial u}{\partial x} + v\frac{\partial u}{\partial y} + w\frac{\partial u}{\partial z}\right) = -\frac{\partial p}{\partial x} + \rho g_x + \mu\left(2\frac{\partial^2 u}{\partial x^2} + \frac{\partial^2 u}{\partial y^2} + \frac{\partial^2 v}{\partial x\partial y} + \frac{\partial^2 w}{\partial x\partial z} + \frac{\partial^2 u}{\partial z^2}\right)$$

$$\Longrightarrow \rho\left(\frac{\partial u}{\partial t} + u\frac{\partial u}{\partial x} + v\frac{\partial u}{\partial y} + w\frac{\partial u}{\partial z}\right) = -\frac{\partial p}{\partial x} + \rho g_x + \mu\left(2\frac{\partial^2 u}{\partial x^2} + \frac{\partial^2 u}{\partial y^2} + \frac{\partial u}{\partial x}\left(\frac{\partial v}{\partial y} + \frac{\partial w}{\partial z}\right) + \frac{\partial^2 u}{\partial z^2}\right)$$

$$\tag{2.149}$$

Note that for a steady state flow of an incompressible fluid, using the continuity Eq. (2.130), we have $\frac{\partial v}{\partial y} + \frac{\partial w}{\partial z} = -\frac{\partial u}{\partial x}$. This implies that Eq. (2.149) becomes:

$$\rho\left(\frac{\partial u}{\partial t} + u\frac{\partial u}{\partial x} + v\frac{\partial u}{\partial y} + w\frac{\partial u}{\partial z}\right) = -\frac{\partial p}{\partial x} + \rho g_x + \mu\left(\frac{\partial^2 u}{\partial x^2} + \frac{\partial^2 u}{\partial y^2} + \frac{\partial^2 u}{\partial z^2}\right) \tag{2.150}$$

With a similar approach, we can find:

$$\rho\left(\frac{\partial v}{\partial t} + u\frac{\partial v}{\partial x} + v\frac{\partial v}{\partial y} + w\frac{\partial v}{\partial z}\right) = -\frac{\partial p}{\partial y} + \rho g_y + \mu\left(\frac{\partial^2 v}{\partial x^2} + \frac{\partial^2 v}{\partial y^2} + \frac{\partial^2 v}{\partial z^2}\right) \tag{2.151}$$

$$\rho\left(\frac{\partial w}{\partial t} + u\frac{\partial w}{\partial x} + v\frac{\partial w}{\partial y} + w\frac{\partial w}{\partial z}\right) = -\frac{\partial p}{\partial z} + \rho g_z + \mu\left(\frac{\partial^2 w}{\partial x^2} + \frac{\partial^2 w}{\partial y^2} + \frac{\partial^2 w}{\partial z^2}\right) \tag{2.152}$$

These sets of partial differential equations (PDEs) of Eqs. (2.150)–(2.152) in addition to the continuity Eq. (2.129) (conservation of mass) are called Navier–Stokes (NS) equations after the French mathematician L.M.H. Navier (1785–1836), and the English engineer G.G. Stokes (1819–1903). These sets of equations can technically form a set of well-posed mathematical system of equations where u, v, w, and p are unknowns of the system. Also, when we have a compressible and non-isothermal flow, two more variables of ρ and T will be added to the system of the equations. In such scenarios, two more sets of equations known as state and energy equations are required to simultaneously be considered with the momentum and continuity equations. While these sets of equations can be solved in some particular and simplified scenarios (e.g. Couette flow, Poiseuille flow, etc.), the non-linear nature of these PDEs makes the exact solutions of the NS equations impossible in many complex cases. Numerical methods are among alternative techniques, enabling us to find solution in many general aspects of the NS equations. Chapters 10–12 comprehensively discuss one of the numerical techniques called finite volume method to solve the NS equations.

Note

Navier-Stokes equation in cylindrical coordinate can be expressed as:

$r - direction$:

$$\rho\left(\frac{\partial v_r}{\partial t} + v_r\frac{\partial v_r}{\partial r} + \frac{v_\theta}{r}\frac{\partial v_r}{\partial \theta} - \frac{v_\theta^2}{r} + v_z\frac{\partial v_r}{\partial z}\right) = -\frac{\partial p}{\partial r} + \rho g_r$$

$$+ \mu\left[\frac{1}{r}\frac{\partial}{\partial r}\left(r\frac{\partial v_r}{\partial r}\right) - \frac{v_r^2}{r^2} + \frac{1}{r^2}\frac{\partial^2 v_r}{\partial \theta^2} - \frac{2}{r^2}\frac{\partial v_\theta}{\partial \theta} + \frac{\partial^2 v_r}{\partial z^2}\right] \tag{2.153}$$

$\theta - direction$:

$$\rho\left(\frac{\partial v_\theta}{\partial t} + v_r\frac{\partial v_\theta}{\partial r} + \frac{v_\theta}{r}\frac{\partial v_\theta}{\partial \theta} + \frac{v_r v_\theta}{r} + v_z\frac{\partial v_\theta}{\partial z}\right) = -\frac{1}{r}\frac{\partial p}{\partial \theta} + \rho g_\theta$$

$$+ \mu\left[\frac{1}{r}\frac{\partial}{\partial r}\left(r\frac{\partial v_\theta}{\partial r}\right) - \frac{v_\theta}{r^2} + \frac{1}{r^2}\frac{\partial^2 v_\theta}{\partial \theta^2} + \frac{2}{r^2}\frac{\partial v_r}{\partial \theta} + \frac{\partial^2 v_\theta}{\partial z^2}\right] \tag{2.154}$$

$z - direction$:

$$\rho\left(\frac{\partial v_z}{\partial t} + v_r\frac{\partial v_z}{\partial r} + \frac{v_\theta}{r}\frac{\partial v_z}{\partial \theta} + v_z\frac{\partial v_z}{\partial z}\right) = -\frac{\partial p}{\partial z} + \rho g_z + \mu\left[\frac{1}{r}\frac{\partial}{\partial r}\left(r\frac{\partial v_z}{\partial r}\right) + \frac{1}{r^2}\frac{\partial^2 v_z}{\partial \theta^2} + \frac{\partial^2 v_z}{\partial z^2}\right]$$

$$\tag{2.155}$$

Example 2.9

Couette flow is a classic shear-driven fluid motion as depicted in Figure 2.E9.1. Two-dimensional viscous and laminar flow occurs amongst two infinite parallel plates with a distance of D while one of the plates moves with a constant relative velocity of U. First, simplify and solve the NS equations for Couette flow. Then, find the velocity profile between the parallel plates assumed as a 2D pipe with a diameter (D) of 0.1m ($\frac{\partial p}{\partial x}$ is constant). The fluid is water at 20 °C.

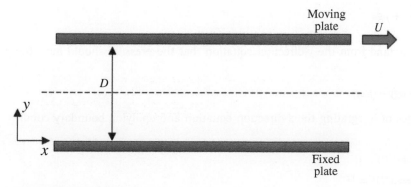

Figure 2.E9.1 Couette flow.

Solution

Assumptions are that the flow is 2D ($w = 0$) and steady, there is no body force, and it is fully developed in the flow x-direction ($\frac{\partial u}{\partial x} = 0$).

We can first rewrite the continuity equation (Eq. (2.129)) as:

$$\frac{\partial u}{\partial x} = 0$$

$$\Longrightarrow \frac{\partial u}{\partial x} + \frac{\partial v}{\partial y} = 0$$

$$\Longrightarrow \frac{\partial v}{\partial y} = 0$$

First, Eqs. (2.150) and (2.151) for a 2D flow (there is no $w\frac{\partial u}{\partial z}$, $w\frac{\partial v}{\partial z}$, $\frac{\partial^2 u}{\partial z^2}$ and $\frac{\partial^2 v}{\partial z^2}$ terms), the velocity component is zero ($v = 0$) in the y-direction ($v\frac{\partial u}{\partial y} = v\frac{\partial v}{\partial y} = u\frac{\partial v}{\partial x} = 0$) and $\mu\left(\frac{\partial^2 v}{\partial x^2} + \frac{\partial^2 v}{\partial y^2}\right) = 0$, and steady ($\frac{\partial u}{\partial t} = \frac{\partial v}{\partial t} = 0$) flow can be simplified to:

$$\rho\left(u\frac{\partial u}{\partial x}\right) = -\frac{\partial p}{\partial x} + \rho g_x + \mu\left(\frac{\partial^2 u}{\partial x^2} + \frac{\partial^2 u}{\partial y^2}\right)$$

$$0 = -\frac{\partial p}{\partial y} + \rho g_y$$

Then, we assume that there is no body force in the x-direction ($\rho g_x = 0$) and the flow is fully developed in the x-direction for the velocity components ($\frac{\partial}{\partial x} = \frac{\partial^2}{\partial x^2} = 0$). Thus, these assumptions further result that Eqs. (2.150) and (2.151) turn to ($g_y = g$):

$$0 = -\frac{\partial p}{\partial x} + \mu\left(\frac{\partial^2 u}{\partial y^2}\right)$$

$$0 = -\frac{\partial p}{\partial y} + \rho g$$

First, we understand from the y-direction equation that the pressure should have the below form:

$$p(y) = -\rho gy + f(x)$$

After two times of integrating the x-direction equation and applying boundary conditions of:

$$@y = 0 \Longrightarrow u(0) = 0$$
$$@y = D \Longrightarrow u(D) = U$$

The velocity can be obtained as:

$$u(y) = \frac{1}{2\mu}\left(\frac{\partial p}{\partial x}\right)(y^2 - Dy) + U\frac{y}{D}$$

Note that $C = \dfrac{\partial p}{\partial x}$ has a constant value as it is only a function of y, and $f(x)$ should be defined as an input to the above equation. If $f(x) = 0$, then the velocity becomes ($\dfrac{\partial p}{\partial x} = 0$):

$$u(y) = U\frac{y}{D}$$

Now, if we would like to keep the flow laminar in the pipe, we should keep the Reynolds number below $Re < 2100$. This implies that if the properties of water at $20°C$ are $\mu = 1.0016 \times 10^{-3}$ Pa/s and $\rho = 0.9982 \times 10^3$ kg/m^3:

$$u_{\text{lam, max}} = \frac{2100\mu}{\rho D} = 2.107 \times 10^{-3} \text{ m/s}$$

We can also assume $C = \pm 0.001$ pa/m and plot the normalized velocity against different Re numbers by changing the value of U as illustrated in Figure 2.E9.2.

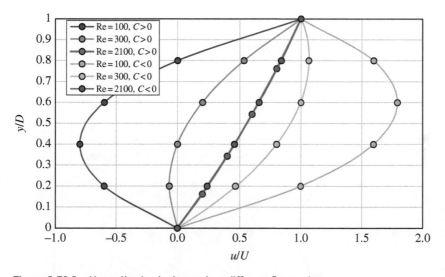

Figure 2.E9.2 Normalized velocity against different Re numbers.

Example 2.10
Consider the 2D parallel plates of Example 2.9 as an infinite pipe again. This time, if the moving plate is also fixed, first simplify, and solve the NS equations for a fully developed and laminar flow. Then, find the acceptable range of $\dfrac{\partial p}{\partial x}$?

Solution

The assumptions and simplifications of NS equations are exactly the same as Example 2.9. The only difference is on the employed boundary conditions as no-slip on the solid surfaces (note that we remove the coordinate to the centre of the pipe):

$$@y = -D/2 \Longrightarrow u(-D/2) = 0$$
$$@y = D/2 \Longrightarrow u(D/2) = 0$$

The velocity can be obtained as:

$$u(y) = \frac{1}{2\mu}\left(\frac{\partial p}{\partial x}\right)\left(y^2 - \frac{D^2}{4}\right)$$

Now, similar to Example 2.9, we would like to keep the flow laminar and thus $u_{\text{lam, max}} = 2.107 \times 10^{-3}$ m/s. This implies that the minimum $\frac{\partial p}{\partial x}$ can be calculated as:

$$\left(\frac{\partial p}{\partial x}\right)_{\min} = \frac{2\mu u_{\text{lam, max}}}{\left(y^2 - \frac{D^2}{4}\right)}$$

$$\Longrightarrow \left(\frac{\partial p}{\partial x}\right)_{\min} = -1.69 \times 10^{-5} \text{ Pa/m}$$

Thus, we can now plot the parabolic velocity profile across the pipe section for various $-1.6897 \times 10^{-5} < \frac{\partial p}{\partial x} < 0$ as shown in Figure 2.E10.

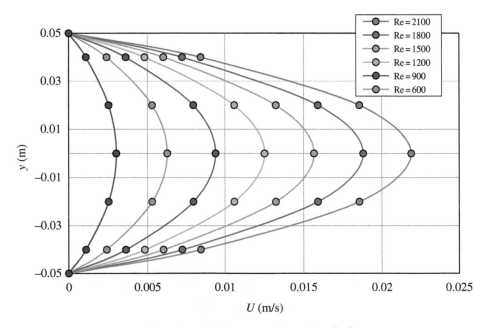

Figure 2.E10 Velocity profile across the pipe section for various ∂p/∂x.

2.7 Dimensionless Analysis

2.7.1 Flow Similarities

There are similarities that exist in different flow problem in the nature. If such similarities can be understood, then a wide range of problems, occurring in different scales and condition, can be experimentally assessed in laboratories transformed to sub-scale models for design purpose or validation of numerical models. The findings then can be extrapolated to the full-scale problems at which an experimental study might be impossible, impractical, monotonous, or expensive.

In building science, there are a wide range of flow problems that can be experimentally assessed in a subscale scenario as explained in Chapter 3. Some famous examples are flow in pipes, natural ventilation through buildings, free and forced convection over surfaces, impinging jet flow from mechanical ventilation, etc. In general, such similarities are explained with ***dimensionless groups*** (***dimensionless products***) associated to a specific flow.

The first step to explore such dimensionless groups is to understand the key parameters in a specific flow, which can be achieved through initial observation, fundamental knowledge, or years of experience about that specific flow. For example, we assume an experiment to calculate pressure drop of a flow in a pipe. As it can be seen in Figure 2.27, it would not be difficult to associate the pressure drop ($\Delta p_L = \frac{p_1 - p_2}{L}$) per unit of length ($L$) of an inviscid flow in a pipe into its physical geometry such as its diameter (D) in addition to the condition of the internal flow such as the mean velocity (V), viscosity (μ), and density (ρ). Thus, one can establish a hypothetical relationship as:

$$\Delta p_L = f(D, V, \mu, \rho) \tag{2.156}$$

Now, we can conduct an experiment and start to alter each parameter one by one while fixing other parameters to be able to observe the associated pressure drops. This process would be monotonous, expensive, and time consuming in most of the scenarios if we would like to generate a general chart for the pressure drop in the pipe flows. For instance, changing pipe diameter to very small or large sizes may make the experiment challenging while we need to also make multiple prototypes for the experiment, which may make it costly. If we change density and viscosity, then we need to use a range of operating fluids while some

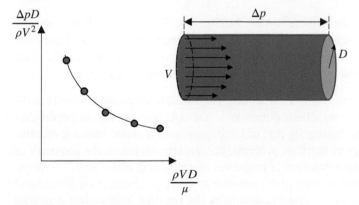

Figure 2.27 Key parameters in pressure drop of flow in pipes.

of them might be hazardous or again expensive to work with. Eventually, we may end up with multiple correlations that it would be difficult to either be employed by end-users or to be merged into a single handy correlation. Thus, grouping of some contributing parameters together and changing them instead in experiments would be an alternative, enabling us to save a series of redundant experiments and post calculations. Through the next section, we will show the associated groups for the later parameters of the pressure drop in the pipes as two dimensionless groups of $\dfrac{D\Delta p_L}{\rho V^2}$ and $\dfrac{\rho VD}{\mu}$. The result can be presented as a single useful and practical correlation as depicted in Figure 2.27. Therefore, with the new grouping technique, there is no further need to change the working fluids or pipe dimensions while we can design a practical experiment with a simple and safe operating fluid.

2.7.2 Buckingham π-Theorem

The number of independent dimensionless groups required in a flow to represents the relation between variables was first formulized by Edgar Buckingham (1867–1940). In his main π-theorem, he described that the number of required independent dimensionless groups to present a meaningful relationship amongst parameters can be found as $n - m$ where n is the number of **key variables** (parameters) and m is the number of **basic dimensions** involved in the variables. Basic dimensions are described earlier in this chapter, usually are referred to mass (M), length (L), and time (T). Thus, in the mathematical terms, we can show that if an equation is defined based on n key variables:

$$y_1 = f(y_2, y_3, ..., y_n) \tag{2.157}$$

Then, it can be rearranged and expressed in terms of $n - m$ independent π-groups:

$$\pi_1 = \phi(\pi_2, \pi_3, ..., \pi_{n-m}) \tag{2.158}$$

For example, for the pressure drop equation in the previous section, we have five key variables $(n = 5)$ as fluid density, fluid viscosity, conduit's diameter, flow velocity, and flow pressure drop. Also, the number of basic dimensions involved are three $(m = 3)$, including M, L, and T. This implies that we can describe the flow characteristics with two $(n - m = 2)$ independent π-groups:

$$\pi_1 = \phi(\pi_2) \tag{2.159}$$

There are several systematic methods to form the dimensionless groups such as method of **repeating variables**, **step-by-step method**, **the exponent method**, etc. Since, explaining all these methods is beyond the scope of this book, here, we use a simplified algebraic approach of repeating variables where we need to follow the main below steps:

i) Identification of n key parameters or variables (y_i) in the expected correlation and rewriting these n variables in terms of basic dimensions. It should be noted that determination of variables is the most challenging part of the process as it is stated before it requires fundamental knowledge of the flow problem, particularly related to the geometry of the system as well as understandings of properties and condition of the flow. Therefore, in many classical experiments such as the pressure drop in pipes, years of experience and trial-and-error processes are involved to determine the required independent π-groups.

ii) At the second step, we must identify basic dimensions (m) involved in the key variables, to be able to calculate the number of expected independent π-groups ($n - m$). Then, we must write the key variables in terms of the involved basic dimensions

iii) At this stage, we select a number of **repeating variables** (y_1, y_2, ..., y_m) equal to the number of involved basic dimensions (m) from the list of key variables (n). Later on, we combine the remaining key variables with these repeating variable $n - m$ to form $n - m$ number of the π-groups. For the selection of the repeating variables, note that (1) all basic dimensions should be included in their combinations, (2) each repeating variable should be independent, meaning that it cannot be reproduced by any combination of other repeating variables, and (3) selection of the dependent variable should be avoided as the repeating variable as it can appear in multiple π-groups.

iv) At the last step, we form the below algebraic equation of π-groups. Here, we put one of the remaining key variables with combination of repeating variables raised with a component:

$$[\pi_i] = \left[y_j\right] \times \left([y_1]^{a_i} \times [y_2]^{b_i} \cdots \times [y_m]^{z_i}\right) \tag{2.160}$$

where y_j are the non-repeating key variables.

Note that to obtain independent π-groups, both sides of Eq. 2.160 should satisfy a same dimension. Therefore, we can identify a system of equation with m equations based on the number of involved basic dimensions. Once we have found the first π-group, then we can proceed to find the rest of them by repeating a same procedure as presented in this section.

Example 2.11

Find the independent π-groups for the laminar pressure drop in pipes.

Solution

We follow the steps introduced in the method of repeating variables:

Step i: we identify the key variables to be Δp_L, D, V, μ, and ρ, so $n = 5$.

Step ii: the number of basic dimensions is $m = 3$ when we scrutinize the key variables and identify M, L, and T to be involved in all the key variables:

$$\Delta p_L = ML^{-2}T^{-2}$$
$$D = L$$
$$V = LT^{-1}$$
$$\mu = ML^{-1}T^{-1}$$
$$\rho = ML^{-3}$$

Now, we can understand the number of independent π-groups to be $n - m = 2$.

Step iii: for the selection of three repeating variables, we can choose D (L), V (LT^{-1}) and ρ (ML^{-3}) since:

1) We can easily see M, L, and T within them,
2) None of these key variables can be reproduced by any combination of other repeating variables as the possible combinations are DV (L^2T^{-1}), $D\rho$ (ML^{-2}), $V\rho$ ($ML^{-2}T^{-1}$), and $DV\rho$ ($ML^{-1}T^{-1}$),
3) $\Delta p_L = ML^{-2}T^{-2}$ and $ML^{-1}T^{-1}$ as dependent variables are not similar to the repeating variables.

Step iv: we can now form $m - n = 2$ sets of equation associated with the number of π-groups as:

$$[\pi_i] = \Theta_i\left(D^{a_i} V^{b_i} \rho^{c_i}\right)$$

where Θ_i are non-repeating variables (here Δp_L and μ).

Or:

$$[\pi_1] = \left[ML^{-2}T^{-2}\right]\left([L]^{a_1} \times [LT^{-1}]^{b_1} \times [ML^{-3}]^{c_1}\right)$$

$$[\pi_2] = \left[ML^{-1}T^{-1}\right]\left([L]^{a_2} \times [LT^{-1}]^{b_2} \times [ML^{-3}]^{c_2}\right)$$

Finally, knowing that each π-group should be equal to $M^0 L^0 T^0$, as they are supposed to be dimensionless, we can form the system of m equations as:

For π_1:

for $M \Longrightarrow 0 = 1 + c_1$

for $L \Longrightarrow 0 = -2 + a_1 + b_1 - 3c_1$

for $T \Longrightarrow 0 = -2 - b_1$

$\Longrightarrow a_1 = 1, b_1 = -2, c_1 = -1$

$\Longrightarrow \pi_1 = \dfrac{\Delta p_L D}{\rho V^2}$

Similarly, for π_2:

for $M \Longrightarrow 0 = 1 + c_2$

for $L \Longrightarrow 0 = -1 + a_2 + b_2 - 3c_2$

for $T \Longrightarrow 0 = -1 - b_2$

$\Longrightarrow a_2 = -1, b_2 = -1, c_2 = -1$

$\Longrightarrow \pi_2 = \dfrac{\mu}{\rho V D}$

Thus, we can express the π-groups as:

$$\frac{\Delta p_L D}{\rho V^2} = \phi\left(\frac{\mu}{\rho V D}\right)$$

We can clearly notice that $\dfrac{\rho D V}{\mu}$ is previously defined as the Reynolds number and the correlation between these two groups can provide us with a same universal graph as discussed in the following sections (also see Figure 2.27).

2.8   Internal Flow

Internal flow is related to the fluid motion in the closed conduits such as pipes and ducts while in most of the cases the flowing fluid is assumed to be incompressible. In general, internal flows can be explained as the circulation of fluid by a pressure gradient imposed through a conduit. It is a valid assumption to neglect the impact of gravity in most of the case studies for further simplification of the analysis.

Figure 2.17 shows a fluid flow in an internal flow. A common classification of internal flows is based on the flow being laminar, transient, or turbulent, depending on the Reynolds number of the conduit calculated using its characteristic length. Each of the mentioned regimes demonstrates different flow mechanism. In a laminar flow, we expect the fluid particles to move in a direction parallel to the conduit's direction while mixing between layers of fluid perpendicular to the flow direction is barely expected to occur. In a transitional regime, an irregular mixing starts between layers of fluid in the normal to the flow direction while this random and irregular motion is highly intensified in a turbulent regime.

Figure 2.28 demonstrates the development of the boundary layer within a conduit in a uniform internal flow. As it can be clearly observed, the boundary layer continues to grow on the conduit's surfaces. The structure of a boundary layer will be explained in more details in the following sections. The region outside of the boundary can be assumed as the irrotational and inviscid flow. Nonetheless, this region diminishes after a certain distance from the entrance of a conduit, which is called the **hydrodynamic entrance region**. As shown in Figure 2.28, in the entrance region, when the uniform flow enters, no-slip boundary layers start to be formed on the conduit's surfaces until colliding to each other at the centre of the conduit at the hydrodynamic entrance region. After this point towards the end of the pipe, the flow becomes **fully developed** where the velocity profile is almost independent of the flow direction, meaning that the profile remains invariant towards the conduit outflow.

Here, to find the shape of a velocity profile in a fully developed region, which is about a major part of a conduit in the engineering problems, we must first clarify if a flow is in the laminar or turbulent regime. In round pipes, if the Re number based on its diameter is calculated as the characteristic length, the flow can be presumed to be laminar if Re is

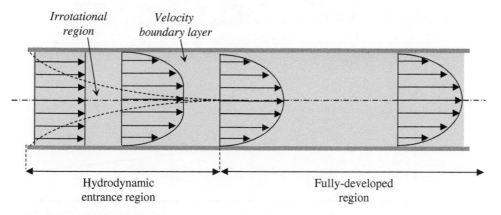

Figure 2.28   Hydrodynamic entrance region and fully developed flow.

approximately below 2100. The flow also can be considered to be turbulent when Re is approximately larger than 4000 while the flow is in the transition between these values, known as the transient flow. For the entrance length of l_e, the below values are suggested:

$$\text{Laminar} \Longrightarrow \frac{l_e}{D} = 0.06 \text{ Re} \tag{2.161}$$

$$\text{Turbulent} \Longrightarrow \frac{l_e}{D} = 4.4 \, (\text{Re})^{1/6} \tag{2.162}$$

This implies that the entrance region can be very short in laminar flows while it can be prolonged in turbulence regimes. The pressure difference (gradient) in horizontal pipes is the sole force that drives a flow. The viscous layer in pipes results in surface frictions or losses, which cause drops in the pressure alongside a pipe. This implies that we can expect a zero pressure drop if a fluid is inviscid. Nonetheless, the pressure drop has a significant effect, which should be carefully considered in the engineering designs. As it can be seen in Figure 2.29, pressure drop is more significant in entrance regions due to the higher viscous effect (shear stress) while it drastically declines to a linear value in the fully developed region. Also, we can describe the friction losses to be a function of laminar and turbulent flow regimes as they effect on the shear stresses on the internal surfaces. Note that the calculation of pressure drops in a turbulent flow regime and within a fully developed region is not a straightforward process while a general expression can be obtained for laminar flows within the fully developed region.

2.8.1 Laminar Internal Flow

The laminar flow in a horizontal enclosed pipe is known as **_Hagen–Poiseuille flow_**. For this flow, we can simplify the Navier–Stokes equations in a cylindrical polar coordinate where z is the flow direction ($g_r = -g \sin \theta$ and $g_\theta = -g \cos \theta$). Under the assumptions

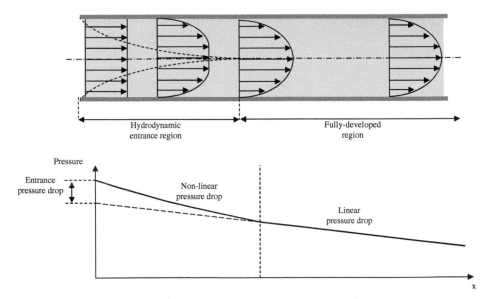

Figure 2.29 Pressure drops alongside entrance and fully developed regions of a horizontal pipe.

of the parallel flow to be one-dimensional ($v_z = v_z(r)$, $v_r = 0$, $v_\theta = 0$) with asymmetrical variation of the velocity in θ-direction ($\frac{\partial v_z}{\partial \theta} = 0$) and a steady-state condition ($\frac{\partial}{\partial t} = 0$), we can derive the NS equations as follows:

From continuity equation:

$$\frac{\partial(v_z)}{\partial z} = 0 \tag{2.163}$$

And, for the momentum equations:

$$r-\text{direction} : 0 = -\frac{\partial p}{\partial r} - \rho g \sin\theta \tag{2.164}$$

$$\theta-\text{direction} : 0 = -\frac{1}{r}\frac{\partial p}{\partial \theta} - \rho g \cos\theta \tag{2.165}$$

$$z-\text{direction} : 0 = -\frac{\partial p}{\partial z} + \mu\left[\frac{1}{r}\frac{\partial}{\partial r}\left(r\frac{\partial v_z}{\partial r}\right)\right] \tag{2.166}$$

From Eqs. (2.164) and (2.165), we can find that the pressure equation that should vary horizontally in the form of:

$$p = -\rho g r \sin\theta + f(z) \tag{2.167}$$

After two times of integration over the x-direction momentum equation (Eq. (2.166)), we can reach to:

$$v_z(r) = \frac{1}{4\mu}\left(\frac{\partial p}{\partial z}\right)r^2 + C_1 lnr + C_2 \tag{2.168}$$

After applying the boundary condition of $\frac{\partial v_z(r = 0)}{\partial r} = 0$, we find $C_1 = 0$. And, after applying the boundary condition of $v_z(r = \mp R) = 0$ for no-slip wall condition (R is the radius of the pipe):

$$C_2 = -\frac{1}{4\mu}\left(\frac{\partial p}{\partial z}\right)R^2 \tag{2.169}$$

Thus, we find a parabolic profile for the velocity of Eq. (2.168):

$$v_z(r) = \frac{1}{4\mu}\left(\frac{\partial p}{\partial z}\right)(r^2 - R^2) \tag{2.170}$$

So, we can now replace the pressure drop in Eq. (2.170) with a linear approximation ($f(z) = \frac{\Delta p}{l}z$) in a laminar flow regime. Thus, using Eq. (2.167) gives:

$$-\frac{\partial p}{\partial z} = -\frac{f(z)}{\partial z} = \frac{\Delta p}{l}$$

$$\Longrightarrow v_z(r) = \frac{\Delta p}{4\mu l}(R^2 - r^2) \tag{2.171}$$

This implies that the maximum velocity (v_c) at the centre line of the pipe can be obtained as:

$$v_c = v_z(r = 0) = \frac{\Delta p R^2}{4\mu l} \tag{2.172}$$

Thus, replacing Eq. (2.172) into Eq. (2.171):

$$v_z(r) = v_c \left[1 - \left(\frac{r}{R} \right)^2 \right]$$

(2.173)

We can further find the flow rate by integration of the velocity on the surface area:

$$Q = \int_0^R v_z(r) dA = \int_0^R v_z(r) 2\pi r dr$$

$$\Longrightarrow Q = \int_0^R \left(\frac{\Delta p}{4\mu l} (R^2 - r^2) \right) 2\pi r dr = \frac{\pi \Delta p R^4}{8\mu l}$$

(2.174)

Also, the mean velocity can be obtained as:

$$V = \frac{Q}{\pi R^2}$$

$$\Longrightarrow V = \frac{\Delta p R^2}{8\mu l} = \frac{v_c}{2}$$

(2.175)

These results confirm that the flow rate is proportional to the pressure drop and pipe diameter (with power of 4) while it is inversely proportional to the pipe length and viscosity.

2.8.2 Turbulent Internal Flow

In many practical engineering problems, the flow within a conduit quickly becomes turbulent. In general, when a flow reaches to a turbulent regime, we can estimate the shear stress to have two contributors as laminar (τ_{lam}) and turbulence (τ_{turb}) parts:

$$\tau = \tau_{lam} + \tau_{turb}$$

(2.176)

Laminar shear stress is following Newton law based on the mean value of the velocity ($\bar{u}$) as described earlier in this chapter:

$$\tau_{lam} = \mu \frac{\partial \bar{u}}{\partial y}$$

(2.177)

On the other hand, the turbulent shear stress is quite depending on the fluctuation of velocity component (u' and v' for a 2D flow), and thus are estimated by the Reynolds stress concept:

$$\tau_{turb} = \rho \overline{u'v'}$$

(2.178)

The key point here is that τ_{lam}, which is only dominant in the sub-viscous layer as will be discussed in detail in the boundary layer structure section, is dependent on the surface roughness. Turbulence can be assumed as a series of random 3D eddies with different sizes that transfer momentum in a fluid. Thus, it was proposed by **Boussinesq** to match turbulent element of the shear stress with a similar concept to the molecular viscosity, known as eddy viscosity (μ_t):

$$\tau_{turb} = \mu_t \frac{\partial \bar{u}}{\partial y}$$

(2.179)

It should be noted that the molecular viscosity is a property of fluid while the eddy viscosity is inherently a property of flow type in addition to the working fluid. With knowing of this background, we can refer the velocity in conduit, for example in a round pipe, to follow physics of shear stress on walls. The uncertainty in accurate understanding of eddy viscosity have resulted in development of many turbulence modes, which is beyond the scope of this book.

2.8.3 Pressure Drop in Conduit

If we use Buckingham π-theorem, we can show that in a turbulent internal flow through a horizontal conduit with a roughness of ε, the total pressure drop (Δp) when the flow is steady and incompressible, can be presented as:

$$\Delta p = f(V, D, L, \mu, \rho, \varepsilon) \tag{2.180}$$

We leave the procedure with the readers to prove that this problem can be represented by four dimensionless groups, which can be written as:

$$\frac{\Delta p}{\frac{1}{2}\rho V^2} = \phi\left(\mathrm{Re}_D, \frac{L}{D}, \frac{\varepsilon}{D}\right) \tag{2.181}$$

where $\dfrac{\varepsilon}{D}$ represents the relative roughness as it was seen to be important parameter in the turbulence flow.

It will not be then difficult to rearrange Eq. (2.181) as:

$$\frac{\Delta p}{\frac{1}{2}\rho V^2} = \frac{L}{D}\psi\left(\mathrm{Re}_D, \frac{\varepsilon}{D}\right)$$

$$\Longrightarrow \frac{\Delta p D}{\frac{1}{2}\rho V^2 L} = \psi\left(\mathrm{Re}_D, \frac{\varepsilon}{D}\right) \tag{2.182}$$

This helps to use a very common term of $\dfrac{\Delta p D}{\frac{1}{2}\rho V^2 L}$ known as **Darcy friction factor** (f) or simply **friction factor**, as it can be found experimentally for different type of conduits, to present Eq. (2.182):

$$f_{turb} = \psi\left(\mathrm{Re}_D, \frac{\varepsilon}{D}\right) \tag{2.183}$$

Also, we can recall Eq. (2.175) in the laminar flow regime, and rewrite it based on the newly defined f as below:

$$V = \frac{\frac{D^2}{4}\Delta p}{8\mu L}$$

$$\Longrightarrow \frac{\Delta p}{\frac{1}{2}\rho V^2 L} = 64\left(\frac{\mu}{\rho D V}\right)\left(\frac{L}{D}\right) \tag{2.184}$$

$$\Longrightarrow f_{lam} = \frac{64}{\mathrm{Re}_D}\left(\frac{L}{D}\right)$$

Friction factors for a range of Reynolds numbers are first measured by [4] as it is known as Moody chart. Based on the energy loss in incompressible flows due to the friction in pipes, it is common to show the **head loss** (h_L) between two sections of a pipe with a distance of L with Darcy–Weisbach equation (for more details see Chapter 3):

$$h_L = f \frac{L}{D} \frac{V^2}{2g} \tag{2.185}$$

Example 2.12

In Example 2.5, calculate the threshold height of a tank (h) to reach a turbulent regime at each cold-water pipe with a minimum flow rate of 0.05 L/s. In the associated turbulent velocity, calculate the pressure loss throughout that pipe between two sinks. The required dimensions are specified in Figure 2.E5.1. We neglect other losses associated with joints and piping elements. Friction factors for ½″ and ¼″ pipes are given as 0.0375 and 0.0312, respectively.

Solution

The schematic diagram of the piping system of Example 2.5 is depicted in Figure 2.E12. At each time, one of Sink-1 to Sink-4 is opened while the rest are closed. Then, h for each case is calculated separately while the threshold of the change of the regime from laminar to turbulence is set as Re $\approx$ 2100. Replacing the required data for water properties at 20°C yields:

$$V_1 = 0.25V_2 = 0.25V_3 = 0.25V_4 = \frac{\text{Re} \times \mu}{\rho D} = \frac{2100 \times 1.0016 \times 10^{-3}}{1000 \times 0.00635} = 0.17 \, \text{m/s}$$

This velocity, however, does not provide the required flow rate as it would be $Q = 0.02$ L/s. Thus, if we increase the velocity to $V_1 = 0.4$ m/s, we can ensure achieving a turbulent regime in the entire system while having a suitable flow rate at the sinks.

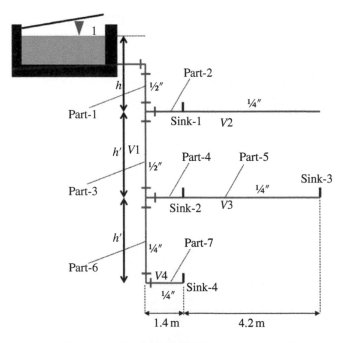

Figure 2.E12 Schematic diagram of the piping system of Example 2.5.

Thus, losses are related to the pipes' diameter, length, water velocity, and friction factor $(h_L = f\frac{l}{D}\frac{V^2}{2g})$.

Name	Length (m)	Diameter (m)	Velocity (m/s)	Friction factor
Part-1	$L_1 = h$	$D_1 = 0.0127$	$V_1 = 0.40$	$f_1 = 0.0375$
Part-2	$L_2 = 1.4$	$D_2 = 0.00635$	$V_2 = 1.58$	$f_2 = 0.0312$
Part-3	$L_3 = 2.8$	$D_3 = 0.0127$	$V_1 = 0.40$	$f_3 = 0.0375$
Part-4	$L_4 = 1.4$	$D_4 = 0.00635$	$V_3 = 1.58$	$f_4 = 0.0312$
Part-5	$L_5 = 4.2$	$D_5 = 0.00635$	$V_3 = 1.58$	$f_5 = 0.0312$
Part-6	$L_6 = 2.8$	$D_6 = 0.0127$	$V_1 = 0.40$	$f_6 = 0.0375$
Part-7	$L_7 = 1.4$	$D_7 = 0.00635$	$V_4 = 1.58$	$f_7 = 0.0312$

For sink-1, we can derive Bernoulli equation between point-1 and sink-1 if the ground floor is the reference point for the height (Part1+Part-2):

$$p_1 + \rho g(h + L_3 + L_6) = p_2 + \rho g(L_3 + L_6) + \frac{1}{2}\rho V_2^2 + \frac{1}{2}\sum_{i=1}^{2} f_i \frac{L_i}{D_i}\rho V_i^2$$

Rearranging the above equation and after simplifications ($p_1 = p_2 = p_{atm}$):

$$h = \frac{V_2^2}{2g} + \left(f_1\frac{L_1}{D_1}\frac{V_1^2}{2g} + f_2\frac{L_2}{D_2}\frac{V_2^2}{2g}\right)$$

Replacing the values:

$$h = 0.1274 + 0.0235h + 0.8761$$
$$\Longrightarrow h = 1.03\,\text{m}$$

Bernoulli between point-1 and sink-2 (Part-1 + Part-3 + Part-4):

$$(h + L_3) = \frac{V_3^2}{2g} + \left(f_1\frac{(h + L_3)}{D_1}\frac{V_1^2}{2g} + f_4\frac{L_4}{D_4}\frac{V_3^2}{2g}\right)$$
$$\Longrightarrow h + 2.8 = 0.1274 + 0.0235h + 09419$$
$$\Longrightarrow h = -0.34\,\text{m}$$

Bernoulli between point-1 and sink-3 (Part-1 + Part-3 + Part-4 + Part-5):

$$(h + L_3) = \frac{V_3^2}{2g} + \left(f_1\frac{(h + L_3)}{D_1}\frac{V_1^2}{2g} + f_4\frac{(L_4 + L_5)}{D_4}\frac{V_3^2}{2g}\right)$$
$$\Longrightarrow h + 2.8 = 0.1274 + 0.0235h + 3.5703$$
$$\Longrightarrow h = 0.92\,\text{m}$$

Bernoulli between point-1 and sink-4 (Part-1 + Part-3 + Part-6 + Part-7):

$$(h + L_3 + L_6) = \frac{V_4^2}{2g} + \left(f_1\frac{(h + L_3 + L_6)}{D_1}\frac{V_1^2}{2g} + f_7\frac{L_7}{D_7}\frac{V_4^2}{2g}\right)$$
$$\Longrightarrow h + 5.6 = 0.1274 + 0.0235h + 1.0077$$
$$\Longrightarrow h = -4.57\,\text{m}$$

Some of the values of the table give negative heights, which states that the velocities in the sinks are already in the turbulent regime while there is no need for the tank to be placed in any height of h to ensure the required flow. Nonetheless, looking at all the found height demands $h = 1.03$ to ensure these conditions in all sinks such as Sink-1. While high velocities were seen exiting from each sink which are not desirable on some occasions. Thus, parameters should be always changed in piping system design to achieve a reasonable water velocity at all taps.

2.9 External Flow

When an object is immersed in a fluid, the flow is known as an external flow. Many engineering problems in nature deal with external flow. Examples are many though we can name external flows related to moving vehicles, airplanes, and ships as objects, which are completely surrounded by one or multiple fluids. We also focus on wind in urban areas in Chapter 3, which is about immersed bodies in external flows. We also discuss wind and water tunnel applications in Chapter 11 as a common technique of investigation of external flows, widely used in hydrodynamic and aerodynamics fields.

It is common to assume bodies immersed in external flows as 2D objects with a similar cross section in shape and size, meaning that their depth dimension has less impact on the wholistic of the flow field. This assumption is valid on many occasions for a large part of an immersed object, for example airfoils. However, 3D effect is way more complex in the end sides of airfoils, which makes the assumption to be very weak in real life studies of aircrafts. Thus, in many scenarios, 3D effect should be taken into account.

2.9.1 Drag and Lift

We have previously discussed that objects will face shear stresses on their skins due to the viscous effect, known as the surface forces. Normal stresses by pressure also act on objects' volumes, known as body forces. As it can be seen in Figure 2.30, the shear stresses on an object skin against the upstream velocity will cause a resultant force, which is called **drag** ($\mathcal{D}$). With a similar definition, but in a direction normal to the upstream velocity, the acting resultant force is called as **lift** ($\mathcal{L}$).

Calculation of drag and lift forces within external flows is extremely crucial to understand the resultant force on the object utilized in many engineering designs from aerospace industry to car manufacturers. Hence, for the immersed body shown in Figure 2.30, we can drive the resultant force as:

$$\mathcal{D} = \int dF_x$$
$$\Longrightarrow \mathcal{D} = \int p \cos \theta dA + \int \tau_w \sin \theta dA \tag{2.186}$$

$$\mathcal{L} = \int dF_y$$
$$\Longrightarrow \mathcal{L} = -\int p \sin \theta dA + \int \tau_w \cos \theta dA \tag{2.187}$$

where, θ is the angle between the normal vector of the surface and upstream velocity direction.

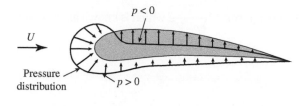

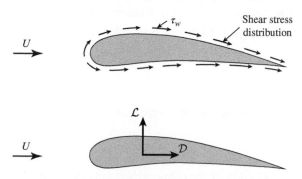

Figure 2.30 Shear and normal stresses as well as drag and lift forces on an immersed body. *Source:* Based on Munson et al. [5].

Calculation of such integration to obtain drag and lift forces is not always theoretically straightforward due to the complexities of many geometries and thus on many occasions experimental study will be employed to find these forces. Theoretically, however, we can parametrize drag and lift of various immerse bodies using Buckingham π-theorem.

For such parametrization, one of the main assumptions is the incompressibility of the flow. Note that parametrization against high-speed flow and compressible flows are beyond the scope of this book although many theoretical and experimental studies can be found in literature [5–7]. Therefore, in our process, we can neglect the impact of surface tension (represented by **Weber number**), gravity (represented by **Froude number**), and compressibility (represented by **Mach number**). We can then justify that the acting force in the upstream direction over an immerse body (the drag force) can be expressed as:

$$\mathcal{D} = f(V, \mu, \rho, \varepsilon, l, l_i) \tag{2.188}$$

where l is the characteristic length of the body and l_i are other key dimension parameters in the drag formation over the immersed body ($n = 7, m = 3$). We leave the procedure to the readers to prove that we can obtain the below independent groups ($n - m = 4$) for the drag force:

$$\frac{\mathcal{D}}{\rho l^2 V^2} = \psi\left(\frac{\rho V l}{\mu}, \frac{\varepsilon}{l}, \frac{l_i}{l}\right) \tag{2.189}$$

The roughness effect is negligible in many external flows. Thus, we can simplify Eq. (2.189) as:

$$\frac{\mathcal{D}}{\rho l^2 V^2} = \phi\left(\mathrm{Re}_l, \frac{l_i}{l}\right)$$

$$\Longrightarrow \frac{\mathcal{D}}{\frac{1}{2}\rho l^2 V^2} = \acute{\phi}\left(\mathrm{Re}_l, \frac{l_i}{l}\right) \tag{2.190}$$

It is very common to reformat Eq. (2.190) in terms of $\dfrac{\mathcal{D}}{\frac{1}{2}\rho l^2 V^2}$, which is known as the ***drag coefficient*** (C_D):

$$C_D = \frac{\mathcal{D}}{\frac{1}{2}\rho l^2 V^2} = \acute{\phi}\left(\text{Re}_l, \frac{l_i}{l} \right) \tag{2.191}$$

As it is obvious from the relation between these dimensionless groups, the drag coefficient is a function of fluid regime appeared in Re_l and the shape of the immerse body embedded in $\frac{l_i}{l}$. l^2 is mainly replaced by a characteristic area ($A \cong l^2$), which is mainly selected to be the frontal area of the immersed body:

$$C_D = \frac{\mathcal{D}}{\frac{1}{2}\rho A V^2} \tag{2.192}$$

A similar approach can be followed to introduce ***the lift coefficient*** (C_L) as below:

$$C_L = \frac{\mathcal{L}}{\frac{1}{2}\rho A V^2} \tag{2.193}$$

As it can be seen from the definitions, selection of the characteristic area as well as the velocity location at the upstream flow requires many considerations and experience, especially in wind tunnel studies.

Example 2.13

Find the lift force acting on a half-cylinder-shaped building with the radius of $r = 10$ m and length of $L = 30$ m as depicted in Figure 2.E13.1 against a uniform wind, ranging from 0.1 to 25 m/s. Assume kinematic viscosity and density of air to be $v_a = 1.0035 \times 10^{-6}$ m^2/s and $\rho_a = 1.23$ kg/m^3, respectively.

Solution

We use Eq. (2.187), and thus with area of $dA = \int_0^\pi Lrd\theta$ for the half cylinder:

$$\mathcal{L} = -\int_0^\pi p \sin\theta Lrd\theta + \int_0^\pi \tau_w \cos\theta Lrd\theta$$

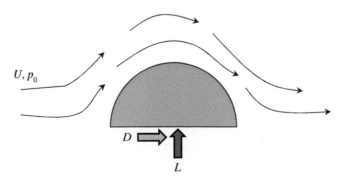

Figure 2.E13.1 Lift force acting on a half-cylinder-shaped building.

Dividing both side by $\frac{1}{2}\rho A U^2$ in accordance with Eq. (2.193) gives C_L (characteristic area is $A = 2Lr$):

$$C_L = -0.5 \int_0^\pi \frac{p \sin \theta}{\frac{1}{2}\rho U^2} d\theta + 0.5 \int_0^\pi \frac{\tau_w \cos \theta}{\frac{1}{2}\rho U^2} Lr d\theta$$

The dimensionless shear stress is also defined as $F(\theta) = \dfrac{\tau_w Re^{0.5}}{\frac{1}{2}\rho U^2}$. Thus, we further can rearrange the above equation as:

$$C_L = -0.5 \int_0^\pi \frac{p \sin \theta}{\frac{1}{2}\rho U^2} d\theta + 0.5 \, Re^{-0.5} \int_0^\pi F(\theta) \cos \theta d\theta$$

Both integral parts can be obtained from numerical techniques in accordance with the flow over a cylinder, which can be found in Fluid Mechanics reference books (see references and suggested reading materials; example is [5]). Therefore, we just provide the values for the half-cylinder as:

$$-0.5 \int_0^\pi \frac{p \sin \theta}{\frac{1}{2}\rho U^2} d\theta = 0.88$$

And:

$$0.5 \, Re^{-0.5} \int_0^\pi F(\theta) \cos \theta d\theta = 1.96 \, Re^{-0.5}$$

Hence, we finally have:

$$C_L = 0.88 + 1.96 \, Re^{-0.5}$$

Now, applying different wind values, we can plot C_L and lift force against Re_r as shown in Figure 2.E13.2.

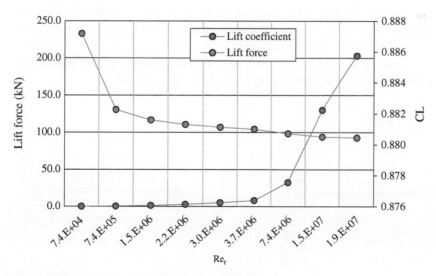

Figure 2.E13.2 Plot of C_L and lift force against Re_r.

2.9.2 Uniform Flow on a Flat Plate

Steady uniform external flow over a flat plate is an interesting fundamental flow that can demonstrate the complexity of the flow even in such a simplistic geometry. Besides, many engineering problems can be, to some extent, simplified to such flows. As it was found in Eq. (2.191), the flow mechanism and drag force on a flat plate is a function of Reynolds number. As it is depicted in Figure 2.31, the flow regime is changing significantly from low to high Reynolds numbers. In the case of laminar flows in the low Reynolds numbers, the viscous effect is dominant, affecting most of the velocity field. Traditionally, a part of the field in which the velocity is more than 99% of the uniform upstream flow is assumed to be far field zone with a negligible effect of viscosity. Thus, although the fluid still has viscosity in the far field region, it is approximated to be inviscid (potential flow) as the shear stresses are extremely low relative to the part that is formed around the object in which velocity gradient are much higher.

In general, besides the inviscid region, we can identify a region stuck to the solid surfaces in which viscosity effect is merely important. This region is known as the boundary layer. Also, there is another region in downstream of the object significantly impacted by the viscosity where the flow leaves the solid surface but continues to rejoin the streamlines after a certain distance. With the knowledge of the concept of boundary layer, it is easier to explain that in the lower Reynolds numbers, the viscous forces are significant in comparison to inertial forces, and thus the boundary layer would be thicker. Nonetheless, in the higher Reynolds numbers, the inertial forces push the viscous region to shrink on the solid surfaces and make the boundary layer thickness (δ) to be thinner.

2.9.3 Boundary Layer Structure

In general, the boundary layer thickness is related to the shape of an object as well as the flow characteristics. For the simplicity purposes, we consider the case of an incompressible steady uniform flow over an infinite flat plate. As we discussed in the previous section, the boundary layer forms due to the existence of a no-slip condition on the solid surfaces. This region is traditionally known as the one in which the velocity is below 99% of the upstream velocity (see Figure 2.32). Beyond the boundary layer region, therefore, we can assume the viscous effect to be negligible and the flow to be inviscid and potential.

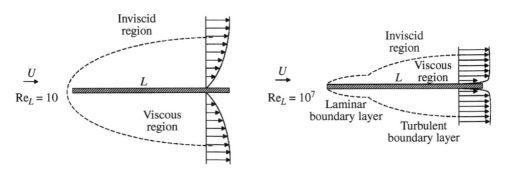

Figure 2.31 Uniform flow around a flat plate in various Reynolds numbers.

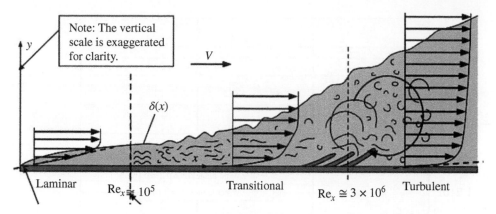

Note: The vertical scale is exaggerated for clarity.

V

$\delta(x)$

Laminar $\mathrm{Re}_x \cong 10^5$ Transitional $\mathrm{Re}_x \cong 3 \times 10^6$ Turbulent

Figure 2.32 Boundary layer over an infinite flat plate. *Source:* Based on Frank et al. [8].

Note that for an infinite plate, we cannot define a characteristic length, thus we calculate the Reynolds number at different locations of the plate as a function of distance from its leading edge (x) presented as:

$$\mathrm{Re}_x = \frac{Ux}{v} \tag{2.194}$$

It can be seen that the flow starts to form a laminar boundary condition right after the location that the flow is impacting the solid surface of the flat plate. As it can be seen, in the boundary layer Figure 2.32, fluid particles are impacted by the no-slip condition of the surface. Moreover, particles face viscous effect when the mixing occurs in a molecular scale. After a certain distance and when the critical Reynolds number on the flat plate is reached to be $2 \times 10^5 < \mathrm{Re}_{xcr} < 3 \times 10^6$, the boundary layer transforms and elevates to a transitional level. Then, after it develops to a turbulent boundary layer where the fluid particles start to move in a random direction and irregular pattern.

Example 2.14
Calculate the range of typical Re number of over a real building with a height of 5m. What would be the Reynolds numbers if two 1:66 and 1:33 building models are placed in a water tunnel and a wind tunnel, respectively. Assume kinematic viscosity of water and air to be $v_w = 1.343 \times 10^{-5}$ m^2/s and $v_a = 1.0035 \times 10^{-6}$m^2/s, respectively.

Solution

If we assume wind to varies from $U < 0.1$ m/s in a typical calm day to about $U > 25$ m/s during a storm (for example see Beaufort Wind Scale), we can plot Re numbers as Figure 2.E14 for the water and wind tunnels.

As it can be seen in Figure 2.E14, in most of the real scenarios for a typical building, the flow is in the turbulence regime though note that the flow in the wind tunnel and water tunnel supposed to remain in the laminar flow regime. However, the definition of critical Reynold (Re_{cr}) as explained in Chapter 3 will help to assume that the flow is turbulent in most of the flow field. Re_{cr} can be chosen to be smaller and easier to achieve in water tunnels

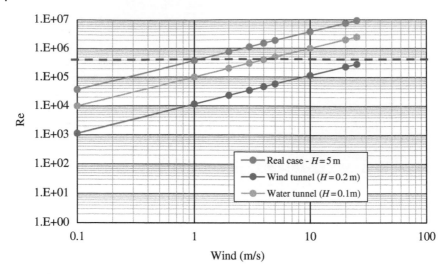

Figure 2.E14 Re numbers for different wind velocities in water and wind tunnels.

although design, operation and maintenance of water tunnels are way more challenging than wind tunnels.

References

1 Nakayama, Y. (2018). Chapter 3 - fluid statics. In: *Introduction to Fluid Mechanics*, 2e (ed. Y. Nakayama), 25–50. Butterworth-Heinemann.

2 Hunt, J.C.R., Abell, C.J., Peterka, J.A., and Woo, H. (1978). Kinematical studies of the flows around free or surface-mounted obstacles; applying topology to flow visualization. *Journal of Fluid Mechanics* 86 (1): 179–200.

3 Wylen, S.-B.-V. (1998). *Fundamentals of Thermodynamics*, 5e. Wiley.

4 Moody, L.F. (1944). Friction factors for pipe flow. *Transactions of the ASME* 66 (8): 671–684.

5 Munson, B.R., Young, D.F., and Okiishi, T.H. (2006). *Fundamentals of Fluid Mechanics*, 6e. Hoboken, NJ: Wiley.

6 Douglas JF. *Fluid Mechanics*: Prentice Hall; 2011.

7 White FM. *Fluid Mechanics* McGraw-Hill, Kogakusha; 1979.

8 Frank, P., Incropera, D.P.D., Bergman, T.L., and Lavine, A.S. (1996). *Fundamentals of Heat and Mass Transfer*, 7e. Wiley.

3

Applications of Fluid Mechanics in Buildings

3   Applications of Fluid Mechanics in Buildings

3.1   Atmospheric Boundary Layer

Boundary layer structure was discussed in Chapter 2. Formation of a boundary layer is a result of molecular and turbulent eddy viscosities. The first one is a result of different mass momentum transport of molecules, generating frictions between them within the layers of a fluid. The latter is related to a transport of momentum by eddies in a turbulent flow where internal friction of molecules is enhanced though in a very larger scale. Surface roughness is also understood to be a major parameter in the formation of boundary layers. In built-up lands in cities, buildings and infrastructures act as an inhomogeneous roughness against the flow, resulting in the formation of **urban boundary layer** (UBL) as depicted in Figure 3.1.

In practice, UBL exists within atmospheric boundary layer (ABL) or planetary boundary layer (PBL), which is itself formed due to earth's land surface roughness. The UBL consists of various layers, including roughness sublayer (RSL) and urban surface layer (USL). Each layer contains other layers categorized based on the streamwise and vertical components of the velocity profile. The height of each layer and ABL in specific are sensitive to the atmospheric stability and urban characteristics. For more detailed understanding of ABL, readers can refer many existing studies and literature (for example see [2]).

3.2   Wind Profile and Directions

Understanding of wind profile within the UBL is important for many applications such as the calculation of wind load over buildings and infrastructure, pedestrian wind, thermal comfort, and building energy calculations (see also Chapters 8 and 9). The wind profile changes in accordance with the type of a land built-up structure. As the built-up structure drastically changes across a city, we expect that the associated wind profile to alter as well (see Figure 3.2). However, it is impractical and, on some occasions, implausible to measure the wind profile across a city and therefore the boundary layer helps to estimate the wind profile in those locations.

Computational Fluid Dynamics and Energy Modelling in Buildings: Fundamentals and Applications, First Edition. Parham A. Mirzaei.
© 2023 John Wiley & Sons Ltd. Published 2023 by John Wiley & Sons Ltd.

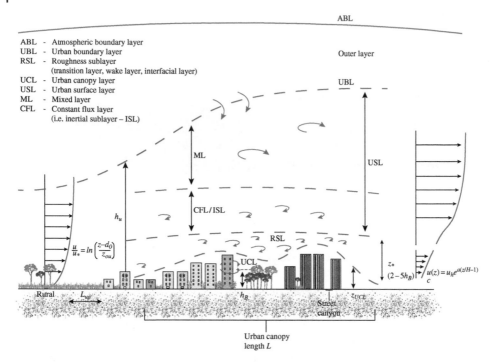

Figure 3.1 Urban surface layer and roughness sublayer within urban boundary layer. *Source:* Modified from Fernando [1]/Annual Reviews.

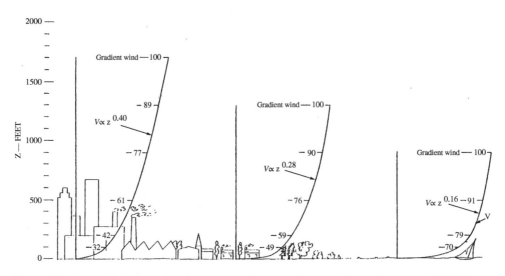

Figure 3.2 Atmospheric boundary layer over various terrain types. *Source:* Davenport [3]/AIVC.

In general, two common models are widely used, developed based on the concept of boundary layer theory, including power-law and log-law, which can account for the alteration of built-up area and wind profile [4]. The velocity profile using the power law can be expressed as:

$$V_z = V_R \left(\frac{z}{z_R} \right)^\alpha \tag{3.1}$$

And, with the log-law as:

$$V_z = \frac{u_*}{k} \left[\ln \frac{z - z_d}{z_0} + \varphi(z, z_0, L) \right] \tag{3.2}$$

where V_R is the reference wind speed at reference height z_R (mainly recorded at a weather station). α is an empirically derived coefficient, depending on the atmosphere. u_* is the friction velocity. k is the Von Kármán constant (~0.41). And, z_0 and z_d are, respectively, the roughness length and zero-plane displacement for air velocity. φ is the universal functions for atmospheric stability, which has different values under stable and unstable weather conditions, depending on L as Obukhov length from Monin–Obukhov similarity theory. Under neutral stability condition, log-law equation reaches to its common application form in urban climate CFD models when $\varphi(z, z_0, L) = 0$.

In general, α and ABL thickness (δ) can be found in Table 3.1. Moreover, the roughness length (z_0), zero-plane displacement (z_d), and normalized friction velocity (u_*/V_{ref}) for some of the rural and urban lands can be found from Table 3.2.

If values are extracted in a reference point such as a meteorological station, they can be extrapolated for another terrain if the land characteristics are known. For a neutral stability condition, for example, the log-law can be written as:

$$V_z = V_{ref} \frac{\ln \dfrac{z - z_d}{z_0}}{\ln \dfrac{z_{ref} - z_{dref}}{z_{0ref}}} \tag{3.3}$$

Furthermore, for the power law, we have:

$$V_z = V_{ref} \left(\frac{\delta_{ref}}{z_{ref}} \right)^{\alpha_{ref}} \left(\frac{z}{\delta} \right)^\alpha \tag{3.4}$$

where index 'ref' implies the associated values to be obtained at the reference point.

Table 3.1 Values of α and atmospheric boundary layer thickness (δ) for different terrain types.

Terrain type	α (–)	δ (m)
Flat, open country	0.14	270
Rough, wooded country, urban, industrial, forest	0.22	370
Towns and cities	0.33	460
Ocean	0.10	210

Source: Modified from [5]/ASHRAE.

Table 3.2 Values of averaged height of roughness element (z_H), roughness length (z_0), zero-plane displacement (z_d), and normalized friction velocity (u_*/u_{ref}).

Terrain type	z_H (m)	z_0 (m)	z_d (m)	u_*/V_{ref}
Farmland, crops	0.4–1	0.05–0.15	0.2–0.7	0.07–0.10 (V_{ref} at 10 m)
Grass, stubble field	0.2–0.5	0.03–0.06	0.1–0.3	0.06–0.07 (V_{ref} at 10 m)
Forest – range from temperate to tropical	12–30	0.8–2	9–24	>0.16 (V_{ref} at 10 m)
Low height and density – houses, gardens, trees; warehouses	5–8	0.3–0.8	2–4	0.09–0.12 (V_{ref} at 30 m)
Medium height and density – row and close houses, town centres	7–14	0.7–1.5	3.5–8	0.11–0.14 (V_{ref} at 30 m)
Tall and high density – less than six floors, row, and block buildings	11–20	0.8–2	7–15	0.13–0.16 (V_{ref} at 30m)
High-rise – office and apartment tower clusters	>20	>2	>12	>0.16 (V_{ref} at 30 m)

Source: Modified from [2]/Cambridge University Press.

Example 3.1

Find the wind profile under a neutrally stable day inside a city with both power law and log-law methods if the reference wind velocity at a height of 10m in a meteorological station, at a nearby airport located in the farmland is recorded to be 3.6 m/s. The average height of buildings in the area is 12 m.

Solution

Using Tables 3.1 and 3.2, we find the required values for the city centre (as 'urban' and 'medium height and density due to $z_H = 12$ m') as below:

$$\delta = 460 \text{ m}$$

$$\alpha = 0.33$$

$$z_0 \approx 1.4 \text{ m}$$

$$z_d \approx 6.5 \text{ m}$$

$$u_*/V_{ref} \approx 0.13 \ (V_{ref} \text{ at } 30 \text{ m})$$

$$\implies u_* \approx 4.7 \text{ m/s}$$

And, for the reference point according to the terrain description of the meteorological station (as 'flat and open country' and 'farmland'), can be found as:

$$\delta_{ref} = 270 \text{ m}$$

$$\alpha_{ref} = 0.14$$

$$z_{0ref} \approx 0.1 \text{ m}$$

$$z_{dref} = 0.45 \text{ m}$$

$$z_{ref} = 10 \text{ m}$$

$$V_{ref} = 3.6 \text{ m/s}$$

Now, we can use Eqs. (3.3) and (3.4) and plot both profiles as shown in Figure 3.E1. Note that for the log-law as expected, the profile has values after $z_d \approx 6.5$ m.

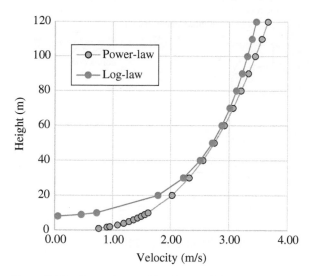

Figure 3.E1 Power-law and log-law profiles.

3.3 Building Aerodynamics

Building aerodynamic plays a crucial role in structure of the flow around a building. Figure 3.3 illustrates an isolated cubic building (a generic bluff body) in which the flow is very complex in terms of separation on its roof as well as the reattachments of flow to the side walls and eventually to the ground. Nonetheless, in the separation regions formed on the buildings' surfaces, the shear stresses are at high levels while the outer flow or the flow outside of these regions can be mainly assumed to be shear free or potential flow (see Chapter 2).

Dissipation of energy from the potential flow in outer regions to the surfaces within boundary layers is caused by molecular and turbulence viscosities, where the latter is due to the movement of eddies within the boundary layer. Whilst the flow velocity decelerates or possibly reaches to a zero value with a negative velocity gradient under an adverse pressure gradient, it can be concluded that the flow has less energy to keep the boundary layer attached to the surface. Eventually, the flow in the boundary layer detaches from the surface and the flow separation occurs. Meanwhile, the adverse pressure gradient strength reduces with an increase of its distance from the initial separation point up to a certain distance, known as reattachment distance, when the flow starts to reattach to the surface.

If a building has a more complex geometry than the shown cuboid in Figure 3.3, the flow field will evidently have more complexities as well. It is noteworthy to highlight that the complexities of flow fields will be consequently more drastic, taking to the account that buildings are in denser and more unstructured urban forms (also known as **sheltered condition**). In such conditions, the impact of neighbouring buildings is even more dominant on the pressure distribution over a building.

Understanding of the flow field hence is very important to envisage building energy performance in terms of convective heat transfer from buildings' surfaces (see Chapters 8 and 9). Furthermore, **wind-driven ventilation** (for example **cross-ventilation** or **single-sided**

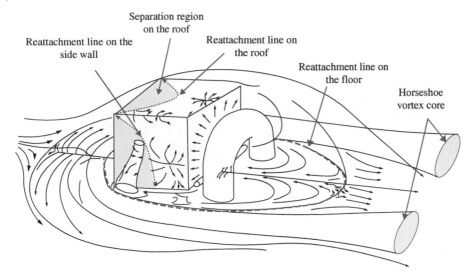

Figure 3.3 Flow field around an isolated building. *Source:* Hunt et al. (1978).

ventilation) in buildings occurs due to a pressure difference between indoor and surrounding environments of a building. This means that scrutinizing the formed negative or positive pressure differences, also known as pressure distributions over a building's surface, dictates the flow direction, which is essential for the design of natural ventilation strategies. It should be noted that the exerted load on building structure is also directly related to the building aerodynamics and their associated pressure differences, which are briefly introduced as drag and lift forces in Chapter 2.

Unlike focusing on the calculation of a resultant pressure force and then finding a unique Drag or Lift coefficient for an object, it is more favourable in building science to employ the ratio of the local pressure to the stagnation pressure, known as the pressure coefficient (C_p). This coefficient, hence, demonstrates a distribution around a building, which reveals how flow is likely to be directed against various approaching winds. Calculation of C_p is not a common practice in real scenarios as the pressure measurement is a sensitive task to the climatic condition and difficult to achieve under rapidly variable atmospheric condition. It is similarly highly associated with the morphological aspects of a building's location while generic conclusions from a specific measurement campaign to be expanded to other locations of a city can be barely envisaged. Wind tunnel studies are in fact a practical approach to control the environment to obtain C_p values as well as a wide range of building shapes and surrounding morphological aspects. Nonetheless, practical difficulties and high expense of experiments are the main limitations of wind tunnel studies as described in Chapter 11. As an alternative, CFD models are commonly used to calculate C_p values although they are bound by their own limitations in turbulence modelling and computational costs, especially when buildings are investigated under high-density urban areas. Interested readers may see these examples for more explanations [5–7].

3.3.1 C_p and Similarity in Buildings

The flow around and inside building models were broadly investigated using atmospheric wind tunnels. When a wind tunnel is equipped with roughness objects (see Figure 3.4) to

(a)

(b)

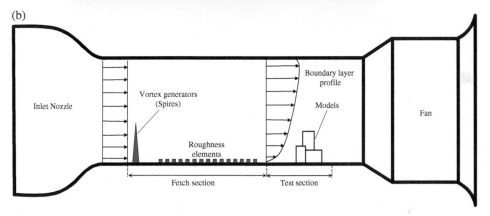

Figure 3.4 (a) Atmospheric wind tunnel with roughness. *Source:* Tominaga. (b) A schematic diagram of a boundary layer wind tunnel.

reshape the uniform approaching wind to a logarithmic profile, it is called **atmospheric wind tunnel** or **boundary layer wind tunnel**.

In general, from a similarity perspective as described in Chapter 2, it is possible to establish similarity between buildings and their physical models, which are mainly in the scale of few centimetres. In addition, flow passing through atmospheric wind tunnels can be assumed in a low Mach number range. In this case, air as the acting fluid is mainly considered to be incompressible. Surface tensions and gravity effect can be also neglected as they will have a marginal effect on the flow. Furthermore, the flow can be focused from a general flow aspect, avoiding studying more advanced effects such as turbulent intensity or vortex shedding problems (represented by **Strouhal Number**). With these assumptions and employed simplifications, any dependent π-term for an arbitrary shape of building, the Buckingham π-theorem explained in Chapter 2 can be exploited. We leave the procedure for the readers as a practice, so that the dependent π-term can be expressed as below:

$$\text{dependent } \pi - \text{term} = \phi\left(\frac{\rho V l}{\mu}, \frac{l_i}{l}, \frac{\varepsilon}{l}, \alpha\right) \tag{3.5}$$

where, the first π-group is Reynolds number, l is the characteristic length of the flow, $\frac{l_i}{l}$ represents other relevant length scale of the investigated building, $\frac{\varepsilon}{l}$ indicates the relative surface roughness, and α is the incidence (orientation towards the wind direction) of a Building. Note that in many scenarios, α is assumed to be fixed while a rotating table in an atmospheric wind tunnel is widely used to place the studied prototype against multiple desired wind directions.

In general, neglecting the impact of α, it can be shown that the π-term serves as the drag force, and it can be shown as:

$$C_D = \frac{D}{\frac{1}{2}\rho V^2 l^2} = \phi\left(\frac{\rho V l}{\mu}, \frac{l_i}{l}, \frac{\varepsilon}{l}\right) \tag{3.6}$$

where l^2 is normally replaced by a characteristic area (A) of the object in drag calculations.

To establish, any kind of similarities for the drag coefficient, we also expect that all three π-groups should be equal between models and real scale buildings (subscript m represents model variables):

$$\frac{\rho V l}{\mu} = \frac{\rho_m V_m l_m}{\mu_m} \tag{3.7}$$

$$\frac{l_i}{l} = \frac{l_{im}}{l_m} \tag{3.8}$$

$$\frac{\varepsilon}{l} = \frac{\varepsilon_m}{l_m} \tag{3.9}$$

It should be noted that due to the practicality and safety issues, air is utilized as the working fluid in many atmospheric wind tunnels while we can assume $\rho = \rho_m$ and $\mu = \mu_m$. Also, roughness effect in many problems is assumed to have a negligible effect on the similarity practices. So, we can ignore Eq. (3.9), and further combine Eqs. (3.7) and (3.8) to:

$$\frac{l}{l_m} = \frac{V_m}{V} = \frac{l_i}{l_{im}} \tag{3.10}$$

Example 3.2

Discuss the similarities between drag coefficient of buildings in a medium-rise neighbourhood and two physical models with the scales of 1 : 66 and 1 : 33 placed in a water tunnel and a wind tunnel, respectively. Assume kinematic viscosity of water and air to be $v_w = 1.343 \times 10^{-5}\,\text{m}^2$ and $v_a = 1.0035 \times 10^{-6}\,\text{m}^2/\text{s}$, respectively. Neglect roughness and orientation towards the wind direction.

Solution

Let us assume that the approaching wind is changing from calm wind of about 0.3 m/s to storm of about 25 m/s, including 0.3, 1, 3, 10, and 25 m/s. We also assume the height variation of the medium-rise neighbourhood to vary from 1 to 5 floors (3–15 m). Thus, we can calculate the range of Re for both wind and water tunnels as shown in Figure 3.E2.

As it can be seen in Figure 3.E2, many of the values are above the dashed line, which means that they are in the turbulent regime and thus the drag coefficient can be assumed to be independent from Re. To establish similarity for those values below the dashed line, based on Eq. (3.10), we must increase the approaching wind velocity to achieve a similar group of dimensionless numbers.

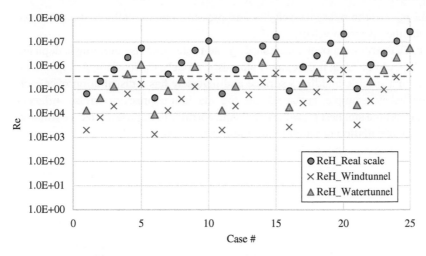

Figure 3.E2 Typical Reynolds number ranges in wind and water tunnels.

As we can see, due to the common size of buildings, their associated Reynolds number is normally greater than 5×10^5. This value is higher than the Reynolds number in which the flow would change to the turbulent regime, and thus we can assume that buildings are most of the times within turbulent flow conditions. As most of the working fluid in atmospheric wind tunnels is air, to achieve similarity in Re numbers with a real case scenario where again air is the working fluid, an extremely higher velocity should be used for the wind tunnel study. This will limit the study when scaling is large; for example, if $\dfrac{H}{H_m} > 66$, then based on Eq. (3.10), we should increase the wind tunnel speed 100 times. So, if we are studying winds around 5 m/s, which is not even very unusual as a strong wind in the built environment, then we must increase the wind tunnel speed to the impossible velocity of 330 m/s. This is a very famous limitation in building science wind tunnel studies, which is tackled by this valid assumption that flow characteristics are not strongly dependent on the Reynolds number higher from a certain number, which is called *critical Reynolds number* ($\mathbf{Re_{cr}}$). The independency of a flow characteristics from Re in its higher values is shown before.

Note, that flow independency condition is not a valid assumption in the thin boundary layers of buildings' surfaces although it is a strong approximation in the rest of the regions when the viscous forces are significantly lower than the inertial forces. In other explanation, we understood from Chapter 2 that drag force has two parts of pressure and friction (shear stress). Thus, when the Re number is large, for example in the flow condition over buildings, the share of the friction drag is less significant whilst the pressure drag is the dominant one.

Using dimensionless analysis approach with almost similar assumptions to the drag equation, we can show that the local pressure coefficient for a specific-shaped building can be expressed as:

$$C_p = \frac{p - p_{\text{ref}}}{\frac{1}{2} \rho V_{\text{ref}}^2} = \phi\left(\frac{\rho V l}{\mu}, \frac{l_i}{l}, \alpha\right) \tag{3.11}$$

where V_{ref} and p_{ref} are the reference velocity and pressure mainly defined at a reference height alongside the upstream velocity.

For example, C_p distribution over a case study gable-roofed low-rise building is depicted in Figure 3.5. As it can be seen, unlike many simplistic models, C_p does not demonstrate a constant value on a specific building's surface and inversely has a distribution over each surface. In the below example, surface-2 has the maximum pressure whilst surface-4 demonstrates almost the smallest C_p values. If a wind corridor to impose a cross-ventilation is designed by the installation of windows on these surfaces, we could expect a strong airflow against $\theta = 30°$ approaching wind (see Figure 3.5).

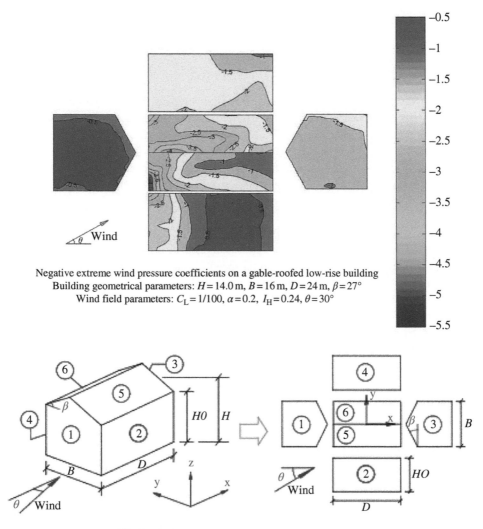

Negative extreme wind pressure coefficients on a gable-roofed low-rise building
Building geometrical parameters: $H = 14.0\,m$, $B = 16\,m$, $D = 24\,m$, $\beta = 27°$
Wind field parameters: $C_L = 1/100$, $\alpha = 0.2$, $I_H = 0.24$, $\theta = 30°$

Figure 3.5 Contours of C_p distribution over a case study gable-roofed low-rise building from a wind tunnel measurement. *Source:* Tamura [8]/Tokyo Polytechnic University.

Example 3.3

Pressure coefficients are measured over the external surfaces of three gable-roof buildings with different pitch angles as seen in Figure 3.E3.1. Calculate the wind pressure over the buildings if the reference velocity is measured as 2.6 m/s at the eave's height (H). The buildings' heights and lengths are 6 and 6.6 m, respectively. The reference pressure is one bar, and the air density is 1.2041 kg/m^3 at 20°C.

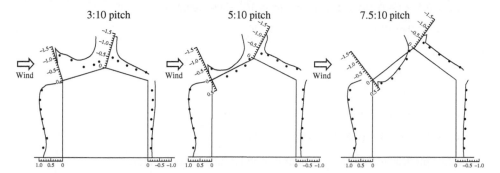

Figure 3.E3.1 Pressure coefficients over the external surfaces of three gable-roof buildings with different pitch angles. *Source:* Tominaga et al. [9]/ With permission of Elsevier.

Solution

Wind velocity is assumed to be constant in the above cases. On each wall an approximate averaged value for C_p is considered using the depicted curves. Then, the pressure and force exerted on each wall are calculated. For each surface first we can obtain the averaged C_p while the walls' length normal to the plane is considered as the unity when the areas are calculated:

Surface	Averaged C_p	Area (m^2)
Top left-angled roof	−0.53	3.45
Top right-angled roof	−0.37	3.45
Right wall	−0.18	6.00
Left wall	0.86	6.00

Detailed calculations are presented for the case of 3 : 10 pitch (Tan(16.7°)) as follows:

Top left-angled roof:

$$C_p = \frac{P_s - P_{ref}}{0.5\rho v_{ref}^2} = -0.53$$

Rewriting the above equation for P_s and inserting the known parameters gives:

$$P_s = 0.5 \times C_p \rho v_{ref}^2 + P_{ref} = 0.5 \times (-0.53) \times 1.2041 \times 2.6^2 + 10^5 = 99,997.84 \text{ Pa}$$
$$F_s = P_s \times A = 99,997.97 \times 3.45 = 344,523.2 \text{ N}$$

Top right-angled roof:

$$C_p = \frac{P_s - P_{ref}}{0.5\rho v_{ref}^2} = -0.37$$

$P_s = 99,997.97 \, \text{Pa}$

$F_s = 344,523.6 \, \text{N}$

Right wall:

$$C_p = \frac{P_s - P_{ref}}{0.5\rho v_{ref}^2} = -0.18$$

$P_s = 0.5 \times (-0.18) \times 1.2041 \times 2.6^2 + 10^5 = 99,999.26 \, \text{Pa}$

$F_s = 99,999.19 \times (6 \times 1) = 599,995.3 \, \text{N}$

Left wall:

$$C_p = \frac{P_s - P_{ref}}{0.5\rho v_{ref}^2} = 0.86$$

$P_s = 0.5 \times 0.86 \times 1000 \times 2.6^2 + 10^5 = 100,003.5 \, \text{Pa}$

$F_s = 100,003.3 \times (6 \times 1) = 600,021.0 \, \text{N}$

For other pitch angle cases, results are provided in the below tables.

Pitch 5 : 10 (26.6°)	Averaged C_p	Pressure (Pa)	Force (N)
Top right-angled roof	0.1	100,000.4	36,9062.8
Top left-angled roof	−0.4	99,998.37	369,055.3
Right wall	−0.3	99,998.78	599,992.7
Left wall	0.8	100,003.3	600,019.5
Pitch 7.5 : 10 (36.9°)	**Averaged C_p**	**Pressure (Pa)**	**Force (N)**
Top right-angled roof	0.2	100,000.8	412,660.1
Top left-angled roof	−0.5	99,997.97	412,648.4
Right wall	−0.4	99,998.37	599,990.2
Left wall	0.8	100,003.3	600,019.5

Therefore, as it can be seen in Figure 3.E3.2, the forces acting on different surfaces of the buildings mainly change when the pitch angle changes.

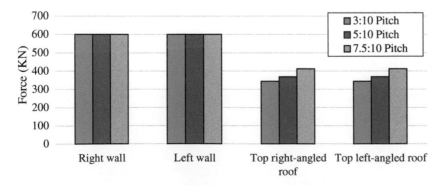

Figure 3.E3.2 Forces acting on surfaces of the gable-roof buildings.

3.3.2 Building Openings

In Chapter 2, we have introduced Bernoulli equation and discussed some of its application in pipes, flow measurement, etc. Similarly, the Bernoulli equation is widely applied to model the airflow between rooms or between indoor and outdoor spaces of a building with a high level of accuracy. We have learned that pressure gradients induced by pumps and fans in pipe, duct, and free jet flows are the driving forces in these flows. In buildings, the pressure gradients are again the driving force behind the airflow throughout the buildings. In fact, these pressure differences are caused either by wind or buoyancy acting though openings, cracks, and gaps of buildings. Hence, the pressure differences result in jets flowing through the opening. Since incompressible, inviscid, steady flows passing through streamlines of openings are not always entirely valid assumptions for many scenarios, the Bernoulli equation can be utilized with some modifications to be applicable for engineering design applications in buildings. Furthermore, similar to flows through sharp areas (see Figure 3.6), the flow mainly face the famous '**vena contracta effect**', which is related to inability of flows to turn over sudden 90°-corners as openings do not have smooth nozzle-shaped areas to direct the flow in a way that the contraction phenomena and loss of energy are avoided.

In other words, if we consider the Bernoulli equation across the streamlines (see Eq. 2.59), we can consider $\mathcal{R} < \infty$ across curved streamlines, which causes a pressure difference through the opening areas. In the case of sharp corners, however, $\mathcal{R} = 0$ and the pressure gradient at that point would be infinite to force the flow to turn 90°. This results in maximum pressure at the centre line and minimum values of zero at the corners, which contradicts the assumption of a constant pressure at the opening area and thus a uniform velocity with straight streamlines. As it can be seen in Figure 3.6, the area with the desired assumptions ($A_j = \dfrac{\pi d^2}{4}$) is relocated outside of the opening location at A-A cross-section where it is called **vena contracta plane**. As this plane is a characteristic of the opening shape, the ratio of its area over the opening area ($A_f = \dfrac{\pi D^2}{4}$) is called as **contraction coefficient** ($C_d = A_j/A_f$) or **discharge coefficient**. Finding this ratio, however, demands a measurement procedure to quantify the place and magnitude of vena contracta plane.

Figure 3.6 Vena contracta effect for a sharp-edged opening.

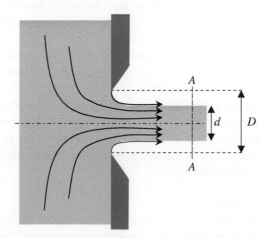

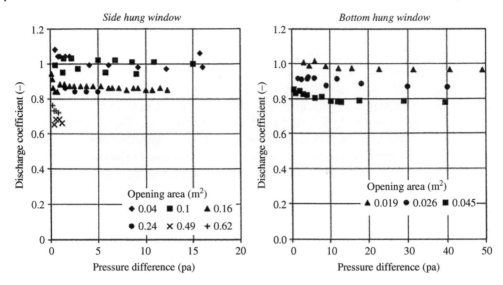

Figure 3.7 C_d for two common window shape opening. *Source:* Heiselberg et al. [10]/ with permission of Elsevier.

Therefore, in many applications, the concept is used with a semi-empirical equation after applying the Bernoulli equation together with the conservation of mass equations between the opening and the enclosed space where the velocity can be assumed to be negligible $(V = \sqrt{\frac{2\Delta p}{\rho}})$:

$$Q = A_j V = C_d A_f V$$

$$\Rightarrow Q = C_d A_f \sqrt{\frac{2\Delta p}{\rho}} = A_{\text{eff}} \sqrt{\frac{2\Delta p}{\rho}} \tag{3.12}$$

where Δp is the value of pressure difference or gradient across the opening, V is the velocity at vena contracta plane, A_f is known as the free area (mainly equal to the cross-sectional area of the opening), and ρ is the air density. Sometimes, A_{eff} as the effective area is used as the product of C_d and A_f.

$C_d = 0.61$ is a widely assumed value for a standard circular sharp-edged orifice although the shape of openings such as windows and door are way more complex than this ideal shape. Thus, extensive experimental works are normally involved to provide discharge coefficient for engineering applications. The equivalent area (A_{eq}) is another popular approach in industry, which is assumed as a hypothetical circular sharp-edged orifice, which equivalently allows air to pass at the same volume flow rate with an identical pressure difference. Figure 3.7 illustrates some example c_d values for two common window shapes.

Example 3.4

If a side hung window with area of 0.18 m^2 is going to be installed at the center of both walls of the buildings of Example 3.3, calculate the volume flow rate through the building.

Solution

C_p distributions depicted on walls of the buildings in Example 3.3 on each wall can be averaged to attain pressure coefficient values as shown before. From Example 3.3, we have obtained the below values for the building with 3 : 10 pitch:

$$P_1 = 100,003.5 \text{ Pa}$$
$$P_2 = 99,999.3 \text{ Pa}$$

There is a direct flow between p_1 and p_2 and thus the velocity is constant through the windows as well as the elevation. So, the Bernoulli equation can be simplified to find the indoor pressure (p_{in}) as an interpolation of pressures exerted on opposite walls p_1 and p_2:

$$P_{in} = 0.5(p_1 + p_2) = 100,001.4 \text{ Pa}$$

Then, the discharge coefficient can be also found $C_d = 0.87$ from side hung window shown in Figure 3.7 while the window area is $A_f = 0.18 \text{ m}^2$. Hence, using Eq. (3.12) results in:

$$Q_{16.7°} = 0.87 \times 0.18 \times \sqrt{\frac{2 \times (100,003.5 - 100,001.4)}{1.2041}} = 0.294 \text{ m}^3/\text{s}$$

Similar calculations can be conducted for 5 : 10 pitch and 7.5 : 10 pitch buildings to find:

$$Q_{26.6°} = 0.302 \text{ m}^3/\text{s}$$

$$Q_{36.9°} = 0.315 \text{ m}^3/\text{s}$$

3.3.3 Wind-Driven Ventilation

As it was discussed before, wind approaching over a building's surfaces cause a pressure gradient between its indoor and outdoor environments and hence drive the mass flow throughout its openings, cracks, and gaps. The pressure difference is understood to be a function of a building's shape as well as wind characteristic (turbulence level, magnitude, and direction). On the other hand, the wind driven through openings, as stated in the previous section, does not fully satisfy the assumptions required to imply the Bernoulli equation and thus should be corrected with the concept of the discharge coefficient. With these key points in mind, we can derive Bernoulli equation between each two points of fully mixed zones of a building (zones are further defined in Chapter 8). For example, for the depicted case of Figure 3.8, we can assume zone-1 of the illustrated building as a fully mixed space and then we can write:

$$\Delta p_w = \left(p_{ww\_1} + \frac{\rho V_{ww\_1}^2}{2} \right) - \left(p_{z_1} + \frac{\rho V_{z_1}^2}{2} \right) + \rho g(z_{ww\_1} - z_{z_1}) \tag{3.13}$$

where ΔP_w is the total pressure difference between two nodes while p_{ww_1} and p_{z_1} are windward and zone-1 pressures. Also, v_{ww_1}, v_{z_1}, z_{ww_1}, and z_1 are windward and zone-1 airflow velocities and elevations, respectively.

If we neglect the effect of elevation variation (hydrostatic pressure) between the upstream wind pressure (p_{ref}) and windward/leeward pressures, then we can derive the Bernoulli

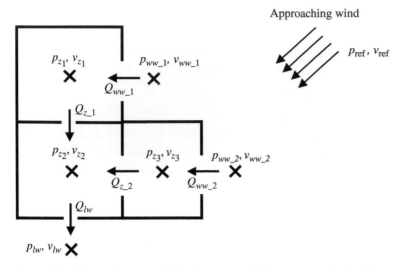

Figure 3.8 Wind-driven ventilation throughout different zones of a building.

equation between them and find pressure at those location using their associated pressure coefficient (see Section 3.1.3.1 for obtaining C_p):

$$C_{pww\_1} = \frac{P_{ww-1} - P_{ref}}{\frac{1}{2}\rho v_{ref}^2} \tag{3.14}$$

$$C_{plw} = \frac{P_{lw} - P_{ref}}{\frac{1}{2}\rho v_{ref}^2} \tag{3.15}$$

The airflow rate passing from the windward zone through the zone-1 and zone-3 are given by Eq. (3.12):

$$Q_{ww\_1} = C_{d\_ww\_1} A_{ww\_1} \sqrt{\frac{2(P_{ww\_1} - P_{z_1})}{\rho}} \tag{3.16}$$

$$Q_{ww\_2} = C_{d\_ww\_2} A_{ww\_2} \sqrt{\frac{2(P_{ww\_2} - P_{z_3})}{\rho}} \tag{3.17}$$

Also, the leaving airflow from zone-2 in Figure 3.8 towards the leeward zone can be written as:

$$Q_{lw} = C_{d\_lw} A_{lw} \sqrt{\frac{2(P_{z_2} - P_{lw})}{\rho}} \tag{3.18}$$

The same concept can be applied to the interzonal airflows:

$$Q_{z_1} = C_{d\_1} A_{z_1} \sqrt{\frac{2(P_{z_1} - P_{z_2})}{\rho}} \tag{3.19}$$

$$Q_{z_2} = C_{d\_3} A_{z_3} \sqrt{\frac{2(P_{z_3} - P_{z_2})}{\rho}} \tag{3.20}$$

Furthermore, the airflow between each two zones can be assumed by the conservation law of the mass and thus we can establish the mass flow balance between all zones to further be able to calculate all the mass flow rates within the building:

$$Q_{ww\_1} = Q_{z_1} \tag{3.21}$$

$$Q_{ww\_2} = Q_{z_2} \tag{3.22}$$

$$Q_{lw} = Q_{z_1} + Q_{z_2} \tag{3.23}$$

One can now solve the non-linear system of equations obtained through Eqs. (3.16) to (3.23) to obtain all the zone pressures and volume flow rates. Further details and application of this approach is explained in Chapter 8.

Example 3.5

Assume the building with $3 : 10$ pitch of Example 3.3 to have an internal layout as illustrated in Figure 3.E5. First, calculate the volume flow rate through each room while all doors have dimensions of $1.98 \text{ m} \times 0.76 \text{ m}$. Now, assume that a mechanical ventilation system within the bedroom imposes $+3 \text{ }pa$ differential pressure with respect to the corridor's pressure and recalculate the volumetric flow rates through the rooms.

Solution

As stated in the example, doors' areas are identical, and density of air is constant throughout the field. Since the difference between p_{ww} and p_{lw} (according to Figure 3.E3.1 in Example 3.3) is less than $5 \text{ }pa$, it is expected that the pressure drops across the rooms inside the building to be a fraction of this overall pressure difference. Moreover, we can further assume similar discharge coefficient values for all the doors.

We can now form the volumetric flow balance as below for internal nodes:

$$Q_{ww} = Q_{\text{Living}\rightarrow\text{Kitchen}}$$

$$Q_{\text{Living}\rightarrow\text{Kitchen}} = Q_{\text{Kitchen}\rightarrow\text{Corridor}}$$

$$Q_{\text{Kitchen}\rightarrow\text{Corridor}} = Q_{\text{Corridor}\rightarrow\text{Bedroom}} + Q_{lw}$$

There is only one door or opening in the bedroom and thus, continuity requires a flow rate:

$$Q_{\text{Corridor}\rightarrow\text{Bedroom}} = 0$$

Note that mass balance equations are written in terms of volumetric flow rates because of constant air density assumption. Rewriting the above equations gives:

$$Q_{ww} = Q_{\text{Living}\rightarrow\text{Kitchen}} = Q_{\text{Kitchen}\rightarrow\text{Corridor}} = Q_{lw}$$

We again apply Eq. (3.12) to resolve each volumetric flow rate passing through each door and form four new sets of equations ($Q = C_d A_f \sqrt{\frac{2\Delta P}{\rho}}$). With further

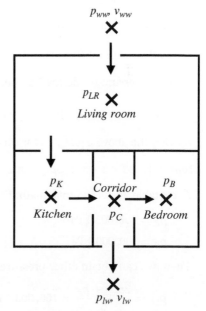

Figure 3.E5 Internal layout of the case study building.

rearrangement of equations, and assuming the same density, door area, and discharge coefficients for all the rooms, we can reach to multiple relationship for the pressure of the zones:

$$p_{ww} - p_{LR} = p_{LR} - p_K = p_K - p_C = p_C - p_{lw}$$

Now, from Example 3.3 for the building case with 3 : 10 pitch:

$$p_{ww} = 100,003.3 \text{ Pa}$$

$$p_{lw} = 99,999.2 \text{ Pa}$$

Solving for other pressures inside the buildings:

$$p_{LR} = \frac{3p_{ww} + p_{lw}}{4} = 100,002.3 \text{ Pa}$$

$$p_K = 2p_{LR} - p_{ww} = 100,001.2 \text{ Pa}$$

$$p_C = 2p_K - p_{LR} = 100,000.2 \text{ Pa}$$

Now each of calculated pressures can be used to obtain the volumetric flow rate:

$$Q_{ww} = 0.87 \times 0.18 \times \sqrt{\frac{2 \times (100,003.3 - 100,002.3)}{1.2041}} = 0.2 \text{ m}^3/\text{s}$$

Now, we assume that the problem is solved when the bedroom is equipped with a mechanical ventilation generating a slight positive pressure. In this case, the volumetric flow rate through the bedroom rooms will be changed to:

$$Q_{ww} = Q_{\text{Living} \to \text{Kitchen}} = Q_{\text{Kitchen} \to \text{Corridor}} = Q_{lw} + Q_{\text{Corridor} \to \text{Bedroom}}$$

Now, recalculating the volumetric flow rates yields:

$$\sqrt{p_{ww} - p_{LR}} = \sqrt{p_{LR} - p_K} = \sqrt{p_K - p_C} = \sqrt{p_C - p_{lw}} - \sqrt{p_B - p_C}$$

Positive pressure inside the bedroom requires that $\sqrt{p_B - p_C} = \sqrt{-3}$, therefore:

$$p_{ww} - p_{LR} = p_{LR} - p_K = p_K - p_C = p_C - p_{lw} + 3 + 2\sqrt{3}\sqrt{p_C - p_{lw}}$$

Solving the above equations for other pressures inside the room gives:

$$16p_C^2 + [-36 \times 3 - 8 \times (3p_{lw} + p_{ww} - 3 \times 3)]p_C + (3p_{lw} + p_{ww} - 3 \times 3)^2 + 36 \times 3p_{lw} = 0$$

Acceptable answer to this quadratic equation will be:

$$p_C = 100,003.1 \text{ Pa}$$

$$p_B = 100,000.1 \text{ Pa}$$

Then, we can obtain other pressures as follows:

$$p_{LR} = \frac{2p_{ww} + p_C}{3} = 100,003.2 \text{ Pa}$$

$$p_K = 2p_{LR} - p_{ww} = 100,003.1 \text{ Pa}$$

Now each of the calculated pressures can be used to obtain the volumetric flow rates:

$$Q_{ww} = Q_{\text{Living}\rightarrow\text{Kitchen}} = Q_{\text{Kitchen}\rightarrow\text{Corridor}}$$

$$= 0.87 \times 0.18 \times \sqrt{\frac{2 \times (100,003.3 - 100,003.2)}{1.2041}} = 0.049 \, \text{m}^3/\text{s}$$

$$Q_{\text{Bedroom}\rightarrow\text{Corridor}} = 0.87 \times 0.18 \times \sqrt{\frac{2 \times (3)}{1.2041}} = 0.35 \, \text{m}^3/\text{s}$$

$$Q_{lw} = Q_{\text{Kitchen}\rightarrow\text{Corridor}} + Q_{\text{Bedroom}\rightarrow\text{Corridor}} = 0.049 + 0.35 = 0.398 \, \text{m}^3/\text{s}$$

3.3.4 Buoyancy-Driven Ventilation

Buoyancy-driven flow, also known as *stack effect*, occurs when the air density in the upper zone of a building is greater than the air density in its lower zone. As shown in Figure 3.9, the stack effect is based on the hydrostatic pressure difference due to the air temperature variation. A warm, humid air column is lighter and less dense than a cold, dry one, tending to rise and be replaced by a cooler one. The stack pressure is also related to the distance or the height difference between the inlet and the outlet openings.

Stack effect can occur unintentionally in the buildings while it is widely designed to intentionally enhance the natural ventilation in buildings. The buoyancy-driven flow is bi-directional in the nature and the amount of the upper flow is equal to the lower flow across the opening. It is also assumed that the maximum buoyancy flow occurs when the pressure difference across the opening due to a forced airflow is zero.

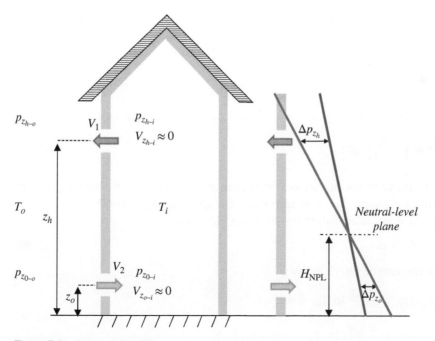

Figure 3.9 Stack ventilation.

Under still-air conditions and when the internal temperature is higher than the external one, the flow moves upward. Conversely, when the external environment is warmer than the internal one, the flow direction is downward. The equation for calculation of the pressure difference due to the stack effect is based on the Bernoulli principle in an isothermal and steady-state condition to be derived for both indoor and outdoor environment. Air movement assumed to be negligible in the spaces, results on a hydrostatic variation of the pressure across the vertical direction in both indoor and outdoor environments, although with two different gradients:

$$dp_i = \rho_i(z_i)gdz_i \qquad (3.24)$$

$$dp_o = \rho_o(z_o)gdz_o \qquad (3.25)$$

where, i and o indices stand for the indoor and outdoor spaces.

As shown in Figure 3.9, the pressure gradient is higher when the outside temperature is cooler. Both pressure gradients intersect each other on a height (H_{NPL}), which is known as the **neutral-level plane** (NPL). This plane is essential in natural ventilation design of buildings. If we assume the behaviour of the air as an ideal gas, then we can refer to the state equation introduced in Chapter 2 as:

$$p = \rho RT \qquad (3.26)$$

A crucial assumption to develop the stack effect equation is that the density variation with elevation around buildings (not a suitable assumption for tall buildings) is relatively low although its product with the gravity (specific weight) cannot be neglected:

$$p/R = \rho T \approx \rho_{atm} T_{atm} = p_{atm}/R \qquad (3.27)$$

Applying Eqs. (3.24) and (3.25) into Eq. (3.27) yields:

$$dp = \frac{p_{atm}}{RT(z)}gdz \qquad (3.28)$$

Now, with integration through the height of openings, we can reach to:

$$\int_{P_{z_0}}^{P_{z_h}} dp = \int_{z_0}^{z_h} \frac{p_{atm}}{RT(z)}gdz \qquad (3.29)$$

Again, it is crucial to the further develop equations to approximate how temperature changes with the elevation. Let us assume that a fair approximation (not the only possible one) would be a linear alteration in height with a slope of α:

$$T(z) = T_{z_0} + \alpha z \qquad (3.30)$$

Hence, we can rewrite Eq. (3.29) as:

$$P_{z_h} - P_{z_0} = \frac{p_{atm}g}{R} \ln\left[T_{z_0} + \alpha z\right]_{z_0}^{z_h}$$

$$\Longrightarrow P_{z_h} - P_{z_0} = \frac{p_{atm}g}{R}\left(\ln\left(T_{z_0} + \alpha z\right) - \ln\left(T_{z_0}\right)\right) \qquad (3.31)$$

The above equation is valid for both indoor and outdoor spaces. Nonetheless, in another assumption, as employed in many problems, the indoor and outdoor temperatures are considered to be constant and homogenous, which implies:

$$T_i(z) = T_i \tag{3.32}$$

$$T_o(z) = T_o \tag{3.33}$$

Therefore, if we account $z_{h-i} - z_{0-i} = z_{h-o} - z_{0-o} = H$, then, we can simplify Eq. (3.29) for both indoor and outdoor spaces as:

$$p_{z_{h-i}} - p_{z_{0-i}} = \frac{p_{atm}g}{RT_i} H \tag{3.34}$$

$$p_{z_{h-o}} - p_{z_{0-o}} = \frac{p_{atm}g}{RT_o} H \tag{3.35}$$

The pressure difference across the opening of a building at a fixed height (H) can be obtained by subtracting Eq. (3.34) from Eq. (3.35), which imposes airflow throughout them (see Figure 3.9) previously introduced as the stack flow:

$$\Delta p_s = \frac{p_{atm}gH}{R}\left(\frac{1}{T_o} - \frac{1}{T_i}\right) \tag{3.36}$$

where Δp_s is the pressure difference due to stack effect at a height of H.

Note that from Eq. (3.26), we can obtain:

$$\rho_i T_i = \rho_o T_o$$

$$\Longrightarrow \left(\frac{\rho_o - \rho_i}{\rho_i}\right) = \left(\frac{T_i - T_o}{T_o}\right) \tag{3.37}$$

And, thus, after applying Eqs. (3.37) to (3.36), we can find another form of the stack pressure at a height of H as (see also Figure 3.9):

$$\Delta p_s = \rho_i g(H_{NPL} - H)\left(\frac{T_i - T_o}{T_o}\right) \tag{3.38}$$

Therefore, the airflow rate at the opening can be calculated as:

$$Q_s = C_d A_h \sqrt{\frac{2\Delta p_s}{\rho}} \tag{3.39}$$

where Q_s is the airflow rate due to the stack effect, A_h is the cross-sectional area of the opening and C_d is its discharge coefficient.

Note that the flow is due to change of static pressures of indoor and outdoor spaces. In reality, still air assumption is not a likely one and wind contribution as well as the pressure difference due to the pressurization of the building (Δp_p) should be integrated with the stack effect to obtain the total pressure difference around an opening:

$$\Delta p = \Delta p_p + \Delta p_s + \Delta p_W \tag{3.40}$$

Example 3.6

We assume a similar building as shown in Figure 3.9 with different indoor and outdoor temperatures of 20° and 0°, respectively, with the same openings' dimensions of 0.3 m × 0.3 m, but at a 10 m height difference. Calculate the mass flow rate due to the stack effect through the openings. The discharge coefficients of both openings are assumed to be 0.7.

Solution

The pressure difference due to the stack effect is given by Eq. (3.36) as below:

$$\Delta p_s = \frac{101,325}{287.18} \times 9.8 \times 10 \times \left(\frac{1}{273.15} - \frac{1}{293.15}\right) = 8.43 \text{ Pa}$$

It can be assumed that the pressure difference for both openings is equal as they have similar dimensions:

$$\Delta p_{\text{opening}} = \frac{\Delta p_s}{2} = 4.21 \text{ Pa}$$

Mass flow rates at both openings can be calculated using Eq. (3.39) multiplied by the average density between two side of the opening ($\rho = \frac{(1.28 + 1.19)}{2} = 1.24 \text{ kg/m}^3$):

$$\dot{m}_{o\_1} = \dot{m}_{o\_2} = \rho Q_s = \rho C_d A_h \sqrt{\frac{2\Delta p_s}{\rho}}$$

$$\Longrightarrow \rho Q_s = 1.24 \times 0.7 \times 0.3 \times 0.3 \sqrt{\frac{2(4.21)}{1.24}} = 0.2 \text{ kg/s}$$

Example 3.7

The illustrated building in Figure 3.E7 is predominantly ventilated using a natural ventilation strategy assisted by wind and buoyancy forces (all dimensions are in mm). Outdoor and indoor temperatures are, respectively 0 and 20°C. Both openings are identical with the dimensions of 0.45 m × 0.45 m while the discharge coefficients are 0.8. Estimate the stack flow throughout the building if the wind-driven flow is neglected during a calm winter day. What would be the density of the leaving air from the building if the entering density is 1.235 kg/m³ while the lower opening is closed.

Solution

According to the building layout, the openings have the heights of $H_1 = 11$ m and $H_2 = 17$ m. Therefore, the stack pressures are calculated at both floors as:

$$\Delta P_{s\_1} = \frac{P_0}{R} g H_1 \left(\frac{1}{T_{\text{out}}} - \frac{1}{T_{\text{in}}}\right) = \frac{101,325}{287.18} \times 9.8 \times 11 \left(\frac{1}{273} - \frac{1}{293}\right) = 9.5 \text{ Pa}$$

$$\Delta P_{s\_2} = \frac{P_0}{R} g H_2 \left(\frac{1}{T_{\text{out}}} - \frac{1}{T_{\text{in}}}\right) = \frac{101,325}{287.18} \times 9.8 \times 17 \left(\frac{1}{273} - \frac{1}{293}\right) = 14.7 \text{ Pa}$$

Now, the mass flow rate entering the building ($\dot{m}_1$) is calculated as follows:

$$\dot{m}_1 = \rho C_d A_f \sqrt{\frac{2\Delta P_{s\_1}}{\rho}} = 1.235 \times 0.8 \times 0.2025 \times \sqrt{\frac{2 \times (9.5)}{1.235}} = 0.785 \text{ kg/s}$$

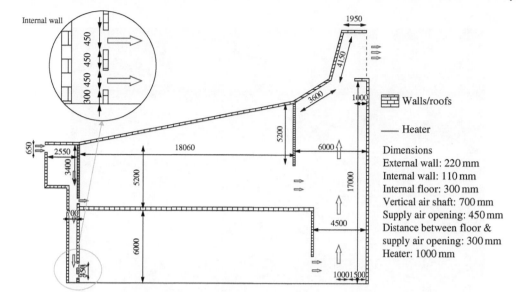

Figure 3.E7 Cross-section layout of a naturally ventilated building.

Due to the continuity within the control volume of the building, we should have $\dot{m}_1 = \dot{m}_2$ while $\dot{m}_2$ is the mass flow rate leaving the building. Thus, we can find the density of the leaving air as:

$$\rho = \left(\frac{C_d A_f \sqrt{2\Delta P_{s\_2}}}{\dot{m}_2} \right)^{1/3} = \left(\frac{0.8 \times 0.2025 \times \sqrt{2 \times (14.7)}}{0.785} \right)^{1/3} = 1.038 \text{ kg/m}^3$$

3.4 Turbulent Jet and Plume

The turbulent jet and plumes are known as a type of the turbulent shear flow, referring to the continuous flows in which a high velocity shear occurs between two fluids when a moving fluid intrudes into a still fluid. When the flow follows an intermittent injection behaviour, it is called puff rather jet or plume. For example, Figure 3.10a shows a turbulent plume generated by chimneys' exhausts of a factory. On the other hand, Figure 3.10b demonstrates an injected turbulent jet throughout another fluid. The fundamental behind the dynamics of jets is amongst the most complex problems of fluid mechanics and is based on decades of research in theoretical and experimental aspects of them, an extensive knowledge is generated in literature.

In building applications, turbulent jets are commonly happening during mechanical or cross-ventilations due to the momentum difference of the intruding fluid while plumes can be generated due to the buoyancy difference of the intruding fluid for instance from respiratory events of occupants or plumes formed in the displacement ventilations. Jets and plumes leaving ducts are most of the time in the turbulent region as they mainly exceed the critical Reynolds number necessary for such flows to enter into the turbulence regime.

While the ventilation strategies, design and best practices are beyond the scope of this book to be scrutinized, we describe the flow mechanism of jets and plumes in further detail

(a)

(b)

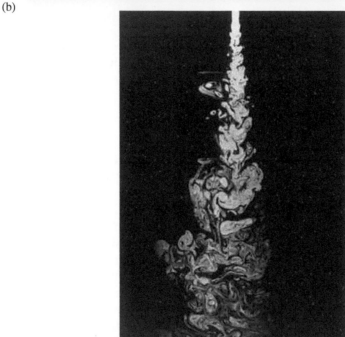

Figure 3.10 (a) Turbulent plume and (b) turbulent jet. *Source:* (a) jwvein/Pixabay. (b) Dimotakis et al. [11].

and from fluid mechanics point of view. Also, we are not focusing on various available shapes of diffusers or openings in this section and instead will assume an orifice shaped opening in which a free jet or plume can be generated through an unconfined space where the impact of proximities (walls and ceiling in buildings) is not yet taken into account.

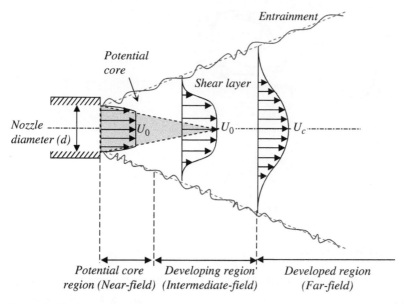

Figure 3.11 Jet structures.

Jets and plume have general characteristics and follow the same fundamental rule in nature. So, when they leave their confined spaces such as ducts, where no-slip boundary condition is attaching the flow particles to the surface, the air suddenly turns into contact with another ambient still air (or with a different velocity or temperature). As it can be seen in Figure 3.11, this still layer at the edge of the jet or plume causes a mixing mechanism in the ***entraining layer*** with a high velocity gradient (shear stress).

Furthermore, the mixing initiates expansion towards the direction of the ambient fluid due to both molecular and turbulence diffusions of the jet or plume. The shear stress from the jet or plume fluid has a dominant direction parallel to the flow, resulting in frictions and eddies in the border of jets or plumes as shown in Figure 3.12. Hence, we can easily understand from the law of the momentum conservation that the jet or plume maximum velocity at its centreline should drop as it progresses towards the flow direction. In other words, the momentum of the flow should remain the same in any arbitrary cross-section of the jet or plume, resulting in a lower velocity when the area becomes larger with distance from the opening of the jet or plume. From the point of view of the law of the mass conservation, the increased mass of the jet or plume with distance from the opening is due to the joining of particles from the quiescent fluid to the entrainment region (see Figure 3.12). In the turbulence regimes, the entrainment layer is stronger than laminar ones due to eddy mixings (turbulent diffusion) rather than a sole molecular diffusion.

3.4.1 Jet Structure

The mechanisms of turbulent jet flow obey the same fundamental rules even though the shape of opening can cause significant differences in their structures. Thus, in this section, we only focus on the round jets acting on Newtonian fluids, which steadily flow through a

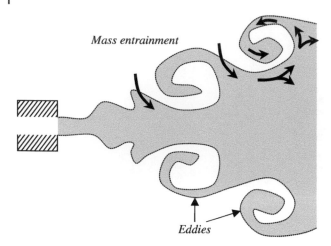

Figure 3.12 Entrainment in shear flow.

nozzle with a diameter d. The flow is expected to be initially uniform. The shape of the jet is almost conical when a moving flow intrudes a quiescent fluid with almost a similar density. The radius of the conical (r) is proportional to the distance from the opening (x) while surprisingly the angle is understood to be universally 23.6° similar in many flows independent from the properties of the moving fluid (e.g. material and velocity) or opening characteristics (e.g., diameter). In general, jets have three main regions (see Figure 3.11), including the near-field, the intermediate-field, and the far-field.

The near-field region is highly impacted by the shape of the opening located within $0 < x/d < 6$. This region contains the potential core region, which is a region starting before the entrance of the opening where the intruding fluid is still not in contact with the ambient fluid. The velocity profile is expected to be uniform at this region similar to the initial velocity of the jet. The far-field is stretched in a range about $x/d > 30$ in a region where the flow is fully-developed, and almost has a similar shape in all jet flows. This implies that jets have almost universal characteristics at this region regardless of the initial conditions imposed in the near-field region. The velocity has been observed to have a bell-shaped (Gaussian) profile with the maximum velocity at the middle (U_{max}), which is stretched across the jet. The intermediate-field region is a transitional region between latter regions. Interestingly, the location of the diffusers based on the chosen ventilation strategy in buildings can result in the ventilated area to be in the far-field zones. On the other hand, near- and intermediate-field parameters (i.e. the exit Reynolds number, nature of exit profiles of the mean velocity, turbulence intensity, nozzle-exit geometric profiles, and aspect ratio in noncircular jets) should be carefully designed as they will define the characteristics of the expected jet in the ventilation applications.

3.4.2 Jet and Plume in Ventilation

Figure 3.13 demonstrates the displacement and entrainment ventilations, which are benefiting from air jets and plumes. Well-mixed condition barely occurs in the displacement ventilation, leaving some of the zone in a stratified air condition. Nonetheless, this ventilation

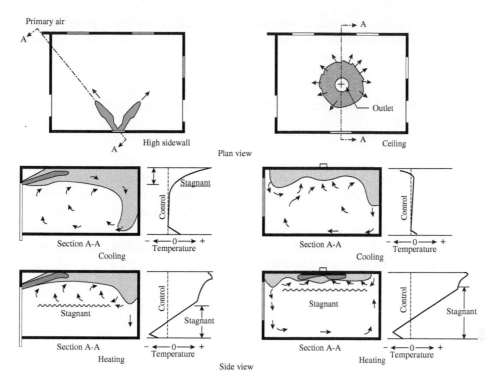

Figure 3.13 Illustrative application of jets and plumes in ventilation. *Source:* [12]/ASHRAE.

strategy is popular when the focus is on pollution removal or air quality enhancement. On the other hand, in the entrainment ventilation, an air jet flow is mainly directed from ceilings or walls while the extracts are installed in a position to promote better mixing of the air. Many effective ventilation systems contain multiple diffusers installed as a single or a combined strategy of the introduced ventilation strategies.

Due to occupant activities, appliances, and room design, most of the time, there is a stratification in the ventilated spaces though an effective ventilation design happens when a well-mixed condition for occupants is provided. Calculation of jet and plume behaviour, however, is way more complex in realistic conditions and not an easy task. Thus, many empirical studies have been conducted to provide guidelines and best practices for the design purposes. In this aspect, designers need to be experienced about many attributes, including geometrical features, climatic conditions, occupant behaviours, costs, environmental considerations, etc. Above this, they should be equipped with a deep understanding of jet and plume mechanisms. For example, the below empirical equation is widely utilized to estimate the maximum velocity ($U_{\max}$) of an isothermal jet within the far-field to be more effective:

$$U_{\max} = \frac{K\dot{Q}_o}{\sqrt{A_o}x} \tag{3.41}$$

where K is the proportionality constant. $\dot{Q}_o$ and A_o are, respectively the airflow rate and area associated to the uniform velocity at the opening.

Besides such empirical equations, as presented in Chapter 12, numerical approaches such as finite volume method can be employed as powerful tools to model complex jets and plumes in different ventilation strategies.

Example 3.8

The velocity profile of a circular jet can be approximated with a nearly Gaussian shape (bell curve) as expressed:

$$u(x, r) = u_{max} e^{\left(-\frac{50r^2}{x^2}\right)}$$

where x is the downstream distance along the jet (counted from the virtual source), r is the cross-jet radial distance from its centerline, and u_{max} is the maximum speed at the centerline.

Also, if Figure 3.E8 demonstrates the relation between the average exit velocity (U) and the orifice diameter (d), then what would be the required flow rate from a circular diffuser ($d = 0.1$ m) to reach a 0.2 m/s velocity at a 2-m distance from it in a displacement ventilation system. What would be the average velocity at this point?

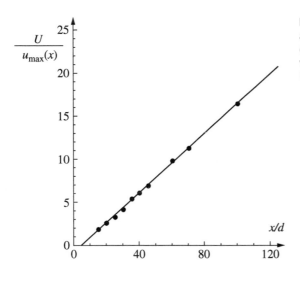

Figure 3.E8 Relation between the average exit velocity (U) and the orifice diameter (d). *Source:* Pope [13]/ with permission of Cambridge University Press.

Solution

It is assumed that $d = 0.1$ m and $x = 2$ m, therefore $x/d = 20$ and in accordance with Figure 3.E8:

$$\frac{U}{u_{max}(x)} = 3$$

As stated in the problem, the velocity at 2 m distance from the nozzle will be 0.2 m/s, which implies that the average velocity can be found as:

$$u(2, 0) = u_{max}(2) = 0.2 \, \text{m/s}$$

$$\Longrightarrow U = 3 \times u_{max}(2) = 3 \times 0.2 = 0.6 \, \text{m/s}$$

Integrating the Gaussian shape velocity profile yields the volumetric flow rate as:

$$Q = \int_0^\infty u2\pi r dr = \frac{\pi}{50} u_{max} x^2 = \frac{\pi}{10} dUx$$

Finally, the flow rate will be calculated as:

$$Q = \frac{\pi}{10} dUx = \frac{\pi}{10} \times 0.1 \times 0.6 \times 2 = 0.038 \text{ m}^3/\text{s}$$

3.5 Wall Effect

In the previous section, we have discussed the application of turbulent shear flow in buildings. Nonetheless, these flows are not as common as flow moving over walls, including internal and external walls as well as roofs and ceilings in buildings. In the previous chapter, we have explained the structure of boundary layer where we understood that the flow regime is almost turbulent over the buildings' surface. Hence, in this section, we further explore the structure of turbulence boundary layer, which can be very essential in the wall modelling stages either empirically or numerically as practiced in Chapter 12.

3.5.1 Inner Layer

In general, the boundary layer can be divided into the ***inner*** and ***outer layers*** or ***regions***. Outer layer is associated with the inertia dominant flow and far from the wall and viscous effects. As it can be seen in Figure 3.14, the inner layer inversely occurs within 10–20% of the

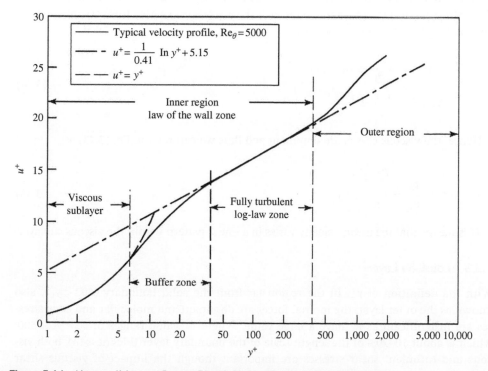

Figure 3.14 Near wall layers. *Source:* Blazek [14]/ with permission of Elsevier.

boundary layer and includes **viscous sublayer**, **buffer layer**, and **fully turbulent layer** or **log-law**. We explore each layer in the following section.

Dimensionless analysis introduced in Chapter 2 is used to explore the impact of solid boundaries on the flow. Thus, we assume that the mean flow over a smooth wall can be impacted by the distance from the wall (y), shear stresses (τ_w), and fluid characteristics of density (ρ) and viscosity (μ). So, we can write the relationship known as **law of the wall** as below:

$$U = f(y, \rho, \tau_w, \mu) \tag{3.42}$$

The dimensionless analysis can be further conducted to reach to:

$$\frac{U}{u_\tau} = f\left(\frac{\rho u_\tau y}{\mu}\right) \tag{3.43}$$

Note that $u_\tau = \sqrt{\tau_w/\rho}$ is defined as the friction velocity and $y^+ = \dfrac{\rho u_\tau y}{\mu}$ represents the Reynolds number in a vertical direction from the solid surface in accordance with the friction velocity. Here, another dimensionless group is also defined as $u^+ = \dfrac{U}{u_\tau}$.

3.5.2 Viscous Sublayer

When $y^+ < 5$ viscous effects are very significant and the influence of mean stream flow is not considerable over this layer, which is called viscous sub-layer (see Figure 3.14). Within the viscous sublayer, flow is almost laminar without any eddy motion, and it is assumed that the shear stress is constant and equal to the wall shear stress (τ_w) across this very thin layer as observed in the experimental studies:

$$\tau(y) = \mu \frac{\partial U}{\partial y} \cong \tau_w$$
$$\Longrightarrow U(y) = \frac{\tau_w y}{\mu} = \frac{\rho u_\tau^2 y}{\mu} \tag{3.44}$$

Hence, the viscous effects are dominant and thus we can rewrite Eq. (3.43) as:

$$u^+ = \frac{U}{u_\tau} = \frac{\frac{\rho u_\tau^2 y}{\mu}}{u_\tau} = y^+ \tag{3.45}$$

This means that the mean velocity varies in a linear pattern within the viscous sublayer.

3.5.3 Log-Law Layer

With the definition of y^+, in the region far from the solid boundary ($500 < y^+$), also known as the outer layer, the inertial forces are dominant and molecular and eddy stresses are insignificant as y^+ have a large value. Where y^+ value ranges of $30 < y^+ < 500$, which is about 10–20% of the length scale of the boundary layer thickness (δ), both viscous and turbulent shear stresses are important though the impact of viscous shear

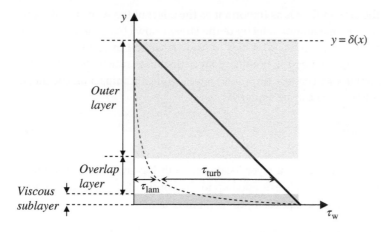

Figure 3.15 Laminar, turbulent, and total shear stresses near a wall.

stress weakens towards the upstream velocity after another layer, which is called buffer layer (see Figure 3.15).

In general, despite a reduction of the total shear stress ($\tau_{tot} = \tau_{turb} + \tau_{lam}$) with distance from the wall, the share of turbulent shear stress (τ_{turb}) over the molecular shear stress (τ_{lam}) becomes extensively dominant towards the upstream mean flow, so one can assume the outer layer as the fully turbulent layer. One of the used assumptions in this region is Prandtl's mixing length model, thus the shear stress can be approximated with the below equation:

$$\tau_w = -\rho\overline{uv} \cong \rho\kappa^2 y^2 \left(\frac{\partial U}{\partial y}\right)^2 \tag{3.46}$$

Back to the dimensionless analysis, with further mathematical and experimental studies applied to Eq. (3.43), the below **log-law equation** was suggested for y^+ for the fully turbulent layer valid for all kind of turbulence flows (**log-law layer**):

$$u^+ = \frac{1}{\kappa}\ln(y^+) + B \tag{3.47}$$

where κ is the von Karman's constant ($\kappa = 0.4$) and B is measured to be 5.5 for smooth walls.

3.5.4 Buffer Layer

This is a blending region where the effect of viscosity and turbulence are equally important normally within the range of $5 < y^+ < 30$. In other words, both molecular and turbulent shear stresses are important and cannot be neglected. In this region, turbulent kinetic energy and dissipation of turbulent energy vividly occurs, which make it to be considered as a turbulent region while the viscous effects are still important.

3.5.5 Outer Layer

Log-law is a valid assumption for the fully turbulent layer normally $y/_\delta < 0.2$. From $y/_\delta > 0.2$ to the thickness of the boundary layer of δ where the flow is not affected only by the wall anymore, the effect of the stream flow is as important as the effect of the wall. The dimensionless analysis based on the maximum velocity of the stream flow (U_{max}), and this time unlike Eq. (3.42) assuming $U = f(y, \rho, \tau_w, \delta)$, yields to **velocity-deficit law**, which implies that the mean velocity is impacted by the boundary layer thickness rather than viscosity itself. Note that still shear stresses (mainly turbulent) have significant impact on the mean velocity. Thus, velocity-deficit law can be expressed as:

$$u^+ = \frac{U_{max} - U}{u_\tau} = g\left(\frac{y}{\delta}\right) \tag{3.48}$$

The **overlap layer** is thus defined as a layer located in both log-law (within the inner layer) and velocity-deficit layer which is extended up to the border of the boundary layer. Thus, it is feasible that we combine Eqs. (3.48) to (3.47) to obtain:

$$u^+ = \frac{U_{max} - U}{u_\tau} = g\left(\frac{y}{\delta}\right) = \frac{1}{\kappa}\ln\left(\frac{y}{\delta}\right) + B \tag{3.49}$$

The above equation is called **law of the wake**. The velocity-deficit layer excluding the overlap layer, thus is called the outer layer where the impact of the viscous effects on the flow can be ignored while the inertial forces enforced by the stream flow are dominant.

Example 3.9

Calculate the thickness of viscous sublayer and log-law layers for a range of airflow over a typical building surface. The boundary layer thickness in a turbulent regime can be calculated as $\dfrac{\delta}{x} = \dfrac{0.37}{Re_x^{1/5}}$ and shear stress can be experimentally estimated as $\tau_w = \dfrac{0.0288\rho U^2}{Re_x^{1/5}}$. Assume air density $\rho_{air} = 1.2047 kg/m^3$ and viscosity $\mu_{air} = 1.8205 \times 10^{-5}$ kg/m s.

Solution

Although the provided equation of the boundary layer may not be accurate for all the buildings' surfaces as this is mainly related to the flat plates, this practice may provide an estimate of the thickness of the mentioned layers on a typical airflow over buildings. Let us assume the velocity to change from 1 to 5 m/s as typical air velocities and surfaces with the size of up to 10 m. Note that those values are only picked to ensure a turbulence regime on the surface ($5 \times 10^5 \leq Re_x$).

First, we calculate the shear stresses using the empirical equation given in the problem. The friction velocity can be resolved by knowing the shear stresses as $u_\tau = \sqrt{\tau_w/\rho}$. Now, for the viscous sublayer, we have $y^+ = \dfrac{\rho u_\tau y}{\mu} < 5$ and for the log-law $30 < y^+ = \dfrac{\rho u_\tau y}{\mu} < 500$ (see Figure 3.14). The plotted thicknesses of the viscous sublayer (δ_s) and log-law layers (δ_l) can be seen in Figure 3.E9.1. Hence, as it can be observed, we can expect from a minimum value of $\delta_s = 0.3$ mm to a maximum value of $\delta_s = 1.7$ mm in the defined range. Obviously, the log-law layer has a larger range from $\delta_l = 3.3$ mm to about $\delta_l = 17$ cm.

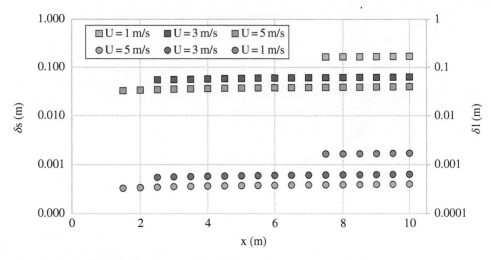

Figure 3.E9.1 Plotted thicknesses of the viscous sublayer (δ_s) and log-law layers (δ_l).

3.6 Piping and Ducting in Buildings

3.6.1 Major Losses

It was shown in Section 2.8.3 that the head loss (h_L) between two sections of a pipe with a distance of L can be obtained by Darcy–Weisbach equation, which is also known as major losses:

$$h_{L\ \text{major}} = f\frac{L}{D}\frac{V^2}{2g} \tag{3.50}$$

As stated in Chapter 2, the friction factor (f) can be found from Moody chart as shown in Figure 3.16. The Moody chart provides f as a function of a flow's Reynolds number within a pipe with a particular relative roughness ($\frac{\varepsilon}{D}$).

3.6.2 Minor Losses

Unlike the friction losses in regular and straight section of pipes, head losses will also occur in the piping systems due to the change in pressure of flow related to the geometrical alteration and thus the flow velocity of the pipe caused by systems' components such as valves, tees, and bends. These losses are known as minor losses ($h_{L\ \text{minor}}$) and in fact can even impose higher losses to a piping system. Due to the complexity in analytical investigations of minor losses in components of piping systems, it is common to write the minor losses in a similar manner as major losses. Hence, from a dimensionless analysis, again, we can show that the pressure drop can be represented as:

$$K_L = \frac{\Delta p}{\frac{1}{2}\rho V^2} \tag{3.51}$$

where K_L is the *loss coefficient* associated to a component.

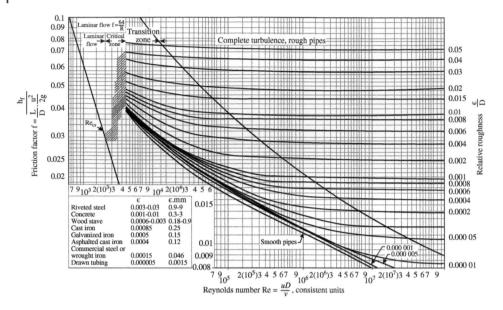

Figure 3.16 Moody chart for a round pipe [15].

With a similar approach to the major head losses ($\Delta p = \gamma h_{L\text{ minor}}$), we can demonstrate that:

$$h_{L\text{ minor}} = K_L \frac{V^2}{2g} \tag{3.52}$$

Equation (3.52) implies that for a value of $K_L = 1$, indicating that the pressure drop is equal to dynamic pressure ($\frac{1}{2}\rho V^2$), $h_{L\text{ minor}}$ is proportional to V^2. On the other hand, K_L itself is a function of flow characteristics and geometry of the component:

$$K_L = \phi(\text{Re}, \text{geomtry}) \tag{3.53}$$

When Re is high enough and the flow is dominated by inertial forces in the components, one can show that K_L is a sole function of geometry ($K_L = \phi(\text{geometry})$) and independent of Re as well as the viscous effect. The latter also results on the major losses to be independent of Re as it can be seen in the Moody chart for the higher Re numbers. On many occasions, to match both major and minor losses together, minor losses are commonly written in terms of an equivalent length (L_{eq}), which imposes a similar major loss within a straight section of a conduit:

$$h_{L\text{ minor}} = h_{L\text{ major}}$$

$$\Longrightarrow K_L \frac{V^2}{2g} = f \frac{L_{\text{eq}}}{D} \frac{V^2}{2g} \tag{3.54}$$

$$\Longrightarrow L_{\text{eq}} = \frac{K_L D}{f}$$

Figure 3.17 illustrates the flow in some sample components while their K_L are also provided mainly by manufacturers. Due to the geometrical changes, the expected pressure

Description	Sketch	Additional data		K
Pipe entrance $h_L = K_c V^2/2g$		r/d 0.0 0.1 >0.2		K_c 0.50 0.12 0.03
Contraction $h_L = K_C V_2^2/2g$		D_2/D_1 0.00 0.20 0.40 0.60 0.80 0.90	K_C $\theta = 60°$ 0.08 0.08 0.07 0.06 0.06 0.06	K_C $\theta = 180°$ 0.50 0.49 0.42 0.27 0.20 0.10
Expansion $h_L = K_E V_1^2/2g$		D_1/D_2 0.00 0.20 0.40 0.60 0.80	K_E $\theta = 20°$ 0.30 0.25 0.15 0.10	K_E $\theta = 180°$ 1.00 0.87 0.70 0.41 0.15
90° miter bend	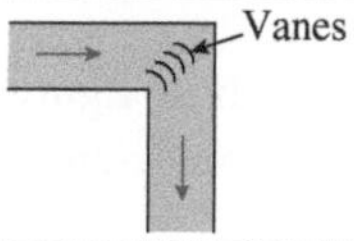	Without vanes	$K_b = 1.1$	
	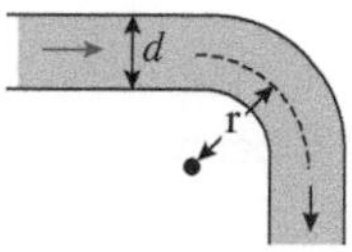	With vanes	$K_b = 0.2$	
90° smooth bend		r/d 1 2 4 6 8 10	$K_b = 0.35$ 0.19 0.16 0.21 0.28 0.32	

Threaded pipe fittings	Globe valve – wide open
	 Angle valve – wide open
	 Gate valve – wide open
	 Gate valve – half open
	 Return bend
	 Tee
	   Straight-through flow
	   Side-outlet flow
	 90° elbow
	 45° elbow
				$K_v = 10.0$
				 $K_v = 5.0$
				 $K_v = 0.2$
				 $K_v = 5.6$
				 $K_b = 2.2$
				
				 $K_t = 0.4$
				 $K_t = 1.8$
				 $K_b = 0.9$
				 $K_b = 0.4$

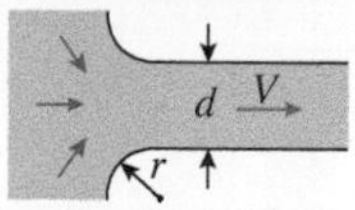

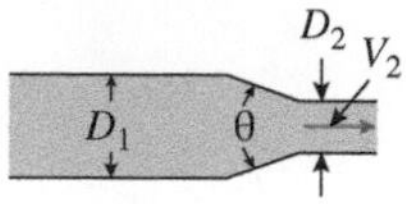

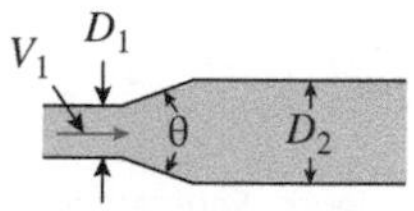

Figure 3.17 Sample pipe components and their loss coefficient. *Source:* Roberson and Chaudhry [16]/ with permission of John Wiley & Sons.

distribution across the components deviates from the ideal condition of Bernoulli equation and thus head losses occur as explained earlier. In other words, flow separations in components or viscous dissipations prevent an efficient exchange of pressure and velocity heads to each other and in fact impose a head loss across the component. Many experimental works have been conducted to obtain efficient design for components. For example, the entrance and exit flow to and from a reservoir to minimize the loss coefficients are examined in many studies as some of the loss coefficients are shown in Figure 3.18. Loss coefficients can be found analytically only in few simplified scenarios such as a sudden expansion case and

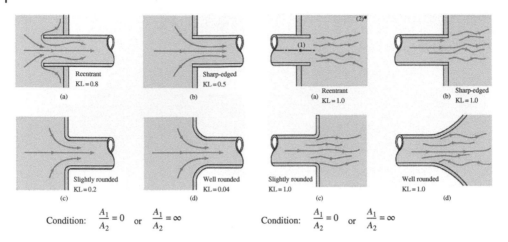

Figure 3.18 Entrance and exit flow and their loss coefficients. *Source:* Munson et al. [17]/ with permission of John Wiley & Sons.

thus in most of the occasions, the loss coefficients from these figures are required for the design of a system. Also, Figure 3.19 demonstrate the loss coefficients of piping components.

3.6.3 Piping System

The total head loss of a piping system is a summation of components and pipe losses. Thus, we can derive the total head loss of a piping system as below:

$$h_L = h_{L\,\text{major}} + h_{L\,\text{minor}}$$

$$\Rightarrow h_L = \sum_{\text{pipes}} f \frac{L}{D} \frac{V^2}{2g} + \sum_{\text{components}} K_L \frac{V^2}{2g} \tag{3.55}$$

$$\Rightarrow h_L = \frac{V^2}{2g} \left(\sum_{\text{pipes}} f \frac{L}{D} + \sum_{\text{components}} K_L \right)$$

Multiple scenarios can determine the design objective of a system of piping in a building, including determination of the total head loss in the system, finding of a desired flow rate, or choosing a suitable diameter for the pipes and components. Each objective, therefore, defines how Eq. (3.55) should be resolved. The solution strategy is, however, beyond the scope of this book. Below example is focused to determinate the total head loss of the system.

Example 3.10

Assume the piping system as depicted in Figure 3.E10.1. A pump installed in the basement is used to supply 30 L/min of cold water with the temperature of $18\,°C$ to a faucet at the first floor. Determine the necessary pressure produced by the pump if:

Part (i) – The minor losses are neglected.

Part (ii) – All losses are included.

All the main pipes are $1''$ up to the first floor where $1/2''$ pipes are used. All pipes have roughness of 0.0015 mm.

a. Elbows

Regular 90°, flanged	0.3
Regular 90°, threaded	1.5
Long radius 90°, flanged	0.2
Long radius 90°, threaded	0.7
Long radius 45°, flanged	0.2
Regular 45°, threaded	0.4

b. 180° return bends

180° return bend, flanged	0.2
180° return bend, threaded	1.5

c. Tees

Line flow, flanged	0.2
Line flow, threaded	0.9
Branch flow, flanged	1.0
Branch flow, threaded	2.0

d. Union, threaded 0.08

*e. Valves

Globe, fully open	10
Angle, fully open	2
Gate, fully open	0.15
Gate, ¼ closed	0.26
Gate, ½ closed	2.1
Gate, ¾ closed	17
Swing check, forward flow	2
Swing check, backward flow	∞
Ball valve, fully open	0.05
Ball valve, ⅓ closed	5.5
Ball valve, ⅔ closed	210

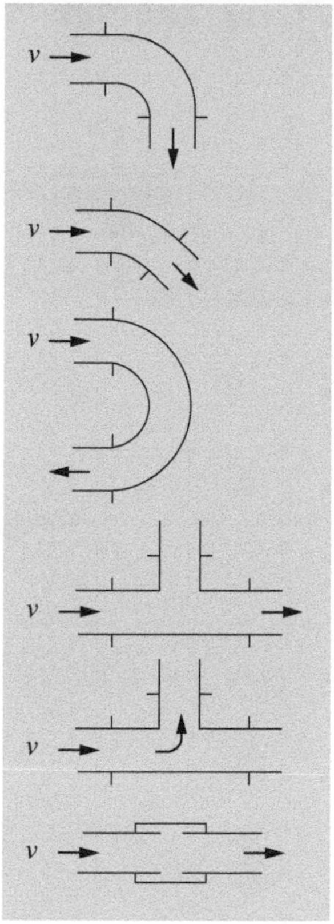

Figure 3.19 Pipe components and their loss coefficients. *Source:* Munson et al. [17]/ with permission of John Wiley & Sons.

Solution

The schematic diagram of the pipeline system is shown in Figure 3.E10.2. Both Point-1 and Point-2 have identical atmospheric pressure.

First, we calculate the velocity at $1''$ pipe as:

$$Q = 30 \frac{L}{min} = 0.0005 \, m^3$$

$$\Rightarrow V_2 = \frac{Q}{A_2} = \frac{0.0005}{\frac{\pi}{4}(0.0127^2)} = 3.95 \, m/s$$

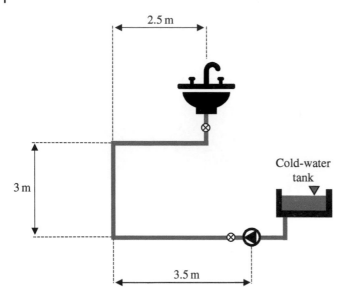

Figure 3.E10.1 A simple cold-water pipeline system.

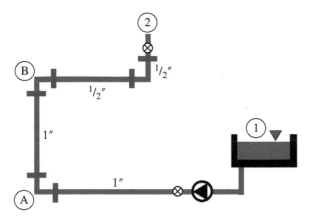

Figure 3.E10.2 Schematic diagram of the case study pipeline system.

Accordingly, the Reynolds number will be:

$$\text{Re}_2 = \frac{\rho V_2 D_2}{\mu} = \frac{998.2 \times 3.95 \times 0.0127}{1.002 \times 10^{-3}} = 49,963$$

Hence, assuming a turbulent flow and the pipe roughness of 0.0015 mm, the friction coefficients can be found from the Moody chart (see Figure 3.16) as $f_2 = 0.0213$. Similarly, for the $1/2''$ pipe, we can find:

$$V_1 = \frac{Q}{A_1} = \frac{0.0005}{\frac{\pi}{4}(0.0254)^2} = 0.99 \text{ m/s}$$

And, the Reynolds number is:

$$\text{Re}_1 = \frac{\rho V_1 D_1}{\mu} = \frac{998.2 \times 0.99 \times 0.0254}{1.002 \times 10^{-3}} = 24,981$$

This implies that the friction coefficient can be found as $f_1 = 0.0246$.

Part (i):

Now, Eq. (3.50) can be directly used to calculate the major losses as:

$$h_{L\,\text{major}} = \sum_{i=1}^{2} f_i \frac{L_i}{D_i} \frac{V_i^2}{2g}$$

$$\Longrightarrow h_{L\,\text{major}} = \frac{0.99^2}{2 \times 9.81} \left(\frac{0.0246 \times (3.5 + 3)}{0.0254} \right) + \frac{3.95^2}{2 \times 9.81} \left(\frac{0.0213 \times 2.5}{0.0127} \right) = 0.31 + 3.33 = 3.65\,\text{m}$$

Now, we can derive Bernoulli equation by considering head losses and pump head as $(P_1 = P_2 = P_{\text{atm}})$:

$$\frac{P_1}{\rho g} + h_1 + h_{\text{pump}} = \frac{P_2}{\rho g} + h_2 + \frac{V_2^2}{2g} + h_{L\,\text{major}}$$

$$\Longrightarrow h_{\text{pump}} = 3 + \frac{3.95^2}{2 \times 9.81} + 3.65 = 7.44\,\text{m}$$

Note that the tank elevation with the main pipe is neglected.

Part (ii)

We know that $h_{L\,\text{minor}} = \sum_{i=1}^{2} \frac{K_{Li} V_i^2}{2g}$. Thus, the minor losses should be added to the Bernoulli equation as:

$$\frac{P_1}{\rho g} + h_1 + h_{\text{pump}} = \frac{P_2}{\rho g} + h_2 + \frac{V_2^2}{2g} + h_{L\,\text{major}} + h_{L\,\text{minor}}$$

K_L for three elbows and the check valve can be found from Figures 3.17 to 3.19.

Element	Position	K_L
Check valve	Pump	2
Elbow	A	1.5
Elbow	B	1.5
Elbow	Point-2	1.5

Note that $P_1 = P_2 = P_{\text{atm}}$ and $h_1 = 0$ as the reference point. Thus, we can find the head of the pump as:

$$h_{\text{pump}} = 3 + \frac{3.95^2}{2 \times 9.81} + 3.65 + (1.5) \times \frac{3.95^2}{2 \times 9.81} + (2 + 1.5 + 1.5) \times \frac{0.99^2}{2 \times 9.81} = 8.88\,\text{m}$$

Hence, neglecting the minor losses can impose about 19% error into the calculations. Furthermore, if no major or minor losses are considered, the required head would be $3.79m$, which is about 134% different from the case where all the losses are taken into the account.

3.6.4 Parallel and Series Piping Systems

Many buildings have multiple piping system set in parallel or series arrangement as illustrated in Figure 3.20. Dealing with such conditions is very similar to Ohm's law in an electrical circuit concept; here, the pressure drop acts as a voltage difference in a circuit between two points, which is causing the flow (current in a circuit) to move from one point to another. The pipe resistance, known as major and minor losses, also can be replicated by a circuit resistance to complete the analogy. Therefore, we need to rewrite the pressure difference in terms of flow rate for a piping system as the below equation:

$$\Delta p = f \frac{L}{D} \frac{V^2}{2g} \tag{3.56}$$

$$\Longrightarrow \Delta p = R_{\text{pipe}} Q^2$$

Using the introduced analogy as presented in Eq. (3.56) enables us to apply Kirchhoff's laws to a system of piping similar to an electrical circuit. So, in the case of a series of pipes installed after each one another, we can assume that a similar flow should pass through each pipe, and thus we can write:

$$Q_i = Q_{i1} = Q_{i2} = ... = Q_{in} \tag{3.57}$$

$$h_i = \sum_{j=1}^{n} h_{ij} \tag{3.58}$$

where Q_i is the flow rate similar in all n sections of the ith pipes in a series arrangement. Also, h_i represents the summation of all losses (h_{ij}) of the ith pipes in a series arrangement.

Similarly, in the case of pipes being installed in a parallel format, the flow is distributed through different branches by different fractions:

$$Q_i = \sum_{j=1}^{n} Q_{ij} \tag{3.59}$$

Figure 3.20 Parallel and series flow in piping systems. *Source:* suparat malipoom/EyeEm/ Adobe Stock.

Furthermore, the same head loss is expected on both ends of each pipe branch as all pipes are connected to each other at these ends:

$$h_i = h_{i1} = h_{i2} = \dots = h_{in} \tag{3.60}$$

Utilizing Eqs. (3.57) through (3.60) in parallel and series piping systems gives extra equations that can be combined with the Bernoulli equation derived at different points of a system in addition to the loss equations to solve the piping system pressure drops, flow rates, or piping dimensions.

Example 3.11

Repeat Example 3.10 to calculate major and minor losses of piping system in a two- storey building as shown in Figure 3.E11.1 if:

Part (i) – in the case of a parallel arrangement, all consumer units are simultaneously open.

Part (ii) – in the case of a series condition in which only one consumer unit is assumed to be open at a time.

Note that in this piping system, only cold-water pipelines are taken into account. All the main pipes are $1''$ while the branch pipes at the consumer units are $1/2''$. All pipes have roughness of 0.0015 mm. The required flow rate of each consumer is also assumed as follow:

$$\text{Point} - 2\ \text{flow rate} = 0.19\ \text{L/s}$$
$$\text{Point} - 3\ \text{flow rate} = 0.33\ \text{L/s}$$
$$\text{Point} - 4\ \text{flow rate} = 0.10\ \text{L/s}$$

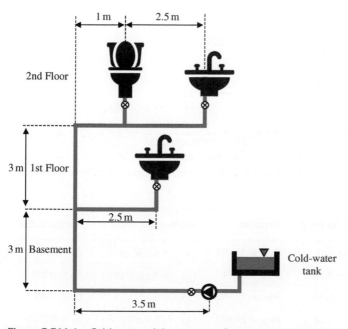

Figure 3.E11.1 Cold-water piping system of a two-story building.

Solution

The schematic diagram of the pipeline system of Figure 3.E11.1 is drawn in Figure 3.E11.2.

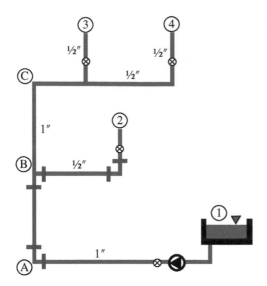

Figure 3.E11.2 Schematic diagram of the case study pipeline system.

(i) Parallel condition

We assume that all consumer units are simultaneously open. First, we calculate the flow rates at the second floor as a parallel combination using Eq. (3.59):

$$Q_{2f} = Q_3 + Q_4 = 0.33 + 0.10 = 0.43\,\text{L/s}$$

$$\Longrightarrow v_{2f} = \frac{Q_{2f}}{A} = \frac{0.43}{1,000 \times \frac{\pi}{4}(0.0254)^2} = 0.85\,\text{m/s}$$

$$\Longrightarrow \text{Re}_{2f} = \frac{\rho V_{2f} D_{2f}}{\mu} = \frac{998.2 \times 0.85 \times 0.0254}{1.002 \times 10^{-3}} = 21,484$$

The relative roughness in this pipe will be $\dfrac{\epsilon}{D} = \dfrac{0.0015}{1000 \times 0.0254} = 5.9 \times 10^{-5}$ and thus using the Moody chart, the friction factor can be found as $f_{2f} = 0.0256$.

With a similar approach, we can find the other flow rates, velocities, and Re numbers at other branches as:

Flow rate (L/s)	Velocity (m/s)	Reynolds	Relative roughness	Friction factor
$Q_{2f} = Q_3 + Q_4 = 0.43$	$V_{2f} = 0.85$	$\text{Re}_{2f} = 21,484$	$\left(\frac{\epsilon}{D}\right)_{2f} = 5.9 \times 10^{-5}$	$f_{2f} = 0.0256$
$Q_{34} = Q_4 = 0.10$	$V_{34} = 0.79$	$\text{Re}_{34} = 9993$	$\left(\frac{\epsilon}{D}\right)_{34} = 1.18 \times 10^{-4}$	$f_{34} = 0.0311$
$Q_{1f} = 0.19$	$V_{1f} = 1.50$	$\text{Re}_{1f} = 18,986$	$\left(\frac{\epsilon}{D}\right)_{1f} = 1.18 \times 10^{-4}$	$f_{1f} = 0.0265$
$Q_{pl} = Q_{1f} + Q_{2f} = 0.62$	$V_{pl} = 1.22$	$\text{Re}_{pl} = 30,977$	$\left(\frac{\epsilon}{D}\right)_{pl} = 5.9 \times 10^{-5}$	$f_{pl} = 0.0235$

The longest path to the pump is Point-4, which will be employed as the criteria for the calculation of the required pump head. Thus, the general form of the Bernoulli equation between points 1 and 4 can be derived as:

$$\frac{P_1}{\rho g} + h_1 + \frac{V_1^2}{2g} + h_{\text{pump}} = \frac{P_4}{\rho g} + h_2 + \frac{V_4^2}{2g} + \sum_{i=1}^{2} f_i \frac{L_i}{D_i} \frac{V_i^2}{2g} + \sum_{i=1}^{2} \frac{K_{Li} V_i^2}{2g}$$

Only major losses:

Simplifying the above Bernoulli equation, for the case that only the major losses are considered. Thus, for the first floor, we have:

$$h_{\text{pump}} = 3 + \frac{1.50^2}{2 \times 9.81} + \frac{1}{2 \times 9.81}\left(0.0235 \times \frac{6.5}{0.0254} \times 1.22^2 + 0.0265 \times \frac{2.5}{0.0127} \times 1.50^2\right)$$

$$\Longrightarrow h_{\text{pump}} = 4.06 \, \text{m}$$

And for the second floor:

$$h_{\text{pump}} = 6 + \frac{0.79^2}{2 \times 9.81} + \frac{1}{2 \times 9.81}\left(0.0235 \times \frac{6.5}{0.0254} \times 1.22^2 + 0.0256\right.$$

$$\left. \times \frac{4}{0.0254} \times 0.85^2 + 0.0311 \times \frac{2.5}{0.0127} \times 0.79^2\right)$$

$$\Longrightarrow h_{\text{pump}} = 6.80 \, \text{m}$$

Major and minor losses:

Loss coefficient of the components can be found as:

Element	Position	K_L
Check valve	Pump	2
Elbow	Point-A	1.5
Tee	Point-B	0.9
Tee turn	Point-B	2
Elbow	Point-C	1.5
Tee	Point-3	0.9
Tee turn	Point-3	2
Elbow	Point-4	1.5
Elbow	Point-2	1.5

Now, replacing the available data where both losses are considered, we can express the Bernoulli equation as the below for the first floor:

$$h_{\text{pump}} = 3 + \frac{0.79^2}{2 \times 9.81} + \frac{1}{2 \times 9.81}\left(0.0235 \times \frac{6.5}{0.0254} \times 1.22^2 + 0.0256 \times \frac{4}{0.0254} \times 0.85^2\right.$$

$$\left. + 0.0311 \times \frac{2.5}{0.0127} \times 0.79^2\right) + \frac{1}{2 \times 9.81}\left((2 + 1.5 + 2) \times 1.22^2 + 2 \times 1.50^2\right)$$

$$\Longrightarrow h_{\text{pump}} = 4.65 \, \text{m}$$

And for the second floor:

$$h_{pump} = 6 + \frac{0.79^2}{2 \times 9.81} + \frac{1}{2 \times 9.81}\left(0.0235 \times \frac{6.5}{0.0254} \times 1.22^2 + 0.0256 \times \frac{4}{0.0254} \times 0.85^2\right.$$

$$\left. + 0.0311 \times \frac{2.5}{0.0127} \times 0.79^2\right) + \frac{1}{2 \times 9.81}\left((2 + 1.5 + 0.9) \times 1.22^2\right.$$

$$\left. + (1.5) \times 0.85^2 + (0.9 + 1.5) \times 0.79^2\right)$$

$$\Longrightarrow h_{pump} = 7.27 \text{ m}$$

(ii) Series condition

In the series condition, only one unit is open at a given point of time while the rest of the consumer's valves are closed. The required pump head can then be calculated separately for points 1, 2, and 3.

Point-2:

$$v_2 = \frac{Q_2}{A} = \frac{0.19}{1000 \times \frac{\pi}{4}(0.0127)^2} = 1.50 \text{ m/s}$$

$$Re_2 = \frac{\rho V_{1f} D_{1f}}{\mu} = \frac{998.2 \times 1.5 \times 0.0127}{1.002 \times 10^{-3}} = 18,986$$

The relative roughness in this pipe is $\frac{\epsilon}{D} = \frac{0.0015}{1,000 \times 0.0127} = 1.18 \times 10^{-4}$ and thus the friction factor can be obtained as $f_2 = 0.0265$. The same flow rate passes through the pump pipeline:

$$v_{pl} = \frac{Q_{pl}}{A} = \frac{0.19}{1000 \times \frac{\pi}{4}(0.0254)^2} = 0.38 \text{ m/s}$$

$$Re_2 = \frac{\rho V_{1f} D_{1f}}{\mu} = \frac{998.2 \times 0.375 \times 0.0254}{1.002 \times 10^{-3}} = 9493$$

Again, the relative roughness and friction factor can be found, respectively as $\frac{\epsilon}{D} = \frac{0.0015}{1,000 \times 0.00254} = 5.9 \times 10^{-5}$ and $f_{pl} = 0.0314$.

Point-2: Only major losses

Now, only considering the major losses for Point-2:

$$h_{pump} = 3 + \frac{1.50^2}{2 \times 9.81} + \frac{1}{2 \times 9.81}\left(0.0265 \times \frac{6.5}{0.0254} \times 0.38^2 + 0.0314 \times \frac{2.5}{0.0127} \times 1.50^2\right)$$

$$\Longrightarrow h_{pump} = 3.77 \text{ m}$$

Point-2: major and minor losses

Again, by adding the minor losses, we can find the head of pump as:

$$h_{pump} = 3 + \frac{1.5^2}{2 \times 9.81} + \frac{1}{2 \times 9.81}\left(0.0265 \times \frac{6.5}{0.0254} \times 0.38^2 + 0.032 \times \frac{2.5}{0.0127} \times 1.5^2\right)$$

$$+ \frac{1}{2 \times 9.81}\left((2 + 1.5 + 0.9) \times 0.375^2 + 1.5 \times 1.50^2\right)$$

$$\Longrightarrow h_{pump} = 3.97 \, m$$

Point-3:

Following a similar procedure for Point-3 and Point-4, we can obtain the related heads.

$$V_3 = \frac{Q_3}{A} = \frac{0.33}{1000 \times \frac{\pi}{4}(0.0254)^2} = 0.65 \, m\,s^{-1}$$

$$Re_3 = \frac{\rho V_{3f} D_3}{\mu} = \frac{998.2 \times 0.65 \times 0.0254}{1.002 \times 10^{-3}} = 16,488$$

The relative roughness is $\frac{\epsilon}{D} = \frac{0.0015}{1,000 \times 0.0254} = 5.9 \times 10^{-5}$ and the friction factor can be obtained as $f_{2f} = 0.0273$.

Point-3: Only major losses

$$h_{pump} = 6 + \frac{0.65^2}{2 \times 9.81} + \frac{1}{2 \times 9.81}\left(0.0273 \times \frac{10.5}{0.0254} \times 0.65^2\right)$$

$$\Longrightarrow h_{pump} = 6.27 \, m$$

Point-3: major and minor losses

$$h_{pump} = 6 + \frac{0.65^2}{2 \times 9.81} + \frac{1}{2 \times 9.81}\left(0.0273 \times \frac{10.5}{0.0254} \times 0.65^2\right)$$

$$+ \frac{1}{2 \times 9.81}\left((2 + 1.5 + 0.9 + 1.5 + 0.9) \times 0.65^2\right)$$

$$\Longrightarrow h_{pump} = 6.41 \, m$$

Point-4:

$$V_4 = \frac{Q_4}{A} = \frac{0.10}{1000 \times \frac{\pi}{4}(0.0127)^2} = 0.79 \, m/s$$

$$Re_{34} = \frac{\rho V_4 D_4}{\mu} = \frac{998.2 \times 0.79 \times 0.0127}{1.002 \times 10^{-3}} = 9993$$

The relative roughness is $\frac{\epsilon}{D} = \frac{0.0015}{1000 \times 0.0127} = 1.18 \times 10^{-4}$ and thus $f_{34} = 0.0311$. The same flow passes through the pump pipeline:

$$V_{pl} = \frac{Q_{pl}}{A} = \frac{0.10}{1000 \times \frac{\pi}{4}(0.0254)^2} = 0.20 \, m/s$$

$$\text{Re}_{pl} = \frac{\rho V_{pl} D_{pl}}{\mu} = \frac{998.2 \times 0.20 \times 0.0254}{1.002 \times 10^{-3}} = 4996$$

The relative roughness is $\dfrac{\epsilon}{D} = \dfrac{0.0015}{1,000 \times 0.0127} = 5.9 \times 10^{-5}$ and thus $f_{34} = 0.0375$.

Point-4: Only major losses

$$h_{\text{pump}} = 6 + \frac{0.79^2}{2 \times 9.81} + \frac{1}{2 \times 9.81} \left(0.0311 \times \frac{2.5}{0.0127} \times 0.79^2 + 0.0375 \times \frac{10.5}{0.0254} \times 0.20^2 \right)$$

$$\Longrightarrow h_{\text{pump}} = 6.26 \, \text{m}$$

Point-4: major and minor losses

$$h_{\text{pump}} = 6 + \frac{0.79^2}{2 \times 9.81} + \frac{1}{2 \times 9.81} \left(0.031 \times \frac{2.5}{0.0127} \times 0.79^2 + 0.038 \times \frac{10.5}{0.0254} \times 0.198^2 \right)$$

$$+ \frac{1}{2 \times 9.81} \left((2 + 1.5 + 0.9 + 1.5) \times 0.20^2 + (0.9 + 1.5) \times 0.79^2 \right)$$

$$\Longrightarrow h_{\text{pump}} = 6.35 \, \text{m}$$

As we can see, the head of the pump can be found for different parallel conditions as:

	h_{pump} (m)	
	Only major loss	**Major and minor losses**
Point-2	3.77	3.97
Point-3	6.27	6.41
Point-4	6.26	6.35

For the selection of the pump, hence, one can consider the worst-case scenario where higher losses should be compensated with the pump head, which implies a value of 6.41 m.

Example 3.12

In Example 2.12 of Chapter 2, calculate the threshold height of a tank (h') if the minor losses are not neglected except in the consumer units. The pipes are of PVC type with the roughness of 0.0015 mm. Loss coefficient for elbows, tee connections, and branches are found as 1.5, 2.0, and 1.0, respectively. All pressure levels are atmospheric and reference frame is at the level of Sink-4 unit.

Solution

Major losses are calculated using the Moody chart and minor losses are also considered at highlighted elbows and tees. Similar to the Example 2.12 of Chapter 2, the velocity to ensure a turbulence regime in the pipes can be calculated as:

$$V_1 = \frac{\text{Re} \times \mu}{\rho D} = \frac{2,100 \times 1.0016 \times 10^{-3}}{998.2 \times 0.0127} = 0.40 \, \text{m/s}$$

Therefore, with the pipe size of $1/4''$ wherever the velocity becomes larger than 1.58 m/s, the flow will be in the turbulence regime. Now, the Bernoulli equation is written between Point-1 and Sink units one by one.

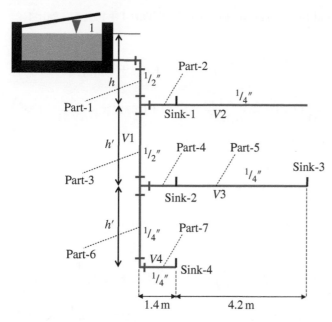

Figure 3.E12 Schematic diagram of the piping system of Example 2.12.

Deriving Bernoulli between Point-1 and Sink-1 gives (Part1 + Part2):

$$h = \frac{V_2^2}{2g} + \left(f_1 \frac{L_1}{D_1} \frac{V_1^2}{2g} + f_2 \frac{L_2}{D_2} \frac{V_2^2}{2g} \right) + \left(K_{L1} \frac{V_1^2}{2g} + K_{L2} \frac{V_2^2}{2g} \right)$$

Continuity requires that $V_1 = 0.25V_2 = 0.25V_3 = 0.25V_4$ while their corresponding Re numbers can be calculated to be 10,000. Therefore, the friction factors for each pipe with the sizes of ½″ and 1/4″ can be found from the Moody chart as $f_1 = 0.0312$ and $f_2 = 0.0375$, respectively. Now, we can find h as:

$$h = \frac{1.58^2}{2 \times 9.81} + \frac{1}{2 \times 9.81} \left(0.0375 \times \frac{h}{0.0127} \times 0.40^2 + 0.0312 \times \frac{1.4}{0.00635} \times 1.58^2 \right)$$

$$+ \frac{1}{2 \times 9.81} \left(1.5 \times 0.40^2 + 2 \times 1.58^2 \right)$$

$$\Longrightarrow h = 0.1274 + 0.0235h + 0.8761 + 0.2070$$

$$\Longrightarrow h = 1.24 \text{ m}$$

Bernoulli between Point-1 and Sink-2 can be found as (Part-1 +Part-3 + Part-4):

$$(2.8 + h) = \frac{V_3^2}{2g} + \left(f_1 \frac{(L_3 + h)}{D_1} \frac{V_1^2}{2g} + f_4 \frac{L_4}{D_4} \frac{V_3^2}{2g} \right) + \left((K_{L1} + K_{L2}) \frac{V_1^2}{2g} + K_{L3} \frac{V_3^2}{2g} \right)$$

$$\Longrightarrow (2.8 + h) = \frac{1.58^2}{2 \times 9.81} + \frac{1}{2 \times 9.81} \left(0.0375 \times \frac{(2.8 + h)}{0.0127} \times 0.40^2 + 0.0312 \times \frac{1.4}{0.00635} \times 1.58^2 \right)$$

$$+ \frac{1}{2 \times 9.81} \left((2 + 1) \times 0.40^2 + 1.5 \times 1.58^2 \right)$$

$$\Longrightarrow h = 0.1274 + 0.0235h + 0.9419 + 0.2149$$

$$\Longrightarrow h = -0.12 \text{ m}$$

Bernoulli between Point-1 and Sink-3 can be expressed as (Part-1 + Part-3 + Part-4 + Part-5):

$$(2.8 + h) = \frac{V_3^2}{2g} + \left(f_1 \frac{(L_3 + h)}{D_1} \frac{V_1^2}{2g} + f_4 \frac{(L_4 + L_5)}{D_4} \frac{V_3^2}{2g} \right) + \left((K_{L1} + K_{L2} + K_{L3}) \frac{V_1^2}{2g} + K_{L4} \frac{V_3^2}{2g} \right)$$

$$\Longrightarrow (2.8 + h) = \frac{1.58^2}{2 \times 9.81} + \frac{1}{2 \times 9.81} \left(0.0375 \times \frac{(2.8 + h)}{0.0127} \times 0.40^2 + 0.0312 \times \frac{5.6}{0.00635} \times 1.58^2 \right)$$

$$+ \frac{1}{2 \times 9.81} \left((2 + 1 + 2) \times 0.40^2 + 1.5 \times 1.58^2 \right)$$

$$\Longrightarrow h = 0.1274 + 0.0235h + 3.5703 + 0.2309$$

$$\Longrightarrow h = 1.16 \, \text{m}$$

Bernoulli between Point-1 and Sink-4 can be written as (Part-1 + Part-3 + Part-6 + Part-7):

$$(5.6 + h) = \frac{V_4^2}{2g} + \left(f_1 \frac{(L_3 + L_6 + h)}{D_1} \frac{V_1^2}{2g} + f_7 \frac{L_7}{D_7} \frac{V_4^2}{2g} \right) + \left((K_{L1} + K_{L2} + K_{L3}) \frac{V_1^2}{2g} + K_{L5} \frac{V_4^2}{2g} \right)$$

$$\Longrightarrow (5.6 + h) = \frac{1.58^2}{2 \times 9.81} + \frac{1}{2 \times 9.81} \left(0.0375 \times \frac{(5.6 + h)}{0.0127} \times 0.40^2 + 0.0312 \times \frac{1.4}{0.00635} \times 1.58^2 \right)$$

$$+ \frac{1}{2 \times 9.81} \left((2 + 2 + 1.5) \times 0.40^2 + 1.5 \times 1.58^2 \right)$$

$$\Longrightarrow h = 0.1274 + 0.0235h + 1.0077 + 0.2348$$

$$\Longrightarrow h = -4.33 \, \text{m}$$

The obtained values specifies that the design should be based on the minimum height that can always guarantee the deemed condition in the system, which is $h = 1.24$ m. As it is expected, adding the minor losses to the calculations in comparison to the case of Example 2.12 of Chapter 2, in which only major losses were considered, results in a few centimetres of head loss due to existence of the components.

3.7 Fan and Pump in Buildings

In the previous section, we have explained how pipes (also conduits in a general aspect) can be modelled as a system. The pumps (fans in the ducting systems) are responsible for providing head for a system to compensate for the pressure loss and in pressurization of a system up to desired levels. Again, an extensive and long-term literature has been developed around design, selection, installation, and maintenance of fans and pumps, which is beyond the scope of this book. On the contrary, we focus on their overall thermofluidic performance in the buildings as we aim to embed them within our general energy and airflow modelling scenarios in future chapters as the source elements.

In a turbomachinery classification, fans and pumps are machines used to move or to increase head (pressure) of a fluid, implying that they increase the energy of a fluid. As shown in Figures 3.21 and 3.22, pumps and fans encompass axial or radial types. Axial machines run the fluid through blades in the same direction as their axis of rotor and

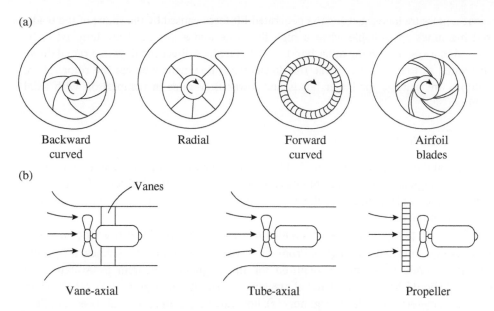

Figure 3.21 (a) Common centrifugal fans and (b) common axial flow fans. *Source:* Blake [18]/ with permission of John Wiley & Sons.

In a *centrifugal pump* flow enters along the axis and is expelled radially. (The reverse is true for a turbine.)

An *axial-flow* pump is like a propeller; the direction of the flow is unchanged after passing through the device.

A *mixed-flow device* is a hybrid device, used for intermediate heads.

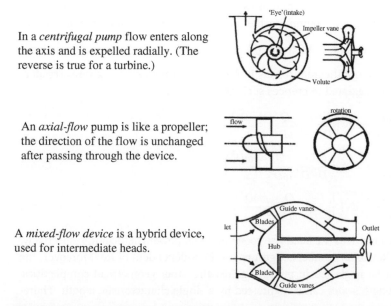

Figure 3.22 Schematic of (a) centrifugal pump and (b) axial-flow pump. *Source:* Munson et al. [17]/ with permission of John Wiley & Sons.

impose a pressure increase using the generated lift force caused by the shape of the blade. Axial machines are suitable when a high flow rate and a low head are targeted to be achieved in a system. On the other hand, radial machines direct the flow in a radial direction related to the rotor axis while the downstream flow axially enters the rotor. These machines are suitable for larger heads and lower flow rates in comparison with axial machines.

3.7.1 Dimensionless Analysis

To understand theoretical background of blades functionality in turbomachines in terms of energy and angular momentum, interested readers are encouraged to refer the suggested references such as [17, 19, 20]. In this section, we only focus on the general performances of fans and pumps.

We start with pumps and thus we can identify dependent variables as head increase (h_a), efficiency (η), and shaft power ($\dot{W}$). From many years of experimental studies, it is found that the dependent parameters required for Buckingham π-theorem as explained in Chapter 2 should be selected as the characteristic diameter of a pump (D), other length related parameters (l_i), surface roughness (ε), flow rate (Q), shaft rotational speed (ω), fluid density (ρ), and fluid viscosity (μ). Thus, the dependent π can be expressed as:

$$\text{dependent } \pi = \phi\left(\frac{l_i}{D}, \frac{\varepsilon}{D}, \frac{Q}{\omega D^3}, \frac{\rho \omega D^2}{\mu}\right) \tag{3.61}$$

First, we assume that the surface roughness as well as other geometrical factors to be not as significant as the diameter of a pump. Further, it is assumed that pumps are mainly running in high Reynolds numbers, which is represented by $\dfrac{\rho \omega D^2}{\mu}$ in these non-dimensional groups. Now, Buckingham π-theorem results in formation of the below π-groups (readers can find details in the suggested references such as [17, 19–21]):

$$\pi_1 = C_H = \frac{gh_a}{\omega^2 D^2} = \phi_1\left(\frac{l_i}{D}, \frac{\varepsilon}{D}, \frac{Q}{\omega D^3}, \frac{\rho \omega D^2}{\mu}\right) \tag{3.62}$$

$$\pi_2 = C_P = \frac{\dot{W}}{\rho \omega^3 D^5} = \phi_2\left(\frac{l_i}{D}, \frac{\varepsilon}{D}, \frac{Q}{\omega D^3}, \frac{\rho \omega D^2}{\mu}\right) \tag{3.63}$$

$$\pi_3 = \eta = \frac{\rho g Q h_a}{\dot{W}} = \phi_3\left(\frac{l_i}{D}, \frac{\varepsilon}{D}, \frac{Q}{\omega D^3}, \frac{\rho \omega D^2}{\mu}\right) \tag{3.64}$$

where C_H and C_P are known as the head rise coefficient and power coefficient, respectively.

Note that in high Reynolds numbers, π-groups can be independent of Re. Moreover, the roughness effect can be neglected in pumps against the other geometrical complexities. Eventually, all the length scales can be expressed by a single characteristic length. Therefore, Eqs. (3.62) through (3.64) can be further simplified to be a function of $C_Q = \dfrac{\rho \omega D^2}{\mu}$ known as the flow coefficient. Hence, we can plot these dimensionless groups against each other and generate general characteristic (performance, efficiency, and brake power) curves for pumps (see example in Figure 3.23).

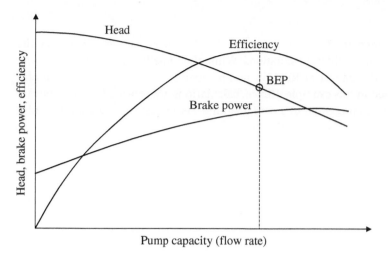

Figure 3.23 A typical pump characteristic curve.

Example 3.13

A performance characteristic curves of a pump with impeller diameter of $D = 6''$ are given for 1750 RPM as shown in Figure 3.E13.1. Determine the performance characteristics of this pump for 2900 RPM.

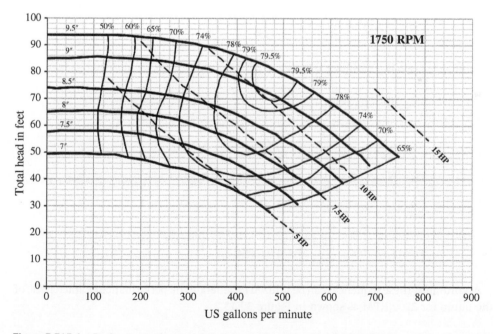

Figure 3.E13.1 Performance characteristic curves of a pump with impeller diameter of $D = 6''$ with the shaft rotational speed of 1750 RPM.

Solution

Similarity law explained in Eqs. (3.62) through (3.64) can be applied to predict the pump's performance under different situations. While the impeller diameter of the pump remains unchanged, the impact of RPM on pump capacity, flow rate, power, and efficiency are investigated in this example. Hence, calculations are provided for an arbitrary point and the results are presented in a new diagram. Let us assume the arbitrary point as below:

$$D_1 = 6''$$

$$h_{a1} = 90 \text{ ft}$$

$$Q_1 = 200 \text{ GPM}$$

$$\eta_1 = 60\%$$

Now, this point will be recalculated with the π-groups assuming that RPM changes from 1750 RPM (C_{H1}, C_{P1}, η_1) to 2900 RPM (C_{H2}, C_{P2}, η_2):

(1) $\qquad C_{H1} = C_{H2} \Longrightarrow \dfrac{gh_{a1}}{\omega_1^2 D^3} = \dfrac{gh_{a2}}{\omega_2^2 D^3}$

$$\Longrightarrow h_{a2} = h_{a1} \times \frac{\omega_2^2}{\omega_1^2} = 90 \times \frac{2900^2}{1750^2} = 247.2 \text{ ft}$$

(2) $\qquad C_{P1} = C_{P2} \Longrightarrow \dfrac{\dot{W}_1}{\rho \omega_1^3 D^5} = \dfrac{\dot{W}_2}{\rho \omega_2^3 D^5}$

$$\Longrightarrow \dot{W}_2 = \dot{W}_1 \times \frac{\omega_2^3}{\omega_1^3} = \dot{W}_1 \times \frac{2900^3}{1750^3}$$

$$\Longrightarrow \dot{W}_2 = 4.55 \dot{W}_1$$

(3) $\qquad \eta_1 = \eta_2 \Longrightarrow \dfrac{\rho g Q_1 h_{a1}}{\dot{W}_1} = \dfrac{\rho g Q_2 h_{a2}}{\dot{W}_2}$

$$\Longrightarrow \frac{Q_2}{Q_1} = \frac{\dot{W}_2}{\dot{W}_1} \times \frac{h_{a1}}{h_{a2}}$$

$$\Longrightarrow \frac{Q_2}{Q_1} = \frac{\omega_2^3}{\omega_1^3} \times \frac{\omega_1^2}{\omega_2^2} = \frac{\omega_2}{\omega_1}$$

$$\Longrightarrow Q_2 = 200 \times \frac{2900}{1750} = 331.4 \text{ GPM}$$

Accordingly, these sequences can be utilized to generate the performance characteristic of the pump for 2900 RPM as shown in Figure 3.E13.2.

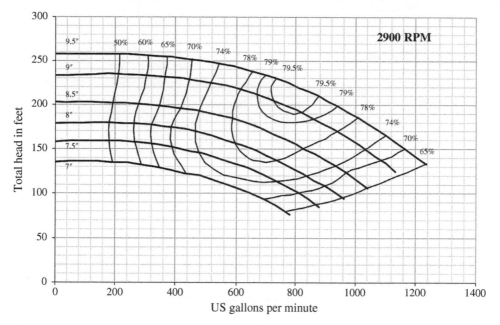

Figure 3.E13.2 Performance characteristic of the case study pump for the shaft rotational speed of 2900 RPM.

3.7.2 System Characteristics and Pumps

We understood that in a piping system with major and minor losses, an actual head produced by a pump is required to compensate the energy and thus sustain the system's flow. In both losses, the velocity was an essential factor as described in the Eq. (3.55). Now, if we convert the velocity to the flow rate as $V = \frac{Q}{A}$, then we can derive the actual head gained from the pump as:

$$h_t = h_{L\text{ major}} + h_{L\text{ minor}} + h_z$$

$$\Rightarrow h_t = \sum_{\text{pipes}} f \frac{L}{D} \frac{Q^2}{2gA} + \sum_{\text{components}} K_L \frac{Q^2}{2gA} + h_z \tag{3.65}$$

$$\Rightarrow h_t = \frac{\left(\sum_{\text{pipes}} f \frac{L}{D} + \sum_{\text{components}} K_L \right)}{2gA} Q^2 + h_z$$

Note that $h_z = z_2 - z_1$ is the elevation or static head due to the height difference between two ends of the system. Now, it is common to represent the geometrical characteristics of the system (including A, D, and L) and the loss coefficients of pipes, fittings, and components (including f and K_L) with an equivalent variable of K. Using Eq. (3.65), the **system equation (curve)**, thus, can be represented as:

$$h_t = KQ^2 + h_z \tag{3.66}$$

We also learned from the previous section about a typical pump characteristic curve in which again the required head of a pump and its efficiency varies due to the flow rate. This implies that in a system of piping, there should be unique point, known as the **operating**

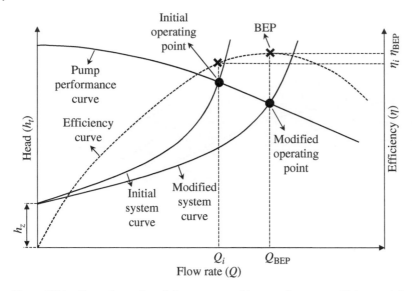

Figure 3.24 Operating point of the system and best performance efficiency of the pump.

point, both the system and pump have a similar flow rate and head. Otherwise, either of system equation (Eq. (3.66)) or pump performance curve (Eqs. (3.62) through (3.64)) contradicts each other, which implies that an operating point should exist where both curves intersect each other as illustrated in Figure 3.24. Another interesting point that can be extracted from colliding both curves on each other is improving the performance of the pump. As it can be seen in Figure 3.24, the pump efficiency at the initial operating point (η_i) with a flow rate of Q_i might not be necessarily associated with the best efficiency point (BEP) of the pump (η_{BEP}).

Hence, if the aim is to achieve the BEP, we must change the system equation. For this purpose, assume that the system equation at the initial state is $h_{ti} = K_i Q_i^2 + h_z$, now we change the system characteristics to reach a modified system equation of $h_{tm} = K_m Q_m^2 + h_z$. It is important that the associated efficiency of the intersection point of the modified system curve and pump performance curve to exactly intercept the BEP point. This implies that $Q_m = Q_{BEP}$. Thus, we can write $h_{tm} = K_m Q_{BEP}^2 + h_z$. We can now conclude from these system equations that:

$$K_m = \frac{h_{tm} - h_z}{Q_{BEP}^2}$$

$$\Longrightarrow K_m = \frac{h_{tm} - \left(h_{ti} - K_i Q_i^2\right)}{Q_{BEP}^2}$$

(3.67)

As it can be seen in Eq. (3.67), all the right-hand side terms can be found from the performance and system curves, which results in finding K_m. Furthermore, as it was explained before, K_m can be changed by alteration of system characteristics such as A, D, L, f, and K_L.

Example 3.14
Assume the piping system of **Example 3.10** with the performance curve presented in Figure 3.E14.1. Now, suggest a modification in the system to maximize the efficiency of the utilized pump.

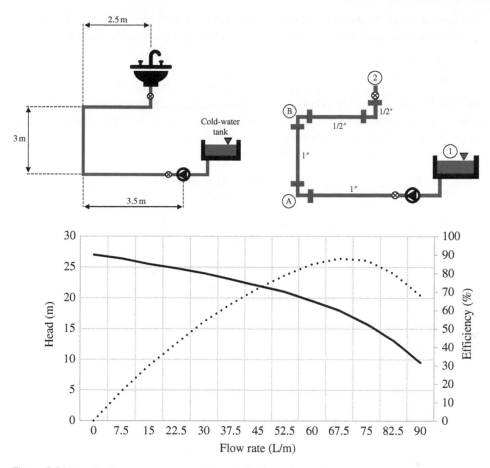

Figure 3.E14.1 Performance curve of Example 3.10.

Solution

We assume that the friction factors of the system to not changing and be fixed as $f = 0.023$. The system equation, thus, can be calculated using Eq. (3.66) as:

$$h_t = \frac{\left(\sum_{\text{pipes}} f \dfrac{L}{D} + \sum_{\text{components}} K_L\right)}{2gA} Q^2 + h_z$$

$$\Rightarrow h_t = \frac{16Q^2}{2 \times 9.81 \times \pi^2 \times 0.0254^4} + \frac{16Q^2}{2 \times 9.81 \times 0.0127^4} \left(\frac{0.023 \times (3.5 + 3)}{0.0254}\right)$$

$$+ \frac{16Q^2}{2 \times 9.81 \times 0.0254^4} \left(\frac{0.023 \times 2.5}{0.0127}\right) + (1.5) \times \frac{16Q^2}{2 \times 9.81 \times 0.0254^4}$$

$$+ (2 + 1.5 + 1.5) \times \frac{16Q^2}{2 \times 9.81 \times 0.0127^4} + 3$$

$$\Rightarrow h_t = 3 + \left(2.45 \times 10^7\right) Q^2$$

The system equation and characteristic pump curve are depicted in Figure 3.E14.2. For this system, we can find the curves' intersection to obtain the operating point as:

$$Q_i = 51 \text{ L/m}$$
$$\Longrightarrow h_{ti} = 20.7 \text{ m}$$

As it can be seen, the associated efficiency of the system is $\eta_i = 78\%$ while $\eta_{BEP} = 87\%$. This implies that the modified curve needs to intersect the line of $y = Q_{BEP}$ point as well ($Q_{BEP} = 70$ L/m). We can read from the pump performance curve that $h_{tm} = 17$ m. Hence, using Eq. (3.67), we can find the modified K_m as:

$$K_m = \frac{h_{tm} - \left(h_{ti} - K_i Q_i^2\right)}{Q_{BEP}^2} = \frac{17 - (20.7 - 2.45 \times 10^7 \times 0.00085^2)}{0.00117} = 1.03 \times 10^7$$

The new performance curve is demonstrated in Figure 3.E14.3.

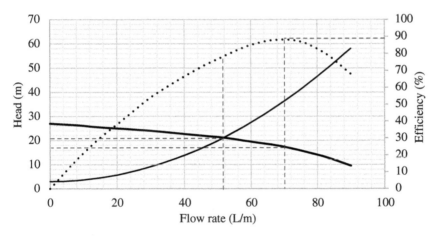

Figure 3.E14.2 System equation and characteristic pump curve of the case study piping system.

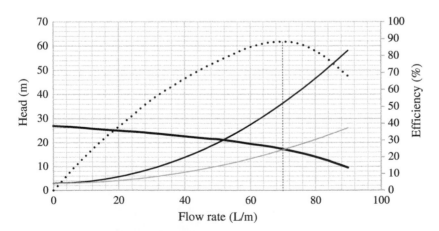

Fig. 3.E14.3 New system equation of the case study piping system.

The procedure to modify a performance curve is, nonetheless, not always a straightforward procedure as it demands many considerations in terms of other factors in the design of the system as well as the overall cost. Here, we just assume two scenarios in which we have changed all pipes diameters to 3/4" in Case 1 and lengths of all pipes to 2 m in Case 2 as it is shown in the below table.

	Pipe diameter (m)			Pipe length (m)		
Pipeline	Base case	Case 1	Case 2	Base Case	Case 1	Case 2
1A	0.0254	0.01905	0.0254	3.5	3.5	2
AB	0.0254	0.01905	0.0254	3	3	3
B2	0.0127	0.01905	0.0127	2.5	2.5	2

The new K_m values are found as 1.15×10^7 for Case 1 and 2.12×10^7 for Case 2. The system equations are also illustrated in Figure 3.E14.4. Therefore, Case 1 is more practical option with only changing the pipe diameters to become closer to the modified case (Case M in Figure 3.E14.4) to elevate the performance of the pump with an efficiency of about 85%, which is very close to its BEP.

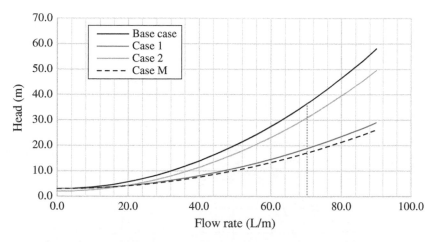

Figure 3.E14.4 System equations of different modification scenarios.

3.7.3 Parallel and Series Pumps

From the previous section, we learned that an operating point is the shared coordinate of head and flow rate amongst both system and pump performance curves. It was also explained that we are always able to change the operating point to a modified one in which the pump has a higher efficiency by changing the shape of the system equation via alteration of its characteristics. The changing of the operating point is also feasible by alteration of a pump performance curve. Aside from the selection of another type of pumps, this change occurs by employment of multiple pumps in the system in a parallel or series format. The reason of utilization of multiple pumps can be due to the functional necessity in the system, nonetheless, it is again a crucial subject of design to ensure that pumps are operating in their higher efficiency ranges.

As demonstrated in Figure 3.25a, if two identical pumps are installed in a series format, the new pumps performance curve will be changed where at any fixed flow rate, the heads will be summed up together as expected. However, the intersection point, as the modified operation point, has not necessary a twice value. In fact, it has coordinates of Q_m and h_{tm} with a new efficiency of η_m. On the other hand, while two identical pumps are installed in a parallel format, the flow rates will be summed together at any fixed actual head. Hence,

(a)

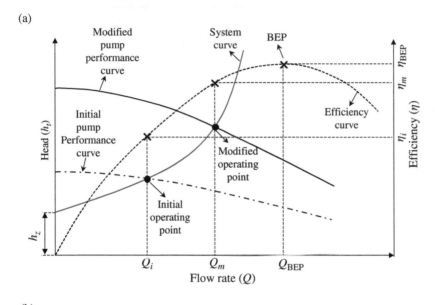

(b)

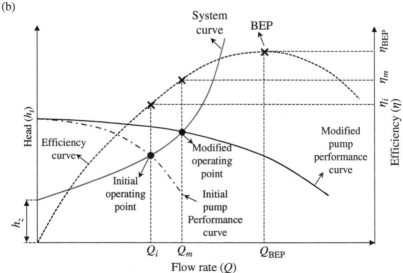

Figure 3.25 Modified pump performance curve of two identical pump installed in a (a) series and (b) parallel format.

again the initial operating point encounters a position change across the system curve to a new position at the intersection with the modified pump performance curve. The coordinates are again named as Q_m and h_{tm} with a new efficiency of η_m.

Example 3.15

Assume a piping system as shown in Figure 3.E15.1. All the main pipes are 1" while the branches to the consumer units are ½". The loss coefficients of all components are given as:

Joints	Loss coefficient
Tee	0.9
Check-valve	2
Elbow	1.5
Tee turn	2

Cold-water temperature is at $18°C$ while the hot water's temperature after leaving the combi boiler is $60°C$. The pressure loss through the boiler is neglected. Cold-water

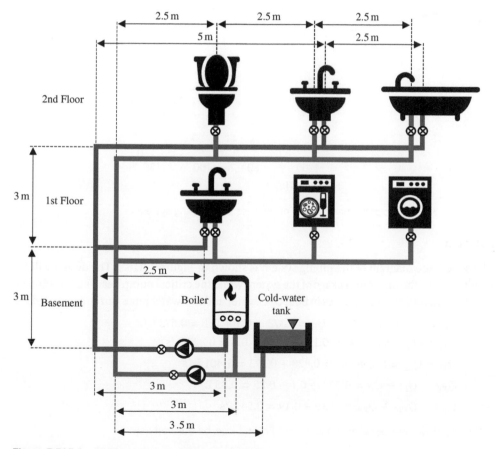

Figure 3.E15.1 Piping system a multi-story building.

consumption of each unit is furthermore provided in accordance with the available standards as below:

Location	Cold-water flow rate (L/s)	Hot-water flow rate (L/s)
A	0.190	0.190
B	0.238	0.100
C	0.400	—
D	0.100	—
E	0.100	—
F	0.373	0.250

First, calculate the system equation. Then, suggest a modification in the pumping of the system to increase the efficiency of the piping system.

The pump performance curve is also given in Figure 3.E15.2.

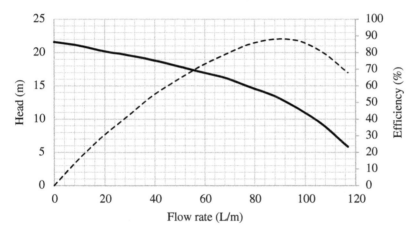

Figure 3.E15.2 Pump performance curve of the case study piping system.

Solution

The schematic diagram of the piping system is shown in Figure 3.E15.3. The design will be conducted for the maximum usage of the system while the critical pump head is at location of Point-F. Therefore, the flow rates in both cold-water and hot-water pipes can be calculated as:

$$Q_{67} = Q_{36} = Q_F + Q_E + Q_D = 0.373 + 0.1 + 0.1 = 0.573 \, \text{L/s}$$
$$Q_{34} = Q_A + Q_B + Q_C = 0.19 + 0.238 + 0.4 = 0.828 \, \text{L/s}$$
$$Q_{23} = Q_{21} = Q_{36} + Q_{34} = 0.828 + 0.573 = 1.401 \, \text{L/s}$$
$$Q_{3'6'} = Q_{F'} + Q_{E'} = 0.373 + 0.1 + 0.25 = 0.35 \, \text{L/s}$$
$$Q_{2'3'} = Q_{3'6'} + Q_{3'A'} = 0.35 + 0.19 = 0.54 \, \text{L/s}$$

So, the flow rate between Point-0 and Point-1 can be derived as:

$$Q_{01} = Q_{23} + Q_{2'3'} = 1.401 + 0.54 = 1.94 \, \text{L s}^{-1}$$

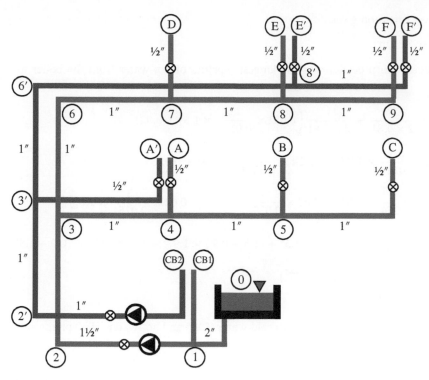

Figure 3.E15.3 Schematic diagram of the case study piping system.

Similar as the previous examples, we first find friction factors through calculated Re and the relative roughness values of each pipe to be able to obtain major losses. As we stated before, the longest cold-water pipeline to the pump is Point-F and thus, first, we derive the Bernoulli equation between these points as follows:

$$h_{\text{pump}} = 6 + \frac{0.74^2}{2 \times 9.81} + \frac{1}{2 \times 9.81} \left(0.026 \times \frac{2.5}{0.0254} \times 0.79^2 + 0.023 \times \frac{2.5}{0.0254} \times 1.26^2 \right.$$

$$+ 0.022 \times \frac{2.5}{0.0254} \times 1.63^2 + 0.026 \times \frac{2.5}{0.0254} \times 0.74^2 + 0.025 \times \frac{2.5}{0.0254} \times 0.93^2$$

$$+ 0.024 \times \frac{2.5}{0.0254} \times 1.13^2 + 0.021 \times \frac{3}{0.0254} \times 1.13^2 + 0.021 \times \frac{3}{0.0381} \times 1.22^2$$

$$\left. + 0.021 \times \frac{3}{0.0381} \times 1.22^2 + 0.021 \times \frac{0.5}{0.0508} \times 0.96^2 \right)$$

$$+ \frac{1}{2 \times 9.81} \left((0.9 + 1.5) \times 0.79^2 + (2 + 0.9) \times 1.26^2 + (2 + 2) \times 1.63^2 \right.$$

$$+ (0.9 + 1.5) \times 0.74^2 + (2 + 0.9) \times 0.93^2 + (2) \times 1.13^2 + (0.9 + 1.5) \times 1.13^2$$

$$\left. + (0.9 + 1.5 + 2) \times 1.23^2 \right)$$

Solving this expression gives:

$$h_{pump} = 7.63 \, m$$

We also repeat the procedure for the longest pipeline of hot-water from the boiler to Point-F':

$$h_{pump} = 6 + \frac{0.74^2}{2 \times 9.81} + \frac{1}{2 \times 9.81}\left(0.022 \times \frac{2.5}{0.0127} \times 1.5^2 + 0.021\right.$$

$$\times \frac{2.5}{0.0127} \times 1.97^2 + 0.022 \times \frac{5}{0.0254} \times 0.69^2 + 0.022$$

$$\left. \times \frac{3}{0.0254} \times 0.69^2 + 0.02 \times \frac{3}{0.0254} \times 1.07^2 + 0.02 \times \frac{3}{0.0254} \times 1.07^2\right)$$

$$+ \frac{1}{2 \times 9.81}\left((2 + 1.5) \times 1.5^2 + (1.5 + 0.9) \times 1.97^2 + (2) \times 0.69^2 + (0.9 + 1.5)\right.$$

$$\left. \times 0.69^2 + (2 + 1.5) \times 1.07^2\right)$$

$$\Longrightarrow h_{pump} = 8.26 \, m$$

Therefore, the system equation for both cold-water (CW) and hot-water (HW) piping system would be:

$$h_{t\_CW} = 6 + \left(4.26 \times 10^5\right)Q^2$$

$$h_{t\_HW} = 6 + \left(7.75 \times 10^6\right)Q^2$$

We can plot both system equations as shown in Figure 3.E15.4 alongside with the pump performance and efficiency curves. As it can be seen, the pumps though do not perform at anywhere near their maximum efficiency (88%) for both cold-water (~73%) and hot-water (~79%) piping systems as shown by blue and red dotted lines, respectively.

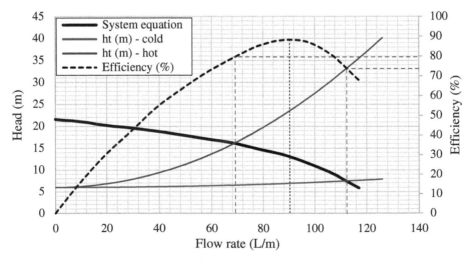

Figure 3.E15.4 System equation of the case study piping system.

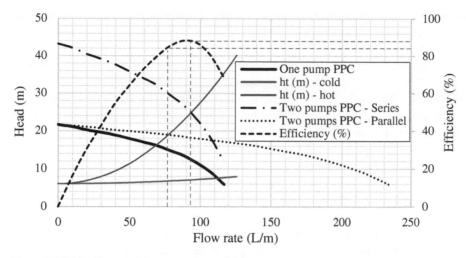

Figure 3.E15.5 New performance curves of the case study dual-pump systems.

Now, let us assume that we have the choice to add another similar pump either in series or parallel arrangement. The new performance curves of the dual-pump system are depicted in Figure 3.E15.5. It is very evident from the graph that two similar pumps installed in a series condition can improve the pump efficiency of the hot-water piping system to ~88%. This value would be about ~84% for the parallel format of two installed pumps. On the other hand, as it can again be seen in Figure 3.E15.5, the cold-water system equation does not intersect any of the new performance curves, which implies a single pump would be a better option though the efficiency is low. In this case, choosing a suitable pump with a less head and higher flow rate is more desirable.

References

1 Fernando, H.J.S. (2010). Fluid dynamics of urban atmospheres in complex terrain. *Annual Review of Fluid Mechanics* 42 (1): 365–389.

2 Oke, T.R., Mills, G., Christen, A., and Voogt, J.A. (2017). *Urban Climates.* Cambridge: Cambridge University Press.

3 Davenport, A. F. (1965). The relationship of wind structure to wind loading. Proceedings of the Conference on Wind Effects on Buildings & Structures, HMSO, 54.

4 Counihan, J. (1975). Adiabatic atmospheric boundary layers: A review and analysis of data from the period 1880-1972. 9: 871.

5 Shirzadi, M., Tominaga, Y., and Mirzaei, P.A. (2019). Wind tunnel experiments on cross-ventilation flow of a generic sheltered building in urban areas. *Building and Environment* 158: 60–72.

6 Shirzadi, M., Tominaga, Y., and Mirzaei, P.A. (2020). Experimental study on cross-ventilation of a generic building in highly-dense urban areas: impact of planar area density and wind direction. *Journal of Wind Engineering and Industrial Aerodynamics* 196: 104030.

7 Shirzadi, M., Tominaga, Y., and Mirzaei, P.A. (2020). Experimental and steady-RANS CFD modelling of cross-ventilation in moderately-dense urban areas. *Sustainable Cities and Society* 52: 101849.

8 Tamura, Y. (2012). Aerodynamic database for low-rise buildings. Global Center of Excellence Program, Tokyo Polytechnic University, Database.

9 Tominaga, Y., Akabayashi, S.-i., Kitahara, T., and Arinami, Y. (2015). Air flow around isolated gable-roof buildings with different roof pitches: wind tunnel experiments and CFD simulations. *Building and Environment* 84: 204–213.

10 Heiselberg, P., Svidt, K., and Nielsen, P.V. (2001). Characteristics of airflow from open windows. *Building and Environment* 36 (7): 859–869.

11 Dimotakis, P.E., Miake-Lye, R.C., and Papantoniou, D.A. (1983). Structure and dynamics of round turbulent jets. *The Physics of Fluids* 26 (11): 3185–3192.

12 N/A (2009). *ASHRAE Handbook - Fundamentals (SI Edition)*. American Society of Heating, Refrigerating and Air-Conditioning Engineers, Inc. (ASHRAE).

13 Pope, S.B. (2000). *Turbulent Flows*. Cambridge: Cambridge University Press.

14 Blazek, J. (2015). *Computational Fluid Dynamics: Principles and Applications*. Elsevier.

15 Moody, L.F. (1944). Friction factors for pipe flow. *Transactions of the ASME* 66 (8): 671–684.

16 Roberson, C. and Chaudhry. (1998). *Hydraulic Engineering*, 2e. Wiley.

17 Munson, B.R., Young, D.F., and Okiishi, T.H. (2006). *Fundamentals of Fluid Mechanics*, 6e. Hoboken, NJ: Wiley.

18 Blake, W.K. (2017). Chapter 6 – noise from rotating machinery. In: *Mechanics of Flow-Induced Sound and Vibration*, 2e, vol. 2 (ed. W.K. Blake), 505–658. Academic Press.

19 Lakshminarayana B. *Fluid Dynamics and Heat Transfer of Turbomachinery*: Wiley; 1995.

20 White, F.M. (1979). *Fluid Mechanics*. Kogakusha: McGraw-Hill.

21 Douglas, J.F. (2011). *Fluid Mechanics*. Prentice Hall.

22 ASHRAE (2019). *ASHRAE Handbook of Fundamentals*. Atlanta, GA: American Society of Heating, Refrigerating and Air Conditioning Engineers.

4

An Overview on Fundamentals of Thermodynamics in Buildings

4   An Overview of Thermodynamics

4.1   Saturation Temperature

Saturation temperature is referred to the temperature in which vaporization in the saturation pressure occurs. The relation between pressure and temperature in water and many substances can be shown as the ***vapour–pressure*** curve as displayed in Figure 4.1.

The substance is a ***saturated liquid*** or ***saturated vapour*** if it exists as either liquid or vapour in the saturation pressure in a given temperature. A substance is also called ***subcooled liquid*** if, as a liquid, it has a lower temperature than the saturation temperature in a given pressure. The liquid substance can be also called ***compressed liquid*** if its pressure is higher than the saturation pressure in a given temperature. Similarly, a vapour substance with a higher temperature than the saturation temperature in a given pressure is called ***superheated vapour.***

A partial existence of a substance in both vapour and liquid phases in a saturation temperature is a presumable situation in which the ratio of the vapour mass to the total mass is known as the ***quality*** and is shown with x or percentile. The quality will be used later to define the moisture content in air. With the definition of the quality, a saturated vapour has a quality value of 1 or 100% and a saturated liquid has a quality value of 0 or 0%.

Another way to present the state of a substances is to plot its volume–temperature relationship. As depicted in Figure 4.2, the pressure-constant lines are parallel lines, crossing saturated liquid points towards saturated vapour points. Thus, the line between two saturations points is a constant temperature process in which a substance in its liquid phase is transformed to its vapour phase. After the saturated vapour point, the vapour is superheated at a constant pressure while both temperature and volume are increasing as expected.

Computational Fluid Dynamics and Energy Modelling in Buildings: Fundamentals and Applications,
First Edition. Parham A. Mirzaei.
© 2023 John Wiley & Sons Ltd. Published 2023 by John Wiley & Sons Ltd.

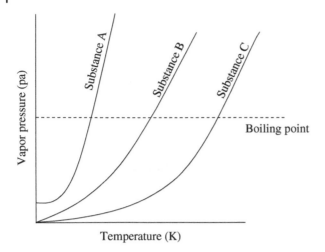

Figure 4.1 Vapour–pressure curve for a pure substance.

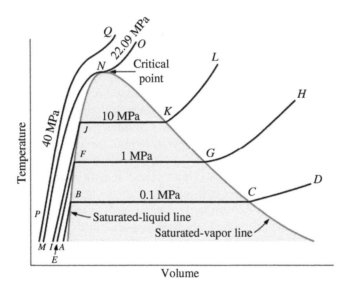

Figure 4.2 Volume–temperature curve for a pure substance.

Example 4.1

An adiabatic room as shown in Figure 4.E1 with the dimensions of 4 m × 3 m × 2.8 m is in equilibrium, containing water vapour at a temperature of 15 °C. (a) What is the pressure of the room as well as the mass of the vapour? (b) What is the mass of the vapour if we assume the water vapour as an ideal gas? (c) Repeat the process if the quality is 70%. (d) What would be the pressure change if we increase the temperature to 25 °C.

Figure 4.E1 Adiabatic case study room.

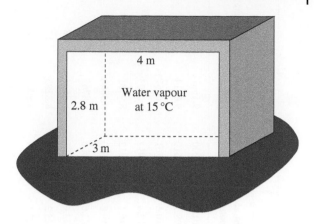

Thermodynamic properties of water vapour for a limited range of temperatures are given in the below Table 4.E1.

Solution

a) First, we calculate the volume as $V = 33.6$ m³. Saturated water vapour at different temperatures can be found from thermodynamic related tables (e.g. [1]). For the temperature of 15 °C, we can find $P_{\text{sat @ 15°C}} = 1.705$ kPa. This also implies that the saturated vapour is $v_g = 77.925$ m³/kg.

Initially, the room has a quality of 100%, thus, we can find the mass of the vapour as:

$$m_g = \frac{V}{v_g} = \frac{33.6}{77.925} = 0.431 \text{ kg}$$

b) Another way is to assume the water vapour as an ideal gas (M_g is water molar mass):

$$R_g = \frac{\overline{R}}{M_g} = \frac{8.314}{18.015} = 0.4615 \text{ kJ/kg K}$$

$$\Longrightarrow m_g = \frac{PV}{R_g T} = \frac{1.7057 \times 33.6}{0.4615 \times 288.15} = 0.431 \text{ kg}$$

Hence, the water vapour can be fairly assumed as an ideal gas.

c) Now if the quality is changed to 70%, we can find the vapour mass as below:

$$x = \frac{m_g}{m_g + m_f}$$

$$\Longrightarrow m_g = 0.7 \times 0.431 = 0.302 \text{ kg}$$

d) Referring again to the related table, we find that $P_{\text{sat @ 25°C}} = 3.169$ kPa. Hence, we expect the pressure change as:

$$\Delta P = 3.169 - 1.706 = 1.463 \text{ kPa}$$

Table 4.E1 Thermodynamics properties of water vapour for a limited range of temperatures.

Temperature (°C)	Pressure bar	Specific volume (m³/kg)		Internal energy (kJ/kg)		Enthalpy (kJ/kg)			Entropy (kJ/kg K)		Temperature (°C)
		Saturated liquid $v_f \times 10^3$	Saturated vapour v_g	Saturated liquid u_f	Saturated vapour u_g	Saturated liquid h_f	Evaporation h_{fg}	Saturated vapour h_g	Saturated liquid s_f	Saturated vapour s_g	
0.01	0.00611	1.0002	206.136	0.00	2375.3	0.01	2501.3	2501.4	0.0000	9.1562	0.01
4	0.00813	1.0001	157.232	16.77	2380.9	16.78	2491.9	2508.7	0.0610	9.0514	4
5	0.00872	1.0001	147.120	20.97	2382.3	20.98	2489.6	2510.6	0.0761	9.0257	5
6	0.00935	1.0001	137.734	25.19	2383.6	25.20	2487.2	2512.4	0.0912	9.0003	6
8	0.01072	1.0002	120.917	33.59	2386.4	33.60	24 82.5	2516.1	0.1212	8.9501	8
10	0.01228	1.0004	106.379	42.00	2389.2	42.01	2477.7	2519.8	0.1510	8.9008	10
11	0.01312	1.0004	99.857	46.20	2390.5	46.20	2475.4	2521.6	0.1658	8.8765	11
12	0.01402	1.0005	93.784	50.41	2391.9	50.41	2473.0	2523.4	0.1806	8.8524	12
13	0.01497	1.0007	88.124	54.60	2393.3	54.60	2470.7	2525.3	0.1953	8.8285	13
14	0.01598	1.0008	82.848	58.79	2394.7	58.80	2468 3	2527.1	0.2099	8.8048	14
15	0.01705	1.0009	77.926	62.99	2396.1	62.99	2465.9	2528.9	0.2245	8.7814	15
16	0.01818	1.0011	73.333	67.18	2397.4	67.19	2463.6	2530.8	0.2390	8.7582	16
17	0.01938	1.0012	69.044	71.38	2398.8	71.38	2461.2	2532.6	0.2535	8.7351	17
18	0.02064	1.0014	65.038	75.57	2400.2	75.58	2458.8	2534.4	0.2679	8.7123	18
19	0.02198	1.0016	61.293	79.76	2401.6	79.77	2456.5	2536.2	0.2823	8.6897	19
20	0.02339	1.0018	57.791	83.95	2402.9	83.96	2454.1	2538.1	0.2966	8.6672	20
21	0.02487	1.0020	54.514	88.14	2404.3	88.14	2451.8	2539.9	0.3109	8.6450	21
22	0.02645	1.0022	51.447	92.32	2405.7	92.33	2449.4	2541.7	0.3251	8.6229	22
23	0.02810	1.0024	48.574	96.51	2407.0	96.52	2447.0	2543.5	0.3393	8.6011	23
24	0.02985	1.0027	45.883	100.70	2408.4	100.70	2444.7	2545.4	0.3534	8.5794	24
25	0.03169	1.0029	43.360	104.88	2409.8	104.89	2442.3	2547.2	0.3674	8.5580	25

26	0.03363	1.0032	40.994	109.06	2411.1	109.07	2439.9	2549.0	0.3814	8.5367	26
27	0.03567	1.0035	38.774	113.25	2412.5	113.25	2437.6	2550.8	0.3954	8.5156	27
28	0.03782	1.0037	36.690	117.42	2413.9	117.43	2435.2	2552.6	0.4093	8.4946	28
29	0.04008	1.0040	34.733	121.60	2415.2	121.61	2432.8	2554.5	0.4231	8.4739	29
30	0.04246	1.0043	32.894	125.78	2416.6	125.79	2430.5	2556.3	0.4369	8.4533	30
31	0.04496	1.0046	31.165	129.96	2418.0	129.97	2428.1	2558.1	0.4507	8.4329	31
32	0.04759	1.0050	29.540	134.14	2419.3	134.15	2425.7	2559.9	0.4644	8.4127	32
33	0.05034	1.0053	28.011	138.32	2420.7	138.33	2423.4	2561.7	0.4781	8.3927	33
34	0.05324	1.0056	26.571	142.50	2422.0	142.50	2421.0	2563.5	0.4917	8.3728	34
35	0.05628	1.0060	25.216	146.67	2423.4	146.68	2418.6	2565.3	0.5053	8.3531	35
36	0.05947	1.0063	23.940	150.85	2424.7	150.86	2416.2	2567.1	0.5188	8.3336	36
38	0.06632	1.0071	21.602	159.20	2427.4	159.21	2411.5	2570.7	0.5458	8.2950	38
40	0.07384	1.0078	19.523	167.56	2430.1	167.57	2406.7	2574.3	0.5725	8.2570	40
45	0.09593	1.0099	15.258	188.44	2436.8	188.45	2394.8	2583.2	0.6387	8.1648	45

Source: Wylen [1], table B.1 – only page 1.

4.2 First Law of Thermodynamics

Work is defined in a quasi-equilibrium process by integration of the product of an intensive property and the change in another intensive property. As a famous example, work, thus, can be represented in the following forms as the integration of the product of pressure and volume change:

$$W_{1-2} = \int_1^2 P dV \tag{4.1}$$

Equation (4.1) indicates that the pressure (P), as the intensive property, is a driving force to cause change in another intensive property volume (V). On the other hand, heat is defined when a change occurs in the system from state-1 to state-2:

$$Q_{1-2} = \int_1^2 dQ \tag{4.2}$$

The first law of the thermodynamics establishes a relation between work and heat during a cycle of a system. From a control mass perspective, the first law of the thermodynamics relates work and heat with another term, which is energy. If in a pressure-volume cycle of a system, as shown in Figure 4.3, the system's state changes from 1 to 2, and returns from state-2 to state-1, then the whole process can be written as:

$$\oint \delta Q = \oint \delta W \tag{4.3}$$

where $\oint \delta$ denotes the cyclic integral for heat and work.

We can further expand the equation for the A and B processes as demonstrated in Figure 4.4:

$$\oint_1^2 \delta Q_A + \oint_2^1 \delta Q_B = \oint_1^2 \delta W_A + \oint_2^1 \delta W_B \tag{4.4}$$

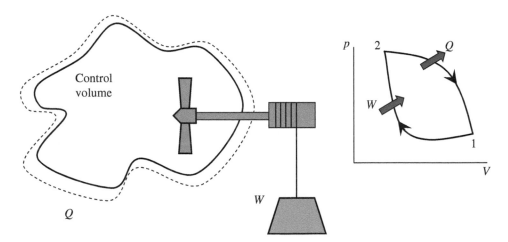

Figure 4.3 System cycle of pressure–volume

The same process can be written for another arbitrary cycle, consisting of a state change during the process C from 1 to 2, and a return during the process of B from state-2 to state-1 (see Figure 4.4):

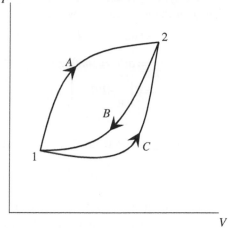

$$\oint_1^2 \delta Q_C + \oint_2^1 \delta Q_B = \oint_1^2 \delta W_C + \oint_2^1 \delta W_B$$

(4.5)

Obviously, if we subtract two cycles of Eqs. (4.4) and (4.5) from each other:

$$\oint_1^2 \delta Q_A - \oint_1^2 \delta Q_C = \oint_1^2 \delta W_A - \oint_1^2 \delta W_C$$

$$\Longrightarrow \oint_1^2 (\delta Q - \delta W)_A = \oint_1^2 (\delta Q - \delta W)_C$$

Figure 4.4 Pressure–volume cycle of a system.

(4.6)

This simply implies that $\delta Q - \delta W$ is independent of any defined path and is only a matter of the initial and final states in any arbitrary cycles. Thus, we can conclude that $\delta Q - \delta W$ is a constant value, which represents energy of the mass:

$$\delta Q - \delta W = dE$$

(4.7)

In an integrated form, we can obtain the below equation from state-1 to state-2:

$$Q_{1-2} - W_{1-2} = E_2 - E_1$$

(4.8)

where E_1 and E_2 are the energy of the control mass in the initial and final states, respectively. Q_{1-2} and W_{1-2} are the heat transferred and work done by the control mass during the process from state-1 to state-2.

The change of energy in a system can be further understood as the change in the kinetic energy of the system (KE), potential energy of the system (PE) and all other forms of the energy, known as the property of the system or the internal energy (U). Thus, Eq. (4.8) can be written as:

$$\delta Q - \delta W = dU + d(PE) + d(KE)$$

$$\Longrightarrow Q_{1-2} - W_{1-2} = (U_2 - U_1) + (KE_2 - KE_1) + (PE_2 - PE_1)$$

(4.9)

A very common form of KE is related to the work done by an external force (F) acting on the control mass in a distance dx:

$$\delta W = -Fdx = -dKE$$

(4.10)

From Eq. (4.10), hence, KE can be derived as below:

$$dKE = (ma)dx = m\frac{d\vec{V}}{dt}dx = m\frac{d\vec{V}}{dx}\frac{dx}{dt}dx = m\vec{V}d\vec{V}$$

$$\Longrightarrow \int_{KE_1}^{KE_2} dKE = \int_{V_1}^{V_2} m\vec{V}d\vec{V}$$

(4.11)

$$\Longrightarrow KE_2 - KE_1 = \frac{1}{2}m\vec{V}_2^{\,2} - \frac{1}{2}m\vec{V}_1^{\,2}$$

where, V is the velocity of the acting external force in the x-direction.

Similarly, it is common to find the potential energy in accordance with the elevation change of the mass control by the gravitational forces (mg) in z-direction:

$$\delta W = -Fdz = -dPE$$

$$\Longrightarrow dPE = (mg)dz$$
$$\Longrightarrow \int_{PE_1}^{PE_2} dPE = \int_{z_1}^{z_2} mgdz \tag{4.12}$$
$$\Longrightarrow PE_2 - PE_1 = mgz_2 - mgz_1$$

Substituting values from Eqs. (4.11) and (4.12) into Eq. (4.9) gives:

$$Q_{1-2} - W_{1-2} = (U_2 - U_1) + \left(\frac{1}{2}m\vec{V}_2^{\,2} - \frac{1}{2}m\vec{V}_1^{\,2}\right) + (mgz_2 - mgz_1) \tag{4.13}$$

The above equation represents the first law of the thermodynamics, and states that the net change of energy in a system is always equal to the net energy transferred from the boundaries in the forms of heat and work. This is another representation form of the conservation of energy from the first law of the thermodynamics as it can be also presented as 'energy can be transformed from one form to another but can be neither created nor destroyed'.

As a rate basis equation, the first law of the thermodynamics can be written in a time interval of δt in which heat transfer of δQ from the boundaries and on which work of δW is done by the control mass. Also, the internal energy of ΔU, kinetic energy of ΔKE and potential energy of ΔPE demonstrate the variation of their values in the system during the time interval of δt. Once again, we can rewrite Eq. (4.9) as:

$$\delta Q - \delta W = \Delta U + \Delta(PE) + \Delta(KE) \tag{4.14}$$

If we divide all terms by δt to obtain the average rate of heat transfer and work in addition to the increase in the energy of the control mass, then we have:

$$\frac{\delta Q}{\delta t} - \frac{\delta W}{\delta t} = \frac{\Delta U}{\delta t} + \frac{\Delta(PE)}{\delta t} + \frac{\Delta(KE)}{\delta t} \tag{4.15}$$

We can now apply the limitation over the time interval for all terms:

$$\lim_{\delta t \to 0} \frac{\delta Q}{\delta t} - \lim_{\delta t \to 0} \frac{\delta W}{\delta t} = \lim_{\delta t \to 0} \frac{\Delta U}{\delta t} + \lim_{\delta t \to 0} \frac{\Delta(PE)}{\delta t} + \lim_{\delta t \to 0} \frac{\Delta(KE)}{\delta t}$$

$$\Longrightarrow \dot{Q} - \dot{W} = \frac{d(PE)}{dt} + \frac{d(KE)}{dt} + \frac{dU}{dt} \tag{4.16}$$

$$\Longrightarrow \dot{Q} - \dot{W} = \frac{dE}{dt}$$

Example 4.2

Assume that the room of Example 4.1 is filled by 10 $^\circ$C air at atmosphere pressure. The heat transfer through walls results the room to reach to another equilibrium state at 25 $^\circ$C. Derive first law of thermodynamics and calculate the change in work, heat transfer, and internal energy of the system.

Thermodynamics properties of air for a limited range of temperatures are given in Table 4.E2.

Table 4.E2 Thermodynamics properties of air for a limited range of temperatures.

Temp K	h kJ/kg	Pr	u kJ/kg	vr	s kJ/kg K	Temp K	h kJ/kg	Pr	u kJ/kg	vr	s kJ/kg K
200	199.97	0.3363	142.56	1707.0	1.29559	580	586.04	14.38	419.55	115.7	2.37348
210	209.97	0.3987	149.69	1512.0	1.34444	590	596.52	15.31	427.15	110.6	2.39140
220	219.97	0.4690	156.82	1346.0	1.39105	600	607.02	16.28	434.78	105.8	2.40902
230	230.02	0.5477	164.00	1205.0	1.43557	610	617.53	17.30	442.42	101.2	2.42644
240	240.02	0.6355	171.13	1084.0	1.47824	620	628.07	18.36	450.09	96.92	2.44356
250	250.05	0.7329	178.28	979.0	1.51917	630	638.63	19.84	457.78	92.84	2.46048
260	260.09	0.8405	185.45	887.8	1.55848	640	649.22	20.64	465.50	88.99	2.47716
270	270.11	0.9590	192.60	808.0	1.59634	650	659.84	21.86	473.25	85.34	2.49364
280	280.13	1.0889	199.75	738.0	1.63279	660	670.47	23.13	481.01	81.89	2.50985
285	285.14	1.1584	203.33	706.1	1.65055	670	681.14	24.46	488.81	78.61	2.52589
290	290.16	1.2311	206.91	676.1	1.66802	680	691.82	25.85	496.62	75.50	2.54175
295	295.17	1.3068	210.49	647.9	1.68515	690	702.52	27.29	504.45	72.56	2.55731
298	298.18	1.3543	212.64	631.9	1.69528	700	713.27	28.80	512.33	69.76	2.57277
300	300.19	1.3860	214.07	621.2	1.70203	710	724.04	30.38	520.23	67.07	2.58810
305	305.22	1.4686	217.67	596.0	1.71865	720	734.82	32.02	528.14	64.53	2.60319
310	310.24	1.5546	221.25	572.3	1.73498	730	745.62	33.72	536.07	62.13	2.61803
315	315.27	1.6442	224.85	549.8	1.75106	740	756.44	35.50	544.02	59.82	2.63280

Source: Wylen [1] with permission of John Wiley & Sons.

Solution

We assume air as an ideal gas, and therefore we can find the mass of air as at $P = 101.325\,\text{kPa}$:

$$m_\text{a} = \frac{PV}{RT}$$

$$\Longrightarrow m_\text{a} = \frac{101.325 \times 33.6}{0.287 \times 283.15} = 41.89\,\text{kg}$$

Now, since the system is in an equilibrium in both states, we can employ Eq. (4.7) as:

$$\delta Q - \delta W = \Delta E$$

The volume of the room is constant ($\Delta V = 0$), thus, the work is zero or $\delta W = \int P dV = 0$. Also, neglecting any change in the kinetic and potential energy, the first law of thermodynamics can be simplified to:

$$\delta Q = \Delta U = m_\text{a}(u_2 - u_1)$$

Now, from the table of ideal-gas properties of air, we can obtain:

$$u_{1@10°\text{C}} = 202.01\,\text{kJ/kg}$$

$$u_{2@25°\text{C}} = 213.04\,\text{kJ/kg}$$

$$\Longrightarrow \delta Q = \Delta U = 41.89 \times (213.04 - 202.01) \approx 462.29\,\text{kJ}$$

4.2.1 Enthalpy

Before defining the specific heat, we introduce the enthalpy for a system. Assume a system under a quasi-equilibrium state is exposed to a constant pressure process with no change in the potential and kinetic energy. So, the work done is only by the boundary movement from an external acting force, which indicates that Eq. (4.8) can be rearranged as:

$$Q_{1-2} - W_{1-2} = (U_2 - U_1)$$

$$\Longrightarrow Q_{1-2} - \int_1^2 P dV = (U_2 - U_1) \tag{4.17}$$

$$\Longrightarrow Q_{1-2} - P(V_2 - V_1) = (U_2 - U_1)$$

$$\Longrightarrow Q_{1-2} = (U_2 + PV_2) - (U_1 + PV_1)$$

So, in this case, we can find the heat transfer from the boundary of the system and during the process by having the value of $U + PV$ at the initial and final states of the process. These values are hence related to the state of the system, which are known as the thermodynamic properties or ***enthalpy*** of the system:

$$H \equiv U + PV \tag{4.18}$$

We can also rewrite Eq. (4.18) per unit of mass:

$$h \equiv u + Pv \tag{4.19}$$

Example 4.3

Use the enthalpy concept and repeat Example 4.2.

Solution

First, we apply the enthalpy equation of Eq. (4.18) into Eq. (4.9) as below:

$$U = H - PV$$
$$\Longrightarrow \Delta E = \Delta U = \Delta(H - PV) = \Delta H - (\Delta P)V - (\Delta V)P$$

Since the volume is not changing ($\Delta V = 0$) and $\delta W = 0$, we can rewrite Eq. (4.8) as:

$$\delta Q = \Delta U = \Delta H - V\Delta P$$

Now, we can find ΔP for air as an ideal gas as ($P_1 = 101.32p$ kPa):

$$P_2 = \frac{T_2}{T_1} P_1$$
$$\Longrightarrow \Delta P = \left(\frac{T_2}{T_1} - 1\right) P_1 = 0.053 P_1 = 5.368 \text{ kPa}$$

Also, from Table 4.E2 of ideal-gas properties of air, we can obtain enthalpies as below:

$$h_1 = 283.29 \text{ kJ/kg}$$
$$h_2 = 298.62 \text{ kJ/kg}$$

From Example 4.2, we have $m = 41.89$ kg. So:

$$\Delta H = m\Delta h = 41.89 \times (298.62 - 283.2863) \approx 642.40 \text{ kJ}$$

Hence, we can find ΔU as:

$$\Delta U = m\Delta h - V\Delta P = 642.40 - 33.6 \times 5.368 = 462.04 \text{ kJ}$$

4.2.2 Specific Heats

The specific heat is defined as the amount of heat necessary to increase the temperature of a unit of mass of a substance by one unit. Knowing this, we assume a system with a simple compressible substance in a quasi-equilibrium process in which the kinetic and potential energies are neglected. The work is defined as $W_{1-2} = \int_1^2 PdV$ and thus we can derive the first law of the thermodynamics using Eq. (4.8) and reach to:

$$\delta Q = dU + PdV \tag{4.20}$$

We can articulate the definition of specific heat for a constant volume as the below mathematical expression:

$$C_v = \frac{1}{m}\left(\frac{\delta Q}{\delta T}\right)_v \tag{4.21}$$

We can assume a control mass with a constant volume ($PdV = 0$). So, Eq. (4.20) becomes $\delta Q = dU$ and we can rewrite Eq. (4.21) as:

$$C_v = \frac{1}{m}\left(\frac{\delta U}{\delta T}\right)_v = \left(\frac{\delta u}{\delta T}\right)_v \tag{4.22}$$

On the other hand, we can redefine the specific heat when the process occurs in a constant pressure as below mathematical expression:

$$C_p = \frac{1}{m}\left(\frac{\delta Q}{\delta T}\right)_p \tag{4.23}$$

Also, we know that $dPV = 0$, so we can expand Eq. (4.18) as:

$$\delta H \equiv dU + dPV + PdV$$
$$\Longrightarrow \delta H \equiv dU + PdV \tag{4.24}$$

Equation (4.24) is equal to Eq. (4.20), which implies that $\delta H = \delta Q$. So, we can rewrite Eq. (4.23) as:

$$C_p = \frac{1}{m}\left(\frac{\delta H}{\delta T}\right)_p = \left(\frac{\delta h}{\delta T}\right)_p \tag{4.25}$$

For the ideal gases, we know that:

$$Pv = RT \tag{4.26}$$

where R is the gas constant number (8.314 J/K/mol).

So, we can rewrite Eq. (4.19) as:

$$h = u + Pv = u + RT \tag{4.27}$$

Thus, h is only a function of temperature since R is a constant value and u is a function of temperature only:

$$C_{po} = \left(\frac{\delta h}{\delta T}\right)_p = \left(\frac{\delta(u + RT)}{\delta T}\right)_p = R + \left(\frac{\delta u}{\delta T}\right)_p \tag{4.28}$$

where C_{po} denotes the specific heat in a constant pressure for an ideal gas.

In the ideal gases, the internal energy is not a function of the system volume and therefore we can find:

$$C_{vo} = \left(\frac{\delta u}{\delta T}\right)_v \tag{4.29}$$

where C_{vo} denotes the specific heat in a constant volume for an ideal gas.

Combining Eqs. (4.28) and (4.29), it is obvious to find the below equation for both specific heat values:

$$C_{po} - C_{vo} = R \tag{4.30}$$

Example 4.4

Estimate the specific heat of vapour in Example 4.1. What is the specific heat if the room is filled by air in a same temperature?

Solution

From Table 4.E1, we find the specific heat at $T = 288.15$ K as:

$$u_v = 2395.5 \, \text{kJ/kg}$$

$$\Rightarrow C_v = \left(\frac{\delta u}{\delta T}\right)_v = \frac{2395.5}{288.15} = 8.31 \, \text{kJ/kg/K}$$

Air is assumed as an ideal gas, so we can use Table 4.E2 to find the specific heat in a constant volume at $T = 288.15$ K:

$$u_a = 205.5854 \, \text{kJ/kg}$$

$$C_{v0} = \left(\frac{\delta u}{\delta T}\right)_v = \frac{205.59}{288.15} = 0.713 \, \text{kJ/kg/K}$$

4.2.3 First Law of Thermodynamics for a Control Volume (Open System)

The first law was presented in Eq. (4.7) for a mass control under a thermodynamic process. Here, we expand the first law of thermodynamics to the concept of control volumes. Assume that we have a control volume as shown in Figure 4.5 where a substance is flowing through it. The control volume has an arbitrary shape while its boundaries are fixed or flexible with contraction and expansion. Mass and energy are conserved in the control volume bounded by the conservation law. Mass in particular cannot be created or destroyed in the control volume. Hence, as shown in the control volume of Figure 4.5, the net increase of the mass and the mass flow entering to or existing out of the control volume can be expressed as:

$$\frac{dm_{C.V.}}{dt} = \sum \dot{m}_{in} - \sum \dot{m}_{out} \tag{4.31}$$

This equation is also known as the **continuity equation** and states that the mass neither can be created nor can be destroyed in a control volume.

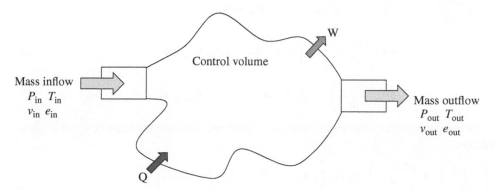

Figure 4.5 Schematic of a control volume in a thermodynamic process.

The mass flow across a control volume can also be described in terms of the volume flow rate while the flow is entering and leaving the control volume with average velocities $\left(\dot{m} = \rho_{avg} \dot{V} = \dfrac{\dot{V}}{v} = \dfrac{\vec{V}A}{v} \right)$:

$$\frac{d\dot{m}_{C.V.}}{dt} = \sum \dot{m}_{in} - \sum \dot{m}_{out} \tag{4.32}$$

To expand the first law for the control volume of Figure 4.5, first we consider the rate of heat transfer and work as before, and now we need to add the rate of energy change, which is due to the rates of energy entering to and leaving from the control volume. To this end, we can notice that the crossing flow in or out from the control volume changes the energy of the control volume per unit of mass in the surfaces as:

$$e = u + \frac{1}{2}\vec{V}^2 + gz \tag{4.33}$$

So, the net change of the energy due to the mass flow in the control volume $\left(\left(\dfrac{dE_{C.V.}}{dt} \right)_1 \right)$ can be presented as:

$$\left(\frac{dE_{C.V.}}{dt} \right)_1 = \dot{m}_{in}e_{in} - \dot{m}_{out}e_{out} \tag{4.34}$$

Now, inserting Eq. (4.33) into Eq. (4.34) gives:

$$\left(\frac{dE_{C.V.}}{dt} \right)_1 = \dot{m}_{in}\left(u_{in} + \frac{1}{2}\vec{V}_{in}^2 + gz_{in} \right) - \dot{m}_{out}\left(u_{out} + \frac{1}{2}\vec{V}_{out}^2 + gz_{out} \right) \tag{4.35}$$

On the other hand, the mass entering or leaving the control volume perform a work, known as $\dot{W}_{flow-in}$, on the surface boundaries. This implies that the mass entering the control volume is pushed by the mass behind the surface boundary due to the pressure of the surrounding exerted on the control volume. In a similar manner, the flow leaving the control volume pushes the surrounding fluid with a work known as $\dot{W}_{flow-out}$. Hence, the net of these works changes the energy of the control volume defined as $\left(\dfrac{dE_{C.V.}}{dt} \right)_2$ and can be expressed as:

$$\left(\frac{dE_{C.V.}}{dt} \right)_2 = \dot{W}_{flow-in} - \dot{W}_{flow-out} = F\vec{V}_{in} - F\vec{V}_{out}$$

$$\implies \left(\frac{dE_{C.V.}}{dt} \right)_2 = \int P_{in}\vec{V}_{in}dA - \int P_{out}\vec{V}_{out}dA \tag{4.36}$$

$$\implies \left(\frac{dE_{C.V.}}{dt} \right)_2 = P_{in}\dot{V}_{in} - P_{out}\dot{V}_{out} = P_{in}v_{in}\dot{m}_{in} - P_{out}v_{out}\dot{m}_{out}$$

The total energy that enters and leaves the control volume with the mass flow can be then written:

$$\left(\frac{dE_{C.V.}}{dt} \right)_1 + \left(\frac{dE_{C.V.}}{dt} \right)_2 = \dot{m}_{in}\left[\left(u_{in} + \frac{1}{2}\vec{V}_{in}^2 + gz_{in} \right) + P_{in}v_{in} \right]$$
$$- \dot{m}_{out}\left[\left(u_{out} + \frac{1}{2}\vec{V}_{out}^2 + gz_{out} \right) + P_{out}v_{out} \right] \tag{4.37}$$

And, thus applying Eq. (4.37) into the first law results in:

$$\frac{dE_{C.V.}}{dt} = \dot{Q}_{C.V.} - \dot{W}_{C.V.} + \left(\frac{dE_{C.V.}}{dt}\right)_1 + \left(\frac{dE_{C.V.}}{dt}\right)_2$$

$$\Longrightarrow \frac{dE_{C.V.}}{dt} = \dot{Q}_{C.V.} - \dot{W}_{C.V.} + \dot{m}_{in}\left[\left(u_{in} + \frac{1}{2}\vec{V}_{in}^2 + gz_{in}\right) + P_{in}v_{in}\right] \tag{4.38}$$

$$- \dot{m}_{out}\left[\left(u_{out} + \frac{1}{2}\vec{V}_{out}^2 + gz_{out}\right) + P_{out}v_{out}\right]$$

We also know that $h \equiv u + Pv$ from Eq. (4.19), hence we can rearrange Eq. (4.38) as:

$$\frac{dE_{C.V.}}{dt} = \dot{Q}_{C.V.} - \dot{W}_{C.V.} + \dot{m}_{in}\left(h_{in} + \frac{1}{2}\vec{V}_{in}^2 + gz_{in}\right) - \dot{m}_{out}\left(h_{out} + \frac{1}{2}\vec{V}_{out}^2 + gz_{out}\right)$$

$$\tag{4.39}$$

For a general case with multiple inlets and outlets, Eq. (4.39) can be further expanded as:

$$\frac{dE_{C.V.}}{dt} = \dot{Q}_{C.V.} - \dot{W}_{C.V.} + \sum_{i=1}^{n} \dot{m}_{in-i}\left(h_{in-i} + \frac{1}{2}\vec{V}_{in-i}^2 + gz_{in-i}\right)$$

$$- \sum_{j=1}^{m} \dot{m}_{out-j}\left(h_{out-j} + \frac{1}{2}\vec{V}_{out-j}^2 + gz_{out-j}\right) \tag{4.40}$$

where n and m are the number of inlets and outlets of the control volume.

Example 4.5

A well-insulated hall room with the dimensions of 5 m × 8 m × 2.8 m contains air at 21°C in the atmosphere pressure as shown in Figure 4.E5. In an equilibrium state, air enters with the velocity of 1 m/s from a window with the area of 0.8 m² at 15°C. On the other hand, air leaves the room through another window with the area of 0.5 m² to the outdoor environment. This cross-ventilation occurs due to a slight pressure difference around the building. Calculate the rate of change in the internal energy of the hall room.

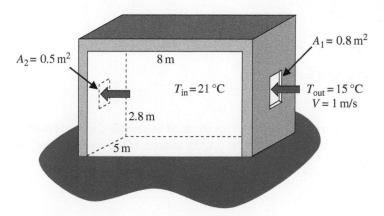

Figure 4.E5 Well-insulated case study hall room.

Solution

If we assume the hall room as the control volume, we can neglect the change in the elevation (potential energy), work on the control volume and any heat transfer from the control volume (isothermal assumption), and hence rewrite Eq. (4.39) as:

$$\frac{dE_{C.V.}}{dt} = \dot{m}_{in}\left(h_{in} + \frac{1}{2}\vec{V}_{in}^2\right) - \dot{m}_{out}\left(h_{out} + \frac{1}{2}\vec{V}_{out}^2\right)$$

We neglect any change in the mass within the control volume and can simplify Eq. (4.31) as:

$$\dot{m}_{in} = \dot{m}_{out}$$

$$\Longrightarrow \dot{m}_{in} = \dot{m}_{out} = \rho\vec{V}_{in}A_{in} = \frac{P}{RT}\vec{V}A_{in} = \frac{101.325}{0.287 \times 294.15} \times 1 \times 0.8 = 0.96\,kg/s$$

If we assume that the density is not changing much, we can now find $\vec{V}_{out}$:

$$\dot{m}_{out} = \rho\vec{V}_{out}A_{out}$$

$$\Longrightarrow \vec{V}_{out} = \frac{0.96}{1.2 \times 0.5} = 1.6\,m/s$$

With the assumption of air as an ideal gas, we can find its properties from Table 4.E2 for the entering air at $15\,°C$ and leaving air at $21\,°C$ as:

$$h_{in} = 288.52\,kJ/kg$$

$$h_{out} = 294.57\,kJ/kg$$

Hence, the first law of thermodynamics can be written as:

$$\frac{dE_{C.V.}}{dt} = 0.96 \times \left(288.52 + \frac{1^2}{2}\right) - 0.96 \times \left(294.57 + \frac{1.6^2}{2}\right) = -6.56\,kJ/s$$

Now, we can estimate the temperature of the hall room ($T_z^{(2)}$) using Eq. (4.22) for ideal gas of air to reach to an equilibrium after time-step of Δt ($C_p \approx 1.007\,kJ/kg/K$):

$$\frac{dE_{C.V.}}{dt} = m_a C_p \frac{\left(T_z^{(2)} - T_z^{(1)}\right)}{\Delta t}$$

$$\Longrightarrow T_z^{(2)} = \frac{-6.56}{1.2 \times 112 \times 1.007}\Delta t + 21$$

For example, if $\Delta t = 10$ s, then we have:

$$\Longrightarrow T_z^{(2)} = 20.5\,°C$$

Note that the above expression can only be used for small Δt as discussed in more details in Chapters 8 and 9. Also, it is very common in the control volume of the rooms (zones) to neglect the kinetic energy:

$$\frac{dE_{C.V.}}{dt} = \dot{m}_{in}h_{in} - \dot{m}_{out}h_{out}$$

Which results in $\frac{dE_{C.V.}}{dt} = -5.81kJ/s$ and for a $\Delta t = 10$ s, we have $T_z^{(2)} = 20.8\,°C$.

4.2.4 Steady-State Steady-Flow (SSSF) Process

When a control volume system is assumed to undergo a steady thermodynamic process for a long period of time, then we can assume the process to be steady state. This means that the transient effects are neglected, and the process is only considered to be independent of time:

$$\frac{dE_{C.V.}}{dt} = 0 \tag{4.41}$$

Another assumption is associated with no variation of mass at any arbitrary point of the control volume, which gives the continuity equation as:

$$\frac{dm_{C.V.}}{dt} = 0 \tag{4.42}$$

$$\Longrightarrow \sum \dot{m}_{in} = \sum \dot{m}_{out} = \sum \dot{m}$$

Then, we can use Eqs. (4.41) and (4.42) and rewrite Eq. (4.40) as a steady-state steady-flow (SSSF) process:

$$\dot{Q}_{C.V.} + \sum_{i=1}^{n} \dot{m} \left(h_{in-i} + \frac{1}{2}\vec{V}_{in-i}^2 + gz_{in-i} \right)$$

$$= \sum_{j=1}^{m} \dot{m} \left(h_{out-j} + \frac{1}{2}\vec{V}_{out-j}^2 + gz_{out-j} \right) + \dot{W}_{C.V.} \tag{4.43}$$

In SSSF processes, it is assumed that the control volume is not moving relative to the reference coordinate, meaning that the relative velocities to the control volume are the same as those of relative to the reference coordinate. Hence, the mass flow states as well as the heat and work rates across the control surface are not varying with time in the SSSF process.

Example 4.6

A dehumidifier similar to the one in Example 2.7 of Chapter 2 is extracting water from the moist air in the room of Example 4.5. The schematic of a dehumidifier system is depicted in Figure 4.E6. In an equilibrium condition, the well-insulated humidifier extracts 0.5 L/h of water while it has a nominal working moist airflow of 165 m^3/h, containing 90% of it as dry air. If the temperature of the leaving moist air, containing 95% of it as dry air, is 19 $^\circ$C, determine the necessary heat that should be extracted from the air.

Solution

The process is SSSF. We consider the control volume around the air and heaters. So, the refrigerant to remove the heat and to reheat the air is excluded from the control volume. Hence, the work done by compressor over the refrigerant is not acting on our presumed control volume and the work is zero. We can also neglect the change in the kinetic and potential energy of the entering air. Equation (4.43) can be written as:

$$\dot{Q}_{C.V.} + \dot{m}_{in}h_{in} = \dot{m}_{out}h_{out}$$

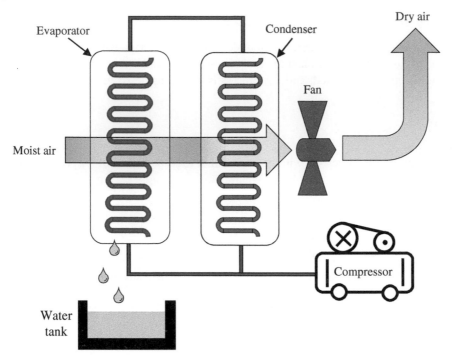

Figure 4.E6 Schematic of a dehumidifier system.

The entering moist air contains dry air ($\dot{m}_{da_{in}}$) and water vapour ($\dot{m}_{v_{in}}$). The leaving air also has dry air ($\dot{m}_{da_{out}}$) and water vapour ($\dot{m}_{v_{out}}$) while it lost some of the moisture through condensations to liquid water ($\dot{m}_{l_{out}}$). From Table E1, we can find $v_{g_{in}} = 61.8 \text{ m}^3/\text{kg}$ and $h_{v_{in}} = 79.82 \text{ kJ/kg}$:

$$\dot{m}_{v_{in}} = (1 - 0.9) \times \frac{Q_{da_{in}}}{v_{g_{in}}} = 0.1 \times \frac{165}{61.8} = 0.267 \text{ kg/h}$$

We also assume the dry air as an ideal gas. Therefore, we can find its mass at the atmosphere pressure as:

$$\dot{m}_{da_{in}} = 0.9 \times Q_{da_{in}} \times \rho_{da} = 0.9 \times 165 \times \frac{P}{RT}$$

$$= 0.9 \times 165 \times \frac{101.325}{0.287 \times 292.15} = 179.54 \text{ kg/h}$$

From the conservation of mass in the control volume (see Eq. (4.42)):

$$\dot{m}_{da_{in}} = \dot{m}_{da_{out}}$$

$$\Longrightarrow 0.9 \rho_{da_{in}} Q_{da_{in}} = 0.95 \rho_{da_{out}} Q_{da_{out}}$$

Assuming no pressure variation through the dehumidifier and a constant temperature for the entering and leaving air, the air density remains unchanged ($\rho_{da_{in}} = \rho_{da_{out}} = 1.21 \text{ kg/m}^3$). So, we can derive:

$$Q_{da_{out}} = \frac{165 \times 0.9}{0.95} = 156.3 \text{ m}^3/\text{h}$$

$$\Rightarrow \dot{m}_{v_{out}} = (1 - 0.95) \times \frac{Q_{da_{out}}}{v_{g_{out}}} = 0.05 \times \frac{156.3}{61.8} = 0.126 \text{ kg/h}$$

Hence, the liquid water flow rate can be found from conservation of mass:

$$\dot{m}_{l_{out}} = \dot{m}_{v_{out}} - \dot{m}_{v_{in}}$$

$$\Rightarrow \dot{m}_{l_{out}} = 0.267 - 0.126 = 0.141 \text{ kg/h}$$

Now, we can calculate the heat that should be removed from the control volume as ($h_f = 79.77$ kJ/kg):

$$\dot{Q}_{C.V.} = (\dot{m}_{da_{out}} - \dot{m}_{da_{in}})C_p T + h_{v_{in}}(\dot{m}_{v_{out}} - \dot{m}_{v_{in}}) + \dot{m}_{l_{out}} h_{l_{out}}$$

$$\Rightarrow \dot{Q}_{C.V.} = 2535.65 \times (0.126 - 0.267) + 0.141 \times 79.77 = 345.1 \text{ kJ/h}$$

Note that we could find h_{fg} of the saturated water at $19°C$ and, as it is explained in the following sections as well as the next Chapter, use the below equation to find the transferred heat:

$$\dot{Q} = \dot{m}_a(\omega_{out} - \omega_{in})h_{fg}$$

where ω_{in} and ω_{out} are the specific humidity of the air–vapour mixture.

4.2.5 Uniform-State Uniform-Flow (USUF) Process

The long-term thermodynamic processes are favourable to be modelled by the SSSF process concept. Nonetheless, thermodynamic processes are frequently subjected to unsteady conditions, which demands the consideration of time-dependent parameters in the modelling approach. In unsteady scenarios, we first assume that the control volume is not moving relative to the reference coordinate. Second, the state of the mass in the control volume assumed to vary in time as below:

$$\frac{dm_{C.V.}}{dt} + \sum \dot{m}_{out} - \sum \dot{m}_{in} = 0 \tag{4.44}$$

Nonetheless, at every time snapshot, the thermodynamic process is assumed to be uniform in the entire control volume. Hence, we can find the mass flow change of $\frac{dm_{C.V.}}{dt}$ during the entire thermodynamic process (t) with the integration of the mass flow variations at each time snapshots:

$$\int_0^t \frac{dm_{C.V.}}{dt} dt = (m_t - m_0)_{C.V.} \tag{4.45}$$

Other terms of Eq. (4.44) as the state of mass crossing each control surface during time can be also expressed as:

$$\int_0^t \left(\sum \dot{m}_{out} \right) dt = \sum m_{out} \tag{4.46}$$

$$\int_0^t \left(\sum \dot{m}_{in} \right) dt = \sum m_{in} \tag{4.47}$$

Thus, we can rearrange Eq. (4.40) using Eqs. (4.45), (4.46) and (4.47) as the uniform-state uniform-flow (USUF) process obtained at each snapshot of time:

$$\frac{d}{dt}\left[m_t\left(u_t + \frac{1}{2}\vec{V}_t^{\,2} + gz_t\right) - m_0\left(u_0 + \frac{1}{2}\vec{V}_0^{\,2} + gz_0\right)\right]_{C.V.} = \dot{Q}_{C.V.} - \dot{W}_{C.V.}$$
$$+ \sum_{i=1}^{n}\dot{m}_{\text{in}-i}\left(h_{\text{in}-i} + \frac{1}{2}\vec{V}_{\text{in}-i}^{2} + gz_{\text{in}-i}\right) - \sum_{j=1}^{m}\dot{m}_{\text{out}-j}\left(h_{\text{out}-j} + \frac{1}{2}\vec{V}_{\text{out}-j}^{2} + gz_{\text{out}-j}\right)$$

(4.48)

Example 4.7

Consider the hall room introduced in Example 4.6. A mechanical ventilation system is installed to provide dry air of $20\,^{\circ}C$ to the room with a velocity of 0.5 m/s through a diffuser with the area of 0.01 m^2 as shown in Figure 4.E7. Calculate the pressure and temperature of the room after 10 minutes. Neglect the moisture content in the room.

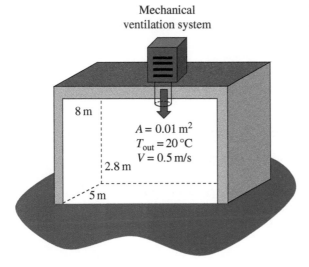

Mechanical ventilation system

$A = 0.01$ m^2
$T_{\text{out}} = 20\,^{\circ}C$
$V = 0.5$ m/s

8 m

2.8 m

5 m

Figure 4.E7 Well-insulated case study hall room of Example 4.5 with a mechanical ventilation system.

Solution

First, we simplify Eq. (4.48) by assuming no work and heat transfer on the control volume as well as neglecting the kinetic and potential energies:

$$\frac{d}{dt}[m_t(u_t) - m_0(u_0)]_{C.V.} = \dot{m}_{\text{in}}h_{\text{in}} - \dot{m}_{\text{out}}h_{\text{out}}$$

Air is considered as an ideal gas while there is no air leaving the control volume. Thus, we can further simplify the first law of thermodynamics as:

$$\frac{m_2 u_2 - m_1 u_1}{dt} = \dot{m}_{\text{in}}C_p T_{\text{in}}$$

Now, finding density at $20\,^{\circ}C$, we can calculate $\dot{m}_{\text{in}}$ as:

$$\dot{m}_{\text{in}} = \rho\vec{V}A_{\text{in}} = 1.205 \times 0.5 \times 0.01 = 0.006 \text{ kg/s}$$
$$\Longrightarrow \dot{m}_{\text{in}}h_{\text{in}} = \dot{m}_{\text{in}}C_p T_{\text{in}} = 1.780 \text{ kJ/s}$$

Note that instead of approximating enthalpy, its value from Table 4.E2 could be found as $h_{in} = 293.55$ kJ/kg. Also, from Table 4.E2 for $T_{in} = 20°$C, we can find $u_1 = 209.88$ kJ/kg. Now, mass of the room at the initial state with $T_1 = 21°$C can be found as:

$$m_1 = \frac{P_1 V}{RT_1} = \frac{101.325 \times 112}{0.287 \times 294.15} = 134.43 \text{ kg}$$

This implies that the mass at the final state can be calculated as:

$$m_2 = m_1 + \dot{m}_{in} \Delta t$$

Hence, after 10 min, we have $\Delta t = 600$ s:

$$m_2 = 134.43 + 0.006 \times 600 = 138.04 \text{ kg}$$

$$\Longrightarrow u_2 = \frac{\dot{m}_{in} C_p T_{in} dt + m_1 u_1}{m_2} = \frac{1.780 \times 600 + 134.43 \times 209.88}{138.04} = 212.07 \text{ kJ/kg}$$

Now, we refer to Table 4.E2, and for $u_2 = 212.07$ kJ/kg, we find $T_2 = 24.1°$C. Thus, the pressure would be:

$$P_2 = \frac{m_2 R T_2}{V} = \frac{138.04 \times 0.287 \times 297.2}{112} = 105.129 \text{ kPa}$$

As we can see the room is unrealistically pressurized and the temperature is considerably elevated though in the reality the control volume encounters considerable losses as discussed in details in Chapter 8.

4.3 Second Law of Thermodynamics and Entropy

Definition and description of second law of thermodynamics are beyond the scope of this book and interested readers are referred to the fundamental thermodynamics' books such as [1–3]. However, we briefly introduce the entropy. The definition of the second law is based on the below *inequality of Clausius* for any reversible or irreversible thermodynamic cycles:

$$\oint \frac{\delta Q}{T} \leq 0 \tag{4.49}$$

To define the entropy (S) of a thermodynamic system as its property, we can develop the below inequality:

$$dS \geq \frac{\delta Q}{T} \tag{4.50}$$

Thus, we can further simplify this equation to:

$$dS = \frac{\delta Q}{T} + S_{gen} \tag{4.51}$$

$\frac{\delta Q}{T} = 0$ is related to the entropy of a reversible process in a system with a similar δQ and T. And, thus we can define $S_{gen} \geq 0$ as the entropy generation of a process due to irreversibilities occurring in a system. Irreversibility itself in a system can occur due to various processes such as friction due to viscosity in fluids.

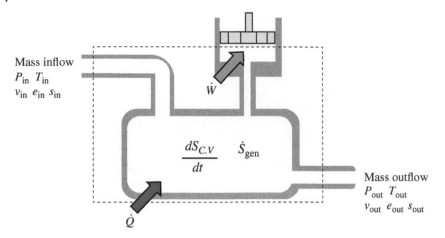

Figure 4.6 Entropy balance in a control volume.

For an example, in the control volume of Figure 4.6, we can further expand Eq. (4.51) as:

$$\frac{dS_{C.V.}}{dt} = \sum \frac{\dot{Q}_{C.V.}}{T} + \dot{S}_{gen} + \sum_{i=1}^{n} \dot{m}_{in-i} s_{in-i} - \sum_{j=1}^{m} \dot{m}_{out-j} s_{out-j}. \tag{4.52}$$

Similarly, we can reach to a SSSF form of Eq. (4.52) when the entropy is not changing at any point of the control volume with time:

$$\sum \frac{\dot{Q}_{C.V.}}{T} + \dot{S}_{gen} = \sum_{i=1}^{n} \dot{m}_{in-i} s_{in-i} - \sum_{j=1}^{m} \dot{m}_{out-j} s_{out-j} \tag{4.53}$$

4.4 Mixture of Ideal Gases

In this section, we study the behaviour of gaseous mixture that can resemble the moisture in buildings. Thus, the fundamental of the mixtures is further scrutinized with the main assumption of the mixture to be an ideal gas. Let us start with the concept of the general mixture with N pure substances as it can be derived as:

$$m_{total} = m_1 + m_2 + ... + m_N = \sum m_i \tag{4.54}$$

$$n_{total} = n_1 + n_2 + ... + n_N = \sum n_i \tag{4.55}$$

where m_i is the mass and n_i is moles of each component. Mass factions (c_i) and mole fractions (y_i) can be also shown as:

$$c_i = \frac{m_i}{m_{total}} \tag{4.56}$$

$$y_i = \frac{n_i}{n_{total}} \tag{4.57}$$

Note that the molecular weight (M_i) can be defined as:

$$M_i = \frac{m_i}{n_i} \tag{4.58}$$

For a mixture, thus, we can find the molecular weight as:

$$M_{\text{mix}} = \frac{m_{\text{total}}}{n_{\text{total}}} = \frac{\sum n_i M_i}{n_{\text{total}}} = \sum y_i M_i \tag{4.59}$$

Dealing with ideal gases, Dalton proposed a model in which pressures of each component separately existing in a same volume and temperature can be superposed in the mixture. This theory is presented in Figure 4.7 when two substances A and B coexist with the **partial pressures** of P_A and P_B, and in a same volume of V and temperature of T:

$$PV = n\overline{R}T \Longrightarrow \begin{matrix} P_A V = n_A \overline{R} T \\ P_B V = n_B \overline{R} T \end{matrix} \tag{4.60}$$

We know from Eq. (4.55) that $n = n_A + n_B$, thus we can rewrite Eq. (4.60) as:

$$\frac{PV}{\overline{R}T} = \frac{P_A V}{\overline{R}T} + \frac{P_B V}{\overline{R}T}$$
$$\Longrightarrow P = P_A + P_B \tag{4.61}$$

Further to Dalton's model, Amagat's model proposes that a superposition can be performed over ideal gases, which coexist in the same temperature of T and pressure of P while in different volumes. For example, two substances A and B exist with the volumes of V_A and V_B, and in a same volume of V and temperate of T:

$$PV = n\overline{R}T \Longrightarrow \begin{matrix} PV_A = n_A \overline{R} T \\ PV_B = n_B \overline{R} T \end{matrix} \tag{4.62}$$

Again, assuming $n = n_A + n_B$ from Eq. (4.55), we can reach to:

$$\frac{PV}{\overline{R}T} = \frac{PV_A}{\overline{R}T} + \frac{PV_B}{\overline{R}T}$$
$$\Longrightarrow V = V_A + V_B \tag{4.63}$$

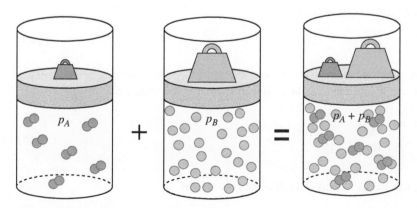

Figure 4.7 Visualization of Dalton model.

Thus, the volume of the mixture with a same temperature and pressure is equal to the sum of the volumes.

4.4.1 Mixture of Air and Vapour

To investigate the air quality and condition in buildings, it is essential to provide a model for the moist air mixture. Since both air and water vapour can be perfectly assumed as ideal gases, we can apply Dalton model to this mixture. Note that the mixture should contain no dissolved gases. Another important assumption is that when equilibrium between liquid and vapour phases are achieved, the impact of other gases on the partial pressure of the vapour is neglected.

4.4.2 Saturated Air, Relative Humidity, and Humidity Ratio

The saturated air–vapour mixture is called **saturated air**. The **relative humidity** (ϕ) is defined as the ratio of mole fraction of the vapour in the mixture (n_v) over the mole fraction of the vapour existing in a saturated mixture (n_g) at the same temperature and pressure:

$$\phi = \frac{n_v}{n_g} \tag{4.64}$$

Note that the mixture is an ideal gas, so the mole fraction can be replaced by partial pressures, or as the ratio of the partial pressure of the mixture (P_v) to the saturation pressure (P_g) in a same temperature:

$$\phi = \frac{n_v}{n_g} = \frac{\left(\dfrac{P_v V}{\overline{R} T}\right)}{\left(\dfrac{P_g V}{\overline{R} T}\right)} = \frac{P_v}{P_g} \tag{4.65}$$

ϕ has therefore a value between 0 and 1. It is also commonly shown in a percentile format. The relative humidity can be also derived in the terms of specific volume:

$$\phi = \frac{P_v}{P_g} = \frac{\rho_v}{\rho_g} = \frac{v_g}{v_v} \tag{4.66}$$

From here, we can now define the **humidity ratio** or **specific humidity** (mass factions) of the air-vapour mixture (ω) as below:

$$\omega = \frac{m_v}{m_{da}} \tag{4.67}$$

where m_v and m_{da} are the mass of water vapour and dry air, respectively. Using the ideal gas assumption, we can again reach to the below equation for the vapour:

$$\frac{P_v V}{\overline{R} T} = n_v = \frac{m_v}{M_v}$$
$$\Longrightarrow m_v = \frac{P_v V M_v}{\overline{R} T} \tag{4.68}$$

Similarly, we can obtain a same equation for the dry air:

$$m_{da} = \frac{P_{da} V M_{da}}{\overline{R} T} \tag{4.69}$$

We can now apply Eqs. (4.68) and (4.69) into Eq. (4.67):

$$\omega = \frac{m_v}{m_{da}} = \frac{\left(\dfrac{P_v V M_v}{\overline{R}T}\right)}{\left(\dfrac{P_{da} V M_{da}}{\overline{R}T}\right)} = \frac{M_v}{M_{da}}\frac{P_v}{P_{da}} \tag{4.70}$$

Knowing the molecular weights of the vapour and air, the relation between the relative humidity and the specific humidity can be found:

$$\omega = 0.622\frac{P_v}{P_{da}}$$

$$\Longrightarrow \omega = \left(0.622\frac{P_g}{P_{da}}\right)\phi \tag{4.71}$$

Example 4.8

Consider an adiabatic condition for the room of Example 4.1, which is in the equilibrium state in the atmosphere pressure and contains water vapour and dry air at a temperature of $15°C$. The humidity ratio is $\omega = 0.008$. What is the mass of the vapour and air?

Solution

At a temperature of $15°C$, we can refer to Table 4.E2 and find:

$$v_g = 77.885\frac{m^3}{kg}$$

$$P_g = 1.706 \text{ kPa}$$

Using Eq. (4.71), we can find the partial pressure of the dry air as:

$$0.008 = 0.622\frac{P_v}{P_{da}}$$

$$\Longrightarrow P_v = 0.0129P_{da}$$

$$\Longrightarrow P_{room} = P_{da} + 0.129P_{da}$$

$$\Longrightarrow P_{da} = \frac{101.325}{1.129} = 100.038 \text{ kPa}$$

$$\Longrightarrow P_v = 101.325 - 100.038 = 1.287 \text{ kPa}$$

Hence, for Eq. (4.66), we can find the relative humidity as:

$$\phi = \frac{P_v}{P_g} = \frac{1.287}{1.706} = 0.754 \ (\cong 75\%)$$

And, also masses can be found:

$$m_{da} = \frac{P_{da}V}{R_{da}T} = \frac{100.038 \times 33.6}{0.287 \times 288.15} = 40.64 \text{ kg}$$

$$m_v = \omega m_{da} = 0.008 \times 40.64 = 0.33 \text{ kg}$$

4.4.3 Dew Point, Dry-Bulb, and Wet-Bulb

The **dew point** of a gas–vapour mixture is the temperature in which the vapour condenses or directly solidifies in a constant-pressure cooling process. The **dry-bulb temperature** is known as the free air temperature measured by a thermometer. The dry-bulb temperature is normally shielded from radiation and moisture. The **wet-bulb temperature** can be defined as the temperature of the air cooled to the saturation by the evaporation of water, which gains its latent heat from the passing air. The device, which provides this mechanism is called **psychrometer**, consisting of a thermometer covered in water-soaked cloth over which air is passing (see Figure 4.8). With this steady flow process, the wet-bulb temperature can be measured, which becomes equal to the dry-bulb air temperature if the relative humidity is 100%. It should be noted that the wet-bulb temperature is the lowest feasible temperature related to the ambient condition caused by the evaporation of water, and therefore, a suitable indicator of measuring moisture of the ambient condition.

More advanced devices are currently used to measure the wet-bulb temperature using other physical concepts such as change in length or electricity capacitance of certain materials when exposed to moisture.

4.4.4 Psychrometric Chart

As we understood that the air–water vapour mixture can be represented with two indicators of dry-bulb and wet-bulb temperatures, we can plot this relation in a chart known as **psychrometric chart** as illustrated in Figure 4.9.

As it can be seen in Figure 4.10, the psychrometric chart is a plot of the humidity ratio as a function of the dry-bulb temperature, the relative humidity, wet-bulb temperature, and enthalpy per mass of the dry-bulb. Lines of constant relative humidity and wet-bulb temperature, enthalpy per mass of dry-bulb, and humidity ratio are clearly sketched in the psychrometric chart. Thus, a certain point in the psychrometric chart provides multiple information associated to these parameters.

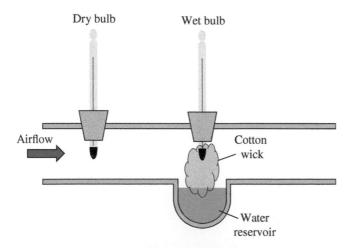

Figure 4.8 Psychrometer for measurement of the wet-bulb temperature.

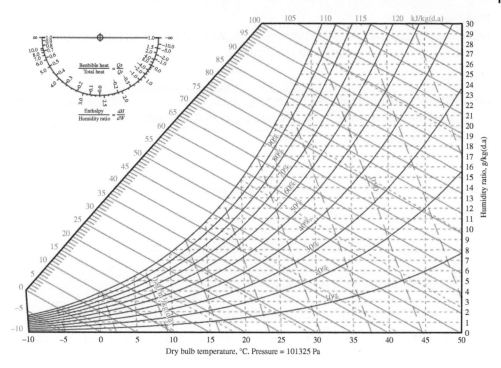

Figure 4.9 Psychrometric chart. *Source:* Wylen [1]/ with permission of John Wiley & Sons.

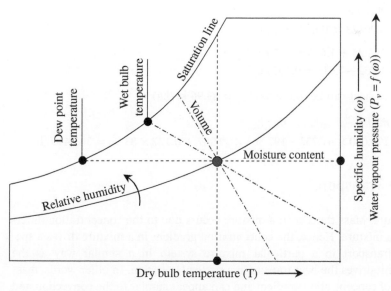

Figure 4.10 Different lines in psychrometric chart.

Example 4.9

Recalculate Example 4.6 using the psychrometric chart if the entering and leaving moist air have 65% and 60% relative humidity, respectively.

Solution

From the psychrometric chart (see Figure 4.9), we can find moist air enthalpies. At the inlet for air with the properties of $T_{in} = 19°C$ and $\phi_{in} = 65\%$, we find:

$$h_{in} = 41.752 \, \text{kJ/kg}$$

$$\omega_{in} = 0.00893$$

We can find the mass flow rate of the entering moist air as:

$$\rho_a Q_{in} = \dot{m}_{da_{in}} + \omega_{in} \dot{m}_{da_{in}}$$

$$\Longrightarrow \dot{m}_{da_{in}} = \frac{\rho_a Q_{in}}{(1 + \omega_{in})} = \frac{1.2 \times 165}{1 + 0.00893} = 196.25 \, \text{kg/h}$$

$$\Longrightarrow \dot{m}_{v_{in}} = 0.00893 \times 196.25 = 1.752 \, \text{kg/h}$$

$$\Longrightarrow \dot{m}_{in} = 196.25 + 1.752 = 198.000 \, \text{kg/h}$$

We can again find the below values for the leaving air at $T_{out} = 19°C$ and $\omega_{out} = 60\%$:

$$h_{out} = 39.987 \, \text{kJ/kg}$$

$$\omega_{out} = 0.00823$$

Since $\dot{m}_{da_{in}} = \dot{m}_{da_{out}}$, we can find the mass of the leaving vapour as:

$$\omega_{out} = \frac{\dot{m}_{v\_out}}{\dot{m}_{da\_out}}$$

$$\Longrightarrow \dot{m}_{v\_out} = 0.00823 \times 196.25 = 1.615 \, \text{kg/h}$$

$$\Longrightarrow \dot{m}_{out} = 196.25 + 1.615 = 197.863 \, \text{kg/h}$$

$$\Longrightarrow \dot{m}_{l_{out}} = 1.752 - 1.615 = 0.137 \, \text{kg/h}$$

Thus, the extracted heat can be found as ($h_{l_{out}} = 83.96 \, \text{kJ/kg}$):

$$\dot{Q}_{C.V.} = \dot{m}_{in} h_{in} - \dot{m}_{out} h_{out} - \dot{m}_{l_{out}} h_{l_{out}}$$

$$\Longrightarrow \dot{Q}_{C.V.} = 198.000 \times 41.752 - 197.863 \times 39.987 + 0.137 \times 83.96 = 343.4 \, \text{kJ/h}$$

4.5 Moisture Transport

Moist air is a mixture. Mass transfer in a mixture occurs due to the concentration difference of species in a mixture. Hence, the concentration gradient in a mixture drives a species (vapour) to transport in a particular mixture system in a similar way as the temperature gradient drives the heat transfer in a particular system. In other words, mass transfer is a result of concentration gradient and can appear similar to the convection and conduction heat transfer mechanisms. Figure 4.11 depicts the convective mass transfer mechanism due to a bulk fluid motion as a result of a concentration gradient as well

Figure 4.11 Convective and diffusive mass transfer mechanisms due to the concentration gradient.

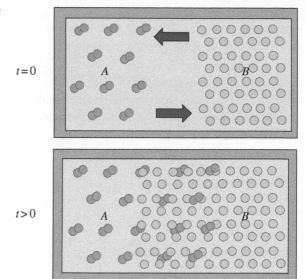

as the diffusive mass transfer mechanism in which the local diffusion of molecules is the acting force.

4.5.1 Mixing

A very classical experiment to understand the mixing mechanism of two different species at a same temperature and pressure, but in two different concentrations, is shown in Figure 4.11. Once the barrier between two species is removed in a sense that a flow motion is not imposed to the fluid, both species start to diffuse towards each other. This implies molecules with higher concentrations from a side of the chamber transport to the opposite side. The process in a sufficient time reaches an equilibrium with a constant concentration regarded to both species at any point of the chamber.

4.5.2 Mass Diffusion Mechanism

It is noteworthy to mention that mass diffusion is highly related to the molecular spacing and thus it is more noticeable in gases in comparison to the liquids and of course solid materials. While the heat transfer via diffusion mechanism is formulized by Fourier equation (see Chapter 6), mass transfer is explained by **Fick's law**, in which species-A in a binary mixture of A and B can be expressed as:

$$j_A = -\rho D_{AB} \nabla c_A = -\rho D_{AB} \nabla \left(\frac{\rho_A}{\rho}\right) \tag{4.72}$$

where j_A is the **diffusive mass flux** of species-A, c_A is the mass fraction, D_{AB} is the mass diffusivity or binary diffusion coefficient, $\rho \approx \rho_A + \rho_B$ is the density of the binary mixture. Another form of the Fick's law can be also driven based on the molar concentration as below:

$$J_A^* = -C D_{AB} \nabla y_A = -C D_{AB} \nabla \left(\frac{C_A}{C}\right) \tag{4.73}$$

where J_A^* is the diffusive molar of species-A and $C \approx C_A + C_B$ is the total molar concentration of the mixture.

Obtaining D_{AB} is not as straight forward as finding of material's conductivity and the related experiments faces various difficulties and limitations. For example, the below equation is being widely used for the ideal gases:

$$D_{AB} \approx P^{-1}T^{3/2} \tag{4.74}$$

Or, as another example for a binary mixture of dry air and vapour (moist air), we can employ the following equation (P is the atmosphere pressure) [4]:

$$D_{AB} \approx 2.26\, P^{-1}\left(\frac{T}{273.15}\right)^{1.81} \tag{4.75}$$

In a binary mixture, we can expect that the summation of both diffusive mass transfer is zero:

$$j_A + j_B = 0 \tag{4.76}$$

We know that $c_A + c_B = 1$, thus $\nabla c_A = -\nabla c_B$ and we can apply Eq. (4.72) into it to further reach to:

$$(-\rho D_{AB}\nabla c_A) + (-\rho D_{BA}\nabla c_B) = 0$$
$$\Longrightarrow D_{AB} = D_{BA} \tag{4.77}$$

Example 4.10
Calculate the mass diffusivity of an air–vapour mixture at the atmosphere pressure and temperature of $T_1 = 20\,^\circ C$.

Solution

From Eq. (4.75), we can find the binary diffusion coefficient as:

$$D_{AB} \approx 2.26\, P^{-1}\left(\frac{T}{273.15}\right)^{1.81} = \frac{2.26}{101325} \times \left(\frac{293.15}{273.15}\right)^{1.81} = 2.53 \times 10^{-5}\ \text{m}^2/\text{s}$$

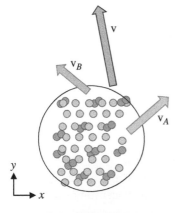

Figure 4.12 Diffusive and advective concentration fluxes in a convective mass transfer.

4.5.3 Mass Convection

In addition to the diffusive mass transfer, the concentration gradient can occur due to the bulk motion of the fluid known as the mass advection. The combination of both mass transfer components is known together as the convective mass transfer (**g**). This implies that while the diffusive mass transfer involves molecular movements of species, if these small motions cause a larger scale movement in the species particles, then we have the **convective mass transfer** or in total the **absolute flux** (diffusive and advective) of the species (see Figure 4.12).

As shown in Figure 4.12, to develop the mass convective diffusion equation, we need to consider a binary species mixture in which the average velocity of all particles of the species-A related to a fixed right-angled coordinate is defined as v_A. So, the absolute mass flux of the species-A can be defined as:

$$g_A = \rho_A v_A \tag{4.78}$$

where g_A is the absolute mass flux of the species-A.

We can now take a similar approach and write the absolute mass flux of the species-B as:

$$g_B = \rho_B v_B \tag{4.79}$$

And, finally the mass-averaged velocity of the binary mixture can be written as:

$$\rho v = \rho_A v_A + \rho_B v_B$$
$$\Longrightarrow v = \frac{g_A + g_B}{\rho} \tag{4.80}$$

Now, we can insert Eqs. (4.78) and (4.79) into Eq. (4.80):

$$v = c_A v_A + c_B v_B \tag{4.81}$$

Note that the velocity of the species-A relative to the mixture mass-averaged velocity is $v_A - v$ and thus we can evidently make the conclusion that the relative mass flux of the species-A is equal to the diffusive mass transfer of the species-A:

$$j_A = \rho_A(v_A - v) \tag{4.82}$$

Now, from Eq. (4.78), we replace $g_A = \rho_A v_A$ into Eq. (4.82):

$$g_A = \rho_A v + j_A \tag{4.83}$$

This equation, therefore, explains the absolute mass transfer of the species-A is equal to a contribution due to the advection of the species-A ($\rho_A v$) and a contribution due to the diffusive mass transfer (j_A). Eq. (4.83) can be further rearranged using Eq. (4.72) as:

$$g_A = \rho_A v - \rho D_{AB} \nabla \frac{\rho_A}{\rho} \tag{4.84}$$

Or, by using Eq. (4.81), we can show Eq. (4.84) as:

$$g_A = \rho_A(c_A v_A + c_B v_B) - \rho D_{AB} \nabla c_A$$
$$\Longrightarrow g_A = \frac{\rho_A}{\rho}(\rho_A v_A + \rho_B v_B) - \rho D_{AB} \nabla c_A \tag{4.85}$$

One can further simplify this Eq. (4.84) in terms of mass transfers of both species:

$$g_A = c_A(g_A + g_B) - \rho D_{AB} \nabla c_A \tag{4.86}$$

Thus, we can send g_A to the left-hand-side of Eq. (4.85) to achieve the below equation:

$$g_A = \frac{c_A g_B - \rho D_{AB} \nabla c_A}{(1 - c_A)} = \frac{c_A g_B - \rho D_{AB} \nabla c_A}{c_B} \tag{4.87}$$

With a similar approach for the species-B, we can obtain:

$$g_B = \frac{c_B g_A - \rho D_{AB} \nabla c_B}{c_A} \tag{4.88}$$

From above equations, we can again simplify the case to which the mass transfer is purely diffusive when the mixture is in the stagnation ($\mathbf{v} = 0$).

4.5.4 Conservation of Mass

Mass transfer is following the conservation laws also known as **conservation of species**. In this section, hence, we develop the equation of the conservation of mass for the stationary medium as it has relevant applications in buildings as we investigate them in the next chapter. For this purpose, similar to any conservation law, we define a control volume and derive the conservation of species (M) similar to other conservation of laws:

$$\dot{M}_{in} - \dot{M}_{out} + \dot{M}_{gen} = \frac{dM}{dt} \tag{4.89}$$

where $\dot{M}_{in}$ and $\dot{M}_{out}$ are the entering and leaving masses of a particular species via diffusion and convection through the surfaces of the control volume. $\dot{M}_{gen}$ represents the generated mass.

In the case of a stationary medium with a binary mixture, we can assume that the advection is absent and therefore rewrite Eq. (4.89) using analogy with heat diffusion fluxes:

$$g_A dydz = g_A dydz + \frac{\partial(g_A dydz)}{\partial x} dx \tag{4.90}$$

$$g_A dxdz = g_A dxdz + \frac{\partial(g_A dxdz)}{\partial y} dy \tag{4.91}$$

$$g_A dxdy = g_A dxdy + \frac{\partial(g_A dxdy)}{\partial z} dz \tag{4.92}$$

If $\dot{g}_A$ is the rate of increase of mass of the species-A per unit volume of the binary mixture, then we have:

$$\dot{M}_{gen} = \dot{g}_A dxdydz \tag{4.93}$$

Also, the rate of change of the species-A within the control volume can be represented as:

$$\frac{dM_A}{dt} = \frac{\partial \rho_A}{\partial t} dxdydz \tag{4.94}$$

Thus, after summing all the associated equations and dividing both sides by $dxdydz$, we can reach to:

$$-\frac{\partial g_A}{\partial x} - \frac{\partial g_A}{\partial y} - \frac{\partial g_A}{\partial z} + \dot{g}_A = \frac{\partial \rho_A}{\partial t} \tag{4.95}$$

$$-\nabla \cdot g_A + \dot{g}_A = \frac{\partial \rho_A}{\partial t}$$

Replacing Eq. (4.84) into Eq. (4.95), when ρ and D_{AB} are constant, we can obtain ($\mathbf{v} = 0$):

$$\nabla^2 \rho_A + \frac{\dot{g}_A}{D_{AB}} = \frac{1}{D_{AB}} \frac{\partial \rho_A}{\partial t} \tag{4.96}$$

References

1 Wylen, S.-B.-V. (1998). *Fundamentals of Thermodynamics*, 5e. Wiley.
2 Moran MJ, Shapiro HN, Boettner DD, Bailey MB. *Fundamentals of Engineering Thermodynamics*: Wiley; 2010.
3 Çengel, Y.A., Boles, M.A., and Kanoğlu, M. (2015). *Thermodynamics: An Engineering Approach. Eighth edition in SI units.* New York: McGraw-Hill Education.
4 Sandall, O.C. (1973). *Transfer Processes* (ed. D.K. Edwards, V.E. Denny, A.F. Mills and R. Holt), 361. New York: Winston, Inc. $15.00. AIChE Journal. 1974;20(2):414.

5

Applications of Thermodynamics in Buildings

5 Introduction

5.1 Human Thermal Comfort

Thermoreceptors on human skins throughout the nervous impulses send the message to hypothalamus in a human brain as shown in Figure 5.1. Hence, in a physiological prospect, **thermal comfort** is defined as the minimum level of signals sent from thermoreceptors. On the other hand, the **thermal sensation** of an environmental temperature are arisen by thermoreceptor sensors of a body in conjunction to the temporal variation of the body's core and skin temperatures [1].

A body can act to adapt itself to a complex thermoregulatory process to keep the body's core temperature in an almost same temperature of 37°C necessary for its physiological functionality.

Body thermoregulation uses four autonomic control mechanisms, including vasoconstriction with narrowing of blood vessels, vasodilatation via widening of the blood vessels, sweating by increase of skin evapotranspiration, and shivering by increase of metabolism rate [1]. Psychological adaption can occur in a long-term sense as a matter of generations through genetic adoption or as a result of short-term acclimatization as a matter of hours in accordance with many environmental parameters.

How these sensory temperature signals on human skins (see Figure 5.1) are translated in the brain through time and expectation can create a **thermal perception**, which is related to a diverse range of environmental factors, known as **thermophysical factors** as well as **psychological factors**. **Thermal comfort** or **thermal acceptability**, thus, can be defined as a state in which mind expresses satisfaction with the thermal environment according to ASHRAE Standard 55 [3]. In other words, reaching to a level of **perception of Thermal comfort** implies an acceptable range by the human body – brain.

Many efforts have been conducted to design scales and measures to evaluate and gauge the thermal sensation and thermal comfort in the past decades, which are beyond the scope of this book and interested readers are suggested to refer the provided reference, e.g. [4, 5]. Nonetheless, the essence of many of these developed models to measure the thermal comfort are based on the first law of thermodynamics (see Chapter 4) applied over a control volume of human body. The heat balance between a human body is related to the exchange

Computational Fluid Dynamics and Energy Modelling in Buildings: Fundamentals and Applications,
First Edition. Parham A. Mirzaei.
© 2023 John Wiley & Sons Ltd. Published 2023 by John Wiley & Sons Ltd.

of convective, radiative, evaporative, and conductive heat transfer between multiple parts of a human body and environment as illustrated in Figure 5.2. The generated heat to maintain the constant core temperature occurs due to the metabolism in addition to conscious and unconscious muscle activities.

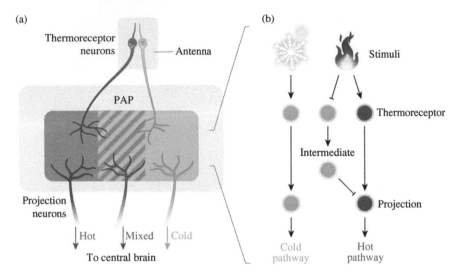

Figure 5.1 Thermoreceptors on human skins. *Source:* Florence and Reiser [2]/with permission of Springer Nature.

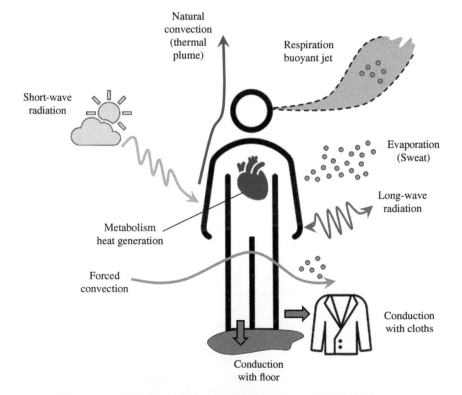

Figure 5.2 Energy balance over the control volume of a human body

An early and fundamental model related to the thermal comfort was proposed by Fanger [6, 7]:

$$PPD = 100 - 95^{\left(-0.03353PMV^4 - 0.2179PMV^2\right)} \tag{5.1}$$

where, the predicted mean vote for thermal comfort (PMV) was calculated from the heat balance within a control volume defined over a human body. Various studies have been conducted to improve, adjust, and criticize PMV while the revised PPD in Eq. (5.1) also have been offered by many researchers [4]. In many of these studies, four environmental parameters have been recognized as dominant factors, including air temperature, relative humidity, mean radiant temperature, and ambient water vapour pressure. Besides, metabolic activity and clothing levels of an individual are found to be significant contributors in thermoregulatory response of human body and thus in thermal comfort.

5.2 Thermal Comfort Measures in Building

It is important to design mechanical heating, ventilation, and air conditioning (HVAC) systems or natural ventilation strategies to provide thermal comfort in buildings. For this purpose, several standards and guidelines have been developed to aid the design, control and maintenance of thermal comfort in buildings. For example, psychrometric chart introduced in Chapter 4 is utilized in ASHRAE Standard 55 to offer a thermal comfort zone in terms of range of operative temperatures for 80% acceptability (see Figure 5.3).

5.3 Thermodynamic Processes in Air-Conditioning Systems

Air–vapour mixture is assumed as an ideal gas. Thus, based on the mixture definition of Dalton's law in the previous chapter, one can treat each component of the mixture separately and identify its partial pressure and thermodynamic properties such as internal energy and enthalpy. We have also explored that the thermodynamic state of a mixture can be embedded in the psychrometric chart. In this section, we will focus on the standard processes in the HVAC designs. These processes are fundamentals to all HVAC systems in controlling and maintaining of a room condition to ensure thermal comfort. Whilst we envisage various processes in this chapter, it is noteworthy to mention that the aim of this chapter is not to focus on the design and control of HVAC system, and interested readers can refer to a diverse, related list of practical resources, e.g. [3, 8, 9].

A variety of HVAC systems is used to change the thermodynamic state of an air–vapour mixture in terms of dry- and/or wet-bulb temperatures. For the purpose of increasing or decreasing a dry-bulb temperature (sensible heating or cooling), heating and cooling processes will be applied to the mixture (e.g. passing the moist air over coils). On the psychrometric chart, as shown in Figure 5.4, these processes are shown by parallel lines with the dry-bulb axes while a heating process would increase the moist air's temperature and a cooling process would decrease it. On the other hand, humidification, or dehumidification processes (e.g. passing the moist air through water sprays or desiccants) are used to increase or decrease the wet-bulb temperature of an air–vapour mixture. In Figure 5.4, these processes

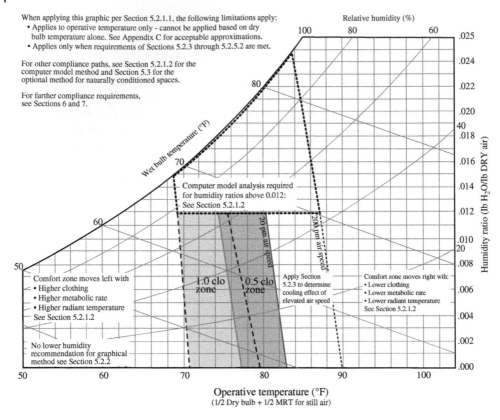

Figure 5.3 Comfort zone in winter and summer seasons. *Source:* ASHRAE [3]/ASHRAE.

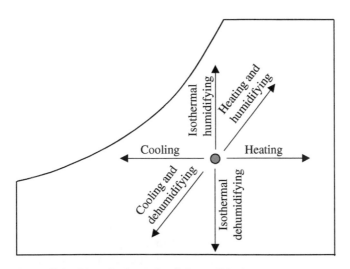

Figure 5.4 General schematic of air conditioning process.

are shown by parallel lines to the vertical axes of the humidity ratio. Again, humidification means an elevation in the humidity ratio or relative humidity while the dehumidification implies a decline in the humidity ratio. We can also see heating and humidification and cooling and dehumidification processes in Figure 5.4 as these processes respectively increase or reduce both dry- and wet-bulb temperatures of the moist air. We can also identify other processes such as adiabatic cooling. In this section, we briefly overview these common existing processes in more details.

5.3.1 Adiabatic Saturation

One of the traditional cooling mechanism in buildings is adiabatic saturation where an external heat extraction from the conditioning system does not occur, assuming that the system is well insulated and has a long enough evaporation section, and instead the evaporative humidification process of the vapour–air mixture can be assumed as a adiabatic one. The supplied air to a conditioned room at the outlet of the conditioning system can be assumed as a saturated moist air with a relative humidity of 100%. As shown in Figure 5.5, the vapour–air mixture is cooled and saturated against a supplied water at a temperature of T_2, and thus the outlet mixture is a saturated one with a similar temperature of $T_3 = T_2$.

If we assume the adiabatic saturation as a SSSF process, then we can derive the first law of thermodynamics for the control volume similar to Eq. (4.43). As shown in Figure 5.5, where there is no work done on the system and the process is adiabatic ($\dot{Q}_{C.V.} = 0$). We also neglect the change in kinetic and potential energies and simplify Eq. (4.43) as:

$$(\dot{m}_{a1}h_{a1} + \dot{m}_{v1}h_{v1}) + (\dot{m}_{l2}h_{l2}) - (\dot{m}_{a3}h_{a3} + \dot{m}_{v3}h_{v3}) = 0 \tag{5.2}$$

From the conservation of mass law in the system, we can derive the below equation from Eq. (4.42) for the dry air:

$$\dot{m}_{a1} = \dot{m}_{a3} = \dot{m}_a \tag{5.3}$$

And, thus, for the vapour and liquid water, the conservation of mass states that:

$$\dot{m}_{v1} + \dot{m}_{l2} = \dot{m}_{v3} \tag{5.4}$$

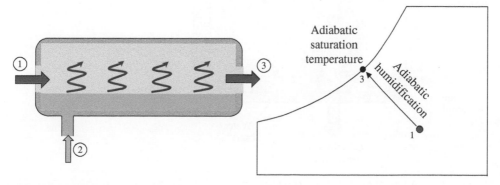

Figure 5.5 Adiabatic saturation or evaporative cooling process.

From the definition of $\omega_1 = \dot{m}_{v1}/\dot{m}_a$ and $\omega_3 = \dot{m}_{v3}/\dot{m}_a$, we can now rewrite Eq. (5.2) as:

$$\left(h_{a1} + \frac{\dot{m}_{v1}}{\dot{m}_a}h_{v1}\right) + \left(\frac{(\dot{m}_{v3} - \dot{m}_{v1})}{\dot{m}_a}h_{l2}\right) - \left(h_{a3} + \frac{\dot{m}_{v3}}{\dot{m}_a}h_{v3}\right) = 0$$

$$\Longrightarrow (h_{a1} + \omega_1 h_{v1}) + ((\omega_3 - \omega_1)h_{l2}) = (h_{a3} + \omega_3 h_{v3})$$

$$\Longrightarrow \omega_1(h_{v1} - h_{l2}) + \omega_3(h_{l2} - h_{v3}) = (h_{a3} - h_{a1})$$

(5.5)

With assuming the dry air as an ideal gas, we can derive the below equation:

$$h_{a3} - h_{a1} = C_{pa}(T_3 - T_1)$$

(5.6)

Moreover, replacing $h_{v3} = h_{v2}$ as $T_3 = T_2$, and since the entering water is a saturated liquid and the leaving vapour is a saturated one, Eq. (5.5) becomes:

$$\omega_1 = \frac{C_{pa}(T_3 - T_1) + \omega_3\left(h_{fg3}\right)}{(h_{v1} - h_{l3})}$$

(5.7)

Example 5.1

In an adiabatic saturation process in an atmospheric pressure, hot air with a dry-bulb temperature of $T_1 = 31°C$ enters the evaporative cooling system with a volume flow rate of 2000 m³/h as depicted in Figure 5.E1.1. If the water at $T_2 = 19°C$ is used in the system, calculate the relative humidity of the moist vapour entering the system.

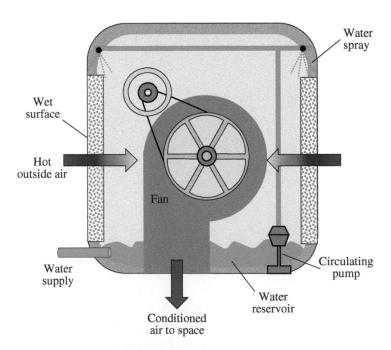

Figure 5.E1.1 A typical evaporative cooling system.

Solution

For the entering air of $T_1 = 31°C$ and leaving air of $T_3 = T_2 = 19°C$ and $\phi_3 = 100\%$, we can either use the saturation table of water or psychrometric chart to find properties of the leaving vapour. From the table, we find p_{v3}:

$$P_{v3} = \phi_3 P_{g3} = 1 \times 2.112 = 2.212 \, \text{kPa}$$

$$\Longrightarrow P_{a3} = 101.325 - 2.12 = 99.113 \, \text{kPa}$$

$$\Longrightarrow \omega_3 = 0.622 \times \frac{2.212}{99.113} = 0.01388$$

And, then from Eq. (5.7), we can find ω_1 as ($C_{pa} = 1.005 \, \text{kJ/kg K}$):

$$\omega_1 = \frac{C_{pa}(T_3 - T_1) + \omega_3 \left(h_{fg3} \right)}{\left(h_{v1} - h_{l3} \right)} = \frac{1.005 \times (19 - 31) + 0.01388 \times 2456.48}{(2558.06 - 79.75)} = 0.00889$$

Now, we can find the vapour partial pressure as ($p_{g1} = 2.212$):

$$\omega_1 = \left(0.622 \frac{P_{v1}}{101.325 - P_{v1}} \right)$$

$$\Longrightarrow P_{v1} = 1.410 \, \text{kPa}$$

$$\Longrightarrow \phi_1 = \frac{P_{v1}}{P_{g1}} = 0.64$$

From psychrometric chart, we can find quite similar values for both entering and leaving air (see Chapter 4, Figure 4.9). First, we identify point-3 as a point with $\phi_3 = 100\%$ at the chart as shown in Figure 5.E1.2. Then, we draw a line parallel to the dry-bulb axes, which intersects a line parallel to the humidity ratio axes starting from $T_1 = 31°C$. The intersection gives the point of interest (point-1).

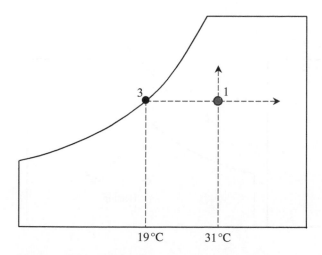

Figure 5.E1.2 Identification of the properties of the entering moist air using psychrometric chart.

5.3.2 Cooling and Heating

Cooling and heating typically occur by directing air though coils installed in HVAC systems. For the sensible cooling purposes, the refrigerant fluid inside the coils is extracting heat from the moist air inside a heat exchanger. Thus, the dry-bulb temperature of the air–vapour mixture drops when it passes through the cool coils as shown in Figure 5.6. It should be noted that if the mixture reaches to a temperature below its dewpoint temperature, then the condensation occurs. Therefore, in the cooling process, the humidity ratio remains constant and the relative humidity increases. It should be mentioned that we will focus on the heat transfer mechanism in exchangers in the following chapters, but here we only follow the thermodynamic process on the control volume of the HVAC system, which only includes the moist air.

Inverse to the sensible cooling, the vapour–air mixture becomes heated up in a heat exchanger during a sensible heating process where the coils are heated by a working fluid or electricity. Figure 5.6 demonstrates a schematic sensible process in which the dry-bulb temperature is increased. Inverse to the sensible cooling process, the relative humidity decreases while the humidity ratio again remains constant as there is no mechanism such as a water spray in the system to control the humidity.

Again, both cooling and heating processes can be presumed to be SSSF while the coils, acting as a heat source or sink, implying a heat generation of $\dot{q}$. We again derive the first law of thermodynamics using Eq. (4.43) as we neglect the change in the kinetic and potential energies (see Figure 5.7):

$$(\dot{m}_{a1}h_{a1} + \dot{m}_{v1}h_{v1}) \mp \dot{Q} - (\dot{m}_{a2}h_{a2} + \dot{m}_{v2}h_{v2}) = 0 \tag{5.8}$$

Note that a negative sign in Eq. (5.8) stands for a cooling process and a positive sign for a heating one. And, the mass balance for both vapour and air (see Eq. 4.42) becomes:

$$\dot{m}_{a1} = \dot{m}_{a2} = \dot{m}_a \tag{5.9}$$

$$\dot{m}_{v1} = \dot{m}_{v2} = \dot{m}_v \tag{5.10}$$

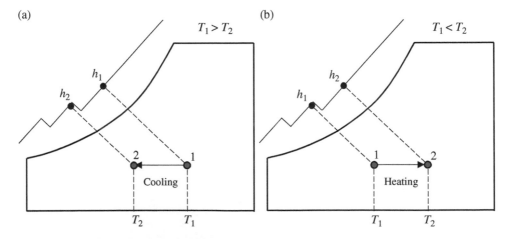

Figure 5.6 Sensible (a) cooling and (b) heating processes.

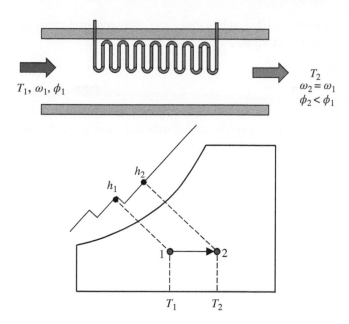

Figure 5.7 Schematic of sensible heating process.

So, we can insert Eqs. (5.9) and (5.10) into Eq. (5.8) to achieve:

$$\left(h_{a1} + \left(\frac{\dot{m}_v}{\dot{m}_a}\right)h_{v1}\right) \mp \dot{q} - \left(h_{a2} + \left(\frac{\dot{m}_v}{\dot{m}_a}\right)h_{v2}\right) = 0$$

$$\Longrightarrow \mp \dot{q} = C_{pa}(T_2 - T_1) + \omega C_{pv}(T_2 - T_1) \tag{5.11}$$

$$\Longrightarrow \mp \dot{q} = \left(C_{pa} + \omega C_{pv}\right)(T_2 - T_1) \approx C_p(T_2 - T_1)$$

Example 5.2

What is the required heat to be extracted though the coils of a chiller to cool down an air initially at $T_1 = 31°C$ and $\phi_1 = 30\%$ to a dry-bulb temperature of 19 °C if the dry-air volume flow rate of the system is 2000 m^3/h.

Solution

We can use the psychrometric chart or saturated water tables to find the humidity ratio as ($C_{pa} = 1.005$ kJ/kg K and $C_{pv} = 1.86$ kJ/kg K):

$$\omega_1 = \omega_2 = 0.00842$$
$$C_{pa} + \omega C_{pv} = 1.005 + 0.00842 \times 1.86 = 1.021 \text{ kJ/kg K}$$

Now, we apply these values into Eq. (5.11):

$$\dot{q} = C_p(T_2 - T_1) = 1.021 \times (19 - 31) = -12.25 \text{ kJ/kg}$$
$$\Longrightarrow \dot{Q} = m_a \dot{q} = \rho Q_a \dot{q} = -1.2 \times 2000 \times 12.25 = -29,383.73 \text{ kJ/h}$$

Another way is to directly extract enthalpy values from the psychrometric chart as $h_1 = 52.724$ kJ/kg(d.a) and $h_2 = 40.471$ kJ/kg(d.a). Hence, we can find:

$$\dot{Q} = m_{\mathrm{a}}(h_2 - h_1) = 29,407.20 \text{ kJ/h}$$

5.3.3 Heating and Humidification

The heating and humidification process employs both heating coils and water sprays (pools) to increase the dry- and wet-bulb temperatures of the vapour–air mixture. The schematic of the process is shown in Figure 5.8. Again, the pathway to reach Point-3 form Point-1 can be conducted with different strategies, depending on the heating and humidification elements used in a HVAC system.

Similar to the previous processes, we can apply the first law of the thermodynamics in addition to the conservation of mass law over the control volume of the flowing moist air to derive the below expression:

$$(\dot{m}_{\mathrm{a1}}h_{\mathrm{a1}} + \dot{m}_{\mathrm{v1}}h_{\mathrm{v1}}) + \dot{Q} + (\dot{m}_{\mathrm{f2}}h_{\mathrm{f2}}) - (\dot{m}_{\mathrm{a3}}h_{\mathrm{a3}} + \dot{m}_{\mathrm{v3}}h_{\mathrm{v3}}) = 0 \tag{5.12}$$

The mass balance can be applied on both dry air and vapour as:

$$\dot{m}_{\mathrm{a1}} = \dot{m}_{\mathrm{a3}} = \dot{m}_{\mathrm{a}} \tag{5.13}$$

$$\dot{m}_{\mathrm{v1}} + \dot{m}_{\mathrm{f2}} = \dot{m}_{\mathrm{v3}} \tag{5.14}$$

Now, using Eqs. (5.13) and (5.14), we can rearrange Eq. (5.12) as:

$$\dot{Q} = C_{\mathrm{pa}}\dot{m}_{\mathrm{a}}(T_3 - T_1) - \dot{m}_{\mathrm{a}}(\omega_3 - \omega_1)h_{\mathrm{f2}} + \dot{m}_{\mathrm{a}}(\omega_3 h_{\mathrm{v3}} - \omega_1 h_{\mathrm{v1}}) \tag{5.15}$$

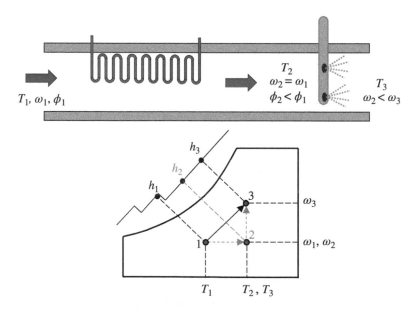

Figure 5.8 Schematic of heating and humidification processes.

Example 5.3

Dry air with the volume flow rate of $Q_a = 0.25$ m³/s in the atmosphere pressure with the relative humidity of $\phi_1 = 20\%$ and dry-bulb temperature of $T_1 = 12°C$ should be conditioned to a comfort range with a temperature of $T_3 = 22°C$ and $\phi_3 = 50\%$. What would be the required heat and water flow rate to achieve the expected air condition in the heating coil and spray system?

Solution

We can find saturation pressures at point-1 and point-2 as $P_{g1} = 1.419$ kPa and $P_{g2} = 2.730$ kPa, respectively. This implies that for Point-1, we can write:

$$P_{v1} = \phi_1 P_{g1} = 0.284 \text{ kPa}$$

$$\Longrightarrow \omega_1 = 0.622 \frac{P_{v1}}{P_{atm} - P_{v1}} = 0.00175$$

$$\Longrightarrow \dot{m}_{a1} = \rho Q_a = \frac{P_{a1}}{RT_1} Q_a = \frac{(P_{atm} - P_{v1})}{RT_1} Q_a = \frac{(101.325 - 0.284)}{0.287 \times 285.15} \times 0.25 = 0.3087 \text{ kg/s}$$

$$\Longrightarrow \dot{m}_{v1} = \dot{m}_{a1}\omega_1 = 0.0005 \text{ kg/s}$$

Also, we can repeat the same procedure for Point-3:

$$P_{v3} = \phi_3 P_{g3} = 1.365 \text{ kPa}$$

$$\Longrightarrow \omega_3 = 0.622 \frac{P_{v3}}{P_{atm} - P_{v3}} = 0.00849$$

From the conservation of mass, we have $\dot{m}_{a1} = \dot{m}_{a3} = \dot{m}_a$:

$$\dot{m}_{v3} = \dot{m}_{a3}\omega_2 = 0.0026 \text{ kg/s}$$

$$\Longrightarrow \dot{m}_{f2} = 0.0021 \text{ kg/s}$$

Now, we can find $h_{f2} = 92.303$ kJ/kg, $h_{v3} = 2541.692$ kJ/kg and $h_{v1} = 2523.408$ kJ/kg, and apply them into Eq. (5.15):

$$\dot{Q} = C_{pa}\dot{m}_a(T_3 - T_1) - \dot{m}_a(\omega_3 - \omega_1)h_{f2} + \dot{m}_a(\omega_3 h_{v3} - \omega_1 h_{v1})$$

$$\Longrightarrow \dot{Q} = 1.005 \times 0.3087 \times (295.15 - 285.15) - 0.3087 \times (0.00849 - 0.00175) \times 92.303$$
$$+ 0.3087 \times (0.00849 \times 2541.692 - 0.00175 \times 2523.408) = 8.21 \text{ kJ/s}$$

From the psychrometric chart, we can find similar values as:

$$h_1 = 16.443 \text{ kJ/kg(d.a)}$$

$$h_3 = 43.109 \text{ kJ/kg(d.a)}$$

$$\Longrightarrow \dot{Q} = \dot{m}_a(h_3 - h_1) = 8.23 \text{ kJ/s}$$

5.3.4 Cooling and Dehumidification

As shown in Figure 5.9, if the air–vapour mixture is cooled below its dewpoint temperature, then the water starts to be condensed. This can result in the dehumidification of the mixture

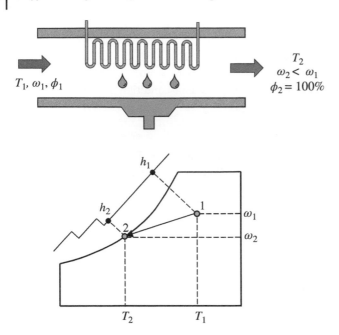

Figure 5.9 Schematic of cooling and dehumidification processes.

at the same time with a decrease in its dry-bulb temperature as illustrated in Psychrometric chart of Figure 5.9. The final relative humidity can be lower than or equal to 100%.

Note that the pathway to reach Point-2 from Point-1 can vary from the one shown in this picture in accordance with the type of coils, surface temperature of coils, condition of the flow, etc. However, we can again assume a SSSF thermodynamic process for the shown cooling and dehumidification process, and derive the first law of thermodynamics and mass conservation law between inlet and outlet points as follows:

$$(\dot{m}_{a1}h_{a1} + \dot{m}_{v1}h_{v1}) - \dot{Q} - (\dot{m}_{f3}h_{f3}) - (\dot{m}_{a2}h_{a2} + \dot{m}_{v2}h_{v2}) = 0 \qquad (5.16)$$

Again, the mass balance becomes:

$$\dot{m}_{a1} = \dot{m}_{a2} = \dot{m}_a \qquad (5.17)$$

$$\dot{m}_{v1} - \dot{m}_{f3} = \dot{m}_{v2} \qquad (5.18)$$

And, with a similar procedure as before, we can find:

$$\dot{Q} = C_{pa}\dot{m}_a(T_1 - T_2) - \dot{m}_a(\omega_1 - \omega_2)h_{f3} + \dot{m}_a(\omega_1 h_{v1} - \omega_2 h_{v2}) \qquad (5.19)$$

Example 5.4

Moist air with the relative humidity of $\phi_1 = 85\%$ and dry-bulb temperature of $T_1 = 27°C$ with a mas flow rate of $\dot{m}_1 = 0.3\,kg/s$ is exposed to a cooling coil system, which extracts 8 kW heat and condensates 0.0002 kg/s water from it. What are the final temperature and relative humidity of the moist air?

Solution

From psychrometric chart, we can find the enthalpy of the mixture for Point-1 as:

$$h_1 = 76.28\,\text{kJ/kg(d.a)}$$

$$\omega_1 = 0.00193$$

Now, we can find the mass flow rates as below:

$$\dot{m}_1 = \dot{m}_{a1} + \dot{m}_{v1} = \dot{m}_{a1}(1 + \omega_1)$$

$$\Longrightarrow \dot{m}_{a1} = \dot{m}_{a2} = \frac{\dot{m}_1}{1 + \omega_1} = \frac{0.3}{1 + 0.00193} = 0.2994\,\text{kg/s}$$

$$\Longrightarrow \dot{m}_{v1} = \dot{m}_{a1}\omega_1 = 0.0006\,\text{kg/s}$$

$$\Longrightarrow \dot{m}_{v2} = \dot{m}_{v1} - \dot{m}_{f3} = 0.0004\,\text{kg/s}$$

We can also calculate the humidity ratio and mass flow rate at Point-2:

$$\omega_2 = \frac{\dot{m}_{v2}}{\dot{m}_{a2}} = 0.00126$$

$$\dot{m}_2 = \dot{m}_{a2} + \dot{m}_{v2} = 0.2998\,\text{kg/s}$$

Thus, the enthalpy at the outlet can be found as:

$$-\dot{Q} = \dot{m}_a(h_2 - h_1)$$

$$\Longrightarrow h_2 = \frac{-\dot{Q} + \dot{m}_a h_1}{\dot{m}_a} = \frac{-8 + 0.2994 \times 76.28}{0.2994} = 49.56\,\text{kJ/kg}$$

Using ω_2 and h_2, we can locate the associated point in the psychrometric chart as shown in Figure 5.E4 with drawing a line from h_2 parallel with the enthalpy lines. Then, we draw a line from ω_2 parallel to the dry-bulb axes. The intersection would locate Point-2, which has a dry-bulb temperature of $T_2 = 17.5°C$ and a relative humidity of $\phi_2 = 99.9\%$.

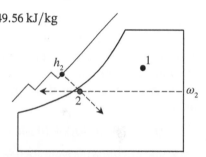

Figure 5.E4 Identification of the properties of the leaving moist air using psychrometric chart.

5.3.5 Adiabatic Humidification

An evaporative cooling process is illustrated in Figure 5.10. As expected, part of this process is similar to the adiabatic saturation in which the dry-bulb temperature drops while the absolute humidity increases. Nonetheless, in this process, the temperature does not necessarily reach its dewpoint. A warmer vapour–air mixture is exposed to a colder water supplied with sprays or pools and thus, the water molecules absorb the mixture energy and change their phases to vapour. This implies that the latent heat and absolute humidity of the vapour–air mixture is increased while the wet-bulb temperature and the total energy remains constant as the sensible heat is only converted to the latent heat.

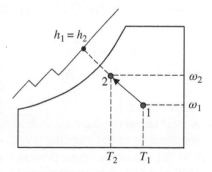

Figure 5.10 Schematic of evaporative cooling process.

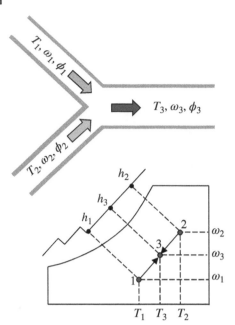

Figure 5.11 Schematic of adiabatic mixing process.

5.3.6 Adiabatic Mixing

In a presumably adiabatic process, different vapour–air mixtures can be mixed together to alter the properties of the mixture to a desired one. This process is mainly effective when a fresh moist air is added to the air mixture returned from the ventilation system of an enclosed space. As illustrated in Figure 5.11, the process can be again expressed in terms of conservation laws. For this purpose, we can derive the below equation for the moist air as:

$$(\dot{m}_{a1}h_{a1} + \dot{m}_{v1}h_{v1}) + (\dot{m}_{a2}h_{a2} + \dot{m}_{v2}h_{v2}) = (\dot{m}_{a3}h_{a3} + \dot{m}_{a3}h_{v3}) \tag{5.20}$$

We can also write the conservation of mass as below for the dry air and water vapour as:

$$\dot{m}_{a1} + \dot{m}_{a2} = \dot{m}_{a3} \tag{5.21}$$

$$\dot{m}_{v1} + \dot{m}_{v2} = \dot{m}_{v3} \tag{5.22}$$

After inserting Eqs. (5.21) and (5.22) into Eq. (5.20) and rearranging it:

$$\frac{h_2 - h_3}{h_3 - h_1} = \frac{\omega_2 - \omega_3}{\omega_3 - \omega_1} = \frac{\dot{m}_{a1}}{\dot{m}_{a2}} \tag{5.23}$$

where $h_i = \dot{m}_{ai}h_{ai} + \dot{m}_{vi}h_{vi}$.

Example 5.5

A mechanical ventilation system in the atmospheric condition supplies air to a room with a combination of 80% from the HVAC system and the rest from the outdoor fresh air of 32°C and 35%. If in an adiabatic process, the air from the mechanical system has the condition of 19°C and 55%, calculate the condition of the supplied air into the room.

Solution

First, we identify both points at psychrometric chart. First, we identify the values of Point-1 with $T_1 = 32°C$ and $\phi_1 = 35\%$:

$$h_1 = 58.89 \text{ kJ/kg}$$

$$\omega_1 = 0.00104$$

For $T_2 = 19°C$ and $\phi_2 = 55\%$, we obtain:

$$h_2 = 38.23 \text{ kJ/kg}$$

$$\omega_2 = 0.00075$$

Note that $\dfrac{\dot{m}_1}{\dot{m}_2} = 0.25$, so we have:

$$\frac{\dot{m}_1}{\dot{m}_2} = \frac{\dot{m}_{a1}(1 + \omega_1)}{\dot{m}_{a2}(1 + \omega_2)} = 0.25$$

$$\Longrightarrow \frac{\dot{m}_{a1}}{\dot{m}_{a2}} = \frac{(1 + \omega_2)}{4(1 + \omega_1)} \cong 0.25$$

Now, we can use Eq. (5.23) to find properties of Point-3:

$$\frac{h_2 - h_3}{h_3 - h_1} = \frac{\omega_2 - \omega_3}{\omega_3 - \omega_1} = \frac{\dot{m}_{a1}}{\dot{m}_{a2}}$$

$$\Longrightarrow h_3 = \frac{h_2 + 0.25h_1}{1.25} = 42.36 \text{ kJ/kg(d.a)}$$

$$\Longrightarrow \omega_3 = \frac{\omega_2 + 0.25\omega_1}{1.25} = 0.00081$$

Again, similar to Example 5.4, we can find the associated temperature and relative humidity of Point-3 as $T_3 = 21.6°C$ and $\phi_3 = 50.4\%$, respectively.

5.4 Moist Air Transport in Buildings

5.4.1 Mass Transport of Moist Air

The fundamentals of the binary mixture of vapour and dry air are described in the previous chapter while the related processes such as evaporative cooling are explained. Other process such as mass transfer through walls is essential in the calculation of heating and cooling loads in building as explained in Chapters 8 and 9. The moisture transport is also essential for the building materials' durability. Hence, in this section, we further focus on the vapour–air mixture to further analyse its transport phenomena in buildings.

Using Eq. (4.84), we can initially define the mass transfer between dry air and vapour as follows:

$$g_v = \rho_v \mathbf{v} - \rho D_{va} \nabla \frac{\rho_v}{\rho} \tag{5.24}$$

where, D_{va} is the binary diffusion coefficient between vapour and dry air.

Under atmospheric condition, we can assume that the mixture is an ideal gas and thus derive the below equation using Eq. (5.24):

$$g_v = \rho_v \mathbf{v} - (P_v/RT + P_a/R_aT)D_{va}\nabla\left(\frac{P_v/RT}{P_v/RT + P_a/R_aT}\right) \tag{5.25}$$

Note that from Eq. (4.80), we can replace $\mathbf{v} = \dfrac{\mathbf{g}}{\rho}$:

$$g_v = \rho_v\frac{\mathbf{g}}{\rho} - (P_v/RT + P_a/R_aT)D_{va}\nabla\left(\frac{P_v/RT}{P_v/RT + P_a/R_aT}\right) \tag{5.26}$$

Moreover, knowing that the total pressure (P) is largely exceeding the vapour pressure (P_v) ($P \gg P_v$), thus, one can assume $P_v/R + P_a/R_a$ to be constant (C) and rewrite Eq. (5.26) as:

$$g_v = \rho_v\frac{\mathbf{g}}{\rho} - \frac{CD_{va}}{RT}\nabla\frac{P_v}{C}$$

$$\Longrightarrow g_v = \rho_v\frac{\mathbf{g}}{\rho} - \frac{D_{va}}{RT}\nabla P_v \tag{5.27}$$

Equation (5.27) is widely used to describe different vapour mass transfer phenomena in buildings. One example is the evaporative cooling where we can assume that the vapour flux is occurring in one direction while the dry air is not transferring in the opposite direction ($g_a = 0$ and $\mathbf{g} = g_v$). Hence, Eq. (5.27) becomes:

$$g_v = \rho_v\frac{g_v}{\rho} - \frac{D_{va}}{RT}\nabla P_v$$

$$\Longrightarrow g_v = \frac{-1}{\left(1 - \dfrac{\rho_v}{\rho}\right)}\frac{D_{va}}{RT}\nabla P_v \tag{5.28}$$

Another important case is associated with the pure diffusion in the case of a stagnant moist air where the mixture has no mass flux ($\mathbf{g} = 0$), which implies $g_v = -g_a$, and Eq. (5.27) can be written as:

$$g_v = -\delta_a\nabla P_v \tag{5.29}$$

where $\delta_a = \dfrac{D_{va}}{RT}$ is the **vapour permeability coefficient** of the moist air mixture.

As it was stated, mass transfer is always associated with the heat transfer. In this case and with the comparison of Eq. (5.29) with Fourier law, δ_a can be linked with the thermal conductivity of materials in the conduction heat transfer (see Chapter 6).

Example 5.6

The moist air with a mass flow rate of $\dot{m}_1 = 0.05$ kg/s at $T_1 = 28°C$ and $\phi_1 = 30\%$ is passing over a water pool with a temperature of $T_2 = 18°C$ in an HVAC unit duct, which is operating at the atmospheric pressure. Calculate the vapour mass flux to the air if the vapour pressure is at its minimum level in the distance of 0.5 m from the pool's surface. What would be the relative humidity of the leaving air if the pool's length and width are 3 m and 2 m, respectively? Assume the binary diffusion coefficient of vapour and air as $D_{va} = 2.36 \times 10^{-5}$ m^2/s.

Solution

First, we find air properties as:

$$\omega_1 = 0.00071$$

$$\rho_{v1} = 0.009 \text{ kg/m}^3$$

$$\boldsymbol{\rho_1} = 1.167 \text{ kg/m}^3$$

$$P_{g1} = 3.782 \text{ kPa}$$

$$\Longrightarrow P_{v1} = P_{g1}\phi_1 = 1.135 \text{ kPa}$$

We use Eq. (5.8) as we can assume that the dry air is not diffused into the water while saturated vapour is transferred to the vapour with the lower partial pressure:

$$g_v = \frac{-1}{\left(1 - \dfrac{\rho_{v1}}{\rho_1}\right)} \frac{D_{va}}{RT_1} \frac{dP_{v1}}{dx} = \frac{-2.36 \times 10^{-5}}{\left(1 - \dfrac{0.023}{1.170}\right) \times 0.287 \times 301.15} \frac{dP_{v1}}{dx} = -2.751 \times 10^{-7} \frac{dP_{v1}}{dx}$$

Note that the variation of the vapour pressure due to change of relative humidity and dry-bulb temperature across the duct cause a very marginal change on the found number for the pressure gradient and, thus, the averaged number is a valid assumption. Hence, we can plot the mass flux from pool's surface as shown in Figure 5.E6.

This implies that after passing the moist air over the pool it can take about $m_{v*} = 9.67 \times 10^{-6}$ kg/s. The initial mass of the moist air is:

$$\dot{m}_{a1} = \frac{\dot{m}_1}{(1 + \omega_1)} = 0.09993 \text{ kg/s}$$

$$\Longrightarrow \dot{m}_{v1} = \dot{m}_{a1}\omega_1 = 0.00007 \text{ kg/s}$$

We know that:

$$\dot{m}_{a1} = \dot{m}_{a2} = 0.09993 \text{ kg/s}$$

$$\dot{m}_{v2} = \dot{m}_{v1} + m_{v*} = 0.00008 \text{ kg/s}$$

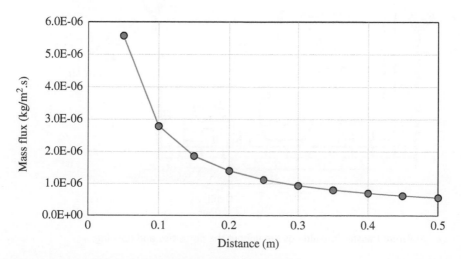

Figure 5.E6 Variation of vapour pressure from pool's surface.

Now, we can calculate the new air properties as:

$$\Longrightarrow \omega_2 = \frac{\dot{m}_{v2}}{\dot{m}_{a2}} = 0.00080$$

Thus, assuming the temperature of leaving air to be at a same temperature as the utilized water temperature ($T_2 = T_3 = 18°C$) in an evaporative cooling, we can find the relative humidity from the psychrometric chart as $\phi_1 = 62.4\%$.

5.4.2 Mass Transport Modes in Buildings

Moist air carries energy in the form of the latent heat from buildings to the surrounding environment or vice versa. When we scrutinize the mass transfer of the moist air in building, two modes of fluxes are more dominant, including throughout the open-pores of buildings' materials such as walls and ceilings, and buildings' openings such as windows, doors, and cracks (see Figure 5.12).

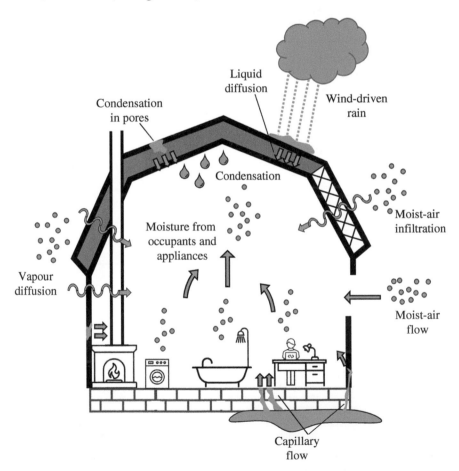

Figure 5.12 Moisture transfer in buildings throughout the porosities and opening.

While a weak insulation can result in a moisture ingress through buildings' surfaces, the water can also flow into the buildings by the capillary effect through the materials' pores. On the other hand, large openings such as windows and door in addition to small cracks, gaps and cavities can carry moisture into a building. One should add moisture released from occupants' activities such as respiration and bathing as well as from appliances such as washing machines and dish washers. It is essential, hence, to manage the wetness of a building in order to prevent plausible condensation and damages to the structure, and harms to occupants' health via mould growth and other respiratory problems.

5.4.3 Pores

In general, all the buildings' skin materials have porosities at a certain level although their porosity level can be very high in concretes to almost zero in moisture insulations. If pores in a material have a certain dimension to allow water molecules to penetrate, the diffusive mass transfer occurs while materials with large pores allow vapour to also be transferred via convection mechanism. Moist air can have both liquid and vapour phases at the same time. Therefore, in accordance with the pores' dimensions, moist air can enter porosities of that material in either of cases. Hence, some materials are inaccessible for liquid water while vapour can pass across them. If the temperature is sub-zero, a solid phase of water (ice) can coexist as well. For example, in the case of condensed vapour at liquid phase or entered liquid water by Capillary effect is frozen. This effect, nonetheless, can cause damages to building materials due to the expansion of the frozen water within pores. As illustrated in Figure 5.13, volume of pores per unit volume of materials are shown by Ψ while Ψ_0 is used for the volume of open-pores per unit volume of materials, which is depending on the fluid type. Therefore, if a pore is filled with the moist air, we have:

$$\Psi = \frac{\rho_s - \rho}{\rho_s - \rho_a} \tag{5.30}$$

where ρ is the density of the material with porosity and ρ_s is the specific material density (without porosity).

Figure 5.13 Pores in porous materials.

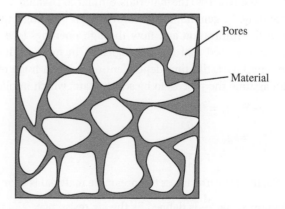

Pores

Material

5.4.4 Air Transport Through Pores

Using Eq. (4.95), we can now derive the conservation of mass for the pores while we assume there is no source and sink term in the equation:

$$-\nabla \cdot g_a = \frac{\partial w_a}{\partial t} \tag{5.31}$$

Here, we use the air content (w_a) instead of the air density for pores, which is another form of density, but defined as the mass of air per unit volume of porous material:

$$w_a = \Psi_0 S_a \rho_a \tag{5.32}$$

where, S_a is the air saturation degree, which represents the ratio of moisture to air in a pore. Thus, applying the ideal gas law for air, we have:

$$-\nabla \cdot g_a = \frac{\partial \left(\frac{\Psi_0 S_a P_a}{R_a T} \right)}{\partial t} \tag{5.33}$$

Solving Eq. (5.33) is a complex process in the realistic scenarios though we can simplify the equation, assuming a constant air saturation degree as well as an isothermal condition:

$$-\nabla \cdot g_a = \frac{\Psi_0 S_a}{R_a T} \frac{\partial (P_a)}{\partial t} = c_a \frac{\partial (P_a)}{\partial t} \tag{5.34}$$

where c_a is named as the isothermal volumetric specific air content.

If we substitute the air flux using **Poiseuille's law**, stating a relationship between air flux and driving forces, we obtain:

$$-\nabla \cdot (-k_a \nabla P_a) = c_a \frac{\partial (P_a)}{\partial t}$$
$$\Longrightarrow \nabla^2 P_a = D_a \frac{\partial (P_a)}{\partial t} \tag{5.35}$$

where, k_a is the air permeability and $D_a = \frac{c_a}{k_a}$ is called the isothermal air diffusivity for an open-pore.

Equation (5.35) resembles the heat diffusion equation and represents air flux through permeable building materials. Likewise, the steady state solution results in a straight line for the pressure across the materials. Similar to U-value, we can also define air resistance of the assembly. In general, the air diffusion in buildings' material is very small in comparison with the infiltration and flow through openings (see Chapter 8) as open pore materials are barely used that can be neglected in most of the scenarios. If permeable materials are employed, the time-dependent term has a huge response as D_a is large and thus we can assume the process to be steady state, which implies that we can rewrite Eq. (5.31) as:

$$-\nabla \cdot g_a = 0$$
$$\Longrightarrow g_a = \frac{\Delta P_a}{\dfrac{d}{k_a}} \tag{5.36}$$

So, the pressure across the preamble material is changed linearly similar to the heat conduction while $\frac{d}{k_a}$ is defined as the air resistance.

5.4.5 Vapour Transport Through Pores

Pores' dimensions play a significant role in the moisture transfer throughout buildings' materials, which is a similar analogy as the previous section, and can help to define the transfer of vapour through the porosity. It should be noted that the vapour permeability of air is very small throughout the pores and can be neglected. Obviously, the analogy is not a binary mixture anymore while there is no opposite direction (porosity does not move), and thus the stagnant condition of the moist air presented in Eq. (5.29) can be applied:

$$g_v = -\delta_m \nabla P_v \tag{5.37}$$

where δ_m is the vapour permeability of the material (porosity), which is larger than the vapour permeability of air (δ_a). This analogy has led to introduction of a parameter related to the pores' size in stagnant moist air condition, known as **vapour resistance factor** (μ):

$$\mu = \frac{\delta_a}{\delta_m} \tag{5.38}$$

Hence, Eq. (5.37) can be written as:

$$g_v = -\frac{\delta_a}{\mu} \nabla P_v \tag{5.39}$$

As seen in Figure 5.14, μ helps to understand the proportion of the vapour permeability of materials and stagnant air at a same temperature and pressure. The vapour permeability of materials is found to be a function of pores' open-area (A), length (l), and deviousness (Ψ) [10]:

$$\mu = f(A, l, \Psi) \tag{5.40}$$

Figure 5.14 Pores' characteristics and vapour resistance factor.

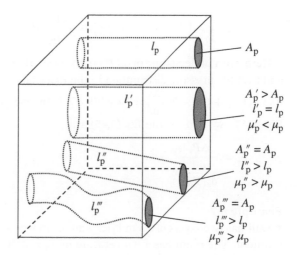

A_p

$A'_p > A_p$
$l'_p = l_p$
$\mu'_p < \mu_p$

$A''_p = A_p$
$l''_p > l_p$
$\mu''_p > \mu_p$

$A'''_p = A_p$
$l'''_p > l_p$
$\mu'''_p > \mu_p$

It should be noted that in small pores close to the size of water molecules, capillary effect is also an essential factor, when the vapour is condensed, that should not be neglected. Interested readers are encouraged to read [10].

Back to a vapour transport in the stagnant condition, we can apply the mass conservation equation for vapour in the non-hygroscopic materials when there is no vapour condensation, evaporation, or sublimations (source/sink term is zero and Ψ_0 is constant) in an isothermal condition as:

$$\nabla \cdot \left(-\frac{\delta_a}{\mu} \nabla P_v \right) = \frac{\Psi_0}{RT} \frac{\partial P_v}{\partial t} \tag{5.41}$$

In a steady-state condition, we can simplify Eq. (5.41) to:

$$\nabla \cdot \left(-\frac{\delta_a}{\mu} \nabla P_v \right) = 0 \tag{5.42}$$

In isothermal conditions, δ_a is a constant number and, thus, one can find the pressure as a straight line ($P_v(x) = Ax + B$) in any layer of the porous materials similar to the temperature distribution due to the conduction. This implies that we can find the vapour mass flux as:

$$-\frac{\delta_a}{\mu} \nabla P_v = 0$$
$$\Longrightarrow g_v = \frac{\Delta P_v}{\frac{\mu d}{\delta_a}} \tag{5.43}$$

In a non-isothermal steady-state condition, nonetheless, we know from the heat diffusion equation that the temperature is linearly changing through a material (without a sink and source). Therefore, δ_a in Eq. (5.42) becomes dependent of temperature ($\delta_a(T(x))$), which is changing in a 1D material with a length of d with two ends' temperatures of T_1 and T_2 as:

$$T(x) = T_1 + (T_2 - T_1)\frac{x}{d} \tag{5.44}$$

After integration of Eq. (5.42) over a 1D thickness, we obtain:

$$-g_v \mu \int_0^d \delta_a \left(T_1 + (T_2 - T_1)\frac{x}{d} \right) dx = \int_{P_1}^{P_2} dP_v \tag{5.45}$$

Here, with understating of $\delta_a(T(x))$ from experiments, we can solve the above integration and find the pressure distribution in the materials. Nevertheless, it is not difficult to prove that $\delta_a \left(T_1 + (T_2 - T_1)\frac{x}{d} \right) \approx T_m$ while T_m is the mean value of T_1 and T_2. Hence, similar to Eq. (5.43), we can rewrite Eq. (5.45) as:

$$g_v = \frac{\Delta P_v}{\frac{\mu d}{\delta_a(T_m)}} \tag{5.46}$$

Example 5.7

A wall has two layers of 0.3 m brick and 0.1 m wood while the brick is exposed to the environment. The room has a temperature of $T_z = 21°C$ and a relative humidity of $\phi_z = 55\%$

while these values are, respectively $T_\infty = 13°C$ and $\phi_\infty = 95\%$ for the outdoor environment with a stagnant condition. Calculate the moisture mass flux through the wall, which is assumed to have an isothermal condition. Assume the wall's surface temperatures at two ends to be $T_{so} = 13.5°C$ and $T_{si} = 20°C$. Material properties are given as below:

Material	Water vapour permeability (kg/m/s/Pa)	Conductivity (W/mK)
Brick (36% porosity)	11.5×10^{-12}	0.80
Wood	6×10^{-12}	0.12

Solution

From the psychrometric chart, we can find the vapour partial pressure as:

$$P_{v_z} = 1373 \text{ Pa}$$

$$P_{v_\infty} = 1428 \text{ Pa}$$

Using Eq. (5.43), we can find the vapour flux as (P_m is the pressure between layers):

$$g_v = -11.5 \times 10^{-12} \times \frac{P_m - 1428}{0.3}$$

$$g_v = -6 \times 10^{-12} \times \frac{1373 - P_m}{0.1}$$

$$\Longrightarrow P_m = 1394 \text{ Pa}$$

$$\Longrightarrow g_v = 1.286 \times 10^{-9} \text{kg/m}^2/\text{s}$$

The temperature distribution within the walls also can be found as (T_m is the temperature between layers):

$$q_{cond} = -0.8 \times \frac{T_m - 13.5}{0.3}$$

$$q_{cond} = -0.12 \times \frac{20 - T_m}{0.1}$$

$$\Longrightarrow T_m = 15.5°C$$

$$\Longrightarrow q_{cond} = -5.38 \text{ W/m}^2$$

If we assume that the vapour permeabilities of the used materials are not changing with the temperature distribution, we can plot the vapour pressure in addition to their associated saturation pressure in accordance with the temperature distribution through the wall as shown in Figure 5.E7.

As it can be seen, the vapour pressure is not exceeding the saturation pressures and thus, we do not expect a condensation within the wall. The design of the thermal and moisture insulations are very crucial in buildings to avoid possible condensation. For further details related to the design of insulation to avoid condensation, transient condition of vapour transport as well as diffusion–convection in open-porous materials refer to [10].

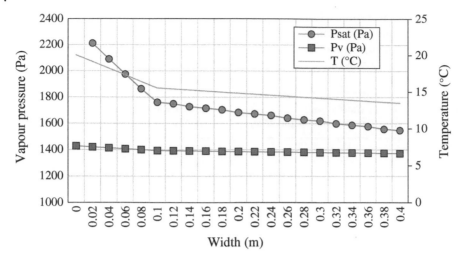

Figure 5.E7 Vapour pressures and their associated saturation pressures in accordance with the temperature distribution through the case study wall.

5.4.6 Mass Transport Through Openings

We discussed the airflow crossing buildings' opening in Chapter 3. When the moisture in air is not neglected, vapour also transports energy to or from buildings. Thus, the equilibrium of moist air in a certain temperature and pressure defines the transported energy to the control volume of a well-mixed room as depicted in Figure 5.15. In this section, we derive the general conservation of mass for the vapour as below:

$$\omega_o \acute{g}_{a_o} + \acute{g}_{gen} + \acute{g}_s - \omega_i \acute{g}_{a_i} = \frac{\partial(\omega_i M_a)}{\partial t} \tag{5.47}$$

where $\omega_o \acute{g}_{a_i}$ and $\omega_i \acute{g}_{a_o}$ denote the vapour mass flow, entering to and leaving from the room with the flowing moist air. $\acute{g}_{gen}$ presents the vapour mass removed or released by equipment, plants, and occupants while $\acute{g}_s$ indicates the vapour transfer from the room's surfaces via drying or condensation. $M_a = \rho_a V$ is the mass of the dry air in the room. Now, we can replace the mass flows ($\acute{g}_{a_o} = \rho_{a_o} nV$) where, n is the ventilation rate and find the below expression:

$$\left(\frac{\rho_{v_o}}{\rho_{a_o}}\right)(\rho_{a_o} nV) + \acute{g}_{gen} + \acute{g}_s - \left(\frac{\rho_{v_i}}{\rho_{a_i}}\right)(\rho_{a_i} nV) = \frac{\partial\left[\left(\frac{\rho_{v_i}}{\rho_{a_i}}\right)\rho_a V\right]}{\partial t} \tag{5.48}$$

And, we can rearrange Eq. (5.48) with the ideal gas law applied to the vapour:

$$\frac{T_i}{T_o}P_o + C\acute{g}_{gen} + C\acute{g}_s - P_i = \frac{1}{n}\frac{\partial P_i}{\partial t} \tag{5.49}$$

where $C = \dfrac{RT_i}{nV}$.

As it can be seen in Eq. (5.49), the moisture similar to the air transport is related to the pressure difference as well as the temperature of the building and the surrounding environment.

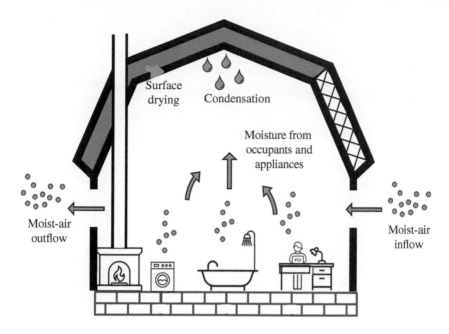

Figure 5.15 Mass flow of vapour in a well-mixed room.

The calculation of these values helps to find the energy demand and thermal comfort of a building as it is discussed in detail in Chapters 8 and 9.

References

1 Wilson, T.E. and Metzler-Wilson, K. (2018). *Autonomic Thermoregulation*. Oxford University Press.

2 Florence, T.J. and Reiser, M.B. (2015). Hot on the trail of temperature processing. *Nature*. 519 (7543): 296–297.

3 ASHRAE (2019). *ASHRAE Handbook of Fundamentals*. Thermal Comfort.

4 Hoof, V. (2008). Forty years of Fanger's model of thermal comfort: comfort for all? *Indoor Air* 18 (3): 182–201.

5 Cheng, Y., Niu, J., and Gao, N. (2012). Thermal comfort models: a review and numerical investigation. *Building and Environment* 47: 13–22.

6 Fanger, P.O. (1967). Calculation of thermal comfort-introduction of a basic comfort equation. *ASHRAE Transacions*. 73.

7 Fanger, P.O. (1970). *Analysis and Applications in Environmental Engineering*. Thermal Comfort.

8 McQuiston, F.C., Parker, J.D., and Spitler, J.D. (2004). *Heating, Ventilating, and Air Conditioning: Analysis and Design*. Wiley.

9 Mitchell JW, Braun JE. *Principles of Heating, Ventilation, and Air Conditioning in Buildings*: Wiley; 2012.

10 Hens, H.S.L.C. (2012). *Building Physics: Heat, Air and Moisture: Fundamentals and Engineering Methods with Examples and Exercises*, 2e. Berlin: Ernst & Sohn.

6

An Overview on Fundamentals of Heat Transfer in Buildings

6 An Overview of Heat Transfer

Heat transfer is thermal energy in transit due to the temperature difference. In general, heat may transfer in a medium or between media with three different mechanisms, including conduction, convection, and radiation (see Figure 6.1). Conduction refers to the diffusion of heat within the molecules of a medium while in a convection mechanism, heat is transferred by movement of the fluid from a location to another one. Finally, in a radiation mode, heat transfer occurs with emitted electromagnetic waves from a surface of a body to another one. We scrutinize each mode of heat transfer in more detail in this chapter.

6.1 Conduction

Conduction refers to the transfer of energy in the molecular form, initiating from molecules with a higher level of activity to the ones with the lower levels. Therefore, the energy transfer is always in a direction in which the temperature shall decrease. As shown in Figure 6.2, the transfer happens with the constant collision of the molecules with their neighbours.

To understand and formulate the conduction heat transfer, we assume a one-dimensional plane wall as shown in Figure 6.3. Hence, the conduction through the plane wall may be represented with **Fourier's law**:

$$q_x'' = -k\frac{dT}{dx} \tag{6.1}$$

where q_x'' (W/m^2) is the conduction heat transfer rate (heat flux) and k (W/m K) is the conductivity, which is a characteristic of a material. The negative '−' sign shows that heat is towards the direction of decreasing temperature and obviously the heat transfer occurs due to the gradient of temperature $\left(\frac{dT}{dx}\right)$.

If the heat conduction is assumed to linearly vary $\left(\frac{dT}{dx} = \frac{T_2 - T_1}{x_2 - x_1} = \frac{T_2 - T_1}{L}\right)$, then we can rewrite the Fourier equation as:

$$q_x'' = -k\frac{T_2 - T_1}{L} = k\frac{T_1 - T_2}{L} = k\frac{\Delta T}{L} \tag{6.2}$$

Computational Fluid Dynamics and Energy Modelling in Buildings: Fundamentals and Applications, First Edition. Parham A. Mirzaei.

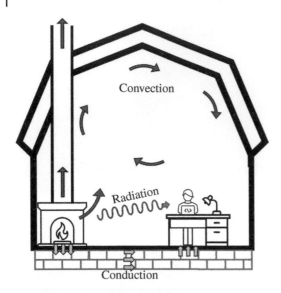

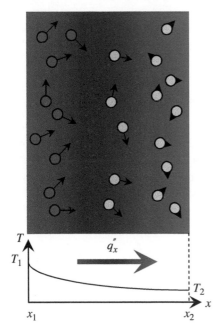

Figure 6.2 Molecular diffusion in the conduction mechanism.

Note that the total conduction heat transfer can be obtained by multiplying the heat transfer rate by the surface area as:

$$q_x = q_x''A$$
$$\Rightarrow q_x = -kA\frac{dT}{dx}$$

(6.3)

Figure 6.3 One-dimensional heat conduction in a wall plane.

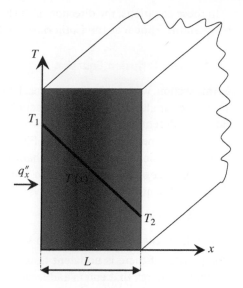

Example 6.1

A building wall with the dimension of 2.8 m × 4 m is modelled as a one-dimensional geometry with a thickness of 20 cm. The indoor and outdoor temperatures are $T_1 = 20°C$ and $T_2 = 5°C$, respectively. Calculate the heat flux through the wall if the thermal conductivity is $k = 1.8$ W/m K.

Solution

We use Eq. (6.2) as below:

$$q''_x = k \frac{\Delta T}{L} = 1.8 \frac{(20-5)}{0.20} = 135 \, \text{W/m}^2$$

The wall area is 11.2 m². Thus, we can find the total conduction as:

$$q_x = 135 \times 11.2 = 1512 \, \text{W}$$

Another key point to be considered in the Fourier equation is that the heat transfer is a directional quantity and hence the heat flux can be written in a general three-dimensional vector forms as expressed as:

$$\vec{q}'' = -k\nabla T = -k\left(\frac{\partial T}{\partial x}\vec{i} + \frac{\partial T}{\partial y}\vec{j} + \frac{\partial T}{\partial z}\vec{k}\right) \tag{6.4}$$

This implies that each component can be represented as:

$$\vec{q}''_x = -k\frac{\partial T}{\partial x}$$

$$\vec{q}''_y = -k\frac{\partial T}{\partial y} \tag{6.5}$$

$$\vec{q}''_z = -k\frac{\partial T}{\partial z}$$

In these equations, the direction of the heat transfer is normal to the surface of constant temperature known as the **isothermal surface**.

6.1.1 Heat Diffusion Equation

In this section, we introduce an approach to find the temperature distribution in a medium. Therefore, similar to the previous practices presented in the earlier chapters, we firstly consider a differential (small) control volume defined in a Cartesian coordinate within a homogenous medium. As shown in Figure 6.4, we assume the heat to be transferred only with the diffusion (conduction) mechanism while there is no advection, and we derive the energy balance on the control volume. Thus, we can write the conservation of energy within the defined control volume as follows:

$$\dot{E}_{in} + \dot{E}_{g} - \dot{E}_{out} = \dot{E}_{storage} \tag{6.6}$$

where, $\dot{E}_{g}$ is the energy source term within the control volume and $\dot{E}_{storage}$ is the energy storage term if there is no latent heat due to the phase change. $\dot{E}_{in}$ and $\dot{E}_{out}$ denote the amount of energy that enters and leaves the control volume from the surfaces.

Below we further describe each term to replace them into Eq. (6.6). First, we assume that the energy source within the control volume is generated with the rate of $\dot{q}$ (W/m³) and hence the rate of thermal energy generation within the control volume is as follows:

$$\dot{E}_{g} = \dot{q}dxdydz \tag{6.7}$$

Moreover, the energy stored within the control volume is related to the thermal mass of the medium due to its material's properties and temperature change in time, which can be written as below:

$$\dot{E}_{storage} = \rho c_{p} \frac{\partial T}{\partial t} dxdydz \tag{6.8}$$

where, $\rho c_{p} \dfrac{\partial T}{\partial t}$ term stands as the time variation rate of the sensible energy within the medium per unit volume.

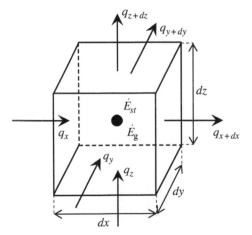

Figure 6.4 Control volume for conduction mechanism.

Referring to Eq. (6.6), the entering and leaving fluxes perpendicularly at the surfaces of the control volume shown in Figure 6.4 in the *x*-direction can be written as:

$$(q_x - q_{x + dx}) = \frac{\partial q_x}{\partial x} dx \tag{6.9}$$

Note that we used the Taylor series $\left(q_{x + dx} = q_x + \frac{\partial q_x}{\partial x} dx + O(\Delta x^2)\right)$ with neglecting the higher order terms to estimate the gradient of temperatures at the surfaces based on their defined values in the centre of the control volume. We can expand this approach to the *y*- and *z*-directions, and rewrite Eq. (6.6) using Eqs. (6.7)–(6.9) as:

$$q_x - q_{x + dx} + q_y - q_{y + dy} + q_z - q_{z + dz} + \dot{q}dxdydz = \rho c_p \frac{\partial T}{\partial t} dxdydz$$

$$\Rightarrow -\frac{\partial q_x}{\partial x} dx - \frac{\partial q_y}{\partial y} dy - \frac{\partial q_z}{\partial z} dz + \dot{q}dxdydz = \rho c_p \frac{\partial T}{\partial t} dxdydz \tag{6.10}$$

Now, replacing Fourier equation (Eq. 6.5) into the gradient terms of Eq. (6.10) and dividing both sides by the volume (*dxdydz*) gives:

$$-\frac{\partial\left(-k_x \frac{\partial T}{\partial x}\right)}{\partial x} - \frac{\partial\left(-k_y \frac{\partial T}{\partial y}\right)}{\partial y} - \frac{\partial\left(-k_y \frac{\partial T}{\partial z}\right)}{\partial z} + \dot{q} = \rho c_p \frac{\partial T}{\partial t} \tag{6.11}$$

If the conductivity is homogenous in the medium, then we can simplify the Eq. (6.11) as:

$$\frac{\partial^2 T}{\partial x^2} + \frac{\partial^2 T}{\partial y^2} + \frac{\partial^2 T}{\partial z^2} + \frac{\dot{q}}{k} = \frac{1}{\alpha} \frac{\partial T}{\partial t} \tag{6.12}$$

where, $\alpha = \dfrac{k}{\rho c_p}$ (m^2/s) is called the thermal diffusivity.

The above equation can be written in the below form in the cylindrical coordinate:

$$\vec{q}'' = -k\nabla T = -k\left(\frac{\partial T}{\partial r}\vec{i} + \frac{1}{r}\frac{\partial T}{\partial \phi}\vec{j} + \frac{\partial T}{\partial z}\vec{k}\right)$$

$$\Rightarrow \frac{1}{r}\frac{\partial}{\partial r}\left(kr\frac{\partial T}{\partial r}\right) + \frac{1}{r^2}\frac{\partial}{\partial \phi}\left(k\frac{\partial T}{\partial \phi}\right) + \frac{\partial}{\partial z}\left(k\frac{\partial T}{\partial z}\right) + \dot{q} = \rho c_p \frac{\partial T}{\partial t} \tag{6.13}$$

Various boundary condition can be applied to the heat diffusion partial differential of Eq. (6.13). These boundaries are demonstrated for the simplified scenario of a one-dimensional heat diffusion transfer as illustrated in Figure 6.5. When a surface has a fixed temperature, the boundary condition is called **Dirichlet**. On the other hand, when a fixed heat flux is assigned to the surface, the boundary condition is named as **Neumann**. For the heat diffusion equation, in the case of a zero Neumann boundary condition, it represents an adiabatic situation or an insulation in buildings. Eventually, the convection boundary condition demonstrates when a heat exchange occurs with a convection mechanism on the surface.

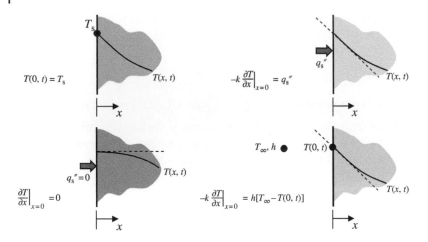

Figure 6.5 Various types of boundary condition for the heat diffusion equation.

Example 6.2

What would be the form of the diffusion equation in 1D wall of Example 6.1 if the conductivity changes as $k_x = \dfrac{c}{x^n}$.

Solution

We first simplify Eq. (6.11) with the condition of 1D and steady-state conditions as:

$$\frac{\partial \left(k_x \dfrac{\partial T}{\partial x} \right)}{\partial x} = 0$$

$$\Longrightarrow \frac{\partial T}{\partial x} = c'x^n$$

$$\Longrightarrow T(x) = c''(n+1)x^{n+1} + d$$

Now, we can apply the boundary conditions as:

$$@x = 0, T_1 = 20\,°\mathrm{C} \Longrightarrow d = 20$$

$$@x = 0.2, T_2 = 5\,°\mathrm{C} \Longrightarrow c'' = \frac{-15}{(n+1)0.2^{n+1}}$$

$$T(x) = -15(5x)^{n+1} + 20$$

Now, we can plot the temperature distribution for different values of n as shown in Figure 6.E2 (we consider $c = 1.8\,\mathrm{W/m\,K}$). Evidently, if $n = 0$, then we reach to the linear temperature variation within the wall ($T(x) = -75x + 20$) independent from the conductivity as expected from Figure 6.3.

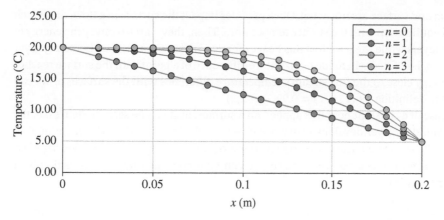

Figure 6.E2 Temperature distribution for different values of n.

6.2 Convection

6.2.1 Thermal Boundary Layer

Similar to the momentum boundary layer, we use a flat plate geometry for explaining the thermal boundary layer. We found in Chapters 2 and 3 that the velocity boundary layer profiles represent the values of velocities, varying in a perpendicular direction to the flow direction. Also, we understood that the **boundary layer thickness** of δ varies in the flow direction as fluid particles have zero velocity on the flat plate surface and these particles themselves tend to retard the motion of other particles in the upper layers. This process continues up to the thickness of δ where the fluid motion is not anymore impacted by the initial obstruction of the particles on the surface, nor by the turbulent stresses. At this location, the fluid velocity is known to have a **free stream velocity**, which is independent from the effect of the wall. Normally, the boundary layer thickness is defined as a depth in the y-direction where the velocity reaches to a value very close to the free stream velocity, mainly assumed as $u = 0.99u_\infty$

A typical thermal boundary layer is shown in Figure 6.6. In a flow over a flat plate or surface, the **thermal boundary layer** forms when the temperatures of both fluids in the

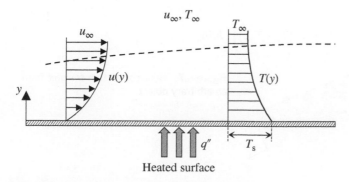

Figure 6.6 Velocity and thermal boundary layers.

free stream and surface are not equal. Once particles impact the surface in a flow, they reach a thermal equilibrium with the surface temperature. Then, they start to convey the energy to the upper layers while the length of this heat exchange is known as thermal boundary layer (δ_t). Again, δ_t is defined where the temperature difference of a point with the surface reaches to a value very close to the free stream's temperature difference with the surfaces or we can express this definition as $(T_s - T) = 0.99(T_s - T_\infty)$

In the case of a constant heat flux applied on a surface, as it can be seen in Figure 6.6, the thermal boundary layer develops with a distance from the leading edge. The equation that represents the associated heat convection on the surface can be represented by Fourier's law as flow has a zero velocity on the surface that can be represented by the conduction heat transfer mechanism:

$$q_s{}'' = -k_f \frac{\partial T}{\partial y}\bigg|_{y=0} \tag{6.14}$$

where, k_f is the fluid conductivity at the surface temperature. On the other hand, we know that the convection from a surface can be represented by the **Newton's law of cooling** as:

$$q_s{}'' = h(T_s - T_\infty) \tag{6.15}$$

One can combine Eqs. (6.14) and (6.15) and find the heat convection coefficient as below:

$$h = \frac{-k_f \dfrac{\partial T}{\partial y}\bigg|_{y=0}}{(T_s - T_\infty)} \tag{6.16}$$

We expect that the heat transfer rate from the wall to vary with the $\dfrac{\partial T}{\partial y}$ term as its value decreases with the increase of the distance from the leading edge, which also results in a reduction in the h value.

6.2.2 Local and Average Convection Coefficients

We assume a convective heat transfer from an arbitrary object as shown in Figure 6.7. When the surface has a constant temperature of T_s, which is different from the flow temperature of T_∞, then, we can define the total surface heat flux with the summation of the heat transfer rates alongside of the object:

$$q_s{}'' = \int_{A_s} q'' dA_s$$
$$\Longrightarrow q_s{}'' = \int_{A_s} h(T_s - T_\infty) dA_s = (T_s - T_\infty)\int_{A_s} h dA_s \tag{6.17}$$

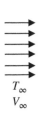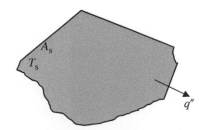

Figure 6.7 Heat convection transfer from an arbitrary object.

Here, we should note that the **local convection heat transfer coefficient** varies alongside the object. Therefore, we can now define the **average convection heat transfer coefficient** as follows:

$$\overline{h} = \frac{1}{A_s} \int_{A_s} h \, dA_s \tag{6.18}$$

Walls are mainly simplified with 1D geometries and mostly assumed as a flat plate. Thus, the average heat transfer coefficient of them can be defined as:

$$\overline{h} = \frac{1}{L} \int_0^L h \, dx \tag{6.19}$$

where, L is the length of the wall and x is the distance from the leading edge.

To find the local convection heat transfer coefficients, many efforts have been conducted, which have resulted in the development of numerous correlations as the convection mechanism in zones (rooms) are not very straightforward to be calculated. The correlations depend on many parameters, including fluid properties, surface geometry and a flow condition as discussed in detail in Chapter 7. Many of the obtained correlations suggest that the heat transfer is a function of Reynolds and Prandtl numbers. The Prandtl number is a dimensionless number, representing the ratio of momentum diffusivity to thermal diffusivity:

$$\text{Pr} = \frac{\upsilon}{\alpha} = \frac{\dfrac{\mu}{\rho}}{\dfrac{k}{\rho c_p}} = \frac{\mu c_p}{k} \tag{6.20}$$

From Eq. (6.16), we know that $\left(h = \dfrac{-k_f \dfrac{\partial T}{\partial y}\big|_{y=0}}{(T_s - T_\infty)} \right)$. Hence, if we define $T^* = \dfrac{(T - T_s)}{(T_\infty - T_s)}$, $x^* = \dfrac{x}{L}$, and $y^* = \dfrac{y}{L}$ as the dimensionless independent variables for the parameters of temperature, x-, and y-direction, respectively, then we can rewrite Eq. (6.16) as:

$$\frac{\partial T}{\partial y}\bigg|_{y=0} = \frac{\partial T^*}{\partial y^*}\bigg|_{y^*=0} \frac{(T_\infty - T_s)}{L} \tag{6.21}$$

$$\Longrightarrow h = \frac{-k_f \dfrac{\partial T^*}{\partial y^*}\big|_{y^*=0}(T_\infty - T_s)}{L(T_s - T_\infty)} = \frac{k_f}{L} \frac{\partial T^*}{\partial y^*}\bigg|_{y^*=0}$$

Here, we define the heat transfer ratio based on the dimensionless number of temperature gradient and the local Nusselt number (Nu) at the surface of an arbitrary object as:

$$h = \frac{k_f}{L} \frac{\partial T^*}{\partial y^*}\bigg|_{y^*=0}$$

$$\Longrightarrow \text{Nu} = \frac{hL}{k_f} = \frac{\partial T^*}{\partial y^*}\bigg|_{y^*=0} \tag{6.22}$$

This equation further implies that Nu can be represented as a function as demonstrated in the below:

$$Nu = f(Re, Pr, x^*) \tag{6.23}$$

Now, with the integration of the local Nu of Eq. (6.22) over the surface of the object, we can obtain the average Nusselt number using Eq. (6.18) as:

$$\overline{Nu} = \frac{1}{A_s} \int_{A_s} \frac{hL}{k_f} dA_s$$

$$\Rightarrow \overline{Nu} = \left(\frac{1}{A_s} \int_{A_s} h dA_s \right) \frac{L}{k_f} = \frac{\overline{h}L}{k_f} \tag{6.24}$$

Thus, $\overline{Nu}$ is independent of the spatial variables, which results in Eq. (6.23) to become:

$$\overline{Nu} = f(Re, Pr) \tag{6.25}$$

Development of Eq. (6.25) helps in many scenarios, including wall problems, to link Nu or $\overline{Nu}$ with the flow and fluid characteristics using empirical or semi-empirical correlations. The required data is mainly obtained by measurement procedures or computational simulations. In all these correlations, the characteristic length (L) and k_f are appeared terms to calculate the local or average heat transfer coefficient.

6.2.3 Convection in External Flows

External flow is associated with the heat and mass transfer from a surface against a flow stream. Building external walls exposed to the outdoor environment are an example of the external flow investigated in many studies. Common ways to measure Nu number for such flows are the empirical methods in which the surface and stream temperature as well as the heat flux exchange between the surface and fluid are simultaneously measured and correlated with each other. In Eq. (6.25), we specified the relation between the average Nu number and dimensionless numbers of Reynolds and Prandtl ($\overline{Nu} = f(Re, Pr)$). In a more general form and based on many studies, the below equation can be suggested:

$$\overline{Nu} = C Re_L^m Pr^n \tag{6.26}$$

where constant values of C, m, and n can be obtained with the empirical methods in accordance with the nature of the flow.

For example, a flat plate in a parallel flow is a practical case of the external flow appeared in many engineering scenarios. To investigate the heat transfer from a flat plate in a parallel flow, first we need to understand the condition of the flow in terms of being in either laminar or turbulent regimes. Recall to the boundary layer development over a flat plate as shown in Figure 6.8. One can solve the **Prandtl boundary layer theorem** using **Blasius solution** in which the flow is steady, incompressible, and laminar with a negligible change in the fluid properties and viscous dissipation while the pressure gradient is zero. We can then obtain the below equation for the boundary layer thickness, the local and average friction coefficients in a laminar flow over a flat plate with a constant temperature as:

$$\delta_x = 5x \, Re_x^{-0.5} \tag{6.27}$$

$$C_{f,x} = \frac{\tau_{s,x}}{\rho u_\infty^2 / 2} = 0.664 \, Re_x^{-0.5} \tag{6.28}$$

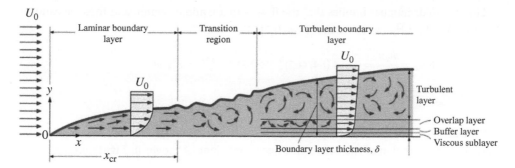

Figure 6.8 Boundary layer over an infinite flat plate. *Source:* Based on Incropera [1] and Frank et al. [2].

$$\overline{C_{f,x}} = 1.328 \text{ Re}_x^{-0.5} \tag{6.29}$$

With a similar approach and solving the thermal boundary layer, the local and average Nusselt numbers can be found as below:

$$\text{Nu}_x = \frac{h_x x}{k} = 0.332 \text{ Re}_x^{0.5} \text{Pr}^{0.33} \quad \text{Pr} \geq 0.6 \tag{6.30}$$

$$\overline{\text{Nu}_x} = \frac{\overline{h_x} x}{k} = 0.664 \text{ Re}_x^{0.5} \text{Pr}^{0.33} \quad \text{Pr} \geq 0.6 \tag{6.31}$$

Example 6.3

In a calm and cold day with the temperature of $5\,°C$, the velocity over a flat roof of a building as shown in Figure 6.E3.1 with the length of 12 m and constant temperature of about $15\,°C$ is 0.1 m/s. Find the average convective heat flux of the roof.

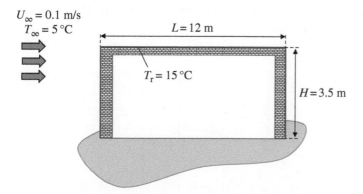

Figure 6.E3.1 Case study building exposed to an approaching wind.

Solution

We can find the air properties at film temperature $\dfrac{5 + 15}{2} = 10\,°C$, which lays around the surface and find Prandtl as $\text{Pr} = 0.71759$, kinematic viscosity as $v = 1.4207 \times 10^{-5}\,\text{m}^2/\text{s}$, and thermal conductivity as $k = 0.024840$ W/m K. Thus, we can obtain Re as:

$$\text{Re} = 84,465$$

The Reynolds number implies that the flow is in a laminar regime and thus we can use Eq. (6.31) to find $\overline{Nu_x}$ as (note that $Pr = 0.71759 > 0.6$):

$$\overline{Nu_x} = \frac{\overline{h_x}x}{k} = 0.664\,(84,\!465)^{0.5} \times 0.71759^{0.33} = 172.96$$

$$\Rightarrow \overline{h_x} = \frac{\overline{Nu_x}k}{x} = \frac{172.96 \times 0.024840}{12} = 0.36\,\text{W/m K}$$

If we consider the validity of Eq. (6.31) up to the transition regime about $Re = 5 \times 10^5$, we may find the variation of the convective heat coefficient as shown in Figure 6.E3.2.

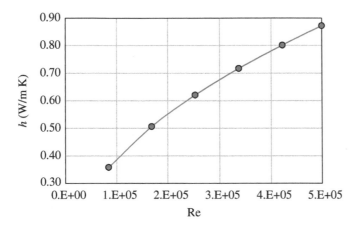

Figure 6.E3.2 Variation of the convective heat transfer coefficient of the case study building's roof against various Re numbers.

Also, we can show the change of Nusselt number over the roof for different Re numbers as illustrated in Figure 6.E3.3 using Eq. (6.30).

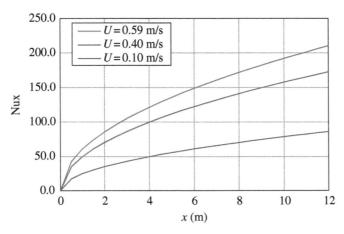

Figure 6.E3.3 Variation of Nusselt number over the case study building's roof against different Re numbers.

If the parallel flow is in turbulence regime over an isothermal flat plate, the calculation of the related parameters is a more complex task due to the random fluctuations of the fluid particles and invalidity of the Prandtl boundary layer theorem. Thus, experimental methods are more favourable in this case. For example, in a high turbulence condition when Re < 10^8, the below equations are found to be reasonable to calculate the local friction coefficient and the boundary layer thickness:

$$C_{f,x} = 0.0592 \ \mathrm{Re}_x^{-0.2} \tag{6.32}$$

$$\delta_x = 0.37x \ \mathrm{Re}_x^{-0.2} \tag{6.33}$$

We should note that the growth in the boundary layer thickness is not dependent on Pr, and thus it can be assumed that the velocity and thermal boundary layer thicknesses are very similar ($\delta_x \approx \delta_t$). Furthermore, we can obtain the local Nusselt number in a turbulent regime as follows:

$$\mathrm{Nu}_x = 0.0296 \ \mathrm{Re}_x^{0.8}\mathrm{Pr}^{0.33} \quad 60 \geq \mathrm{Pr} \geq 0.6 \tag{6.34}$$

If the flow over the flat plate is turbulent from the beginning, then we can show with an approach similar to the laminar flow that the average Nusselt number can be calculated as:

$$\overline{\mathrm{Nu}_x} = 0.037 \ \mathrm{Re}_x^{0.8}\mathrm{Pr}^{0.33} \quad 60 \geq \mathrm{Pr} \geq 0.6 \tag{6.35}$$

Note that in many cases, the turbulence flow occurs in a parallel flow after a certain length of the flow being in a laminar regime. In this case, a mixed Nusselt number should be calculated, which implies that the heat transfer in the laminar regime should be resolved followed by the heat transfer in the turbulent regime after a critical length (x_{cr}):

$$\overline{\mathrm{Nu}_L} = \left(0.037 \ \mathrm{Re}_L^{0.8} - C\right)\mathrm{Pr}^{0.33} \quad 60 \geq \mathrm{Pr} \geq 0.6$$
$$C = 0.037 \ \mathrm{Re}_{xcr}^{0.8} - 0.664 \ \mathrm{Re}_{xcr}^{0.5} \tag{6.36}$$

If the boundary condition of the flat plate exposed to a parallel flow is changed to a uniform surface heat flux condition, then different expressions for the discussed parameters can be found. In a laminar condition, the Nusselt number can be expressed as:

$$\mathrm{Nu}_x = 0.453 \ \mathrm{Re}_x^{0.5}\mathrm{Pr}^{0.33} \quad \mathrm{Pr} \geq 0.6 \tag{6.37}$$

And, when the flow is turbulent, we can obtain the Nusselt number as:

$$\mathrm{Nu}_x = 0.0308 \ \mathrm{Re}_x^{0.8}\mathrm{Pr}^{0.33} \quad 60 \geq \mathrm{Pr} \geq 0.6 \tag{6.38}$$

Example 6.4
Find the convective heat transfer coefficient in Example 6.3 when the velocity on the roof exceeds to a maximum value of 2 m/s.

Solution

Above the velocity of 0.6 m/s, the flow becomes turbulent in a part of the roof. In the velocity of 2 m/s, the critical distance is 3.6 m. Moreover, the local Nu number also is calculated for both laminar and turbulent flows using Eqs. (6.30) and (6.34), respectively, as depicted in Figure 6.E4.1.

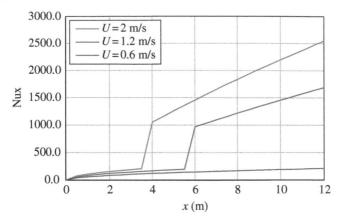

Figure 6.E4.1 Variation of the local Nu numbers.

Also, the average Nusselt number of the building roof as x_{cr} changes with the alteration of the velocity up to 2 m/s after utilization of Eq. (6.36) as shown in Figure 6.E4.2.

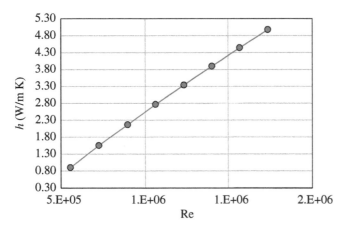

Figure 6.E4.2 Variation of the average Nusselt number of the case study building's roof against different Re numbers.

As we discussed before, the rate of convection heat transfer is highly related to the flow structure and geometry, in the specific way that boundary layer develops over a surface. Many studies have been conducted to measure the local and average Nusselt number for different types of flows. Example is the external flow over a cylindrical object. Nusselt number over a cylinder varies in accordance with the boundary layer thickness, the fluid regime (laminar or turbulent) and the separation angle, which promotes more heat transfer from the surface due to the formation of eddies (see Figure 6.9).

The local and average Nusselt numbers are also investigated in a wide range of engineering problems. For example, Figure 6.10 presents the average Nusselt number for an external flow over circular and noncircular cross sections in a format of:

$$\overline{Nu_D} = C\ Re_D^{\ m}\ Pr^{0.33} \tag{6.39}$$

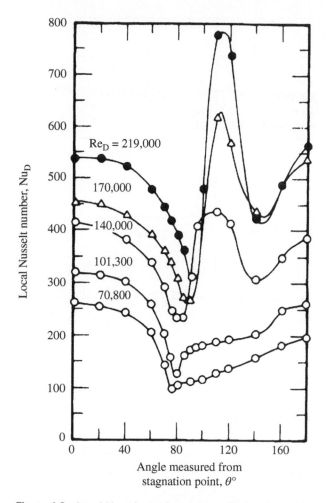

Figure 6.9 Local Nusselt number over a cylinder. *Source:* Based on Incropera [1].

6.2.4 Convection in Internal Flow

In many engineering applications, the heat transfer in internal flows is as important as the external flows. Unlike internal flows, the surface boundary layer in external flows can be developed without any external constraint. Nevertheless, in internal flows the boundary layer growth is limited to the geometry of the surface, e.g. in pipes or ducts. Internal flow was introduced in Chapters 2 and 3 where the flow characteristics were discussed. Here, we further focus on the thermal boundary layer and the associated heat transfer in internal flows. First condition to understand the heat transfer is again to scrutinize the flow regime being in either laminar or turbulent conditions. In a laminar flow in a circular tube with a uniform inflow velocity, we could realize that the boundary layer on the surface starts to develop in the *hydrodynamic entrance region* until reaching the *fully developed region* where boundary layers merge with each other (see Chapter 2, Figure 2.28). Note that the region out of the boundary layer in the hydrodynamic entrance region can be assumed as the irrotational or inviscid flow region. At the fully developed region, the flow is viscous

Cross-section of the cylinder	Fluid	Range of Re	Nusselt number
Circle 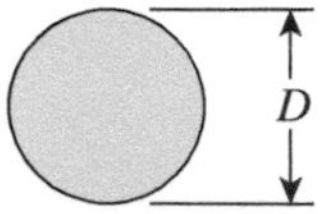	Gas or liquid	0.4–4 4–40 40–4000 4000–40,000 40,000–400,000	$Nu = 0.989 Re^{0.330} Pr^{1/3}$ $Nu = 0.911 Re^{0.385} Pr^{1/3}$ $Nu = 0.683 Re^{0.466} Pr^{1/3}$ $Nu = 0.193 Re^{0.618} Pr^{1/3}$ $Nu = 0.027 Re^{0.805} Pr^{1/3}$
Square 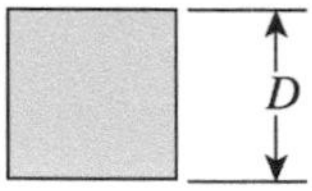	Gas	3900–79,000	$Nu = 0.094 Re^{0.675} Pr^{1/3}$
Square (tilted 45°) 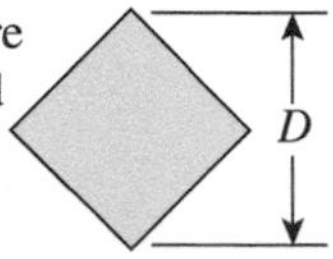	Gas	5600–111,000	$Nu = 0.258 Re^{0.588} Pr^{1/3}$
Hexagon 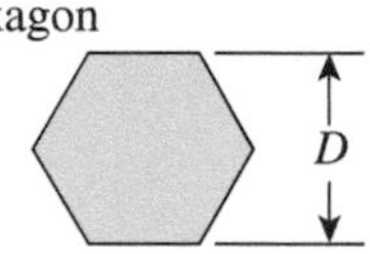	Gas	4500–90,700	$Nu = 0.148 Re^{0.638} Pr^{1/3}$
Hexagon (tilted 45°) 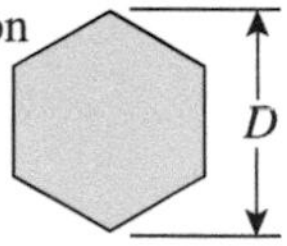	Gas	5200–20,400 20,400–105,000	$Nu = 0.162 Re^{0.638} Pr^{1/3}$ $Nu = 0.039 Re^{0.782} Pr^{1/3}$
Vertical plate 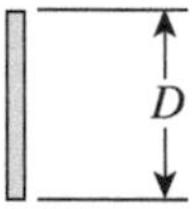	Gas	6300–23,600	$Nu = 0.257 Re^{0.731} Pr^{1/3}$
Ellipse 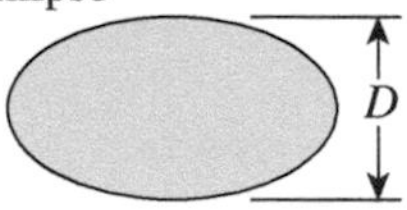	Gas	1400–8200	$Nu = 0.197 Re^{0.612} Pr^{1/3}$

Figure 6.10 Average Nusselt number coefficients for circular and noncircular cross sections. *Source: Based on Incropera [1].*

across the section of the tube while the velocity profile remains unchanged along the flow direction with a parabolic profile in the laminar flows and a flatter one in the turbulent flows due to the turbulent mixing in the radial direction. The indicator to understand the flow regime condition is Reynolds number defined as:

$$Re_D = \frac{u_{mean} D}{v} \tag{6.40}$$

where, u_{mean} is the mean velocity of the flow across the tube and D is the diameter of the tube.

It is very common to assume the critical Reynolds number where an internal flow become turbulent to be $Re_{D-critical} \approx 2100-2300$. It would be in the transition regime when $2100-2300 < Re_{D-critical} < 4000$. The fully developed condition to be achieved though requires larger $Re_{D-critical} \approx 10,000$. The **hydrodynamic entry length** $(x_{e,f})$ can be also found from the below expression for a laminar flow and with a uniform inflow as:

$$\left(\frac{x_{e,f}}{D}\right)_{laminar} \approx 0.05 \, Re_D \tag{6.41}$$

The turbulent hydrodynamic entry length is found to be an independent number from Reynolds number, which is also a controversial value to be exactly specified and can vary as:

$$10 \leq \left(\frac{x_{e,f}}{D}\right)_{turbulence} \leq 60 \tag{6.42}$$

Finding the mean velocity of the flow across a conduit (u_{mean}) is critical in calculation of the mass flow rate as it can be seen in the below example. Although the analytical calculation of u_{mean} is not a straightforward task in the hydrodynamic entrance region and the turbulence regime, it can be mathematically found in the fully developed region of a laminar flow for an incompressible fluid with constant properties crossing a circular tube with radius r_0 as (see also Example 2.9 in Chapter 2):

$$u_r = -\frac{1}{4\mu}\left(\frac{dp}{dx}\right)r_0^2\left[1-\left(\frac{r}{r_0}\right)^2\right] \tag{6.43}$$

where $\frac{dp}{dx}$ is the axial pressure gradient, and $u_r = u(x, r)$ is the axial velocity, which only depends on radius r. Therefore, the mean velocity can be obtained by integration of the found parabolic axial velocity across the section of the circular tube $(A = \pi r_0^2)$:

$$u_{mean} = \frac{\int_A \rho u(x,r)dA}{\rho A} = \frac{-\frac{2\rho\pi}{4\mu}\left(\frac{dp}{dx}\right)\int_0^{r_0}\left(r_0^2\left[1-\left(\frac{r}{r_0}\right)^2\right]\right)rdr}{\rho\pi r_0^2} = -\frac{r_0^2}{8\mu}\frac{dp}{dx} \tag{6.44}$$

Pressure drop in conduits such as pipes using the mean flow velocity were comprehensively discussed in Chapter 2.

6.2.4.1 Thermally Fully Developed Condition

Similar to the velocity boundary layer, the **thermal boundary layer** develops alongside of an internal flow if the surface temperature is higher than the fluid temperature. Hence, we can again reach to a **thermally fully developed condition** if we assume a uniform temperature for the inflow while the surface temperature is maintained with either a uniform temperature (T_s) or heat flux (q''_s). As it can be observed in Figure 6.11, the profile of the

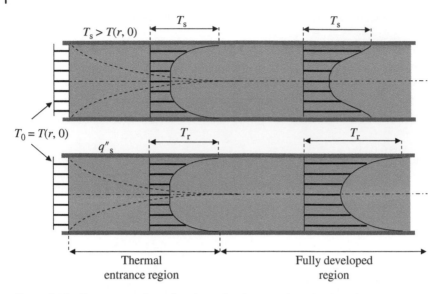

Figure 6.11 Temperature boundary layer development in a circular tube.

thermally fully developed temperature can be varied in either scenario. Again, in case of having a laminar regime, the **thermal entry length** $(x_{e,t})$ can be calculated as:

$$\left(\frac{x_{e,t}}{D}\right)_{laminar} \approx 0.05 \, \mathrm{Re}_D \mathrm{Pr} \tag{6.45}$$

This value can be approximated with the below equation for turbulent flows where the thermal entry length is independent from the Prandtl number:

$$\left(\frac{x_{e,t}}{D}\right)_{turbulence} \geq 10 \tag{6.46}$$

It should be noted here that temperature difference between tube surface and internal flow is causing the heat transfer, thus the thermally fully developed condition is not defined as $\frac{\partial T}{\partial x} \neq 0$, dissimilar to the velocity fully developed condition $\frac{\partial u}{\partial x} = 0$, but it is defined using the dimensionless temperature as defined before:

$$\frac{\partial \left[\dfrac{T_s(x) - T(r,x)}{T_s(x) - T_{mean}(x)} \right]_{e,t}}{\partial x} = 0 \tag{6.47}$$

where, $T(r, x)$ is the local temperature and $T_s(x)$ is the tube surface temperature. The mean temperature is indicated with $T_{mean}(x)$ and is defined as the bulk temperature of the fluid across the section of the circular tube.

Unlike the external flows, $T_{mean}(x)$ is not constant alongside the parallel flow and varies $(\frac{\partial T_{mean}}{\partial x} \neq 0)$ to maintain the heat transfer between the surface and fluid particles. The mean temperature can be calculated based on the rate of enthalpy advection integrated across the circular section using the velocity profile for an incompressible fluid:

$$\dot{m}c_p T_{mean}(x) = \int_A \rho u c_p T dA$$

$$\Longrightarrow T_{mean}(x) = \frac{\int_A \rho u c_p T dA}{\dot{m}c_p} \tag{6.48}$$

For a circular tube, we can use the mean velocity ($\dot{m} = \rho A u_m = \rho \pi r_0^2 u_m$) to reduce Eq. (6.48) to:

$$T_{mean}(x) = \frac{\int_A \rho u c_p T dA}{\dot{m}c_p} = \frac{\rho c_p \int_r u T (2\pi r dr)}{\rho \pi r_0^2 u_m c_p}$$

$$\Longrightarrow T_{mean}(x) = \frac{2}{u_m r_0^2} \int_0^{r_0} u T r dr \tag{6.49}$$

Now, back to the independent thermally fully developed condition from x-direction, we can assume two conditions for the conduit surface, including a uniform surface heat flux or a uniform temperature. In both cases, the derivative of the dimensionless temperature with respect to r as presented in Eq. (6.47) can lead to:

$$\frac{\partial \left[\frac{T_s(x) - T(r,x)}{T_s(x) - T_{mean}(x)} \right]_{r=r_0}}{\partial r} = \frac{\left[-\frac{\partial T}{\partial r} \right]_{r=r_0}}{T_s - T_{mean}} \neq f(x) \tag{6.50}$$

The right-hand side term is not a function of x and we can conclude from the Fourier's conduction and Newton's cooling laws on the conduit surface that:

$$q''_s = k \frac{\partial T}{\partial r} \bigg|_{r=r_0} = h(T_s - T_{mean}) \tag{6.51}$$

And replacing in the thermally fully developed condition of Eq. (6.50):

$$\frac{h}{k} \neq f(x) \tag{6.52}$$

This draws an important conclusion, implying that the local convection coefficient in a flow in which fluid properties are not changing, remains independent from x.

6.2.4.2 Mean Temperature at Internal Flows

To understand the mean temperature in an internal flow, we assume a control volume within a conduit as illustrated in Figure 6.12. We can now apply the energy conservation law to this control volume, considering a steady-state condition, negligible dissipation due

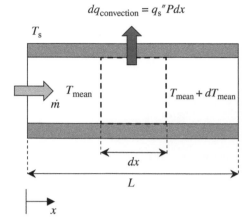

$dq_{\text{convection}} = q_s'' P dx$

Figure 6.12 Control volume in an internal flow.

to viscosity, incompressible condition for the fluid, and a negligible pressure change. Thus, we can assume that the amount of energy that the fluid mass carries through the control volume is equal to the one passing through the conduit surface while the temperature varies alongside the flow direction with dT_{mean}:

$$dq_{\text{convection}} = q_s'' P dx = [h(T_s - T_{\text{mean}})]Pdx$$

$$dq_{\text{convection}} = \dot{m}c_p[(T_{\text{mean}} + dT_{\text{mean}}) - T_{\text{mean}}] \qquad (6.53)$$

$$\Longrightarrow \frac{dT_{mean}}{dx} = \frac{q_s'' P}{\dot{m}c_p} = \frac{Ph}{\dot{m}c_p}(T_s - T_{\text{mean}})$$

where, P is the conduit perimeter (here $2\pi r_0$ for a circular tube).

The shape of the above equation is a first order ordinary differential equation (ODE) and its solution is down to the assumption made for each of the right-hand side terms. For example, the perimeter (P) is mainly considered to be fixed when there is no change in the geometry of the tube, or h is assumed to be constant when the flow is analyzed in the thermally fully developed region as discussed before. In this case, the solution is only a parameter of the tube surface condition. If it is assumed with a constant surface heat flux of q_s'', then we can integrate equation across the entire tube as:

$$\frac{dT_{\text{mean}}}{dx} = \frac{q_s'' P}{\dot{m}c_p}$$

$$\Longrightarrow \int_{T_{\text{mean,inlet}}}^{T_{\text{mean,outlet}}} dT_{\text{mean}} = \int_0^L C dx \qquad (6.54)$$

$$\Longrightarrow T_{\text{mean}}(x) = C q_s'' x + T_{\text{mean,inlet}}$$

where, $C = \dfrac{P}{\dot{m}c_p}$ is a constant value. Thus, the result is a linear variation in the temperature between inflow and outflow sections as shown in Figure 6.13 (red region).

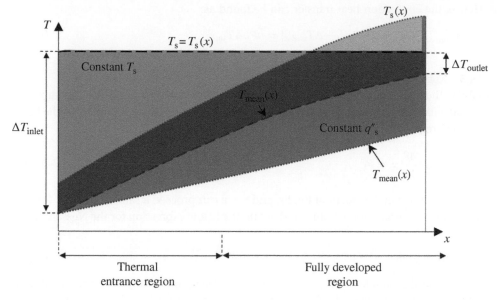

Figure 6.13 Mean temperature variation in a tube with constant heat flux and constant surface temperature.

Inversely, when a constant temperature of T_s is assigned to the conduit's surface, then we can rewrite the equation by integration of the Eq. (6.53) from the inlet to the outlet as (see Figure 6.13, blue region):

$$\frac{dT_{mean}}{dx} = Ch(T_s - T_{mean})$$

$$\Longrightarrow \int_{\Delta T_{inlet}}^{\Delta T_{outlet}} \frac{d(\Delta T)}{\Delta T} = -\int_0^L Ch\,dx \tag{6.55}$$

$$\Longrightarrow \ln \frac{\Delta T_{outlet}}{\Delta T_{inlet}} = -CL\left(\frac{1}{L}\int_0^L h\,dx\right)$$

where, $\Delta T = T_s - T_{mean}$. From Eq. (6.19), we understood that $\bar{h} = \left(\frac{1}{L}\int_0^L h\,dx\right)$. Hence, we can rewrite Eq. (6.55) as:

$$\ln \frac{\Delta T_{outlet}}{\Delta T_{inlet}} = -CL\bar{h}$$

$$\Longrightarrow \frac{\Delta T_{outlet}}{\Delta T_{inlet}} = e^{\left(-CL\bar{h}\right)} \tag{6.56}$$

In many problems, the log-mean temperature difference is used to calculate the total heat flux:

$$\Delta T_{lm} = \frac{(\Delta T_{outlet} - \Delta T_{inlet})}{\ln\left(\dfrac{\Delta T_{outlet}}{\Delta T_{inlet}}\right)} \tag{6.57}$$

Hence, the convection heat transfer can be found as:

$$q_{\text{convection}} = \dot{m}c_{\text{p}}[\Delta T_{\text{inlet}} - \Delta T_{\text{outlet}}] = \overline{h}PL\Delta T_{\text{lm}} \tag{6.58}$$

6.2.4.3 Nusselt Number of Internal Flows

Applying the discussed concept for a circular conduit while the flow is laminar, incompressible, and with constant properties, one can theoretically find the Nusselt number in the fully developed region and the case of constant surface heat flux as [1]:

$$\text{Nu}_D \equiv \frac{48}{11} = 4.36 \tag{6.59}$$

While the Nu is independent of Re, Pr, and x, we can proceed a same procedure for the case of constant surface temperature and obtain the below expression for the Nusselt number [3]:

$$\text{Nu}_D \equiv 3.66 \tag{6.60}$$

Nusselt number should be calculated in the entry region if the fully developed condition in either of constant surface heat flux or temperature are not satisfied. The condition to solve the general equation for fully developed regions contain many assumptions, which are not valid in the entry region (e.g. $v \neq 0$, or temperature and velocity depend on x and r), thus, making the analytical solution very complicated. Further details of empirical studies or analytical solutions for specific cases such as fully developed velocity profile or combined entry length problem can be found in fundamental textbooks such as [4, 5].

If the fully developed conditions are in turbulent regime, the empirical approaches are mainly preferred due to the complexity of the governing equations. For example, [6] explained a correlation for the latter case related to smooth circular tubes ($0.6 \leq \text{Pr} \leq 160, 10^4 \leq \text{Re}_D, 10 \leq \frac{L}{D}$):

$$\text{Nu}_D \equiv 0.023 \, \text{Re}_D^{0.8} \, \text{Pr}^n \tag{6.61}$$

where, n is 0.4 for heating ($T_{\text{s}} > T_{\text{mean}}$) and 0.3 for cooling ($T_{\text{s}} < T_{\text{mean}}$) cases. The temperate difference of $T_{\text{s}} - T_{\text{mean}}$ is small or moderate while properties are calculated at T_{mean}. In case of having a large temperature difference, [7] proposed the below equation while properties except μ are calculated at T_{mean} ($0.7 \leq \text{Pr} \leq 16,700, 10^4 \leq \text{Re}_D, 10 \leq \frac{L}{D}$):

$$\text{Nu}_D \equiv 0.027 \, \text{Re}_D^{0.8} \, \text{Pr}^{0.33} \left(\frac{\mu}{\mu_{\text{s}}}\right)^{0.14} \tag{6.62}$$

Further correlations are proposed in fundamental textbooks and literature. In case of having other shapes for cross section of the conduit rather than a circular one, Figure 6.14 demonstrates some of the found Nusselt numbers in the fully developed laminar regime based on the ***effective diameter*** or ***hydraulic diameter*** ($D_{\text{h}} = \frac{4A}{P}$) as the characteristic length for noncircular conduits. D_{h} is used to calculate the associated Re_D and Nu_D.

$$\mathrm{Nu_D} \equiv \frac{hD_\mathrm{h}}{k}$$

Cross section	$\dfrac{b}{a}$	(Uniform q_s'')	(Uniform T_s'')	$f\,\mathrm{Re_{D_h}}$
⬤	—	4.36	3.66	64
a ◻ b	1.0	3.61	2.98	57
a ▭ b	1.43	3.73	3.08	59
a ▭ b	2.0	4.12	3.39	62
a ▭ b	3.0	4.79	3.96	69
a ▭ b	4.0	5.33	4.44	73
▭ b	8.0	6.49	5.60	82
▭ Heated	∞	8.23	7.54	96
▭ Insulated	∞	5.39	4.86	96
△	—	3.11	2.49	53

Figure 6.14 Nusselt number and friction factor for different conduits. *Source:* Based on Incropera [1].

Example 6.5

Air in a solar chimney with a length of 3 m and a circular section of $D = 0.3m$ is shown in Figure 6.E5.1. While the solar heat flux is assumed to impact the wall with 300 W/m², calculate the air and surface temperatures at the exit of the chimney if the entering air has the temperature of 22°C and velocity of 1 m/s induced by a fan.

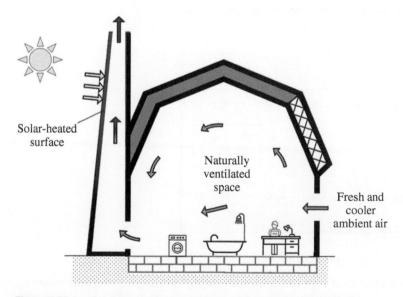

Figure 6.E5.1 The case study house with an integrated solar chimney.

Solution

Assuming that the air properties are constant through the solar chimney, we use the entrance temperature to find the properties as $\rho_a = 1.18$ kg/m^3, $\nu_a = 15.52 \times 10^{-6}$ m^2/s, $c_{pa} = 1.007$ kJ/kg K, $k_a = 25.83 \times 10^{-3}$ W/mK, and Pr $= 0.714$.

We first determine the Reynolds number for $r = 0.15$ m, assuming that the section is not changing:

$$\mathrm{Re}_H = \frac{UD}{\nu_a} = 19,330$$

Assuming a fully developed condition, the flow is thus turbulent, and we use Eq. (6.61) ($\frac{L}{D} = 10$ and $n = 0.4$ for heating) to calculate the Nusselt number as:

$$\mathrm{Nu}_D = 0.023(19,330)^{0.8}(0.714556)^{0.4} = 54.0$$

We then calculate $\dot{m} = \rho_a v_a \pi r^2 = 0.083$ kg/s and then we apply the energy balance equation on the control volume of the chimney as (see Eq. (6.53)):

$$q'' A_e = \dot{m} c_p (T_{\mathrm{mean}}(L) - T_{\mathrm{mean}}(0))$$
$$\Longrightarrow T_{\mathrm{mean}}(L) = \frac{q'' \pi D}{2 \dot{m} c_p} x + T_{\mathrm{mean}}(0)$$

Note that the solar ration is assumed only to incident half of the chimney ($P = \frac{\pi D L}{2}$), which may contradict the condition for a uniform heat flux. If this effect is neglected, then we can find $T_{\mathrm{mean}}(L) = 27°$C. We can find the linear variation of the mean temperature as below and plot it as Figure 6.E5.2:

$$T_{\mathrm{mean}}(x) = \frac{q'' \pi D}{2 \dot{m} c_p} x + T_{\mathrm{mean}}(0)$$

Also, we can use the Nusselt number to find the local h as:

$$h = \frac{\mathrm{Nu}_D k}{D} = 4.65 \, \mathrm{W/m^2 \, K}$$

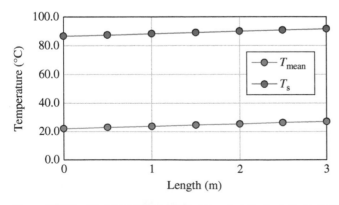

Figure 6.E5.2 Linear variation of the mean temperature through the length of the solar chimney.

Furthermore, this implies that we can find the surface temperature at the outlet $(L = 3\,\text{m})$ as:

$$q'' = h(T_s(L) - T_{\text{mean}}(L))$$

$$\implies T_s(x) = \frac{q''}{h} + T_{\text{mean}}(L) = 91.6^{\circ}\text{C}$$

As it could be noticed, finding the convective heat transfer coefficient is the key factor in the calculation of the air and surface temperatures. As it will be discussed in the following chapters, correlations or advanced numerical methods are widely adopted to find more accurate estimation for Nusselt numbers and hence convective heat transfer coefficients.

6.2.5 Free Convection

When a fluid particle is pushed by an external force, for example caused by a fan or a pump, then we could expect forced convection occurs. Without the presence of such external forces, the convection can still occur if the buoyancy force (body force) is acting on the fluid particle as the external force. This type of convection is known as *free* or *natural convection*, which is on the basis of the density change causes a buoyant force. Although free convection results in a smaller heat transfer in most of the cases, it is a dominant effect in many engineering problems, especially building related ones when the fluid density alters due to a temperature variation. The existence of the free convection does need a temperature gradient mainly on the surfaces and depends on the geometry of the associated problem. The example is **stable** and **unstable** conditions of flow between two large horizontal parallel plates as depicted in Figure 6.15. This phenomenon is broadly happening in rooms when there are no external forces on fluid and air can be presumed to be still. As it can be seen in Figure 6.15, in an unstable condition, if the temperature of the bottom plate is higher than the top one, then the density of fluid particles become lower as they are exposed to a plate with a higher temperature and start to rise while the heavier fluid particles exposed to the top plate with a lower temperature begin to descend and replace the lighter particles due to the gravity effect if this force can dominate the initial viscous forces. Inversely, in a stable

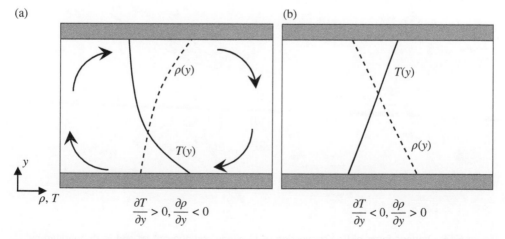

(a)   (b)

$$\frac{\partial T}{\partial y} > 0, \frac{\partial \rho}{\partial y} < 0 \qquad\qquad \frac{\partial T}{\partial y} < 0, \frac{\partial \rho}{\partial y} > 0$$

Figure 6.15 (a) Unstable and (b) stable condition of fluid between two large horizontal parallel plates.

condition, this phenomenon is not expected if the temperature of the top plate is higher than the bottom one as the density will be stratified from the bottom plate toward the top one. Although in an unstable condition, the flow circulation results in a considerable convection heat transfer, in a stable condition, a dominant air movement does not happen, implying that the prevailing heat transfer mechanism is only a weak conduction within the fluid.

Thus, before starting to investigate the free convection, it is important to identify the geometry condition of a problem, specifically in the building related problems. If the free convection flow is not bounded on top, then it can be in a form of **plume** (see Figure 6.16a), or **buoyant jet** (see Figure 6.16b). Examples of plume are flow in a chimney or above a city known as **urban plume**, which is formed on some occasions in atmospheric boundary layer. Example of a buoyant jet is horizontal flow discharged out from a diffuser into a room with still (quiescent) and lower temperature air while the jet is risen by the buoyancy effect. The buoyant jet is further discussed in detail in Chapter 3.

In this chapter, nonetheless, we focus mainly on the free convection scenarios formed over flat plates. In this case, a temperature difference between the fluid and plate, if $T_\infty > T_s$, results in an downward flow (see Figure 6.16c). Note that the flow would be upward if $T_\infty < T_s$. In this case, the velocity of fluid particles is zero at $y = 0$ while it gradually increases upwards. The exposed still air to the vertical plate is continuously heated, becomes lighter and rises upward as its density is decreased. At this stage, a boundary layer is formed around a vertical or inclined plate, which can be again laminar or turbulence in accordance with the geometry of the plate, fluid properties, and the temperature difference between the fluid and surface. The analytical approach can be utilized to simplify the Navier–Stokes equation and derive laminar boundary layer equation when the flow is

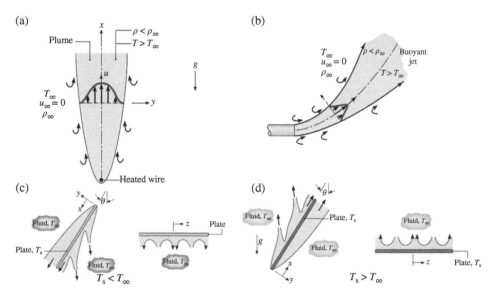

Figure 6.16 Various forms of free convection (a, b) plume and buoyant jet and (c, d) bounded by vertical, horizontal, or inclined plate. *Source:* Based on Incropera [1].

two-dimensional, steady, and incompressible as described in [8], which is beyond the scope of this book. Here, we introduce the simplified and transformed equation in which the nature of the free convection is described by the dimensionless number of Grashof (Gr), which is presenting the ratio of buoyancy forces to viscous forces in a flow:

$$\text{Gr} = \frac{g\beta(T_s - T_\infty)L^3}{\nu^2} \tag{6.63}$$

where, L is the length scale, and β is the thermal expansion and defined as:

$$\beta = -\frac{1}{\rho}\left(\frac{\partial\rho}{\partial T}\right)_p \tag{6.64}$$

Note that in the solution of the mentioned laminar boundary layer problem, the thermal expansion is used to relate the density of fluid with only the temperature difference, which is known as **Boussinesq approximation**:

$$\beta = -\frac{1}{\rho}\left(\frac{\partial\rho}{\partial T}\right)_p \approx -\frac{1}{\rho}\frac{\rho_\infty - \rho}{T_\infty - T} \tag{6.65}$$

$$\Longrightarrow (\rho_\infty - \rho) = \beta(T - T_\infty)$$

For an ideal gas, we can further assume $\rho = \dfrac{p}{RT}$, so we can derive Eq. (6.64) as:

$$\beta = -\frac{1}{\rho}\left(\frac{\partial\rho}{\partial T}\right)_p$$

$$\Longrightarrow \beta = -\frac{1}{\rho}\left(\frac{\partial\left(\frac{p}{RT}\right)}{\partial T}\right)_p = -\frac{1}{\rho}\frac{p}{RT^2} = -\frac{1}{\frac{p}{RT}}\frac{p}{RT^2} = \frac{1}{T} \tag{6.66}$$

where, T is the absolute temperature.

The role of Grashof number in identification of a flow characteristics in free convections is essential and similar to the role of Reynolds number in forced convections. Thus, in a similar manner, we expect the Nusselt number to be a function of Gr and Pr numbers in free convections. In many realistic flows such as mixed ventilation in buildings, however, a combination of both forced and free convections is the effective means of heat transfer. This implies that correlations or advanced numerical technique are required to address both convection factors. The ratio of $\dfrac{\text{Gr}}{\text{Re}^2}$, known as Richardson number (Ri), is a suitable index to understand the significance of each convection type. In general, when $\text{Ri} \approx 1$, then both convection parts are similarly important in a particular flow. When $\text{Ri} \ll 1$, then forced convection is dominant and inversely it can be neglected when $\text{Ri} \gg 1$. To identify the flow regime in a free convection, an indicator similar to Reynolds number in forced convection is usually employed, known as Rayleigh number (Ra), which is a product of Grashof and Prandtl numbers:

$$\text{Ra}_x = \text{Gr}_x\text{Pr} = -\frac{g\beta(T_s - T_\infty)x^3}{\nu\alpha} \tag{6.67}$$

The critical value of $\text{Ra}_{x,c}$ to reach a turbulence free convection is accounted when $\text{Ra}_{x,c} \approx 10^9$.

Table 6.1 Nusselt numbers in some of the free convection scenarios.

Value	Flow condition	Case
$\overline{\mathrm{Nu}_L} = 0.59\mathrm{Ra}_L^{0.25}$ [9]	Laminar flow: $10^4 \leq \mathrm{Ra}_L \leq 10^9$	Vertical flat plate
$\overline{\mathrm{Nu}_L} = 0.10\mathrm{Ra}_L^{0.33}$ [10]	Turbulent flow: $10^9 \leq \mathrm{Ra}_L \leq 10^{13}$	
$\overline{\mathrm{Nu}_L} = \left\{0.825 + \dfrac{0.387\mathrm{Ra}_L^{1/6}}{\left[1 + (0.429/\mathrm{Pr})^{9/16}\right]^{8/27}}\right\}^2$ [11]	Wide range of Ra_L	Vertical flat plate
$\overline{\mathrm{Nu}_L} = 0.68 + \left\{\dfrac{0.670\mathrm{Ra}_L^{1/4}}{\left[1 + (0.429/\mathrm{Pr})^{9/16}\right]^{4/9}}\right\}$ [11]	Laminar flow: $\mathrm{Ra}_L \leq 10^9$	Vertical flat plate

6.2.5.1 Empirical Correlations for Vertical Surfaces

Many empirical correlations are developed to represent the heat transfer in laminar and turbulence flows related to a free convection flow. In the case of having a heated vertical plate, the format of the obtained correlations is mainly in the below form (fluid properties are mainly defined with the **film temperature** defined as $T_f = \dfrac{T_s + T_\infty}{2}$):

$$\overline{\mathrm{Nu}_L} = \frac{\overline{h}L}{k} = C\mathrm{Ra}_L^n \tag{6.68}$$

where, $n = \dfrac{1}{4}$ when the flow is laminar and $n = \dfrac{1}{3}$ when it is turbulent. Examples of some correlations are provided in Table 6.1.

Example 6.6

Calculate the average convective coefficient for the vertical windward wall of Figure 6.E6, which has a $H = 2.4$ m height and a constant surface temperature of $T_{WW} = 21^\circ$C. The approaching wind has the temperature of $T_\infty = 45^\circ$C and velocity of $U_\infty = 0.1$ m/s.

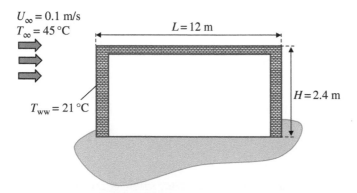

$U_\infty = 0.1$ m/s
$T_\infty = 45\,^\circ$C

$L = 12$ m

$H = 2.4$ m

$T_{ww} = 21\,^\circ$C

Figure 6.E6 Case study building exposed to an approaching wind.

Solution

First, we obtain the air properties for the film temperature of $\dfrac{T_s + T_\infty}{2} = 33°C$ as $\rho_a = 1.14$ kg/m^3, $\nu_a = 16.55 \times 10^{-6}$ m^2/s, $c_{pa} = 1.007$ kJ/kgK, $k_a = 26.64 \times 10^{-3}$ W/mK, $\alpha_a = 232.764 \times 10^{-7}$ m^2/s, $\beta_a = 3.27916 \times 10^{-3}$ 1/K and Pr = 0.713. Then, we calculate the Ra number over the windward wall (H) to explore the flow regime using Eq. (6.67):

$$\text{Ra}_H = -\frac{9.81 \times 3.27916 \times 10^{-6} \times (21 - 45) \times 2.4^3}{16.55 \times 10^{-6} \times 232.764 \times 10^{-7}} = 2.77 \times 10^{10}$$

Hence, the flow is turbulent, and we can use the below equation in Table 6.1 to find the Nusselt number:

$$\overline{\text{Nu}_L} = \left\{ 0.825 + \frac{0.387 \text{Ra}_L^{1/6}}{\left[1 + (0.429/\text{Pr})^{9/16} \right]^{8/27}} \right\}^2 = \left\{ 0.825 + \frac{0.387 \times (2.77 \times 10^{10})^{1/6}}{\left[1 + (0.429/0.713)^{9/16} \right]^{8/27}} \right\}^2 = 348.84$$

Therefore, the average convective coefficient for the windward wall can be expressed as:

$$\overline{h} = \frac{\overline{\text{Nu}_L} k}{L} = \frac{348.84 \times 26.64 \times 10^{-3}}{2.4} = 3.87 \text{ W/m}^2\text{K}$$

In the calculation of $\overline{h}_{\text{conv}}$ related to the forced convection, the approaching wind can be used as demonstrated earlier to calculate Nu. However, the local wind on each surface of the building can be significantly different from U_∞. To resolve such shortcomings and also to improve the empirical equations, many other correlations are developed for the building internal and external surfaces as presented in Chapter 7. In many of those correlations though the local velocity can be more accurately calculated with the numerical techniques as demonstrated in Chapter 12.

6.2.5.2 Empirical Correlations for Horizontal and Inclined Surfaces

For a horizontal or inclined plate, the buoyancy force, which is in the direction of gravity force, is not necessarily parallel to the surface angle. Thus, the fluid particle velocity parallel to the plate will be impacted by the force perpendicular to the plate. The result of interference of both buoyancy components on fluid particle is formation of a three-dimensional flow over the plate. As demonstrated in Figure 6.17, in these cases we can simultaneously have both two-dimensional and three-dimensional flows in either side of a plate.

For an inclined plate, for the case of a downward buoyancy exposed to a cold plate ($T_s < T_\infty$) as shown in Figure 6.16c, the fluid particles at the top of the plate are impacted by x-component of the gravity ($g \cos \theta$) while y-component ($g \sin \theta$) drag the boundary layer down to the top surface; the whole process remains as a three-dimensional airflow and the overall heat transfer drops as the fluid particles' velocity is reduced at the top of the surface. For the bottom surface (see Figure 6.16c), the y-component just move the fluid particles from the surface to the warmer environment, forming a three-dimensional flow between the cold plate and the surrounding warmer air. Three-dimensional process as shown in Figure 6.16c for the z-direction of the hotter air movement enhances the heat transfer in total. A same argue can be conducted for a hot plate scenario as shown in Figure 6.16d.

(a)   (b)

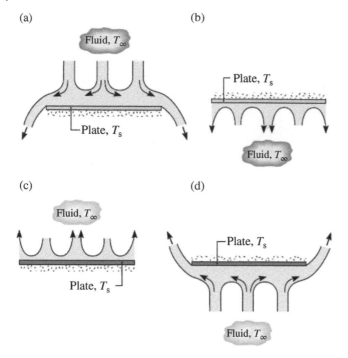

(c)   (d)

Figure 6.17 Free convection heat transfer from downward/upward facing cold/hot plates: (a, b) $T_s < T_\infty$ and (c, d) $T_s > T_\infty$ (*source:* and Fig. 9.7).

In the empirical calculation of Nusselt number for inclined plate scenarios, in either top or bottom surfaces of cold or hot plates, it is recommended to replace g by $g \sin \theta$ only if $0° \le \theta \le 60$ in the calculation of Ra_L number, and then to utilize the related correlation for the vertical surfaces with the new Ra_L in Table 6.1 [12]. For the bottom and top surfaces in the cold and hot plates, it is also suggested to use other empirical correlations [13].

In case of having a horizontal plate, the heat transfer rate is directly influenced by the facing of the plate (downward or upward) and the temperature condition of the surface (cold or hot) as stated before (see Figure 6.17). For the upper surface of a hot plate or lower surface of a cold plate, the average Nusselt numbers can be found as [14]:

$$\overline{Nu_L} = 0.54 \, Ra_L^{0.25} \left(10^4 \le Ra_L \le 10^7, 0.7 \le Pr\right) \tag{6.69}$$

$$\overline{Nu_L} = 0.15 \, Ra_L^{0.33} \left(10^7 \le Ra_L \le 10^{11}, \text{ all } Pr\right) \tag{6.70}$$

While for the lower surface of a hot plate or upper surface of a cold plate, we can derive the average Nusselt number as below [15]:

$$\overline{Nu_L} = 0.52 \, Ra_L^{0.2} \left(10^4 \le Ra_L \le 10^9, 0.7 \le Pr\right) \tag{6.71}$$

The heat transfer calculation is particularly important in the building applications when the convective heat transfer from internal and external surface are aimed as discussed in

Chapters 8 and 9. If the plate does have various shapes, then the length scale should be modified before applying the introduced correlations:

$$L \equiv \frac{A_s}{P} \tag{6.72}$$

where, A_s is the plate surface area and P is the plate perimeter.

Example 6.7

Calculate the average convective coefficient for the horizontal surfaces within the building of Figure 6.E7 if the room temperature is $T_r = 19°C$ while the ceiling and floor have temperatures of $T_c = 40°C$ and $T_f = 15°C$, respectively. Recalculate the values if the temperatures are inversed.

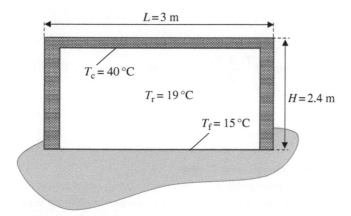

Figure 6.E7 Case study building with nonuniform internal surface temperatures.

Solution

For the first sets of temperatures, we expect a stable condition in room as also shown in Figure 6.15. First, we calculate Ra on the ceiling and floor as shown in the below table:

T_f	ρ_a (kg/m³)	c_{pa} (kJ/kg K)	k_a (W/m K) × 10^{-3}	β_a (1/K) × 10^{-3}	α_a (m²/s) × 10^{-7}	ν_a (m²/s) × 10^{-6}	Pr	Ra
29.5°C	1.15	1.007	26.39	3.316	227.84	16.23	0.713	5.0×10^{10}
17°C	1.20	1.007	25.46	3.462	210.88	15.04	0.715	1.2×10^{10}

Thus, we can use Eq. (6.70) with a film temperature of $\dfrac{T_c + T_r}{2} = 29.5°C$ to find the Nusselt numbers for the ceiling as a hot-plate upper surface:

$$\overline{Nu_L} = 0.15\,Ra_L^{0.33} = 508.63$$

$$\Longrightarrow \overline{h} = 5.59\,W/m^2K$$

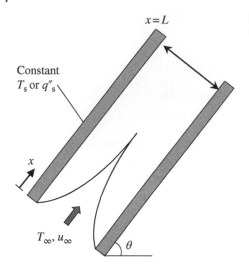

$x = L$

Constant
T_s or q''_s

x

T_∞, u_∞

θ

Figure 6.18 Free convection heat transfer in a channel flow.

Also, for the floor with a film temperature of $\dfrac{T_f + T_r}{2} = 17\degree\,\text{C}$, we can again use Eq. (6.70) as a cold-plate lower surface:

$$\overline{\text{Nu}_L} = 0.15\,\text{Ra}_L^{0.33} = 314.00$$

$$\Longrightarrow \overline{h} = 0.33\,\text{W/m}^2\,\text{K}$$

6.2.5.3 Empirical Correlations for Channel Flows and Cavities

Free convection within parallel plate channels also has some engineering applications. In particular, it can be expanded to chimneys in buildings. As it can be seen in Figure 6.18, different boundary conditions can be considered for either side of a channel, consisting of two parallel plates. In general, the width to length (S/L) ratio of the channel dictates the flow regime in terms of being laminar or turbulence in addition to reaching or not reaching to a fully developed condition. Further details of calculation of the Nusselt number in each of these scenarios can be found in literature [16–18].

Free convection in cavities (see Figure 6.19) are very complex and highly depending on surface conditions of a cavity. Many studies suggest various Nusselt numbers under different conditions to estimate free convection in cavities [19–21]. The free convection in rooms can be expanded as a free convection in cavities in case of the absence of forced convection although the heat transfer in rooms is mainly more complex than the cases with uniform temperatures on opposite walls (bottom/top or side walls) as shown in Figure 6.19. The utilized correlations hence are mainly not representing accurate results, justifying the implementation of simplistic correlations with a same order of magnitude of the accuracy or the utilization of more advanced methods like finite volume method as described in the following chapters.

6.3 Radiation

The major difference in radiation with convection and conduction is related to the condition in which the existence of a medium to transfer the heat is not necessary. Hence, radiation can bring energy from a higher surface temperature to a lower one while there is no matter between them. As shown in Figure 6.20, a basic radiation heat transfer mechanism can be

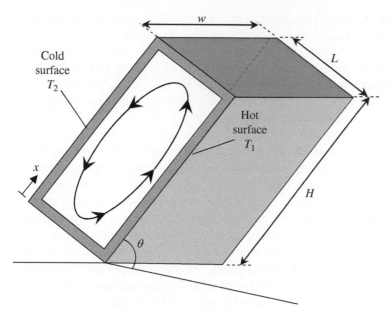

Figure 6.19   Free convection heat transfer in cavities.

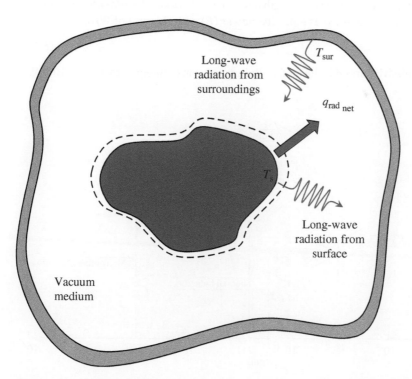

Figure 6.20   Radiation mechanism between a solid object and its surrounding environment.

explained with a solid object with temperature of T_s surrounded by an enclosed environment with temperature of T_{sur} within a vacuum condition to ignore other heat transfer mechanisms, including conduction and convection. This condition also helps to neglect the interference of the medium in radiation in terms of reflection or absorption. As it can be seen in Figure 6.20, the surface temperatures are set as $T_{sur} < T_s$, thus a radiation heat transfer from the object toward the surrounding environment ($q_{rad_{net}}$) can be expected until both surfaces reach an equilibrium condition.

In general, the radiation emission from a surface includes various wavelengths from very short such as X-rays to very long such as microwaves as seen in Figure 6.21. However, only a range of the spectrum contributes to changing the temperature of the impacted matter, including UV, visible, and infrared as highlighted in Figure 6.21. An obvious example of **short-wave** or **solar radiation** is the one emitted from the sun. Also, in building science the **long-wave radiation** on exterior surfaces of buildings with respect to the environment is significant and should be considered in its energetic analysis as discussed in Chapter 9. In an indoor condition, multiple longwave radiations between all objects with various surface temperatures occur although in many circumstances, these interactions are neglected to reduce the complexity of the modelling as well as its related computational cost. Some elements such as radiator heaters, however, cannot be neglected and should be considered in the calculations. In general, emitted radiation is related to the **spectral distribution** in which it is generated and the **directional distribution** in which it is directed.

In an ideal condition, when a surface homogenously emits radiation fluxes per unit of area to all directions, and over all the wavelengths, it has a behaviour of a **blackbody,** and the radiation emission rate is defined as **emissive power** (E) with the unit of W/m²:

$$E = \varepsilon \sigma T_s^4 \tag{6.73}$$

where, ε is the emissivity and σ is the Stefan–Boltzmann constant equal to 5.670×10^{-8} W/m². K^4.

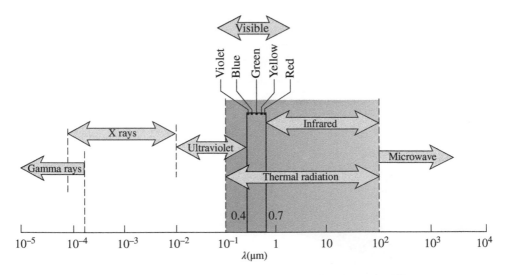

Figure 6.21 Electromagnetic radiation spectrum. *Source:* Based on Incropera [1].

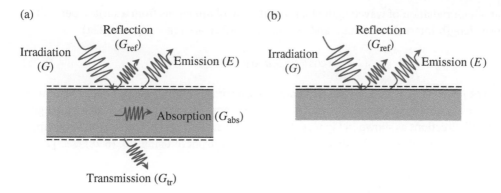

Figure 6.22 Radiation energy balance upon (a) semi-transparent and (b) opaque surfaces.

On the other side, when the radiation is incident upon a surface per unit of area from all directions and over all wavelengths, it is defined as ***irradiation*** (G) with the unit of W/m². Moreover, radiosity (J) with the unit of W/m² is defined as the radiant energy leaving a surface. As it can be imagined, the irradiation can incident upon an ***opaque*** or ***transparent*** medium. As illustrated in Figure 6.22, the incident upon an opaque surface result in absorption or reflection by the surface though it can be also transmitted through the surface if it is transparent in addition to a part of it being reflected and a part being absorbed. Hence, for transparent surfaces, we can write the below expression:

$$\rho + \alpha + \tau = 1 \tag{6.74}$$

where, ρ is the reflectivity, α is absorptivity, and τ is the transmissivity.

Moreover, in the case of opaque surfaces ($\tau = 0$), Eq. (6.74) becomes:

$$\rho + \alpha = 1 \tag{6.75}$$

The energetic balance upon an opaque surface can be written as (see Figure 6.22):

$$J = E + G_{ref} = E + \rho G \tag{6.76}$$

Note that for transparent surfaces, radiosity has two parts, leaving from top and bottom surfaces. The net radiative heat flux (q''_{rad}) can be also defined as irradiation subtracted from radiosity:

$$q''_{rad} = J - G$$
$$\Longrightarrow q''_{rad} = (E + \rho G) - G \tag{6.77}$$
$$\Longrightarrow q''_{rad} = \varepsilon \sigma T_s^4 - (\rho - 1)G = \varepsilon \sigma T_s^4 - \alpha G$$

6.3.1 Total Emission, Irradiation, and Radiosity

Since radiation propagates in all directions and all wavelengths from a surface, we can define ***spectral, hemispherical emissive power*** (E_λ) with the unit of W/m² μm as the rate at which radiation of wavelength of λ is emitted to all directions from a surface per unit of wavelength interval $d\lambda$ about λ and per unit of surface area (see Figure 6.23a):

$$E_\lambda(\lambda) = \int_0^{2\pi} \int_0^{\pi/2} I_{\lambda,e}(\lambda,\theta,\phi) \cos\theta \sin\theta \, d\theta d\phi \tag{6.78}$$

The **total hemispherical emissive power** (E), or simply known as **total emissive power**, can be also defined as the rate at which radiation is emitted per unit area at all wavelengths and all directions as shown in Figure 6.23b. The total emissive power can be calculated from integration of all emitted radiation from dA_1 to hemisphere using Eq. (6.78) (see [1] for details):

$$E = \int_0^\infty E_\lambda(\lambda) d\lambda$$

$$\Longrightarrow E = \int_0^\infty \int_0^{2\pi} \int_0^{\pi/2} I_{\lambda,e}(\lambda,\theta,\phi) \cos\theta \sin\theta d\theta d\phi \, d\lambda \tag{6.79}$$

Now, by assuming that the emitted radiation from a particular surface is independent of directions ($I_{\lambda,e}(\lambda,\theta,\phi) = I_{\lambda,e}(\lambda)$), the surface can be defined as a **diffuse emitter,** and we can simplify Eq. (6.78) to find the total emissive power as:

$$E_\lambda(\lambda) = \pi I_{\lambda,e}(\lambda)$$

$$\Longrightarrow E = \int_0^\infty E_\lambda(\lambda) d\lambda = \pi I_e \tag{6.80}$$

where, I_e is the **total intensity** of the emitted radiation.

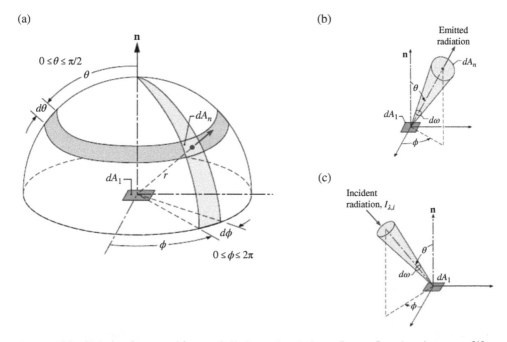

Figure 6.23 Emission from a sold area of dA_1 into a hemisphere. *Source:* Based on Incropera [1].

We can also define **spectral irradiation** (G_λ) with the unit of W/m² μm as the rate at which radiation of wavelength λ is incident on a surface per unit area of surface and per unit wavelength interval $d\lambda$ about λ (see Figure 6.23c):

$$G_\lambda(\lambda) = \int_0^{2\pi} \int_0^{\pi/2} I_{\lambda,i}(\lambda, \theta, \phi) \cos\theta \sin\theta d\theta d\phi \tag{6.81}$$

Similarly, we can define **total irradiation** (G) with the unit of W/m², which is the rate at which radiation is incident per unit area from all directions and all wavelengths as:

$$G = \int_0^\infty G_\lambda(\lambda) d\lambda$$

$$\Longrightarrow G = \int_0^\infty \int_0^{2\pi} \int_0^{\pi/2} I_{\lambda,i}(\lambda, \theta, \phi) \cos\theta \sin\theta d\theta d\phi d\lambda \tag{6.82}$$

With a same approach used before to define a diffuse emitter, we can also define **diffuse incident irradiation** as:

$$G_\lambda(\lambda) = \pi I_{\lambda,i}(\lambda)$$

$$\Longrightarrow G = \int_0^\infty G_\lambda(\lambda) d\lambda = \pi I_i \tag{6.83}$$

We can also define spectral radiosity (J_λ) with the unit of W/m² μm as the rate at which radiation of wavelength λ leaves a unit area of the surface per unit wavelength interval $d\lambda$ about λ:

$$J_\lambda(\lambda) = \int_0^{2\pi} \int_0^{\pi/2} I_{\lambda,e+r}(\lambda, \theta, \phi) \cos\theta \sin\theta d\theta d\phi \tag{6.84}$$

Here, the radiosity, as defined before, is a combination of the reflection and emission intensities. Hence, with a similar approach, the **total radiosity** (J) with the unit of W/m² can be represented as:

$$J_\lambda(\lambda) = \pi I_{\lambda,e+r}(\lambda)$$

$$\Longrightarrow J = \int_0^\infty J_\lambda(\lambda) d\lambda = \pi I_{e+r} \tag{6.85}$$

6.3.2 Black and Grey Bodies

In the calculation of the total emission, irradiation, and radiosity of a real opaque surface, the spectral forms represented in Eqs. (6.78), (6.81), and (6.84) were integrated over the hemisphere. To further define the diffuse characteristics, the assumption of spatial independency from θ and ϕ was utilized to reach directionally independent equations as presented in Eqs. (6.80), (6.83), and (6.85). Now, the definition of a **blackbody** helps as a benchmark to provide a better quantification of a realistic surface, emitting under any circumstances. Hence, we can define the blackbody concept to obey from the below rules:

1) A blackbody absorbs all incident radiation, regardless of wavelength and direction.
2) Although the radiation emitted from a blackbody is a function of wavelength and temperature, it is independent of direction. On the other words, a blackbody is a diffuse emitter.
3) For a prescribed temperature and wavelength, no surface can emit more energy than a blackbody.

The shown cavity of Figure 6.24a is a close representative of a black body, which is a perfect absorber and emitter. Radiation encounters multiple reflections to be completely absorbed, assuming the aperture is small enough to satisfy a blackbody condition. Once the blackbody is in an equilibrium temperature of T as presented in Figure 6.24b, the possible emission due to this temperature from the aperture of the blackbody should be independent of direction and should be diffuse ($I_{\lambda, e} = I_{\lambda, b}$). Also, any small surface that can be positioned anywhere in the cavity receives diffuse irradiation (G_λ) as shown in Figure 6.24c equal to the diffuse emission from the blackbody ($G_\lambda = E_{\lambda, b}$).

Using the definition of a blackbody, we can now explain **emissivity** or **hemispherical emissivity** over all wavelengths and all directions as a property of a particular surface to be the ratio of the total emissive power of that surface ($E(T)$) to the total emissive power of a blackbody ($E_b(T)$) at the same temperature of T. Thus, a similar expression as: Eq. (6.73) can be found as:

$$\varepsilon(T) \equiv \frac{E(T)}{E_b(T)} = \frac{E(T)}{\sigma T^4} \tag{6.86}$$

6.3.3 View Factor

Radiation exchange between two surfaces is a significant mechanism of heat transfer in the nature and engineering problems. The main factors in the radiation exchange, as shown before, are the geometry of the surfaces, their temperatures, and their radiative

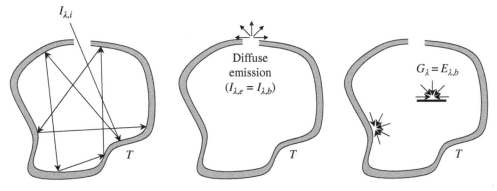

Figure 6.24 Definition of an isothermal blackbody cavity.

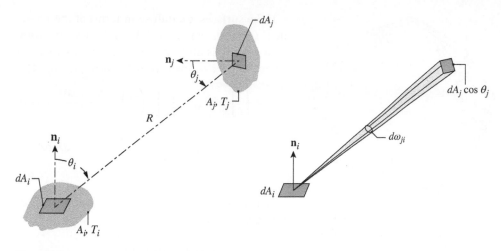

Figure 6.25 View factor of F_{ij}. *Source:* Based on Incropera [1].

properties. Another important factor is recognized as the relative orientation of surfaces toward each other, which is known as **view factor** (VF). As shown in Figure 6.25, a fraction of the radiation emitted from a surface i to intercept surface j is defined as the view factor F_{ij}:

$$F_{ij} = \frac{q_{i \rightarrow j}}{A_i J_i} \tag{6.87}$$

The view factor is calculated based on an integration of all radiations, leaving from elemental areas in the surface i, which are intercepting elemental areas in the surface j:

$$q_{i \rightarrow j} = \int_{A_i} \int_{A_j} dq_{i \rightarrow j} \tag{6.88}$$

where A_i and A_j are the areas. We can also derive $dq_{i \rightarrow j}$ as the radiation intensity, which leaves dA_i and reaches dA_j from Eq. (6.85) as:

$$dq_{i \rightarrow j} = I_{e+r,i} \cos \theta_i \, dA_i d\omega_{j-i} \tag{6.89}$$

The solid angle $d\omega_{j-i}$, as an angle of which dA_j can be seen from dA_i, can be also replaced by:

$$d\omega_{j-i} = \frac{\cos \theta_j dA_j}{R^2} \tag{6.90}$$

Thus, we can rewrite Eq. (6.89) using Eq. (6.90) as:

$$dq_{i \rightarrow j} = I_{e+r,i} \cos \theta_i dA_i \frac{\cos \theta_j dA_j}{R^2} \tag{6.91}$$

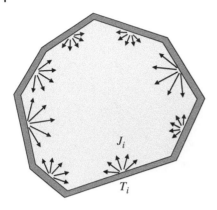

Figure 6.26 View factor in a diffuse emitter and reflector enclosure with uniform radiosities.

If the surfaces are diffuse in terms of the reflection (r) and emission (e) while the radiosity ($J_i = \pi I_{e+r, i}$) is uniform over the surfaces, then we can obtain the below expression using Eq. (6.91):

$$q_{i \to j} = \int_{A_i} \int_{A_j} dq_{i \to j} = \int_{A_i} \int_{A_j} \left(\frac{J_i}{\pi} \right) \frac{\cos \theta_i \cos \theta_j}{R^2} dA_j dA_i$$

$$\Longrightarrow q_{i \to j} = J_i \int_{A_i} \int_{A_j} \frac{\cos \theta_i \cos \theta_j}{\pi R^2} dA_j dA_i$$

(6.92)

Hence, we can combine Eqs. (6.87) and (6.92) to find the view factor as:

$$F_{ij} = \frac{1}{A_i} \int_{A_i} \int_{A_j} \frac{\cos \theta_i \cos \theta_j}{\pi R^2} dA_j dA_i \qquad (6.93)$$

We can similarly define the view factor F_{ji} as the radiation emitted from a surface j and intercepting surface i:

$$F_{ji} = \frac{1}{A_j} \int_{A_i} \int_{A_j} \frac{\cos \theta_i \cos \theta_j}{\pi R^2} dA_j dA_i \qquad (6.94)$$

As it can be seen from Eqs. (6.93) and (6.94), we can obtain the below relation, known as the **reciprocity relation**, between view factors in the case of having two surfaces in the enclosure:

$$F_{ij} A_i = F_{ji} A_j \qquad (6.95)$$

Equation 6.95 can be expanded to an enclosure with multiple surfaces of N as depicted in Figure 6.26. Therefore, one can develop the below conservation equation for the view factors, known as the **summation rule**:

$$\sum_{j=1}^{N} F_{ij} = 1 \qquad (6.96)$$

This equation implies that in an enclosed system, when all leaving radiations from a surface should intercept other surfaces, we will have N^2 view factors. Reciprocity relation equations give $N(N-1)/2$ equations while the summation rule provides N more equations. The rest of needed equations, nevertheless, should be determined with the mathematical definition of the view factor as described earlier in Eq. (6.94). These mathematical expressions are numerically or analytically solved for many cases as provided in the related tables such as Figure 6.27 for some of two-dimensional geometries and Figures 6.28 and 6.29 for some of three-dimensional geometries.

Geometry	Relation
Parallel plates with midlines connected by perpendicular 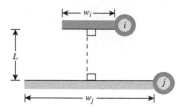	$$F_{ij} = \frac{[(W_i + W_j)^2 + 4]^{1/2} - [(W_j + W_i)^2 + 4]^{1/2}}{2W_i}$$ $$W_i = w_i/L, \; W_j = w_j/L$$
Inclined parallel plates of equal width and a common edge 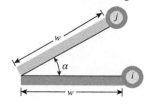	$$F_{ij} = 1 - \sin\left(\frac{\alpha}{2}\right)$$
Perpendicular plates width a common edge 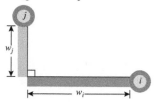	$$F_{ij} = \frac{1 + (w_j/w_i) - [1 + (w_j/w_i)^2]^{1/2}}{2}$$
Three-sided enclosure 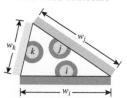	$$F_{ij} = \frac{w_i + w_j - w_k}{2w_i}$$

Figure 6.27 View factors for some of two-dimensional geometries. *Source:* Based on Incropera [1].

Example 6.8

Calculate the view factor of the 2D building surface A_1 in Figure 6.E8.1 in accordance with other surfaces.

Solution

We assume a unit depth for all the dimensions to calculate the areas. We also define the street canyon aspect ratio as $AR = \dfrac{L}{H}$. We use Figure 6.27 to find F_{12} and F_{13}. For this purpose, perpendicular plates with a common edge are used to resolve F_{12}:

Geometry	Relation
Aligned parallel rectangles (Figure 13.4) 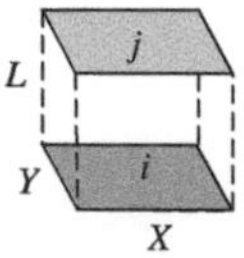	$\bar{X} = X/L,\ \bar{Y} = Y/L$ $F_{ij} = \dfrac{2}{\pi \bar{X}\bar{Y}} \left\{ \ln \left[\dfrac{(1 + \bar{X}^2)\,(1 + \bar{Y}^2)}{1 + \bar{X}^2 + \bar{Y}^2} \right]^{1/2} \right.$ $+ \bar{X}(1 + \bar{Y}^2)^{1/2} \tan^{-1} \dfrac{\bar{X}}{(1 + \bar{Y}^2)^{1/2}}$ $\left. + \bar{Y}(1 + \bar{X}^2)^{1/2} \tan^{-1} \dfrac{\bar{Y}}{(1 + \bar{X}^2)^{1/2}} - \bar{X} \tan^{-1} \bar{X} - \bar{Y} \tan^{-1} \bar{Y} \right\}$
Coaxial parallel disks (Figure 13.5) 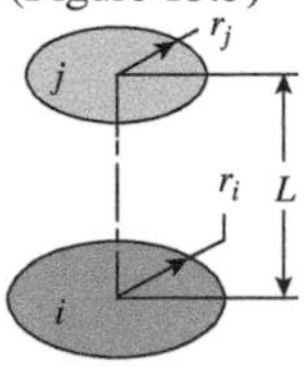	$R_i = r_i/L,\ R_j = r_j/L,$ $S = 1 + \dfrac{1 + R_j^2}{R_i^2}$ $F_{ij} = \dfrac{1}{2} \{ S - [S^2 - 4\,(r_j/r_i)^2]^{1/2} \}$
Perpendicular rectangles with a common edge (Figure 13.6) 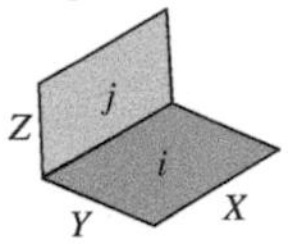	$H = Z/X,\ W = Y/X$ $F_{ij} = \dfrac{1}{\pi W} \left(W \tan^{-1} \dfrac{1}{W} + H \tan^{-1} \dfrac{1}{H} \right.$ $- (H^2 + W^2)^{1/2} \tan^{-1} \dfrac{1}{(H^2 + W^2)^{1/2}}$ $+ \dfrac{1}{4} \ln \left\{ \dfrac{(1 + W^2)(1 + H^2)}{1 + W^2 + H^2} \left[\dfrac{W^2(1 + W^2 + H^2)}{(1 + W^2)(W^2 + H^2)} \right]^{W^2} \right.$ $\left. \left. \times \left[\dfrac{H^2(1 + H^2 + W^2)}{(1 + H^2)(H^2 + W^2)} \right]^{H^2} \right\} \right)$

Figure 6.28 View factors for some of three-dimensional geometries.

$$F_{21} = \frac{1 + (H/L) - \left[1 + (H/L)^2\right]^{0.5}}{2}$$

Also, we have from reciprocity ($F_{21}L = F_{12}H$). Thus, we can find F_{12} as:

$$F_{12} = AR\left\{\frac{1 + AR^{-1} - \left[1 + AR^{-2}\right]^{0.5}}{2}\right\}$$

Moreover, parallel plates with midlines connected by perpendicular are used to find F_{13} as follows:

$$W_1 = W_2 = \frac{H}{L} = AR^{-1}$$

$$F_{13} = \frac{\left[4AR^{-2} + 4\right]^{0.5} - 2}{2AR^{-1}}$$

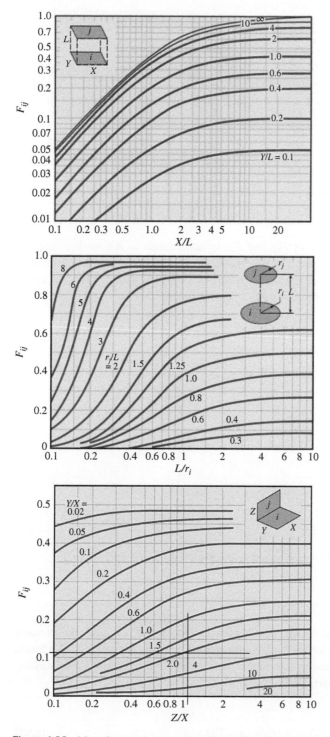

Figure 6.29 View factors for parallel and perpendicular surfaces.

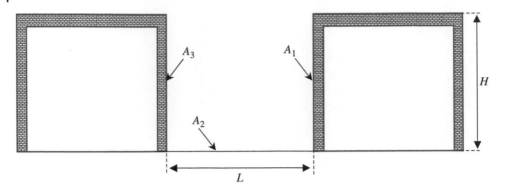

Figure 6.E8.1 Case study street canyon.

Now, we can plot the variation of F_{12} and F_{13} in accordance to AR as illustrated in Figure 6.E8.2. As expected, with the increase of AR, F_{12} elevates to about 0.475 due to a larger A_2, which can receive more irradiation from A_1. Inversely, F_{13} significantly drops as the distance between A_1 and A_3 increases.

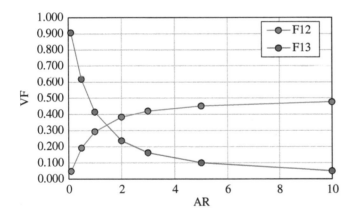

Figure 6.E8.2 Variation of view factors in accordance to the street canyon's aspect ratio.

6.3.4 Radiation Exchange at Surfaces

Understanding of the net rate at which radiation leaves a surface is very essential in many engineering systems. As we understood it from the earlier sections, the net radiative flux is equivalent to the difference between radiosity ($J_i = E_i + \rho_i G_i$) and irradiation on a surface. Thus, we can rewrite it for a particular surface of i as:

$$q_i = A_i(J_i - G_i)$$
$$\implies q_i = A_i([E_i + \rho_i G_i] - G_i) = A_i(E_i + [\rho_i - 1]G_i) \tag{6.97}$$

We know that for the opaque surfaces $\alpha_i = 1 - \rho_i$, so we can modify Eq. (6.97) as:

$$q_i = A_i(E_i - \alpha_i G_i) \tag{6.98}$$

Also, we can replace the emissive power of the surface based on the value for a blackbody found as $\alpha_i = \varepsilon_i$ and $E_i = \varepsilon_i E_{bi}$, so we can rewrite $J_i = E_i + \rho_i G_i$ as below:

$$J_i = \varepsilon_i E_{bi} + (1 - \varepsilon_i)G_i$$

$$\Longrightarrow G_i = \frac{J_i - \varepsilon_i E_{bi}}{(1 - \varepsilon_i)} \tag{6.99}$$

Now, applying Eq. (6.99) into Eq. (6.97) ($q_i = A_i(J_i - G_i)$) gives:

$$q_i = A_i\left(J_i - \left[\frac{J_i - \varepsilon_i E_{bi}}{1 - \varepsilon_i}\right]\right) \tag{6.100}$$

Or, Eq. (6.100) can be rearranged in the form of:

$$q_i = \frac{E_{bi} - J_i}{\left(\dfrac{1 - \varepsilon_i}{\varepsilon_i A_i}\right)} \tag{6.101}$$

This equation represents a driving potential of $E_{bi} - J_i$ between two surfaces while a resistance of $R_{ri} = \left(\dfrac{1 - \varepsilon_i}{\varepsilon_i A_i}\right)$ occurs, which is called ***surface radiative resistance***. A useful conclusion from this net heat radiative equation is that for a large surface of A_i, one can assume that $A_i \rightarrow \infty$, and thus, the resistance (R_{ri}) would be zero, meaning that the surface can be assumed and treated as a blackbody with $\varepsilon_i = 1$ since $E_{bi} = J_i$. This is a practical assumption in the calculation of radiative heat fluxes between surfaces of buildings and their surrounding environments as it will be discussed in further details in Chapter 8.

6.3.5 Radiation Network

The net radiative heat flux on a surface can be written in terms of other radiative fluxes within an enclosure. This leads to form a network of radiative fluxes between all the participating surfaces of a particular enclosure. For this purpose, we can start to find all N irradiations reaching a specific surface-i as:

$$A_i G_i = \sum_{j=1}^{N} F_{ji} A_j J_i \tag{6.102}$$

These irradiations should be used with the radiosity from surface-i to form the net radiative flux (see Eq. (6.97)). Also, after applying the reciprocity relation ($F_{ij}A_i = F_{ji}A_j$) shown in Eq. (6.95), we can rewrite Eq. (6.102) as:

$$q_i = A_i\left(J_i - \sum_{j=1}^{N} F_{ij} J_i\right) \tag{6.103}$$

(a)

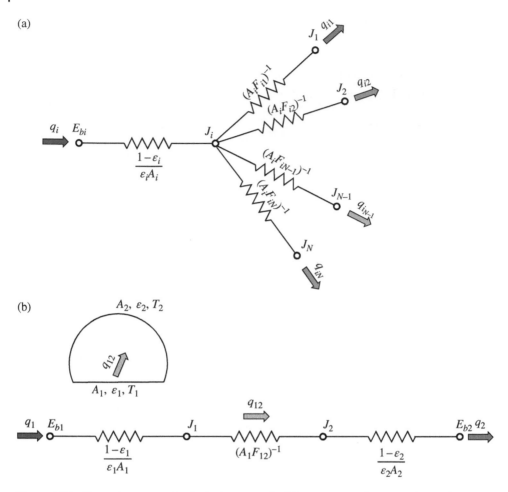

(b)

Figure 6.30 Network concept of radiative heat transfer (a) between a surface and other surfaces in an enclosure, and (b) for an enclosure with two surfaces.

Now, we apply the summation rule introduced in Eq. (6.96) ($\sum_{j=1}^{N} F_{ij} = 1$) into Eq. (6.103) to further reach to:

$$q_i = A_i \left(\sum_{j=1}^{N} F_{ij} J_i - \sum_{j=1}^{N} F_{ij} J_j \right)$$

$$\Rightarrow q_i = \sum_{j=1}^{N} F_{ij} A_i (J_i - J_j) = \sum_{j=1}^{N} q_{ij}$$

(6.104)

Equation (6.104) can be explained as a network concept as illustrated in Figure 6.30a in which q_i reaches from the surface A_i to the other surfaces through multiple components of q_{ij}. Thus, we can replace each component with the network concept in which $J_i - J_j$ is the

Large (innite) parallel planes

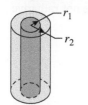

$A_1 = A_2 = A$
$F_{12} = 1$

$$q_{12} = \frac{A\sigma(T_1^4 - T_2^4)}{\dfrac{1}{\varepsilon_1} + \dfrac{1}{\varepsilon_2} - 1}$$

Long (innite) concentric cylinders

$\dfrac{A_1}{A_2} = \dfrac{r_1}{r_2}$

$F_{12} = 1$

$$q_{12} = \frac{\sigma A_1(T_1^4 - T_2^4)}{\dfrac{1}{\varepsilon_1} + \dfrac{1 - \varepsilon_2}{\varepsilon_2}\left(\dfrac{r_1}{r_2}\right)}$$

Concentric spheres

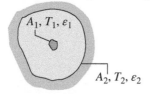

$\dfrac{A_1}{A_2} = \dfrac{r_1^2}{r_2^2}$

$F_{12} = 1$

$$q_{12} = \frac{\sigma A_1(T_1^4 - T_2^4)}{\dfrac{1}{\varepsilon_1} + \dfrac{1 - \varepsilon_2}{\varepsilon_2}\left(\dfrac{r_1}{r_2}\right)^2}$$

Small convex object in a large cavity

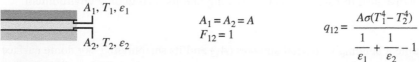

$\dfrac{A_1}{A_2} \approx 0$

$F_{12} = 1$

$q_{12} = \sigma A_1 \varepsilon_1 (T_1^4 - T_2^4)$

Figure 6.31 Radiative heat transfer in two-surface enclosures. *Source:* Based on Incropera [1].

driving potential and $\dfrac{1}{F_{ij}A_i}$ is the radiative resistance, e.g., a case for a two-surface enclosure is shown in Figure 6.30b. Combining Eqs. (6.101) and (6.104), thus, we can write:

$$\frac{E_{bi} - J_i}{\left(\dfrac{1 - \varepsilon_i}{\varepsilon_i A_i}\right)} = \sum_{j=1}^{N} \frac{J_i - J_j}{\left(\dfrac{1}{F_{ij}A_i}\right)} \tag{6.105}$$

Combining Eqs. (6.104) and (6.105), and using Eq. (6.86) for the blackbody total emissive power ($E_{bi} = \sigma T_i^4$), we can demonstrate the below expression for the net radiative exchanges between two surfaces:

$$q_1 = -q_2 = q_{12} = \frac{\sigma(T_1^4 - T_2^4)}{\dfrac{1 - \varepsilon_1}{\varepsilon_1 A_1} + \dfrac{1}{A_1 F_{12}} + \dfrac{1 - \varepsilon_2}{\varepsilon_2 A_2}} \tag{6.106}$$

Figure 6.31 demonstrates the radiative heat exchange for a number of two-surface enclosure cases.

Example 6.9

Find the radiative heat transfer between a building and its surrounding environment.

Solution

We can assume a building's external surfaces (A_1) and its surrounding sky dome surface (A_2) to form a two-surface enclosure similar to a small convex object in a large cavity. We can also assume that $A_1/A_2 \approx 0$. Now, we can use Eq. (6.106) to find the net radiative heat transfer as:

$$q_{12} = \frac{\sigma\left(T_1^4 - T_2^4\right)}{\dfrac{1-\varepsilon_1}{\varepsilon_1 A_1} + \dfrac{1}{A_1 F_{12}} + \dfrac{1-\varepsilon_2}{\varepsilon_2 A_2}}$$

$$\Longrightarrow q_{12} = \frac{\sigma\left(T_1^4 - T_2^4\right)}{\dfrac{1}{A_1}\left(\dfrac{1-\varepsilon_1}{\varepsilon_1} + \dfrac{1}{F_{12}} + \dfrac{1-\varepsilon_2}{\varepsilon_2 \dfrac{A_2}{A_1}}\right)}$$

Since, we have $\dfrac{A_2}{A_1} = \infty$, we can find $\dfrac{1-\varepsilon_2}{\varepsilon_2 \dfrac{A_2}{A_1}} = 0$. Thus, we can simplify the above equation

as:

$$q_{12} = \frac{\sigma\left(T_1^4 - T_2^4\right)}{\dfrac{1}{A_1}\left(\dfrac{1-\varepsilon_1}{\varepsilon_1} + \dfrac{1}{F_{12}}\right)}$$

Let us assume that the building surfaces can barely see each other, implying that with a high level of accuracy, we have $F_{11} = 0$. So, from the summation rule, we can conclude $F_{12} = 1$. We can then further simplify q_{12} as below:

$$q_{12} = \frac{\sigma\left(T_1^4 - T_2^4\right)}{\dfrac{1}{A_1}\left(\dfrac{1-\varepsilon_1}{\varepsilon_1} + 1\right)}$$

$$\Longrightarrow q_{12} = \frac{\sigma\left(T_1^4 - T_2^4\right)}{\dfrac{1}{\varepsilon_1 A_1}\left(1 - \varepsilon_1 + \varepsilon_1\right)}$$

$$\Longrightarrow q_{12} = \sigma\varepsilon_1 A_1\left(T_1^4 - T_2^4\right)$$

In this two-surface enclosure, hence, we can find q_{21} as:

$$q_{21} = \sigma\varepsilon_2 A_2\left(T_2^4 - T_1^4\right)$$

As it is shown in Chapter 8, this approach is used to approximate the long-wave radiation exchanges between buildings and their surrounding environments.

References

1 Incropera, F.P. (2006). *Fundamentals of Heat and Mass Transfer*. Wiley.

2 Frank, P., Incropera, D.P.D., Bergman, T.L., and Lavine, A.S. (1996). *Fundamentals of Heat and Mass Transfer*, 7e. Wiley.

3 Kays, W.M. and Whitelaw, J.H. (1967). Convective heat and mass transfer. *Journal of Applied Mechanics* 34: 254.

4 Kays, W.M. (1955). Numerical solution for laminar-flow heat transfer in circular tube. *Transactions of the ASME* 77: 1265–1274.

5 Shah, R.K. (1975). Laminar flow friction and forced convection heat transfer in ducts of arbitrary geometry. *International Journal of Heat and Mass Transfer* 18 (7): 849–862.

6 Winterton, R.H.S. (1998). Where did the Dittus and Boelter equation come from? *International Journal of Heat and Mass Transfer* 41 (4): 809–810.

7 Sieder, E.N. and Tate, G.E. (1936). Heat transfer and pressure drop of liquids in tubes. *Industrial & Engineering Chemistry* 28 (12): 1429–1435.

8 Schlichting, H. (1982). *Boundary-Layer Theory*, 843. Braun, Karlsruhe, Germany, F.R.: Grenzschicht-Theorie.

9 Warner, C.Y. and Arpaci, V.S. (1968). An experimental investigation of turbulent natural convection in air at low pressure along a vertical heated flat plate. *International Journal of Heat and Mass Transfer*. 11 (3): 397–406.

10 McAdams, W.H. (1954). *Heat Transmission* (ed. W.H. McAdams). New York: McGraw-Hill.

11 Churchill, S.W. and Chu, H.H.S. (1975). Correlating equations for laminar and turbulent free convection from a vertical plate. *International Journal of Heat and Mass Transfer* 18 (11): 1323–1329.

12 Rich, B. (1953). An investigation of heat transfer from an inclined flat plate in free convection. *Transacions of the ASME*. 75: 489–499.

13 Jiji, L.M. (2009). *Free Convection. Heat Convection*, 2e, 259–292. Berlin, Heidelberg: Springer Berlin Heidelberg.

14 Lloyd, J.R. and Moran, W.R. (1974). Natural convection adjacent to horizontal surface of various planforms. *Journal of Heat Transfer*. 96 (4): 443–447.

15 Radziemska, E. and Lewandowski, W.M. (2001). Heat transfer by natural convection from an isothermal downward-facing round plate in unlimited space. *Applied Energy*. 68 (4): 347–366.

16 Elenbaas, W. (1948). The dissipation of heat by free convection from vertical and horizontal cylinders. *Journal of Applied Physics*. 19 (12): 1148–1154.

17 Bar-Cohen, A. and Rohsenow, W.M. (1984). Thermally optimum spacing of vertical, natural convection cooled, parallel plates. *Journal of Heat Transfer*. 106 (1): 116–123.

18 Azevedo, L.F.A. and Sparrow, E.M. (1985). Natural convection in open-ended inclined channels. *Journal of Heat Transfer*. 107 (4): 893–901.

19 Ostrach, S. (1988). Natural convection in enclosures. *Journal of Heat Transfer*. 110 (4b): 1175–1190.

20 Ostrach, S. (1972). Natural convection in enclosures. In: *Advances in Heat Transfer*, vol. 8 (ed. J.P. Hartnett and T.F. Irvine), 161–227. Elsevier.

21 Catton I. (1978). Natural Convection in Enclosures.

   ... ...

7

Applications of Heat Transfer in Buildings

7   Introduction

7.1   Conduction in Walls

The simplest form of the general heat equation as presented in Chapter 6 (Eq. 6.11) is its steady state one-dimensional form while there is no heat generation inside the medium. Hence, we can rewrite Eq. (6.11) in the Cartesian coordinate as below:

$$\frac{\partial}{\partial x}\left(k\frac{\partial T}{\partial x}\right) = 0 \tag{7.1}$$

A resemblance of the above equation in buildings is a 1D plane wall without heat generation as shown in Figure 7.1. In a case that the wall conductivity is homogenous in x-direction, then we can solve the equation with double integration as shown before:

$$\frac{\partial^2 T}{\partial x^2} = 0$$
$$\Longrightarrow \int \frac{\partial T}{\partial x} dx = C_1 \tag{7.2}$$
$$\Longrightarrow T(x) = C_1 x + C_2$$

Therefore, the temperature distribution is a line in the case of a simplified wall modeling. Note that the coefficients C_1 and C_2 can be obtained by applying the boundary conditions. For example, for a Dirichlet boundary condition of $T(0) = T_1$ and $T(L) = T_2$, we can obtain:

$$(\text{at } x = 0, \ T(0) = T_1) \Rightarrow T(0) = C_1 \times 0 + C_2 = T_1$$
$$\Rightarrow C_2 = T_1$$
$$(\text{at } x = L, \ T(L) = T_2) \Rightarrow T(L) = C_1 \times L + T_1 = T_2 \tag{7.3}$$
$$\Rightarrow C_1 = \frac{T_2 - T_1}{L}$$
$$\Rightarrow T(x) = \frac{T_2 - T_1}{L} x + T_1$$

Computational Fluid Dynamics and Energy Modelling in Buildings: Fundamentals and Applications, First Edition. Parham A. Mirzaei.
© 2023 John Wiley & Sons Ltd. Published 2023 by John Wiley & Sons Ltd.

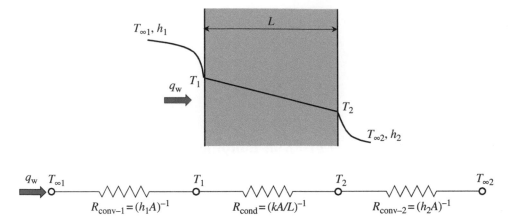

Figure 7.1 One-dimensional conduction in a plane wall.

Example 7.1

Assume a 1D wire with a length of $L = 150$ cm and a constant thermal conductivity of $k = 10$ W/m K as illustrated in Figure 7.E1. Find the temperature profile within the wire if the temperatures at the warm and cold sides of the wire are set as 100 and 30 °C, respectively.

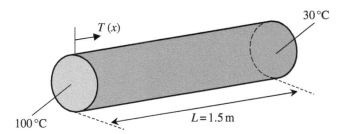

Figure 7.E1 One-dimensional wire with constant temperatures on the boundary conditions.

Solution

Using the obtained general Eq. (7.3), we have:

$$T(x) = \frac{T_2 - T_1}{L} x + T_1$$

$$\Longrightarrow T(x) = \frac{(30 - 100)}{1.5} x + 100 = -46.7x + 100$$

Example 7.2

Assume in the previous Example 7.1, as illustrated in Figure 7.E2, that we assign a constant heat flux to the cold side in Scenario 1 and then in the next scenario, insulate the cold side (Neumann condition). Find the temperature profiles within the wire in both scenarios.

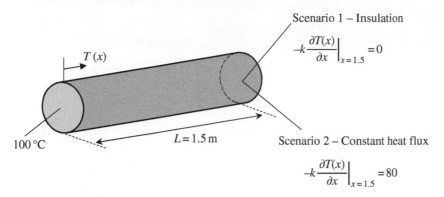

Figure 7.E2 One-dimensional wire with a constant temperature and a constant heat flux on the boundary conditions.

Solution

Scenario 1 – the left-hand side boundary obeys the same approach as before:

$$\left(\text{at } x = 0 \text{ m}, \quad T(0) = 100^\circ\text{C}\right)$$

$$\Rightarrow T(0) = C_1 \times 0 + C_2 = 100$$
$$\Rightarrow C_2 = 100$$

Now, we apply the Neumann boundary at the right-hand side:

$$\left(\text{at } x = 1.5 \text{ m}, \quad -k\frac{\partial T(x)}{\partial x}\bigg|_{x=1.5} = 0\right)$$

$$\Rightarrow \frac{\partial T(x)}{\partial x} = \frac{\partial(C_1 x + C_2)}{\partial x} = 0 = C_1$$

$$\Rightarrow T(x) = 100$$

Scenario 2 – $C_2 = 100$ can be found in a similar way as before. Now, we apply the Neumann boundary at the right-hand side:

$$\left(\text{at } x = 1.5 \text{ m}, \quad -k\frac{\partial T(x)}{\partial x}\bigg|_{x=1.5} = 80\right)$$

$$\Rightarrow -k\frac{\partial T(x)}{\partial x} = -k\frac{\partial(C_1 x + C_2)}{\partial x} = -10C_1 = 80$$

$$\Rightarrow T(x) = -8x + 100$$

Example 7.3

Now solve Example 7.1 with assigning a convection boundary condition to the cold side and find the temperature profile within the wire ($T_\infty = 25°C$ and $h = 10$ W/m² K) as shown in Figure 7.E3.

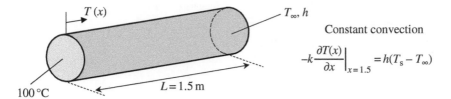

Figure 7.E3 One-dimensional wire with a constant temperature and a constant convection on the boundary conditions.

Solution

$C_2 = 100$ can be found in a similar way as before. The temperature of the cold side of the wire is not given as we have:

$$(\text{at } x = 1.5\,\text{m}, \quad T(1.5) = T_s)$$
$$\Rightarrow T_s = 1.5C_1 + 100$$

However, an extra equation can be applied at this point as using the convective flux equation:

$$\left(\text{at } x = 1.5\,\text{m}, \ -k\frac{\partial T(x)}{\partial x}\bigg|_{x=1.5} = h(T_s - T_\infty)\right)$$
$$\Rightarrow -k\frac{\partial T(x)}{\partial x} = -k\frac{\partial(C_1 x + C_2)}{\partial x} = -10C_1 = 10 \times (T_s - 25)$$
$$\Rightarrow T_s = 25 - C_1$$

Thus, we combine both above equations to achieve:

$$(25 - C_1) = 1.5C_1 + 100$$
$$\Rightarrow C_1 = -30$$

So, we obtain the general equations as:

$$T(x) = -30x + 100$$

7.2 Thermal Resistance Analogy

We obtained the 1D equation for the heat conduction without source/sink generation within a plane wall presented as Eq. (7.3). Here, a particular **nodal analogy** can be defined between the heat diffusion and electricity charge or in other words between Fourier's law and Ohm's law. Ohm's law states that the electric current (I) conducted through two points of a medium is directly proportional to the voltage difference ($\Delta V = V_1 - V_2$) across those two different points. To further relate these parameters to each other, we introduce the electricity resistance (R) of the medium. So, we can write the Ohm's law as:

$$V = RI \tag{7.4}$$

It can be similarly justified that the temperature difference between two different points $(\Delta T = T_1 - T_2)$ can be associated with the voltage while the flux between them (q_x) can be linked with the electric current. In this case, the **conduction thermal resistance** (R_{TC}) of the medium can be defined similar to the electricity resistance:

$$q_{\text{cond}} = \frac{kA}{L}(T_1 - T_2)$$
$$\Longrightarrow T_1 - T_2 = R_{\text{TCond}}q_{\text{cond}} \tag{7.5}$$

Therefore, we can define the conduction thermal resistance with the unit of K/W as:

$$R_{\text{TCond}} = \frac{L}{kA} \tag{7.6}$$

This analogy can be also seen in Figure 7.1. The same concept can be applied to the heat convection equation (more details are provided in Chapter 8 related to the convection):

$$q_{\text{conv}} = hA(T_1 - T_2)$$
$$\Longrightarrow T_1 - T_2 = R_{\text{TConv}}q_{\text{conv}} \tag{7.7}$$

where, $R_{\text{TConv}} = \dfrac{1}{hA}$ denotes the convection thermal resistance with the unit of K/W.

Therefore, using the Ohm's law analogy or nodal analogy helps to first simplify the heat transfer in walls and other buildings' elements as discussed in Chapter 8. In addition, we can consider a similar approach and apply series and parallel circuits' rules to the heat transfer resistances within buildings' elements such as walls in accordance with Kirchhoff's current law.

For example, in a series circuit as shown in Figure 7.1, the currents passing through all of the elements in a particular electricity (heat transfer) pathway are the same $(I_1 = I_2 = I_3 = \dots = I_n)$ and therefore we can conclude that the total resistance is equal to the sum of their individual resistances at this pathway:

$$R_{\text{total}} = \sum_{k=1}^{n} R_k \tag{7.8}$$

Similar rules can be applied for the combination of resistances related to convective and conductive fluxes. For example, for a series of convection and conduction resistances in a wall as shown in Figure 7.1, we can write:

$$T_1 - T_2 = R_{\text{TCond}}q_w$$
$$\Longrightarrow T_1 - T_2 = (R_{\text{TConv1}} + R_{\text{TCond1}} + R_{\text{TConv2}})q_w$$
$$\Longrightarrow T_{\infty 1} - T_{\infty 2} = \left(\frac{1}{h_1 A} + \frac{L}{kA} + \frac{1}{h_2 A}\right)q_w \tag{7.9}$$

Also, for a multi-layer wall, for example as demonstrated in Figure 7.2, we can similarly obtain the below equation:

$$T_{\infty 1} - T_{\infty 2} = \left(\frac{1}{h_1 A} + \frac{L_1}{k_1 A} + \frac{L_2}{k_2 A} + \frac{1}{h_2 A}\right)q_w \tag{7.10}$$

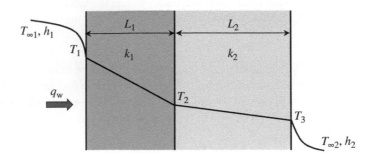

Figure 7.2 One-dimensional series conduction in a multi-layer plane wall.

We can also find the specific temperature of each node using the similar analogy of the series circuits rule with applying the condition of $I_1 = I_2 = I_3 = ... = I_n$:

$$q_1 = q_2 = q_3 = ... = q_x$$
$$\Rightarrow q_w = \frac{T_{\infty 1} - T_1}{\frac{1}{h_1 A}} = \frac{T_1 - T_2}{\frac{L_1}{k_1 A}} = \frac{T_2 - T_3}{\frac{L_2}{k_2 A}} = \frac{T_3 - T_{\infty 2}}{\frac{1}{h_2 A}}$$

(7.11)

In the case of parallel resistances as depicted in Figure 7.3, we can use the nodal analogy and justify that the voltages (temperature differences) across two shared ends of parallel electricity (heat transfer) pathways are similar ($V_1 = V_2 = V_3 = ... = V_n$). On the other hand,

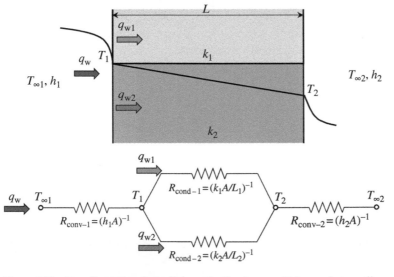

Figure 7.3 One-dimensional parallel conduction in a multi-layer plane wall.

Table 7.1 U-value for some of the common walls.

Wall type	U-value (W/m² K)
Solid brick wall	2
Cavity wall with no insulation	1.5
Insulated wall	0.18
Single glazing	4.8–5.8
Double glazing	1.2–3.7
Triple glazing	Below 1
Solid timber door	3

the total current at these ends is the sum of the currents passing through each individual electricity (heat transfer) pathways. Thus, the total resistance can be calculated as:

$$\frac{1}{R_{\text{total}}} = \sum_{k=1}^{n} \frac{1}{R_k} \tag{7.12}$$

In the case of a multi-layer wall, it is very common in the building science terminology to define the ***overall heat transfer coefficient*** or ***U-value*** with the unit of W/m² K as follows, which implies that for a 1D wall $U = 1/RA$:

$$q_{\text{w}} = UA\Delta T \tag{7.13}$$

where, ΔT is the overall temperature difference between two points of interest. Table 7.1 demonstrates U-value for some of the common wall materials and components. Note that R-value with the unit of m² K/W can be defined as the conduction thermal resistance divided by area.

Example 7.4
Apply and sketch the nodal analogy to the one-dimensional wall of Figure 7.E4.1. Find and if the boundary values are given as $T_{\infty 1} = 25°C$, $T_{\infty 2} = 15°C$ and $h_1 = 9\,\text{W/m}^2\,\text{K}$, and $h_2 = 4\,\text{W/m}^2$ K. Furthermore, the conductivity and lengths of layers are $k_1 = 10\,\text{W/mK}$, $k_2 = 5\,\text{W/mK}$, $k_3 = 7\,\text{W/mK}$, $k_4 = 12\,\text{W/mK}$, $L_1 = 0.2\,\text{m}$, $L_2 = L_3 = 0.1\,\text{m}$, and $L_4 = 0.4\,\text{m}$.

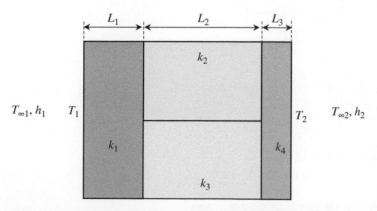

Figure 7.E4.1 One-dimensional wall with multi-layer materials.

Solution

We can find the nodal system of the wall of Figure 7.E4.1 in two different ways. In the first one as illustrated in Figure 7.E4.2a, we can consider that the convective heat flux is transferred from environment-1 (∞_1) to the layer-1 and then it is passing through layer-2 and layer-3 in a parallel format as it reaches layer-4. Then, it is transferred with a convective flux to the environment-2 (∞_2). In another approach, it is possible to vertically split layer-1 and layer-4 and hence assume that the heat between ∞_1 and ∞_2 is initially transferred throughout two different parallel pathways in the wall. As shown in Figure 7.E4.2b, this implies that the resistances associated with layer-1 and layer-4 in the parallel pathways ($\left(\dfrac{k_1 A}{2L_1}\right)^{-1}$ and $\left(\dfrac{k_4 A}{2L_4}\right)^{-1}$) have values twice of those in the previous case ($\left(\dfrac{k_1 A}{L_1}\right)^{-1}$ and $\left(\dfrac{k_4 A}{L_4}\right)^{-1}$).

First, we can find the overall resistance of the wall in each scenario using Eqs. (7.8) and (7.12) as:

Scenario-1:

$$R_{total} = \sum_{k=1}^{n} R_k = \left(\frac{1}{h_1 A} + \frac{L_1}{k_1 A} + \left(\frac{1}{\dfrac{L_2}{k_2 A/2}} + \frac{1}{\dfrac{L_3}{k_3 A/2}} \right)^{-1} + \frac{L_4}{k_4 A} + \frac{1}{h_2 A} \right) = 0.431\,\text{K/W}$$

So, from Eq. (7.10), we can find:

$$q_w = \frac{T_{\infty 1} - T_{\infty 2}}{R_{total}} = \frac{25 - 15}{0.431} = 23.20\,\text{W}$$

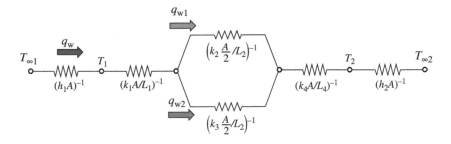

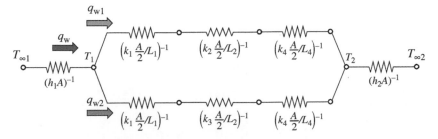

Figure 7.E4.2 Nodal system of the case study multi-layer wall.

Now, from Eq. (7.11), T_1 can be obtained as:

$$q_w = \frac{T_{\infty 1} - T_1}{\dfrac{1}{h_1 A}} = \frac{25 - T_1}{0.11}$$

$$\Longrightarrow T_1 = 25 - 23.20 \times 0.11 = 22.4°\text{C}$$

Scneario-2:

$$R_{total} = \sum_{k=1}^{n} R_k = \left(\frac{1}{h_1 A} + \left(\frac{1}{\left(\dfrac{L_1}{k_1 A/2} + \dfrac{L_2}{k_2 A/2} + \dfrac{L_4}{k_4 A/2} \right)} + \frac{1}{\left(\dfrac{L_1}{k_1 A/2} + \dfrac{L_3}{k_3 A/2} + \dfrac{L_4}{k_4 A/2} \right)} \right)^{-1} + \frac{1}{h_2 A} \right)$$

$$= 0.431 \, \text{K/W}$$

Therefore, the same procedure as Scenario-1 can be followed to find a similar value of $T_1 = 22.4°\text{C}$.

Example 7.5

Calculate the conduction through the multi-layer wall of the building presented in Figure 7. E5. The wall is exposed to indoor and outdoor temperatures of $T_{room} = 18°\text{C}$ and $T_{out} = 8°\text{C}$, respectively, while the heat convection coefficients are $h_{in} = 3 \, \text{W/m}^2$ and $h_{out} = 6 \, \text{W/m}^2$ for the internal and external walls. Assume the heat transfer to be steady state with no heat generation inside the wall. Moreover, the radiation fluxes are neglected in this case study. The properties of the wall's layers are provided in the below table.

Wall layer	Thickness (m)	Conductivity (w/m K)	Density (kg/m³)	Specific heat capacity (J/kg K)
Reinforced concrete	0.3	2.3	2,300	1,000
Insulation	0.2	0.025	20	1,030
Plasterboard	0.1	0.21	700	1,000

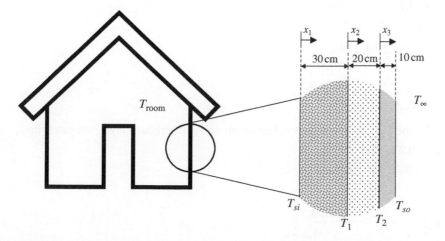

Figure 7.E5 A building wall with three layers of reinforced concrete, insulation, and plasterboard.

Solution

We know from Example 7.3 that the steady-state conduction distribution with convection boundary conditions has a linear form as $T(x) = cx + d$. Hence, for three different layers, we can obtain the below equations:

$T_1(x_1) = ax_1 + b$ for reinforced concrete layer;
$T_2(x_2) = cx_2 + d$ for the insulation layer;
$T_3(x_3) = ex_3 + f$ for plasterboard layer

First, we derive the energy balance over the control volume of the external surface where the convective flux from the environment is conducted through the wall (T_{so} is the external surface temperature):

$$h_{out}AT_{out} - T_{so}| = -kA\frac{dT_3}{dx_3}\Big|_{x_3=0.1}$$
$$\Rightarrow 6(8 - T_{so}) = -0.21e$$

We can similarly derive an energy balance equation over the interior surface between the conducted heat through the wall and the heat transferred to the room with the convection (T_{si} is the internal surface temperature):

$$h_{in}A(T_{si} - T_{room}) = -kA\frac{dT_1}{dx_1}\Big|_{x_1=0}$$
$$\Rightarrow 3(T_{si} - 18) = -2.3a$$

In addition to the above equations, we can apply the boundary conditions into each linear equation:

(at $x_1 = 0$)
$$\Rightarrow T_1(0) = T_{si} = 0 \times a + b = b$$
(at $x_3 = 0.1$)
$$\Rightarrow T_3(0.1) = T_{so} = 0.1e + f$$

For the face, which connects the reinforced concrete and insulation layers, we should have a similar temperature:

$$T_1(x_1 = 0.3) = T_2(x_2 = 0)$$
$$\Rightarrow 0.3a + b = d$$

As well as a similar heat flux:

$$-k_1\frac{dT_1}{dx_1}\Big|_{x_1=0.3} = -k_2\frac{dT_2}{dx_2}\Big|_{x_2=0}$$
$$\Rightarrow 2.3a = 0.025c$$

Similarly, for the face, which connects insulation and plasterboard layer, we can write:

$$T_2(x_2 = 0.2) = T_3(x_3 = 0)$$
$$\Rightarrow 0.2c + d = f$$

And:

$$-k_2\frac{dT_2}{dx_2}\Big|_{x_2=0.2} = -k_3\frac{dT_3}{dx_3}\Big|_{x_3=0}$$
$$\Rightarrow 0.025c = 0.21e$$

Thus, we could derive eight equations while the variables are also eight, including a, b, c, d, e, f, T_{si}, and T_{so}. The system of equations can be solved to find the linear equations as: below:

Reinforced concrete layer $T(x_1) = -0.48x_1 + 17.63$

Insulation layer $T(x_2) = -43.92x_2 + 17.49$

Plasterboard layer $T(x_3) = -5.23x_3 + 8.71$

7.3   Walls with Heat Generation

The general 1D steady-state heat conduction equation with a uniform and constant source/sink ($\dot{q}$) can be written as below:

$$\frac{\partial}{\partial x}\left(k\frac{\partial T}{\partial x}\right) + \dot{q} = 0 \tag{7.14}$$

If the conductivity is homogenous across the solid material, then we can simplify the equation to:

$$\frac{\partial^2 T}{\partial x^2} + \frac{\dot{q}}{k} = 0 \tag{7.15}$$

And, after twice of integration over the temperature, we can reach:

$$\frac{\partial T}{\partial x} = -\frac{\dot{q}}{k}x + C_1$$
$$\Longrightarrow T(x) = -\frac{\dot{q}}{k}x^2 + C_1 x + C_2 \tag{7.16}$$

Coefficient C_1 and C_2 can be obtained after applying the boundary conditions, e.g. for the constant temperatures on both walls ($T(-L) = T_1$ and $T(L) = T_2$) if the origin of x-coordinate is set to be in the middle of the plane wall, we can find:

$$(\text{at } x = -L, \ \ T(-L) = T_1)$$

$$\Rightarrow T(-L) = -\frac{\dot{q}}{k}(-L)^2 + C_1 \times -L + C_2 = T_1$$

$$(\text{at } x = L, \ \ T(L) = T_2) \tag{7.17}$$

$$\Rightarrow T(L) = -\frac{\dot{q}}{k}(L)^2 + C_1 \times L + C_2 = T_2$$

$$\Rightarrow C_1 = \frac{T_2 - T_1}{2L}$$

And:

$$C_2 = \frac{\dot{q}}{2k}L^2 + \frac{T_2 + T_1}{2L} \tag{7.18}$$

Thus, the general form of the temperature distribution is a polynomial as follows:

$$T(x) = -\frac{\dot{q}L^2}{2k}\left(1 - \frac{x^2}{L^2}\right) + \left(\frac{T_2 - T_1}{2L}\right)x + \left(\frac{T_2 + T_1}{2L}\right) \tag{7.19}$$

7.4 Convective Heat Transfer Coefficient of Exterior Walls

Heat transfer from the exterior skin of building plays a pivotal role in the energy demand calculation of buildings as it is comprehensively discussed in Chapters 8 and 9. To resolve a reasonable and accurate heat transfer due to convection, it is crucial that we present an accurate convective heat transfer coefficient (h-values) for the external and internal surfaces of buildings. These values were explored in Chapter 6 to be significantly related to Re in the forced convection. Note that Pr for air can be assumed to stay in a similar range for most of the related problems. In the free convection, however, we explored that the convective heat transfer coefficient is highly dependent on Ra. Presented empirical values, thus, can be applied to some specific cases of building surfaces to find h-values. However, there are a wide range of scenarios that these ideal correlations are not providing accurate values, which resulted in many other correlations to be developed during the past decades to estimate the convective heat transfer coefficients at internal and external surfaces as explained in the following sections.

7.4.1 Wind on Buildings' Exterior Surfaces

The wind velocity at external surfaces is the key parameter in many of the correlations to find convective heat transfer coefficients. The wind profiles presented in Chapter 3 can attain the wind value only over specific regions and not over specific surfaces. While advanced numerical techniques as presented in Chapters 10–12 are capable of providing accurate local values of wind, the application is not an easy task for many scenarios. Hence, simplified correlations are preferred methods to attain the local wind velocities as the following section explain a common one developed based on the power-law wind profile.

In this simplified model, the local wind velocity at the centroid of an external surface (V_z) is calculated based on the values converted from the power-law wind velocity profile (see Chapter 3) recorded at a proximity weather station to the place of a surface:

$$V_z = U_R \left(\frac{\delta_R}{z_R}\right)^{\alpha_R} \left(\frac{z}{\delta}\right)^{\alpha} \tag{7.20}$$

where, z is the centroid of the investigated surface. δ and δ_R are the terrain boundary layer thicknesses at the position of the surface and weather station (reference), respectively. α and α_R are also the terrain exponents at the position of the surface and weather station, respectively. U_R is the recorded wind velocity at the height of z_R, mainly in a proximity weather station. The coefficients can be found in Tables 3.1 and 3.2.

Example 7.6

Calculate the local wind velocity at the surfaces of the case study building as shown in Figure 7. E6, located in a city centre if the wind velocity and direction at the closest airport, located in an open area is recorded (at the height of 10 m) to be 3.6 m/s and North–South, respectively.

Solution

We can assume that the airport is in a 'flat and open country' area and therefore according to Tables 3.1 and 3.2, the terrain description of the meteorological station can be found as:

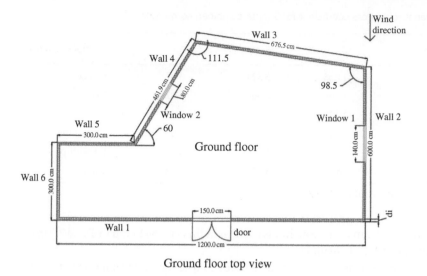

Ground floor top view

Figure 7.E6 A building layout exposed to the local wind.

$$\alpha_{met} = 0.14$$
$$V_{met} = 3.6 \, \text{m/s}$$
$$z_{met} = 10 \, \text{m}$$
$$\delta_R = 270 \, \text{m}$$

According to the wall geometry and building location in 'town and cities' category, the wind velocity at the middle of their heights can be also calculated as:

$$z = \frac{3.4}{2} = 1.7 \, \text{m}$$

$$\delta = 460 \, \text{m}$$

$$\alpha = 0.33$$

Hence, the wind velocity from Eq. (7.20) can be calculated as:

$$V_z = V_{met} \left(\frac{\delta_{met}}{z_{met}}\right)^{\alpha_{met}} \left(\frac{z}{\delta}\right)^{\alpha} = 3.6 \times \left(\frac{270}{10}\right)^{0.14} \left(\frac{1.7}{460}\right)^{0.33} = 0.9 \, \text{m/s}$$

7.4.2 Simple-Combined Correlation

This scheme is a simplified correlation to estimate the convective heat transfer coefficient at external surfaces. This method combines the radiation and convection effect together and is defined as follows:

$$h_{SC} = D + EV_z + FV_z^2 \tag{7.21}$$

Table 7.2 Material roughness coefficient for Simple-combined correlation.

Roughness index	D	E	F	Example material
1 (very rough)	11.58	5.894	0.0	Stucco
2 (Rough)	12.49	4.065	0.028	Brick
3 (medium rough)	10.79	4.192	0.0	Concrete
4 (medium smooth)	8.23	4.0	−0.057	Clear pine
5 (Smooth)	10.22	3.1	0.0	Smooth plaster
6 (very smooth)	8.23	3.33	−0.036	Glass

Source: Modified from ASHRAE [3]/ASHRAE.

where, D, E, and F are the material roughness coefficient that can be found in Table 7.2 from AHRAE Handbook of Fundamentals.

Note that this correlation is only linked to the force convection and is not taking the natural convection to the account similar to some earlier correlation such as McAdams presented as below:

$$h_{MA} = 5.7 + 3.8 V_z \tag{7.22}$$

7.4.3 TARP Correlation

In Thermal Analysis Research Program (TARP) model, the convective heat transfer coefficient is split into two components associated with forced (h_f) and natural (h_n) convection. The general TARP correlation is defined as below:

$$h_{TARP} = h_f + h_n \tag{7.23}$$

The forced component of Eq. (7.23) can be found as:

$$h_f = 2.537 W_f R_f \left(\frac{PV_z}{A}\right)^{0.5} \tag{7.24}$$

where, P is the perimeter and A is the area of the focused surface. W_f is the wind direction modifier and has a value of $W_f = 1.0$ for a windward surface and $W_f = 0.5$ for a leeward one (see Figure 7.4). R_f is the surface roughness multiplier, which can be found in Table 7.3.

For the natural component, TARP correlates the convective heat transfer coefficient to the surface type and its temperature difference with the surrounding environment (see Chapter 6). When there is no temperature difference between a surface and its surrounding environment or a surface is vertical, the following correlation can be used to calculate the natural convection component:

$$h_n = 1.31 |\Delta T|^{\frac{1}{3}} \tag{7.25}$$

When $\Delta T < 0$ and the surface has an upward facing, or $\Delta T > 0$ and the surface has a downward facing (see Figure 7.4), a modified natural convection correlation can be employed as expressed below:

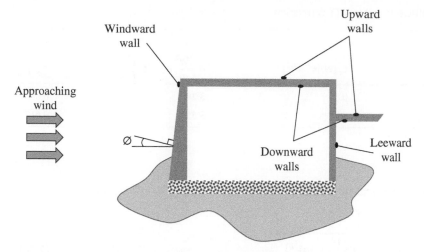

Figure 7.4 Windward, leeward, upward, and downward faces.

Table 7.3 Material roughness coefficient for TARP correlation.

Roughness index	R_f	Example material
1 (very rough)	2.17	Stucco
2 (rough)	1.67	Brick
3 (medium rough)	1.52	Concrete
4 (medium smooth)	1.13	Clear pine
5 (Smooth)	1.11	Smooth plaster
6 (very smooth)	1.00	Glass

Source: Hittle and Lawrie [10]/Public Domain

$$h_n = \frac{9.482|\Delta T|^{\frac{1}{3}}}{7.283 - |\cos \emptyset|} \tag{7.26}$$

Note that $\emptyset$ is the surface tilt angle as shown in Figure 7.4 and $\Delta T = T_\infty - T_{so}$. Finally, when $\Delta T > 0$ and the surface has an upward facing, or $\Delta T < 0$ and the surface has a downward facing, the modified natural convection correlation has the below form:

$$h_n = \frac{1.810|\Delta T|^{\frac{1}{3}}}{1.328 + |\cos \emptyset|} \tag{7.27}$$

7.4.4 MoWiTT Correlation

The Mobile Window Thermal Test (MoWiTT) model is based on the measurements conducted at the MoWiTT facility by Yazdanian and Klems (1994). The correlation applies

Table 7.4 Coefficients in MoWiTT correlation.

Wind direction	C_t (W/m² K⁴/³)	$a \left(\dfrac{W^2}{m} K(m/s) \right)^b$	b (−)
Windward	0.84	3.26	0.89
Leeward	0.84	3.55	0.617

Source: Based on [11, 12].

to the very smooth vertical surfaces (e.g. window glass) in the low-rise buildings, which has the general form as:

$$h_{\text{MoWiTT}} = \sqrt{\left[C_t (\Delta T)^{\frac{1}{3}} \right]^2 + \left[a V_z^b \right]^2}$$ (7.28)

where, C_t is the turbulent natural convection constant while a and b are constant values that are given in Table 7.4.

Note that this correlation is not a suitable match for rough surfaces and high-rise surfaces

7.4.5 DOE-2 Correlation

DOE-2 is a combination of the MoWiTT and Building Loads Analysis and System Thermodynamics (BLAST) correlations and is calculated by the following equation:

$$h_{\text{DOE}-2} = \sqrt{\left[h_n \right]^2 + \left[a V_z^b \right]^2}$$ (7.29)

The natural convection component (h_n) is calculated as described in TARP correlation. a and b can be again found from Table 7.4. When the surface is not smooth and has a roughness, the following modification can be applied to Eq. (7.29) (R_f is given in Table 7.3):

$$h_{\text{DOE}-2-\text{modified}} = h_n + R_f (h_{\text{DOE}-2} - h_n)$$ (7.30)

7.4.6 Adaptive Correlations

Some of the developed models to estimate the convection heat transfer coefficients benefit from a range of correlations in accordance with the surface stability in terms of free convection (see Chapter 6) and the surface sheltering condition against wind. Example is the presented adaptive correlation model in Table 7.5. Taking the vertical wall in a leeward condition as an example, h_f can be selected from TARP, MoWiTT, DOE2, Emmel Vertical, Nusselt Jurges, McAams, and Mitchel model while h_n Models can be chosen from ASHRAE Vertical Wall, Alamdari Hammond Vertical Wall, Fohanno Polidor Vertical Wall, and ISO 15099 Windows model.

Table 7.5 Adaptive convection algorithm details.

#	Surface classification	Heat flow direction	Wind direction	h_f models	h_n models
1	Roof stable	Down	Any	• TARPWindward • MoWiTTWindward • DOE2Windward • NusseltJurges • BlockenWindward • EmmelRoof • ClearRoof	• WaltonStableHorizontal • TiltAlamdariStableHorizontal
2	Roof unstable	Up	Any	• TARPWindward • MoWiTTWindward • DOE2Windward • NusseltJurges • BlockenWindward • EmmelRoof • ClearRoof	• WaltonUnstableHorizontal • TiltAlamdariUnstableHorizontal
3	Vertical wall windward	Any	Windward	• TARPWindward • DOE2Windward • MoWiTTWindward • NusseltJurges • McAdams • Mitchell • BlockenWindward • EmmelVertical	• ASHRAEVerticalWall • AlamdariHammondVerticalWall • FohannoPolidoriVerticalWall • ISO15099Windows
4	Vertical wall leeward	Any	Leeward	• TARPLeeward • MoWiTTLeeward • DOE2Leeward • EmmelVertical • NusseltJurges • McAdams • Mitchell	• ASHRAEVerticalWall • AlamdariHammondVerticalWall • FohannoPolidoriVerticalWall • ISO15099Windows

Source: Based on Beausoleil-Morrison [13].

Example 7.7

Assume the building layout of Figure 7.E7.1. Wind is approaching from north-south direction with a velocity of 3.6 m/s. The wall surface (smooth) temperature is $T_{so} = 12°C$ and the outdoor temperature is $T_\infty = 10°C$. All the window surface conditions are assumed to be the same as the wall surface. Calculate the external convection heat transfer coefficients of 12 external walls using the below correlations:

1) Simple-Combined
2) TARP
3) MoWiTT
4) DOE-2

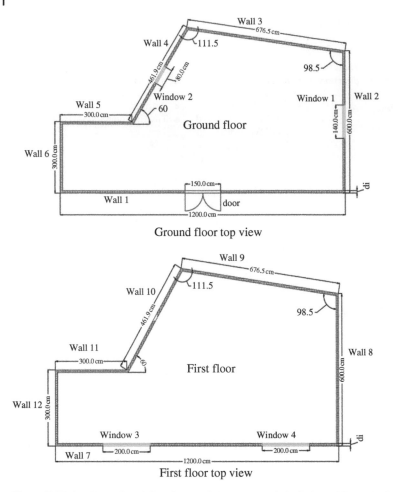

Figure 7.E7.1 Ground and first floors of the case study building of Example 7.6.

Solution

Velocity at the building walls can be calculated similar to Example 7.6. Taking Wall-1 as an example, the detailed calculations are shown below.

(1) Simple-Combined

As the external layer material of the walls are smooth plaster, the material roughness coefficient from ASHRAE Handbook can be found in the coefficients from Table 7.2 as $D = 10.22; E = 3.1; F = 0$. This implies that the velocities at the center of surfaces can be found as:

Wall ID	Material roughness coefficient				
	Material	D	E	F	V_z (m/s)	*h* (W/m² K)
Wall_1, Wall_2, Wall_3, Wall_4, Wall_5, Wall_6	Plaster	10.22	3.1	0	0.90	13.01
Wall_7, Wall_8, Wall_9, Wall_10, Wall_11, Wall_12	Plaster	10.22	3.1	0	1.29	14.23

Hence, the convective heat coefficient according to eq. 7.21 can be found as:

$$h_{SC} = 10.22 + 3.1 \times 0.9 + 0 = 13.01 \text{ W/m}^2\text{K}$$

(2) TARP

Wall-1 is a leeward surface with a smooth material, so from Table 7.3, we have $W_f = 0.5$ and $R_f = 1.11$. Perimeter and area of Wall-1 are calculated as:

$$P = (12 + 3.4) \times 2 = 30.8 \text{ m}$$

$$A = 12 \times 3.4 = 40.8 \text{ m}^2$$

Hence, the forced-convection component according to Eq. (7.24) can be calculated as below:

$$h_f = 2.537 W_f R_f \left(\frac{PV_z}{A}\right)^{1/2} = 2.537 \times 0.5 \times 1.11 \times \left(\frac{30.8 \times 0.9}{40.8}\right)^{1/2} = 1.16 \text{ W/m}^2\text{K}$$

Wall-1 is a vertical wall, therefore, Eq. (7.25) is applied to resolve the natural convection part ($\Delta T = T_\infty - T_{so} = 2°C$):

$$h_n = 1.31|\Delta T|^{\frac{1}{3}} = 1.31 \times |2|^{\frac{1}{3}} = 1.65 \text{ W/m}^2\text{K}$$

The total h-value of Wall-1, hence, can be expressed as:

$$h_{TARP} = h_f + h_n = 1.16 + 1.65 = 2.81 \text{ W/m}^2\text{K}$$

With the same approach, the total convection heat coefficient from TARP correlation for the rest of surfaces can be calculated as shown in the below table.

Wall ID	Length (m)	Width (m)	Perimeter (m)	Area (m^2)	Surface direction	W_f	R_f	Material	h_f (W/m^{2}K)	h_n (W/m^{2}K)	h_{TARP} (W/m^{2}K)
Wall_1	12.0	3.4	30.8	40.8	Leeward	0.5	1.11	Plaster	1.16	1.65	2.81
Wall_2	6.0	3.4	18.8	20.4	Parallel	0.75	1.11	Plaster	1.92	1.65	3.57
Wall_3	6.8	3.4	20.3	23.0	Windward	1	1.11	Plaster	2.51	1.65	4.16
Wall_4	4.6	3.4	16.0	15.7	Windward	1	1.11	Plaster	2.70	1.65	4.35
Wall_5	3.0	3.4	12.8	10.2	Windward	1	1.11	Plaster	2.99	1.65	4.64
Wall_6	3.0	3.4	12.8	10.2	Parallel	0.75	1.11	Plaster	2.24	1.65	3.89
Wall_7	12.0	3.4	30.8	40.8	Leeward	0.5	1.11	Plaster	1.39	1.65	3.04
Wall_8	6.0	3.4	18.8	20.4	Parallel	0.75	1.11	Plaster	2.31	1.65	3.96
Wall_9	6.8	3.4	20.3	23.0	Windward	1	1.11	Plaster	3.01	1.65	4.66
Wall_10	4.6	3.4	16.0	15.7	Windward	1	1.11	Plaster	3.24	1.65	4.89
Wall_11	3.0	3.4	12.8	10.2	Windward	1	1.11	Plaster	3.59	1.65	5.24
Wall_12	3.0	3.4	12.8	10.2	Parallel	0.75	1.11	Plaster	2.69	1.65	4.34

(3) MoWiTT

Wall-1 is a leeward surface; thus, we can find the coefficients from Table 7.4:

$$C_t = 0.84 \frac{\text{W}}{\text{m}^2\text{K}^{\frac{4}{3}}}$$

$$a = \frac{3.55 \text{W}}{\text{m}^2 \text{K} \left(\dfrac{\text{m}}{\text{s}}\right)^b}$$

$$b = 0.617$$

Hence, according to Eq. (7.28), we can derive the h-value as ($\Delta T = T_{\infty} - T_{so} = 2°\text{C}$):

$$h_{\text{MoWiTT}} = \sqrt{\left[C_t(\Delta T)^{\frac{1}{3}}\right]^2 + \left[aV_z^b\right]^2} = \sqrt{\left[0.84 \times (2)^{\frac{1}{3}}\right]^2 + [3.55 \times 0.9^{0.617}]^2} = 3.49 \,\text{W/m}^2\text{K}$$

Wall ID	Surface direction	Convection constant (C_t)	a	b	$h_{\text{MoWiTT}} \left(\dfrac{\text{W}}{\text{m}^2\text{K}}\right)$
Wall_1	Leeward	0.84	3.55	0.62	3.49
Wall_2	Parallel	0.84	3.41	0.75	3.32
Wall_3	Windward	0.84	3.26	0.89	3.15
Wall_4	Windward	0.84	3.26	0.89	3.15
Wall_5	Windward	0.84	3.26	0.89	3.15
Wall_6	Parallel	0.84	3.41	0.75	3.32
Wall_7	Leeward	0.84	3.55	0.62	4.29
Wall_8	Parallel	0.84	3.41	0.75	4.26
Wall_9	Windward	0.84	3.26	0.89	4.23
Wall_10	Windward	0.84	3.26	0.89	4.23
Wall_11	Windward	0.84	3.26	0.89	4.23
Wall_12	Parallel	0.84	3.41	0.75	4.26

(4) DOE-2

Wall-1 is a vertical and leeward surface, and we can use Eq. (7.29) with similar coefficients as the previous correlation ($a = 3.55 \,\text{W/m}^2\text{K} \left(\dfrac{\text{m}}{\text{s}}\right)^b, b = 0.617$) to find the corresponding h-value:

$$h_{\text{DOE}-2} = \sqrt{[h_n]^2 + [aV_z^b]^2} = \sqrt{\left[1.31 \times (2)^{\frac{1}{3}}\right]^2 + [3.55 \times 0.9^{0.617}]^2} = 3.71 \,\text{W/m}^2 \text{K}$$

Again, all the h-values can be presented as the below table.

Wall ID	Surface direction	a	b	$h_n \left(\dfrac{\text{W}}{\text{m}^2\text{K}}\right)$	$h_{\text{DOE}-2}$ (W/m^2 K)
Wall_1	Leeward	3.55	0.62	1.65	3.71
Wall_2	Parallel	3.41	0.75	1.65	3.55
Wall_3	Windward	3.26	0.89	1.65	3.40
Wall_4	Windward	3.26	0.89	1.65	3.40
Wall_5	Windward	3.26	0.89	1.65	3.40
Wall_6	Parallel	3.41	0.75	1.65	3.55
Wall_7	Leeward	3.55	0.62	1.65	4.47
Wall_8	Parallel	3.41	0.75	1.65	4.45
Wall_9	Windward	3.26	0.89	1.65	4.42
Wall_10	Windward	3.26	0.89	1.65	4.42
Wall_11	Windward	3.26	0.89	1.65	4.42
Wall_12	Parallel	3.41	0.75	1.65	4.45

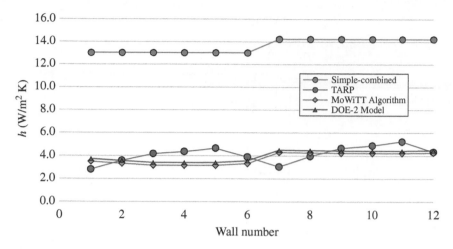

Figure 7.E7.2 External convective heat transfer coefficients over the case study building's walls using various schemes.

The obtained h-value from all the schemes for 12 walls can be shown in Figure 7.E7.2. As it can be seen, these values significantly vary from correlation to correlation and hence the justification of their selection is paramount of importance in the building-related problems in accordance with the given conditions.

7.5 Convection on Interior Walls

Wind effect is mainly absent in indoor spaces, which makes the free convection as the dominant factor in many cases, especially when there is no mechanical ventilation in a room. Some of the common correlation applied in many building energy simulation tools to calculate the interior convection heat transfer coefficients are presented in this section.

7.5.1 Khalifa Correlation

This correlation is developed based on the temperature difference for various types of surfaces as presented in Figure 7.5. The values can be, hence, represented in Table 7.6; where, ΔT is the surface temperature between wall surface and air temperature.

7.5.2 Walton Correlation

Walton correlations have been adopted for different surface types from ASHRAE, which is based on ΔT and $\emptyset$ (tilt angle). The utilized equations are presented equation in Table 7.7. An unstable condition indicates the direction of flow relative to the surfaces as discussed in Chapter 6. In an unstable condition, the warmer air on the lower surfaces or colder air on the upper surfaces can leave the associated surfaces to promote the air circulation. On the other hand, in a stable condition the surfaces are mainly blocking the movement of warmer air on the upper surfaces or colder air on the lower surfaces and thus minimize a potential air circulation.

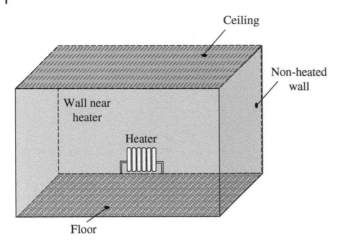

Figure 7.5 Different surface types in a regular room.

Table 7.6 Different surface types in Khalifa correlation.

Surface type	Equation	
Wall away from heat	$h = 2.07|\Delta T|^{0.23}$	(7.31)
Ceiling away from heat	$h = 2.72|\Delta T|^{0.13}$	(7.32)
Wall near heater	$h = 1.98|\Delta T|^{0.32}$	(7.33)
Non-heated walls	$h = 2.30|\Delta T|^{0.24}$	(7.34)
Ceiling	$h = 3.10|\Delta T|^{0.17}$	(7.35)

Source: Based on Khalifa and Marshall [14].

Table 7.7 Different surface types in Walton correlation.

Surface type	Equation	
Vertical wall for natural convection	$h = 3.10|\Delta T|^{\frac{1}{3}}$	(7.36)
Unstable horizontal or tilt	$h = \dfrac{9.482|\Delta T|^{\frac{1}{3}}}{7.283 - |\cos \emptyset|}$	(7.37)
Stable horizontal or tilt	$h = \dfrac{1.810|\Delta T|^{\frac{1}{3}}}{1.283 + |\cos \emptyset|}$	(7.38)

Source: Based on Walton GNUSNBoSUSDoEBESB [5]

Table 7.8 Different surface types in Alamdari–Hammond correlation.

Surface type	Equation	
Vertical wall	$h = \left\{ \left[1.5\left(\frac{\|\Delta T\|}{H}\right)^{\frac{1}{4}} \right]^6 + \left[1.23\|\Delta T\|^{\frac{1}{3}} \right]^6 \right\}^{1/6}$	(7.39)
Unstable horizontal	$h = \left\{ \left[1.4\left(\frac{\|\Delta T\|}{D_h}\right)^{\frac{1}{4}} \right]^6 + \left[1.63\|\Delta T\|^{\frac{1}{3}} \right]^6 \right\}^{1/6}$	(7.40)
Stable horizontal	$h = 0.6\left(\frac{\|\Delta T\|}{D_h}\right)^{\frac{1}{5}}$	(7.41)

Source: Based on Alamdari and Hammond [15].

7.5.3 Alamdari–Hammond Correlation

This correlation is developed based on the temperature difference between a surface and its surrounding air (ΔT) in addition to the hydraulic diameter of the space ($D_h = \frac{4A}{P}$). The associated equations are presented in Table 7.8.

7.5.4 Awbi–Hatton Heated Floor Correlation

This correlation is similarly developed based on ΔT and hydraulic diameter as the related equations in accordance with different surface types are presented in Table 7.9.

7.5.5 Fisher–Pedersen Correlation

This model defines the h-value based on the air exchange rate (ACH) through a space. The related equations are presented in Table 7.10.

7.5.6 Goldstein–Novoselac Correlation

Some of the correlations are developed to be even more specific to a particular design. For example, in the case of a bare windows within perimeter zones with highly glazed spaces served by overhead slot-diffuser based air systems, we can demonstrate the related equations as presented in Table 7.11. Note that WWR is the window to wall ratio, L is the length

Table 7.9 Different surface types in Awbi–Hatton heated floor correlation.

Surface type	Equation	
Heated floor	$h = \dfrac{2.175\|\Delta T\|^{0.308}}{D_h^{\,0.076}}$	(7.42)
Heated wall	$h = \dfrac{1.823\|\Delta T\|^{0.293}}{D_h^{\,0.076}}$	(7.43)

Source: Based on Awbi and Hatton [16].

Table 7.10 Different surface types in Fisher–Pedersen correlation.

Surface type	Equation
Diffuser walls	$h = 1.208 + 1.012 \times \text{ACH}^{0.604}$ (7.44)
Diffuser ceiling	$h = 2.234 + 4.099 \times \text{ACH}^{0.503}$ (7.45)
Diffuser floor	$h = 3.873 + 0.082 \times \text{ACH}^{0.98}$ (7.46)

Source: Based on Fisher and Pedersen [17].

Table 7.11 Different surface types in Goldstein–Novoselac correlation.

Surface type	Equation
WWR < 50% with window in upper part of wall	$h = 0.117\left(\dfrac{\dot{V}}{L}\right)^{0.8}$ (7.47)
WWR < 50% with window in lower part of wall	$h = 0.093\left(\dfrac{\dot{V}}{L}\right)^{0.8}$ (7.48)
WWR > 50%	$h = 0.103\left(\dfrac{\dot{V}}{L}\right)^{0.8}$ (7.49)
Walls located below a window	$h = 0.063\left(\dfrac{\dot{V}}{L}\right)^{0.8}$ (7.50)
Walls located above a window	$h = 0.093\left(\dfrac{\dot{V}}{L}\right)^{0.8}$ (7.51)
Ceiling diffuser floor	$h = 0.048\left(\dfrac{\dot{V}}{L}\right)^{0.8}$ (7.52)

Source: Based on Goldstein and Novoselac [18].

of exterior wall with glazing in the zone, and $\dot{V} = \text{ACH} \times \forall$ is the air flow rate of the system ($\forall$ is the volume).

7.5.7 Fohanno–Polidori Correlation

some of the correlations are highly related to the Rayleigh number ($\text{Ra}_H{}^*$) as explained in Chapter 6. For example, the correlation by Fohanno–Polidori can be presented as below [1]:

$$h = \begin{cases} 1.332\left(\frac{|\Delta T|}{H}\right)^{1/4}, & \text{Ra}_H{}^* \leq 6.3 \times 10^9 \\ 1.235 e^{0.0467H}|\Delta T|^{0.316}, & \text{Ra}_H{}^* > 6.3 \times 10^9 \end{cases} \tag{7.53}$$

Example 7.8

There is a heater and a diffuser placed on the Wall-1 of the layout as illustrated in Figure 7. E8.1. The outdoor temperature is given as 18°C while the indoor temperature is 20°C (celling temperature is 15°C). Moreover, a diffuser is installed on Wall-1 with a constant ACH = 0.15. Calculate the convection heat transfer coefficient for all the internal walls and ceiling with the below correlations:

1) Khalifa
2) Walton
3) Alamdari–Hammond
4) Awbi–Hatton heated floor
5) Fisher–Pedersen correlations

Solution

1) Khalifa

Wall-1 is near the heater; therefore, we can use Eq. (7.33):

$$h_{\text{Wall\_1}} = 1.98|\Delta T|^{0.32} = 1.98 \times |18 - 20|^{0.32} = 2.47\,\text{W/m}^2\,\text{K}$$

As Wall -2 and Wall-6 are non-heated walls, we can find h-value using Eq. (7.34):

$$h_{\text{Wall\_2}} = h_{\text{Wall\_6}} = 2.30|\Delta T|^{0.24} = 2.30 \times |18 - 20|^{0.24} = 2.72\,\text{W/m}^2\,\text{K}$$

Wall-3, Wall-4, and Wall-5 are located away from the heater; thus, we can use Eq. (7.31):

$$h_{\text{Wall\_3}} = h_{\text{Wall\_4}} = h_{\text{Wall\_5}} = 2.07|\Delta T|^{0.23} = 2.07 \times |18 - 20|^{0.23} = 2.43\,\text{W/m}^2\,\text{K}$$

Eventually, we can calculate the h-value of the ceiling using Eq. (7.35):

$$h_{\text{ceiling}} = 3.10|\Delta T|^{0.17} = 3.1 \times |15 - 20|^{0.17} = 4.08\,\text{W/m}^2\,\text{K}$$

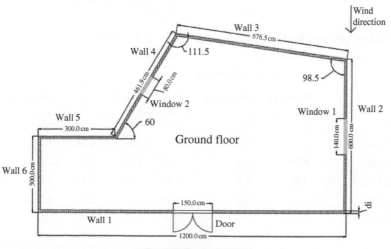

Ground floor top view

Figure 7.E8.1 Ground floor of the case study building.

2) Walton model

For the vertical walls of Wall-1 to Wall-6 in a natural convection, we can use Eq. (7.36):

$$h_{\text{Wall\_1-6}} = 3.10|\Delta T|^{\frac{1}{3}} = 3.10 \times |18 - 20|^{\frac{1}{3}} = 3.91 \text{ W/m}^2 \text{ K}$$

For the ceiling, as it is an unstable horizontal surface, therefore, we can employ Eq. (7.37) as ($\emptyset = 0°$):

$$h_{\text{ceiling}} = \frac{9.482|\Delta T|^{\frac{1}{3}}}{7.283 - |\cos\emptyset|} = \frac{9.482 \times |18 - 20|^{\frac{1}{3}}}{7.283 - |\cos 0|} = 2.58 \text{ W/m}^2 \text{ K}$$

3) Alamdari–Hammond

For the vertical Wall-1 to Wall-6, we can use Eq. (7.39):

$$h_{\text{Wall\_1-6}} = \left\{ \left[1.5 \left(\frac{|\Delta T|}{H} \right)^{\frac{1}{4}} \right]^6 + \left[1.23|\Delta T|^{\frac{1}{3}} \right]^6 \right\}^{1/6}$$

$$= \left\{ \left[1.5 \times \left(\frac{|18 - 20|}{3.4} \right)^{\frac{1}{4}} \right]^6 + \left[1.23 \times |18 - 20|^{\frac{1}{3}} \right]^6 \right\} = 1.63 \text{ W/m}^2 \text{ K}$$

For the ceiling, again, as it is an unstable horizontal surface, Eq. (7.40) can be utilized after the calculation of D_h as:

$$D_h = \frac{4A}{P} = \frac{4 \times 64.05}{35.384} = 7.24$$

$$h_{\text{ceiling}} = \left\{ \left[1.4 \left(\frac{|\Delta T|}{D_h} \right)^{\frac{1}{4}} \right]^6 + \left[1.63|\Delta T|^{\frac{1}{3}} \right]^6 \right\}^{1/6}$$

$$= \left\{ \left[1.4 \times \left(\frac{|15 - 20|}{7.24} \right)^{\frac{1}{4}} \right]^6 + \left[1.63 \times |18 - 20|^{\frac{1}{3}} \right]^6 \right\} = 2.79 \text{ W/m}^2 \text{K}$$

Details of required parameters for each wall are presented in the below table.

| | Alamdari–Hammond | | | | | | |
ID	Type	Width (m)	Height (m)	P (m)	A (m²)	D_h	h (W/m² K)
Wall-1	Vertical wall	12	3.4	30.80	40.8	5.30	1.63
Wall-2	Vertical wall	6	3.4	18.80	20.4	4.34	1.63
Wall-3	Vertical wall	6.76	3.4	20.33	23.0	4.53	1.63
Wall-4	Vertical wall	4.62	3.4	16.04	15.7	3.92	1.63
Wall-5	Vertical wall	3	3.4	12.80	10.2	3.19	1.63
Wall-6	Vertical wall	3	3.4	12.80	10.2	3.19	1.63
Ceiling	Unstable horizontal	—	—	35.38	64.1	7.24	2.79

4) Awbi–Hatton heated floor

Wall-1 is being actively heated; thus, we can use Eq. (7.43):

$$D_h = \frac{4A}{P} = \frac{4 \times 40.8}{30.8} = 5.30$$

$$h_{wall\_1} = \frac{1.823|\Delta T|^{0.293}}{D_h^{0.076}} = \frac{1.823 \times |18 - 20|^{0.293}}{5.299^{0.076}} = 1.97\,\text{W/m}^2\,\text{K}$$

5) Fisher–Pedersen

It is assumed that a diffuser is installed on Wall-1. Hence, using Eq. (7.44) with ACH = 0.15, we can find the h-value as follows:

$$h_{Wall\_1} = 1.208 + 1.012 \times \text{ACH}^{0.604} = 1.208 + 1.012 \times 0.15^{0.604} = 1.53\,\text{W/m}^2\,\text{K}$$

For walls without diffusers (ACH = 0), we can find h-value as:

$$h_{Wall\_2-6} = 1.208 + 1.012 \times \text{ACH}^{0.604} = 1.21\,\text{W/m}^2\,\text{K}$$

Eventually, for the ceiling without a diffuser, one can find the h-values as:

$$h_{ceiling} = 2.234 + 4.099 \times \text{ACH}^{0.503} = 2.23\,\text{W/m}^2\,\text{K}$$

Now, we can plot all the convection heat transfer coefficients for different walls as depicted in Figure 7.E8.2. as it can be seen, the choice of the model can significantly alter the h-value.

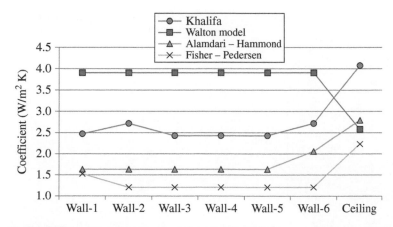

Figure 7.E8.2 Internal convective heat transfer coefficients over the case study building's walls using various schemes.

Example 7.9

A highly glazed room with the dimension of 4m × 4m × 4m is illustrated in Figure 7.E9, which has a ceiling diffuser system with a supply airflow rate of 50 m³/h. The room has a window with a height of 1.5 m and width of 4 m. Calculate the convection heat transfer coefficient for the wall containing the window.

Solution

Since the ACH is given and the room is highly glazed, we can select Goldstein-Novoselac correlation to calculate the h-value. First, we calculate the window to wall ratio as:

$$\text{WWR} = \frac{1.5 \times 4}{4 \times 4} = 37.5\%$$

Thus, we can use Eq. (7.47) as WWR<50% and the window is in the upper part of the wall to calculate the h-value:

$$h_1 = 0.117 \left(\frac{\dot{V}}{L}\right)^{0.8} = 0.117 \times \left(\frac{50}{4}\right)^{0.8} = 0.88 \,\text{W/m}^2\,\text{K}$$

For the wall located below the window, Eq. (7.50) can be used:

$$h_2 = 0.063 \left(\frac{\dot{V}}{L}\right)^{0.8} = 0.063 \times \left(\frac{50}{4}\right)^{0.8} = 0.48 \,\text{W/m}^2\,\text{K}$$

Note that the area weighted average can be used if the calculation of the overall h-value of the surface is required.

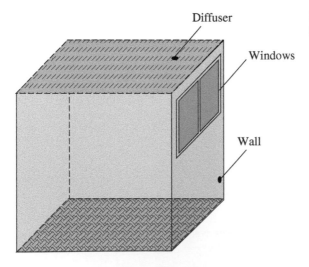

Diffuser

Windows

Wall

Figure 7.E9 A highly glazed case study room.

7.6 Radiations

The concept of a blackbody is explained in Chapter 6. Sun can be assumed as a blackbody with the surface temperature of about 5800 K. As it is shown in Figure 7.6, the reason that sun radiation is occasionally called as **short-wave radiation** is due to the fact that it is concentrated in the range of short wavelengths. As it was explained before, the irradiation incident on the atmosphere of earth is about $I = 1368 \pm 0.65$ W/m^2. The decrease in the solar intensity is a function of the ratio of sun radius (r_s) over the distance between sun and earth's atmosphere (r_d):

$$I = \left(\frac{r_s}{r_d}\right)^2 \tag{7.54}$$

When sun irradiation reaches the atmosphere a part of it is reflected by atmosphere and clouds while another part is absorbed by Ozone (O_3) (mainly UV and visible regions), H_2O (mainly Infrared region), O_2, CO_2, dusts, and aerosols in the atmosphere. As illustrated in Figure 7.6, the part which penetrates through the atmosphere either directly impacts the earth surface, known as the **direct radiation**, or is scattered by **Rayleigh** and **Mie** mechanisms [2], known as the **diffuse radiation** since it is almost independent of direction. The diffuse radiation can vary from 10% to 100% of the total irradiation incident on the earth surface in accordance with the cloud and particle level over a terrain. A part of the solar irradiation can be also reflected in addition to a part being absorbed by the earth surfaces.

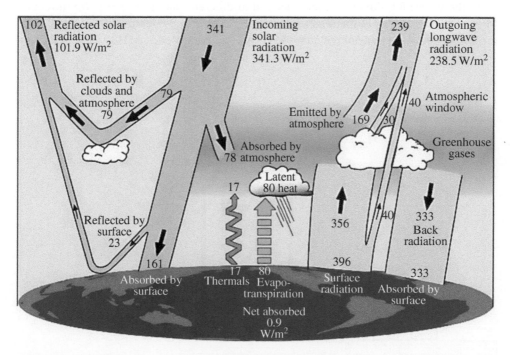

Figure 7.6 Solar intensity incident on the outer atmosphere. *Source:* Trenberth et al. [9]/American Meteorological Society.

Earth surfaces in turn have certain temperatures, which vary in accordance with their location and season of a year. Surface temperatures are different from sky and cloud temperatures, resulting in an upward or downward radiation, kwon as **long-wave radiation**, as it is mainly in the long wavelength ranges of the light spectrum. A part of this long-wave radiation is again absorbed by clouds and atmosphere while the penetrating part emits to the sky as scattered or direct rays. The long-wave radiation between the earth surface and sky can be calculated as (see Chapter 6):

$$E = \varepsilon\sigma\left(T^4 - T_{sky}{}^4\right) \tag{7.55}$$

where, T is the earth surface temperature and ε is the earth surface emissivity. T_{sky} is the **effective sky temperature** and can typically vary from 230 K in cold and clear condition to 285 K in warm and cloudy condition.

The downward short-wave and upward long-wave radiations are normally in a balance during a day cycle, preventing the earth temperature to be varied although a change in this equilibrium can occur due to artificial or external factors and result in phenomena such as global warming.

7.6.1 Solar Radiation on Building Surfaces

As it was explained, solar radiation (q''_{sw}) consists of direct, diffuse and reflecting solar irradiation, which can also reach a building exterior surface as shown in Figure 7.7. For the calculation of the direct solar radiation, there are different models that can be used to estimate the values for different surfaces located in different places on earth. We introduce three models to estimate the direct part of solar radiation, which are widely applied in the building related problems including ASHRAE Clear Sky Model, ASHRAE Revised Clear

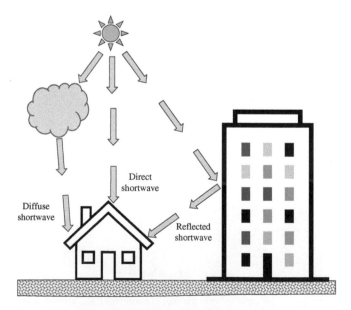

Direct
shortwave

Diffuse
shortwave

Reflected
shortwave

Figure 7.7 Solar radiation in buildings' exterior surfaces.

Sky Model, and Zhang–Huang Model. For the diffuse part of the solar radiation, again many studies have proposed various methods though we introduce only one diffuse model used in many related studies.

7.6.2 ASHRAE Clear Sky Model

This model estimates the hourly clear-day solar radiation for any month of the year in both Northern and Southern hemispheres. At the earth's surface on a clear day, the direct normal irradiation can be represented by the below equation [3]:

$$I_0 = \frac{A}{\exp(B/\sin\beta)} \tag{7.56}$$

where, A is the apparent solar irradiation at the relative optical air mass of zero ($m = 0$) and B is the atmospheric extinction coefficient while both of these coefficients can be found in Table 7.12. Also, β is the solar altitude or elevation (degree) as illustrated in Figure 7.8. As it can been, elevation and azimuth of sun can be employed to track its position in the sky in a specific location on earth and at a particular time of a day. Note that ASHRAE clear sky model is mainly for horizontal surfaces and is also overestimating solar radiation values since the air mass is presumed to be zero.

7.6.3 ASHRAE Revised Clear Sky Model

The model is a revised version of the clear sky model based on the location-specific optical depths for direct and diffuse radiation. The direct (beam) normal irradiance component (I_b) can be calculated by following equation (7.16):

Table 7.12 Extraterrestrial Solar Irradiance and Related Data Note: Data are for 21st day of each month during the base year of 1964.

	I_0 (W/m^2)	Equation of time (minutes)	Declination (degrees)	A (W/m^2)	B (−)	C (−)
Jan	1416	−11.2	−20.0	1202	0.141	0.103
Feb	1401	−13.9	−10.8	1187	0.142	0.104
Mar	1381	−7.5	0.0	1164	0.149	0.109
Apr	1356	1.1	11.6	1130	0.164	0.120
May	1336	3.3	20.0	1106	0.177	0.130
Jun	1336	−1.4	23.45	1092	0.185	0.137
Jul	1336	−6.2	20.6	1093	0.186	0.138
Aug	1338	−2.4	12.3	1107	0.182	0.134
Sep	1359	7.5	0	1136	0.165	0.121
Oct	1380	15.4	−10.5	1166	0.152	0.111
Nov	1405	13.8	−19.8	1190	0.144	0.106
Dec	1417	1.6	−23.45	1204	0.141	0.103

Source: Modified from ASHRAE [3]/ASHRAE.

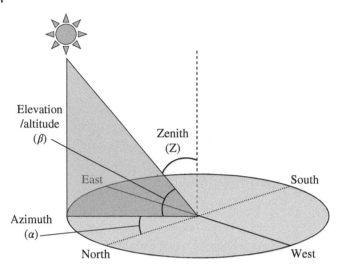

Figure 7.8 Sun position in the sky presented by its elevation, and azimuth.

$$I_b = I_0 \exp\left(-\tau_b \bullet m^{ab}\right) \tag{7.57}$$

Also, the diffuse horizontal irradiance (I_d) can be expressed as:

$$I_d = I_0 \exp\left(-\tau_d \bullet m^{ad}\right) \tag{7.58}$$

where, I_0 is the extraterrestrial normal irradiance, which can be found from Table 7.12 or Table 7.13. τ_b and τ_d are the direct and diffuse optical depths and can be found in the related tables (see example for Atlanta, GA, USA as shown in Table 7.14). Furthermore, ab and ad are the direct and diffuse air mass exponents and can be calculated as follows:

$$ab = 1.219 - 0.043\tau_b - 0.151\tau_d - 0.204\tau_b\tau_d \tag{7.59}$$

$$ad = 0.202 + 0.852\tau_b - 0.007\tau_d - 0.357\tau_b\tau_d \tag{7.60}$$

Furthermore, the relative optical air mass (m) can be calculated as follows:

$$m = 1/\left[\sin\beta + 0.50572 \bullet (6.07995 + \beta)^{-1.6364}\right] \tag{7.61}$$

Table 7.13 Approximate Astronomical Data for the 21st Day of Each Month.

Month	Jan	Feb	Mar	Apr	May	Jun	Jul	Aug	Sep	Oct	Nov	Dec
Day of year	21	52	80	111	141	172	202	233	264	294	325	355
Io, Btu/h·ft2	447	443	437	429	423	419	420	424	430	437	444	447
Equation of time (ET), min	− 10.6	− 14.0	− 7.9	1.2	3.7	− 1.3	− 6.4	− 3.6	6.9	15.5	13.8	2.2
Declination, degrees	− 20.1	− 11.2	− 0.4	11.6	20.1	23.4	20.4	11.8	− 0.2	− 11.8	− 20.4	− 23.4

Source: Modified from 14 AC [19]/ASHARE.

Table 7.14 Design conditions for Atlanta, GA, USA.

Clear sky solar irradiance	Annual	Jan	Feb	Mar	Apr	May	Jun	Jul	Aug	Sep	Oct	Nov	Dec
τ_b		0.325	0.349	0.383	0.395	0.448	0.505	0.556	0.593	0.431	0.373	0.339	0.32
τ_d		2.461	2.316	2.176	2.175	2.028	1.892	1.779	1.679	2.151	2.317	2.422	2.514
I_{bn}		282	285	282	283	268	252	238	227	264	273	273	277
I_{dn}		30	37	45	47	55	63	70	76	46	37	31	28

Source: Based on 169-2006 BAAbtAAS [20].

7.6.4 Zhang–Huang Model

The Zhang–Huang solar model is a regression-based model, which uses a series of parameters to estimate the direct and diffuse component of the solar radiation. Thus, the total (global horizontal) solar radiation can be estimated by the below equation:

$$I = \frac{\left[I'' \sin(\beta)\left(c_0 + c_1 CC + c_2 CC^2 + c_3(T_n - T_{n-3}) + c_4\varphi + c_5 V_w\right) + d\right]}{k} \tag{7.62}$$

where, I'' is the global solar constant (1355W/m^2 as an annual average of I_0), CC is the cloud cover, φ is the relative humidity, V_w is the wind speed, and $T_n - T_{n-3}$ is the dry-bulb temperature at hours n (current) and $n-3$, respectively. Moreover, $c_1, c_2, c_3, c_4, c_5, d,$ and k are the regression coefficeints, which are defined as $c_0 = 0.5598$, $c_1 = 0.4982$, $c_2 = -0.6762$, $c_3 = 0.02842$, $c_4 = -0.00317$, $c_5 = 0.014$, $d = -17.853$, $k = 0.843$.

7.6.5 Diffuse Solar Radiation Model

The model is developed based on the study by [4]. In the absence of shadowing, the sky diffuse irradiance on a tilted surface is expressed to include three components of dome (I_{dome}), circumsolar ($I_{\text{circumsolar}}$), and horizon (I_{horizon}):

$$I_{\text{sky}} = I_{\text{dome}} + I_{\text{circumsolar}} + I_{\text{horizon}} = \frac{I_h(1-F_1)(1+\cos S)}{2} + \frac{I_h F_1 a}{b} + I_h F_2 \sin S \tag{7.63}$$

where, I_h is the horizontal solar irradiance (see I_{dn} in Table 7.14), F_1 is the circumsolar brightening coeffcient, F_2 is the horizon brightening coeffcient, and S is the surface tilt (radians) from the horizon where a horizontal surface has $S = 0°$ and a vertical wall surface has $S = 90°$. Moreover, $a = \max(0, \cos\alpha)$ and $b = \max(0.087, \cos Z)$. Z is the solar zenith angle (radians) and α denotes the incidence angle of sun on the surface (radians).

The circumsolar brightening coefficient F_1 and the horizon brightening coeffcient F_2 can be calculated with the below equations:

$$F_1 = F_{11}(\varepsilon) + F_{12}(\varepsilon)\Delta + F_{13}(\varepsilon)Z \tag{7.64}$$

$$F_2 = F_{21}(\varepsilon) + F_{22}(\varepsilon)\Delta + F_{23}(\varepsilon)Z \tag{7.65}$$

where, $F_{11}, F_{12}, F_{13}, F_{21}, F_{22},$ and F_{23} can be also obtained from Table 7.15. ε is the sky cleaness factor and is defined as:

$$\varepsilon = \frac{(I_h + I)/I_h + 1.041 Z^3}{1 + 1.041 Z^3} \tag{7.66}$$

where, I is the direct normal soalr irradiance (I_{bn} in Table 7.14), Δ is a coefficient that can be calculated by the below equation:

$$\Delta = \frac{I_h m}{I_0} \tag{7.67}$$

Table 7.15 F_{ij} factors as a function of sky clearness range.

ε Range	1.000–1.065	1.065–1.230	1.230–1.500	1.500–1.950	1.950–2.800	2.800–4.500	4.500–6.200	>6.200
F_{11}	−0.0083117	0.1299457	0.3296958	0.5682053	0.873028	1.1326077	1.0601591	0.677747
F_{12}	0.5877285	0.6825954	0.4868735	0.1874525	−0.3920403	−1.2367284	−1.5999137	−0.3272588
F_{13}	−0.0620636	−0.1513752	−0.2210958	−0.295129	−0.3616149	−0.4118494	−0.3589221	−0.2504286
F_{21}	−0.0596012	−0.0189325	0.055414	0.1088631	0.2255647	0.2877813	0.2642124	0.1561313
F_{22}	0.0721249	0.065965	−0.0639588	−0.1519229	−0.4620442	−0.8230357	−1.127234	−1.3765031
F_{23}	−0.0220216	−0.0288748	−0.0260542	−0.0139754	0.0012448	0.0558651	0.1310694	0.2506212

Source: Based on Perez et al. [4].

Example 7.10

Calculate short-wave radiation for a building with vertical walls located at Atlanta, GA, USA (33.75N, 84.39W) on 21st October 2016 (8:00–17:00) using the below direct methods:

1) ASHRAE clear sky model
2) ASHRAE revised clear sky model
3) Zhang–Huang Model

Furthermore, utilize the presented diffuse solar model and find the disused part of the solar radiation. The environment conditions of Atlanta are given as below:

Hour	Elevation (degree)	Cloud cover (%)	Relative humidity φ (%)	Dew point (°C)	Air temperature (°C)	Surface temperature (°C)	Wind velocity (m/s)	Tn-Tn-3 (°C)
06:00:00	\	\	\	\	16.1	17	\	\
07:00:00	\	\	\	\	16.1	17.1	\	\
08:00:00	\	\	\	\	15.6	17.2	\	\
09:00:00	13.44	9.00	64.00	8.9	16.7	17.3	6.2	0.6
10:00:00	24.4	4.00	53.00	7.2	17.2	17.4	9.3	1.1
11:00:00	33.94	2.00	54.00	7.8	16.7	17.5	7.7	1.1
12:00:00	41.17	10.00	56.00	7.8	17.8	17.6	7.2	1.1
13:00:00	44.92	9.00	48.00	6.7	18.9	17.7	6.7	1.7
14:00:00	44.31	10.00	42.00	5.6	20	17.8	7.7	3.3
15:00:00	39.51	10.00	37.00	5	19.4	17.9	5.7	1.6
16:00:00	31.55	10.00	37.00	4.4	19.4	18	9.8	0.5
17:00:00	21.55	10.00	34.00	3.3	19.4	18.1	7.2	−0.6

Solution

1) ASHRAE Clear Sky Model

Let us find the value at 10:00am as the sample calculation. In October, the solar altitude is $\beta_{at10:00} = 24.4°$ and the coefficients can be found as $A = 1166$ W/m²and $B = 0.152$. Therefore, the direct normal irradiation is:

$$I_0 = \frac{1166}{\exp\left(0.152/\sin\left(24.4\right)\right)} = 807.05 \text{ W/m}^2$$

Similarly, the direct normal radiation can be found for other time steps as below:

	ASHRAE clear sky model			
Hour	A (W/m²)	B	Solar altitude (β)	I_0 (W/m²)
09:00:00	1166	0.152	13.44	606.29
10:00:00	1166	0.152	24.4	807.05
11:00:00	1166	0.152	33.94	888.11
12:00:00	1166	0.152	41.17	925.59
13:00:00	1166	0.152	44.92	940.18
14:00:00	1166	0.152	44.31	937.99
15:00:00	1166	0.152	39.51	918.20
16:00:00	1166	0.152	31.55	872.04
17:00:00	1166	0.152	21.55	770.87

2) ASHRAE revised clear sky model

We again derive a sample calculation at 10:00am. From Table 7.14, the beam and diffuse optical depths for the clear sky can be obtained as $\tau_b = 0.373$ and $\tau_d = 2.317$, respectively. Therefore, the associated beam and diffuse air mass exponents can be calculated as:

$$ab = 1.219 - 0.043 \times 0.373 - 0.151 \times 2.317 - 0.204 \times 0.373 \times 2.317 = 0.6768$$

$$ad = 0.202 + 0.852 \times 0.373 - 0.007 \times 2.317 - 0.357 \times 0.373 \times 2.317 = 0.1950$$

Hence, at 10:00am ($\beta_{at10:00} = 24.4°$), the relative optical air mass can be found from Eq. (7.61):

$$m_{@10:00} = 1/\left[\sin(24.4) + 0.50572 \times (6.07995 + 24.4)^{-1.6364} = 2.41\right.$$

At October, $I_0 = 437$ Btu/h $\cdot ft^2 = 1378.6$ W/m^2 from Table 7.12. Hence, the beam normal irradiance and diffuse horizontal irradiance from Eqs. (7.57) and (7.58) can be calculated as:

$$I_b = 1378.6 \times \exp\left[-0.373 \times 2.41^{0.6768}\right] = 700.9 \text{ W/m}^2$$

$$I_d = 1378.6 \times \exp\left[-2.317 \times 2.41^{0.195}\right] = 88.1 \text{ W/m}^2$$

The result of other time steps can be similarly shown as below:

Hour	τ_b	τ_d	ab	ad	m	I_0 (W/m^2)	I_b (W/m^2)	I_d (W/m^2)
09:00:00	0.373	2.317	0.6768	0.1950	4.23	1378.60	512.19	64.01
10:00:00	0.373	2.317	0.6768	0.1950	2.41	1378.60	700.93	88.08
11:00:00	0.373	2.317	0.6768	0.1950	1.79	1378.60	793.35	102.92
12:00:00	0.373	2.317	0.6768	0.1950	1.52	1378.60	840.75	111.68
13:00:00	0.373	2.317	0.6768	0.1950	1.41	1378.60	860.18	115.55
14:00:00	0.373	2.317	0.6768	0.1950	1.43	1378.60	857.22	114.95
15:00:00	0.373	2.317	0.6768	0.1950	1.57	1378.60	831.13	109.83
16:00:00	0.373	2.317	0.6768	0.1950	1.91	1378.60	774.03	99.60
17:00:00	0.373	2.317	0.6768	0.1950	2.71	1378.60	663.24	82.71

3) Zhang–Huang Model

At 10:00am, the solar altitude, cloud cover, and relative humidity are $\beta_{@10:00} = 24.4°$, $CC_{@10:00} = 4\%$, and $\varphi_{@10:00} = 53\%$, respectively. Also, from the climatic information, we have:

$$T_{@10:00} - T_{@7:00} = 17.2 - 16.1 = 1.1°C$$

$$V_{w@10:00} = 9.3 \text{ m/s}$$

Therefore, the estimated hourly solar radiation using Eq. (7.62) can be found as below:

$$I = \frac{[1355\sin(24.4)(0.5598 + 0.4982\times0.04 - 0.6762\times0.04^2 + 0.02842\times1.1 - 0.00317\times0.53 + 0.014\times9.3) - 17.853]}{0.843}$$

$$= 484.29 \text{ W/m}^2$$

Again, the result of other time steps can be found as below:

Hour	I_o (W/m²)	sin (h)	I (W/m²)
09:00:00	1355	0.23	243.13
10:00:00	1355	0.43	484.29
11:00:00	1355	0.59	651.84
12:00:00	1355	0.72	825.59
13:00:00	1355	0.78	911.05
14:00:00	1355	0.77	977.15
15:00:00	1355	0.69	784.59
16:00:00	1355	0.55	645.39
17:00:00	1355	0.38	393.27

Diffuse solar radiation

We again perform a sample calculation at 10:00am. From Table 7.14, the diffuse horizontal and direct normal solar irradiance at October are as below:

$$I_h = 30 \text{ Btu/h} \cdot ft^2 = 94.6 \text{ W/m}^2$$

$$I = 282 \text{ Btu/h} \cdot ft^2 = 889.6 \text{ W/m}^2$$

As all of the external walls of the building are vertical, the surface tilt and the incidence angle of the sun on the surface can be calculated as:

$$S_{@10:00} = \frac{\pi}{2}$$

$$\alpha_{@10:00} = \frac{24.4 \times \pi}{180} = \frac{61}{450}\pi$$

At 10:00am, the solar zenith angle is:

$$Z_{@10:00} = \frac{\pi}{2} - \frac{24.4 \times \pi}{180} = \frac{82}{225}\pi$$

This implies that a and b can be found as:

$$a_{@10:00} = \max(0, \cos\alpha) = \cos\left(\frac{61}{450}\pi\right) = 0.91$$

$$b_{@10:00} = \max(0.087, \cos Z) = \cos\left(\frac{82}{225}\pi\right) = 0.41$$

The relative optical air mass is therefore calculated similar to the previous section:

$$m_{@10:00} = 1/\left[\sin(24.4) + 0.50572 \times (6.07995 + 24.4)^{-1.6364}\right] = 2.41$$

Therefore, the sky cleanness factor from Eq. (7.66) can be derived as:

$$\varepsilon = \frac{(I_h + I)/I_h + 1.041Z^3}{1 + 1.041Z^3} = \frac{\dfrac{94.6 + 889.6}{94.6} + 1.041 \times \left(\dfrac{82}{225}\pi\right)^3}{1 + 1.041 \times \left(\dfrac{82}{225}\pi\right)^3} = 4.67$$

And, Δ could be calculated from Eq. (7.67):

$$\Delta_{@10:00} = \frac{I_h m}{I_0} = \frac{94.6 \times 2.41}{1353} = 0.168$$

As $2.8 < \varepsilon < 4.5$, F_{ij} factors as a function of the sky clearness range can be expressed as below:

Hour	F_{11}	F_{12}	F_{13}	F_{21}	F_{22}	F_{23}
10:00:00	1.060159	-1.59991	-0.3589221	0.26421	-1.127234	0.1310694

Hence, the circumsolar brightening coefficient from Eq. (7.64) and horizon brightening coefficient from Eq. (7.65) can be obtained as:

$$F_1 = F_{11}(\varepsilon) + F_{12}(\varepsilon)\Delta + F_{13}(\varepsilon)Z = 1.060159 - 1.59991 \times 0.168 - 0.3589221 \times \frac{82}{225}\pi = 0.38$$

$$F_2 = F_{21}(\varepsilon) + F_{22}(\varepsilon)\Delta + F_{23}(\varepsilon)Z$$

$$= 0.26421 - 1.127234 \times 0.168 + 0.1310694 \times \frac{82}{225}\pi = 0.225$$

Eventually, the sky diffuse irradiance from Eq. (7.63) can be found as:

$$I_{sky} = \frac{I_h(1-F_1)(1+\cos S)}{2} + I_h F_2 \sin S + \frac{I_h F_1 a}{b} = \frac{94.6 \times (1-0.38)(1+\cos\frac{\pi}{2})}{2}$$
$$+ 94.6 \times 0.225 \times \sin\left(\frac{\pi}{2}\right) + \frac{94.6 \times 0.38 \times 0.91}{0.41} = 130\,\text{W/m}^2$$

The result of other time steps can be also shown as below:

Hour	I_h	I	Surface tilt S	Solar zenith angle Z	Incidence angle	a	b	ε	I_0 (W/m²)	Δ
09:00:00	94.6	889.6	1.57	1.34	0.23	0.97	0.23	3.70	1353	0.30
10:00:00	94.6	889.6	1.57	1.14	0.43	0.91	0.41	4.67	1353	0.17
11:00:00	94.6	889.6	1.57	0.98	0.59	0.83	0.56	5.76	1353	0.12
12:00:00	94.6	889.6	1.57	0.85	0.72	0.75	0.66	6.72	1353	0.11
13:00:00	94.6	889.6	1.57	0.79	0.78	0.71	0.71	7.24	1353	0.10
14:00:00	94.6	889.6	1.57	0.80	0.77	0.72	0.70	7.15	1353	0.10
15:00:00	94.6	889.6	1.57	0.88	0.69	0.77	0.64	6.49	1353	0.11
16:00:00	94.6	889.6	1.57	1.02	0.55	0.85	0.52	5.47	1353	0.13
17:00:00	94.6	889.6	1.57	1.19	0.38	0.93	0.37	4.39	1353	0.19

(Continued)

Hour	F_{11}	F_{12}	F_{13}	F_{21}	F_{22}	F_{23}	F_1	F_2	I_{sky} (W/m²)
09:00:00	1.132608	−1.23673	−0.4118	0.287781	−0.82304	0.05587	0.21641	0.11894	133.983
10:00:00	1.060159	−1.59991	−0.3589	0.264212	−1.12723	0.13107	0.37966	0.22436	129.742
11:00:00	1.060159	−1.59991	−0.3589	0.264212	−1.12723	0.13107	0.50905	0.2516	118.579
12:00:00	0.677747	−0.32726	−0.2504	0.156131	−1.3765	0.25062	0.42961	0.22373	94.6169
13:00:00	0.677747	−0.32726	−0.2504	0.156131	−1.3765	0.25062	0.44834	0.21718	89.1701
14:00:00	0.677747	−0.32726	−0.2504	0.156131	−1.3765	0.25062	0.44533	0.21837	90.0491
15:00:00	0.677747	−0.32726	−0.2504	0.156131	−1.3765	0.25062	0.42116	0.22594	97.0674
16:00:00	1.060159	−1.59991	−0.3589	0.264212	−1.12723	0.13107	0.48076	0.24768	122.062
17:00:00	1.132608	−1.23673	−0.4118	0.287781	−0.82304	0.05587	0.40658	0.19879	144.268

We can now compare and plot the calculated direct and diffuse radiations by different models in Figure 7.E10. As it can be seen, the choice of a solar radiation model is also a crucial factor in the energy balance within a building as discussed in detail in Chapter 8.

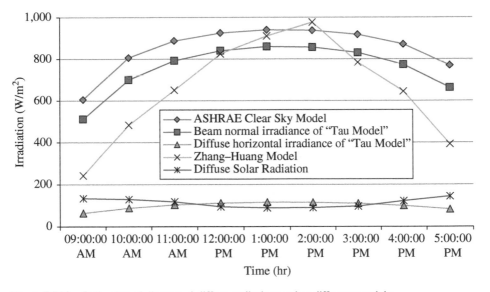

Figure 7.E10 Daily plot of direct and diffuse radiations using different models.

7.7 Long-wave Radiation on Building Surfaces

In Chapter 6, the fundamentals of long-wave radiation between surfaces as well as the importance of view factors were explained. While the interior surfaces have mainly similar temperatures, which turns long-wave radiation exchanges to be negligible in many occasions, in a surrounding environment around a specific building, long-wave exchanges become important due to the significant temperature differences during a day and night cycle. In particular, the long-wave radiation exchanges (q''_{LWR}) are more significant between

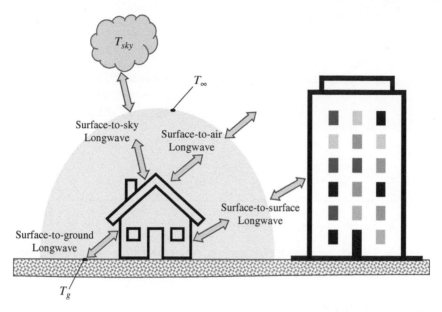

Figure 7.9 Long-wave radiation between buildings' exterior surfaces and surrounding environment.

building surfaces to each other (q_{bs}'') and between them and the vicinity ground surface (q_{gnd}''), the sky dome (q_{sky}''), and the surrounding air (q_{air}'') (Figure 7.9):

$$q_{LWR}'' = q_{gnd}'' + q_{sky}'' + q_{air}'' + q_{bs}'' \tag{7.68}$$

Equation (7.68) resulted in development of different models to estimate any of these radiation exchanges. Since calculation of irradiation exchanges between building surfaces to each other is a difficult task, for now, we can assume that the building is a standalone one ($q_{bs}'' = 0$). The first step in the modelling, thus, is to present these surrounding environment components with a bulk temperature and then utilizing the Stefan–Boltzmann law as:

$$q_{LWR}'' = \varepsilon_g \sigma F_g \left(T_g^4 - T_{so}^4\right) + \varepsilon_{sky} \sigma F_{sky} \left(T_{sky}^4 - T_{so}^4\right) + \varepsilon_\infty \sigma F_\infty \left(T_\infty^4 - T_{so}^4\right) \tag{7.69}$$

where, ε_g, ε_{sky}, and ε_∞ are the ground surface, sky, and air emissivities. σ is the Stefan–Boltzmann constant ($\sigma = 5.67 \times 10^{-8}$ W/m^2 K^4). F_g, F_{sky}, and F_∞ are the view factors of the studied wall in respect to the ground surface, sky, and air temperature, respectively. T_{so} denotes the outside surface temperature while T_g, T_{sky}, and T_∞ are the ground surface, sky, and air bulk temperatures, respectively.

7.7.1 View Factors of Surrounding Environment

Finding view factors is one of the most difficult tasks in the calculation of long-wave radiations in Eq. (7.69). Various approaches are feasible to find these values, which is beyond the scope of this book. Here, we refer to a simplified model by [5] in which these values can be roughly estimated as below:

$$F_g = 0.5(1 - \cos \emptyset) \tag{7.70}$$

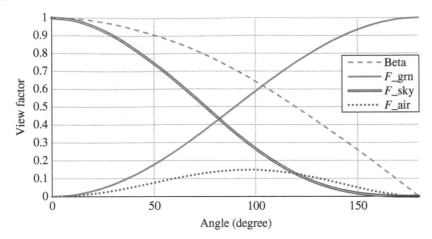

Figure 7.10 Variation of view factors of building surfaces in respect to their surrounding ground, sky, and air. *Source:* Based on [5].

$$F_{sky} = 0.5(1 + \cos\emptyset) \times \beta \tag{7.71}$$

$$F_{\infty} = F_{sky} \times \frac{1-\beta}{\beta} \tag{7.72}$$

$$\beta = \sqrt{0.5(1 + \cos\emptyset)} \tag{7.73}$$

where, $\emptyset$ is the tilt angle of the building surface.

Figure 7.10 demonstrates the variation of these view factors of building surfaces in respect to their surrounding ground, sky, and air.

7.7.2 Emissivity of Surrounding Environment

Another challenge in obtaining the long-wave radiation exchanges in Eq. (7.69) is finding the emissivity values for the sky. Again, different values are offered in this respect though we only present an equation for the calculation of the air emissivity (ε_{air}) proposed by [5, 6]:

$$\varepsilon_{sky} = \left(0.787 + 0.764 \cdot \ln\left(\frac{T_{dp}}{273.15}\right)\right)\left(1 + 0.0224N - 0.0035N^2 + 0.0028N^3\right) \tag{7.74}$$

where, T_{dp} is the dew point temperature (K) and N is the opaque sky cover (tenths).

Note that the emissivity of the surrounding air and ground material can be found in the related handbooks and tables.

7.7.3 Bulk Temperature of Surrounding Environment

Air temperature is a parameter required for the calculation of the convective and radiative fluxes. This value can be normally converted using the recorded air temperature from the vicinity weather stations. Recent studies, however, demonstrate that the recorded air temperatures can be considerably variable within a city due to urban heat island phenomenon

[7, 8]. There are many studies in literature to develop models to estimate the air temperature around a particular building using local weather stations. Also, the surrounding ground temperature can be related to the soil and greening around a particular building, which is again widely being investigated in literature. Here, in this book, we assume these values to be given as boundary conditions to building energy systems as explained in Chapters 8 and 9. For the calculation of the sky temperature, nonetheless, we can use the horizontal irradiation intensity (I_0) at a surface, and the dry-bulb temperature of the air (T_{db}) to find T_{sky}:

$$I_0 = \varepsilon_{sky} \sigma T_{db}^4$$

$$\Longrightarrow T_{sky} = \left(\frac{I_0}{\sigma}\right)^{0.25} \tag{7.75}$$

Example 7.11

Consider the building of Example 7.10 and calculate the long-wave radiation on its surfaces. Assume the soil and air emissivity as $\varepsilon_{gnd} = 0.93$ and $\varepsilon_{air} = 0.75$, respectively. The ground temperature around the building is $T_{gnd} = 5°C$.

Solution

As all the walls are vertical, the surface tilt angle is $\emptyset = 90°$. Therefore, the view factors can be calculated using Eqs. (7.70)–(7.73) as below (Figure 7.10 can be also used):

$$\beta = \sqrt{0.5(1 + \cos\emptyset)} = 0.707$$

$$F_{ground} = 0.5(1 - \cos 90°) = 0.50$$

$$F_{sky} = 0.5(1 + \cos\emptyset) \times \beta = 0.35$$

$$F_{air} = F_{sky} \times \frac{1 - \beta}{\beta} = 0.146$$

Note that for the calculation of the long-wave radiation at 10:00am, we use the values for $h_{r, gnd}$, $h_{r, air}$, and $h_{r, sky}$ from the previous time-step (9:00am):

$$T_{dp@9:00} = 8.9°C = 282.05 \text{ K}$$

$$T_{dp@10:00} = 7.2°C = 280.35 \text{ K}$$

$$T_{air@9:00} = T_{db} = 16.7°C = 289.85 \text{ K}$$

$$T_{air@10:00} = T_{db} = 17.2°C = 290.35 \text{ K}$$

$$T_{so@9:00} = 17.3°C = 290.45 \text{ K}$$

$$T_{so@10:00} = 17.4°C = 290.55 \text{ K}$$

$$N_{cc@9:00} = 0.09$$

$$N_{cc@10:00} = 0.04$$

Moreover, the sky emissivity for two consecutive time steps can be calculated from Eq. (7.74):

$$\varepsilon_{sky@9:00} = \left(0.787 + 0.764 \cdot \ln\left(\frac{282.05}{273.15}\right)\right)$$

$$(1 + 0.0224 \times 0.09 - 0.0035 \times 0.09^2 + 0.0028 \times 0.09^3) = 0.813$$

$$\varepsilon_{sky@10:00} = \left(0.787 + 0.764 \cdot \ln\left(\frac{280.35}{273.15}\right)\right)$$
$$\left(1 + 0.0224 \times 0.04 - 0.0035 \times 0.04^2 + 0.0028 \times 0.04^3\right) = 0.808$$

Now, we can employ Eq. (7.75) to calculate the sky temperatures:

$$I_{0@9:00} = \varepsilon_{sky}\sigma T^4_{dry-bulb} = 0.813 \times 5.67 \times 10^{-8} \times 289.85^4 = 325.54 \text{ W/m}^2$$

$$\Longrightarrow T_{sky@9:00} = \left(\frac{325.54}{5.67 \times 10^{-8}}\right)^{0.25} - 273.15 = 2.16°\text{C} = 275.27 \text{ K}$$

$$I_{0@10:00} = \varepsilon_{sky}\sigma T^4_{dry-bulb} = 0.808 \times 5.67 \times 10^{-8} \times 290.35^4 = 325.58 \text{ W/m}^2$$

$$\Longrightarrow T_{sky@10:00} = \left(\frac{325.58}{5.6697 \times 10^{-8}}\right)^{0.25} - 273.15 = 2.13°\text{C} = 275.28 \text{ K}$$

Now, we can calculate each radiative flux as below:

$$h_{r,gnd@9:00} = \frac{\varepsilon\sigma F_{gnd}\left(T^4_{gnd@9:00} - T^4_{surf@9:00}\right)}{T_{gnd@9:00} - T_{surf@9:00}}$$
$$= \frac{0.93 \times 5.67 \times 10^{-8} \times 0.5 \times (278.15^4 - 290.45^4)}{278.15 - 290.45} = 2.424 \text{ W/m}^2\text{K}$$

$$h_{r,air@9:00} = \frac{\varepsilon\sigma F_{air}\left(T^4_{air@9:00} - T^4_{surf@9:00}\right)}{T_{air@9:00} - T_{surf@9:00}}$$
$$= \frac{0.75 \times 5.67 \times 10^{-8} \times 0.14645 \times (289.85^4 - 290.45^4)}{289.85 - 290.45} = 0.608 \text{ W/m}^2\text{K}$$

$$h_{r,sky@9:00} = \frac{\varepsilon\sigma F_{sky}\left(T^4_{sky@9:00} - T^4_{surf@9:00}\right)}{T_{sky@9:00} - T_{surf@9:00}}$$
$$= \frac{0.814 \times 5.6697 \times 10^{-8} \times 0.35355 \times (275.275^4 - 290.45^4)}{275.275 - 290.45} = 1.477 \text{ W/m}^2\text{K}$$

Therefore, the total long-wave radiation can be found from Eq. (7.69) as:

$$q''_{LWR} = h_{r,gnd@9:00}\left(T_{gnd@10:00} - T_{surf@10:00}\right) + h_{r,air@9:00}\left(T_{air@10:00} - T_{surf@10:00}\right)$$
$$+ h_{r,sky@9:00}\left(T_{sky@10:00} - T_{10:00}\right) = 2.424 \times (278.15 - 290.55)$$
$$+ 0.608 \times (290.35 - 290.55) + 1.477 \times (275.28 - 290.55) = -52.7 \text{ W/m}^2$$

Now, we can repeat the calculations for other time steps to find all the long-wave radiations as presented in the below table.

Hour	T_{so}	T_{dp}	ε_{sky}	T_{sky}	$h_{r, gnd}$	$h_{r, air}$	$h_{r, sky}$	q''_{LWR} (W/m²)
09:00:00	290.45	282.05	0.814	275.27	2.424	0.608	1.477	—
10:00:00	290.55	280.35	0.808	275.28	2.426	0.610	1.468	−52.7
11:00:00	290.65	280.95	0.809	274.92	2.427	0.609	1.469	−53.9
12:00:00	290.75	280.95	0.811	276.08	2.428	0.613	1.481	−52.0
13:00:00	290.85	279.85	0.808	276.85	2.430	0.617	1.482	−50.8
14:00:00	290.95	278.75	0.805	277.65	2.431	0.620	1.483	−49.4
15:00:00	291.05	278.15	0.803	276.94	2.432	0.619	1.476	−51.4
16:00:00	291.15	277.55	0.801	276.80	2.434	0.619	1.472	−51.9
17:00:00	291.25	276.45	0.798	276.53	2.435	0.620	1.466	−52.7

References

1 Fohanno, S. and Polidori, G. (2006). Modelling of natural convective heat transfer at an internal surface. *Energy and Buildings.* 38 (5): 548–553.

2 Lockwood, D.J. (2016). Rayleigh and Mie scattering. In: *Encyclopedia of Color Science and Technology* (ed. M.R. Luo), 1097–1107. New York, NY: Springer New York.

3 ASHRAE (1989). *ASHRAE Handbook – Fundamentals: American Society of Heating, Refrigerating, and Air-Conditioning Engineers.* ASHARE.

4 Perez, R., Ineichen, P., Seals, R. et al. (1990). Modeling daylight availability and irradiance components from direct and global irradiance. *Solar Energy.* 44 (5): 271–289.

5 Walton GNUSNBoSUSDoEBESB (1983). *Thermal Analysis Research Program Reference Manual.* Washington, DC; Springfield, VA: U.S. Dept. of Commerce, National Bureau of Standards; [National Technical Information Service, Distributor].

6 Clark G, Allen C, editors. The estimation of atmospheric radiation for clear and cloudy skies. Proc 2nd National Passive Solar Conference (AS/ISES); 1978.

7 Mirzaei, P.A. (2015). Recent challenges in modeling of urban heat island. *Sustainable Cities and Society.* 19: 200–206.

8 Mirzaei, P.A. and Haghighat, F. (2010). Approaches to study urban heat island–abilities and limitations. *Building and Environment.* 45 (10): 2192–2201.

9 Trenberth, K., Fasullo, J., and Kiehl, J. (2009). Earth's global energy budget. *Bulletin of the American Meteorological Society.* 90.

10 Hittle, D.C. and Lawrie, L. (1978). *The Building Loads Analysis and System Thermodynamics Program (BLAST). Release Nnumber 1. Software.* Champaign, IL: Army Construction Engineering Research Lab. Report No.: AD-A-056226 United States NTIS HEDB English.

11 Mehry, Y. and Joseph, H.K. (1993). Measurement of the exterior convective film coefficient for windows in low-rise buildings. *ASHRAE Transactions.* 100: Part 1.

12 Booten, C., Kruis, N., and Christensen, C. (2012). *Identifying and Resolving Issues in EnergyPlus and DOE-2 Window Heat Transfer Calculations.* National Renewable Energy Laboratory.

13 Beausoleil-Morrison, I. (2002). The adaptive simulation of convective heat transfer at internal building surfaces. *Building and Environment.* 37 (8): 791–806.

14 Khalifa, A.J.N. and Marshall, R.H. (1990). Validation of heat transfer coefficients on interior building surfaces using a real-sized indoor test cell. *International Journal of Heat and Mass Transfer.* 33 (10): 2219–2236.

15 Alamdari, F. and Hammond, G.P. (1983). Improved data correlations for buoyancy-driven convection in rooms. *Building Services Engineering Research and Technology.* 4 (3): 106–112.

16 Awbi, H.B. and Hatton, A. (1999). Natural convection from heated room surfaces. *Energy and Buildings.* 30 (3): 233–244.

17 Fisher, D. E. and Pedersen, C. O. (1997). Convective heat transfer in building energy and thermal load calculations. Conference: American Society of Heating, Refrigerating and Air-Conditioning Engineers (ASHRAE) annual meeting, Boston, MA (United States), 28 Jun – 2 Jul 1997; Other Information: PBD: 1997; Related Information: Is Part Of ASHRAE transactions: Technical and symposium papers, 1997 Volume 103, Part 2; PB: 1072 p; United States: American Society of Heating, Refrigerating and Air-Conditioning Engineers, Inc., Atlanta, GA (United States); p. Medium: X; Size: p. 137.48.

18 Goldstein, K. and Novoselac, A. (2010). Convective heat transfer in rooms with ceiling slot diffusers (RP-1416). *HVAC&R Research.* 16 (5): 629–655.

19 14 AC (2009). *ASHRAE Handbook – Fundamentals: American Society of Heating, Refrigerating, and Air-Conditioning Engineers.* ASHRAE.

20 169-2006 BAAbtAAS (2012). *Proposed Addendum b to Standard 169, Climatic Data for Building Design Standards.* ASHRAE.

8

Fundamental of Energy Modelling in Buildings

8   Introduction

Buildings can be assumed as an energy system and thus all the fundamentals introduced in the earlier chapters can be applied to these systems. From a thermodynamic point of view, buildings can be modelled to some extent similar to ***quasi-equilibrium*** uniform-state unsteady-flow processes (see Chapter 4) as the process relatively happens slowly enough within the building systems to reach an internal equilibrium. With a quasi-equilibrium definition, thus, we expect that the system changes from one equilibrium in a specific time-step to another one in the next time step. Obviously, this slow process is relative to the process of monthly or annual building energy calculations where the timescale is considered to be large enough related to a single time step. Inversely, if we are solemnly focused on phenomena with smaller timescales such as details in convection between a heated wall and indoor air, the process barely can be assumed to comply the quasi-equilibrium condition. It should be noted that we always have internal heat generation in the buildings related to human, other bio sources and appliances, which will be explained in the following sections.

From a heat transfer perspective, the whole transient energetic balance mechanism in a building energy system is broken down to steady-state milestones (similar to equilibrium states) in which the system characteristics are assumed to be invariant during that specific time step. This assumption is known as ***quasi-steady*** energy balance approach. Nonetheless, both prospects are interpreted from the conservation law of energy (see Chapter 4) which implies that energy can neither be created nor destroyed, and only can be transformed from one form to another.

8.1   Definition of a Zone

Any energy system, including a building energy system, requires a closed boundary to be firstly defined. We shall define zones in a building envelope as control volumes where we can later derive the balance of the energy on them. As it can be seen in Figure 8.1, various zones (boundaries) can be defined in a building based on the desired level of information.

From a thermodynamic point of view, buildings' energy process is an isochoric one where the system volume is constant as we do not expect the boundaries of the control volume (zones) to become shrunk or expanded. The choice of splitting a building energy system

Computational Fluid Dynamics and Energy Modelling in Buildings: Fundamentals and Applications,
First Edition. Parham A. Mirzaei.
© 2023 John Wiley & Sons Ltd. Published 2023 by John Wiley & Sons Ltd.

(a)

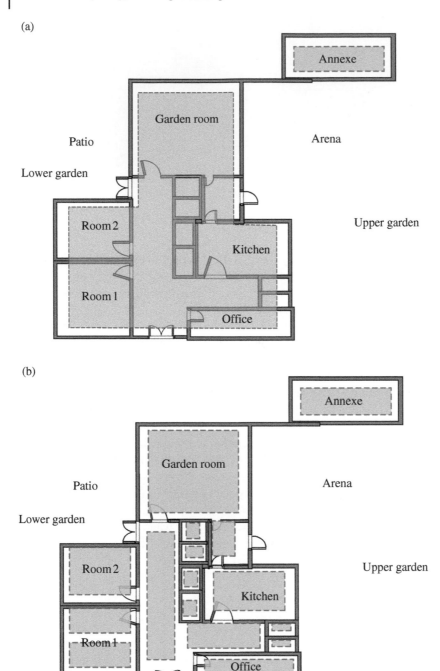

(b)

Figure 8.1 Zoning in buildings.

to multiple zones is not always straight forward, nor limited to the number of rooms. In general, a zone should represent the characteristics of one or similar thermal spaces in terms of conditioning requirements or energetic performance. If two rooms need to be mechanically ventilated with a similar standard effective temperature (SET) point, then merging both into a single zone can be a viable option. On the contrary, a large space with various types of usage and occupant behaviour shall be separated to multiple zones to accommodate the design and calculation purpose of the modelling.

Cluster of zones or energy systems together define our building energy system. Adding one more zone will increase one degree of freedom to the building energy system. Nonetheless, increasing the number of zones may inversely increase the calculation costs and necessity to insert more inputs to the system. For example, definition of two zones (Figure 8.1a) will provide only two temperatures assigned to each zone while defining 16 zones (Figure 8.1b) will allocate 16 temperatures, one to each room (space), in the shown building. Therefore, the first step in deriving the conservation law equations is to identify the spaces with the same thermal characteristics, which can be identified as control volumes or zones.

8.2 Conservation Law in Buildings

In Chapter 4, we have investigated the conservation laws of energy and mass. For building energy system, which is about flow of moist air within the zones, again we can employ the conservation of energy law demonstrated as the first law of thermodynamics and the conservation of mass exploited as a mass continuity in within the control volume (zone). It should be noted that the mass continuity should be applied to each mass species that is considered in the system. As we can see in the proceeding sections, beside air as the main mass species in the building energy systems, moisture or water vapour might be included or not into the account depending on the level of accuracy of the modelling. Likewise, other mass species such as CO_2 can be taken into the consideration when the aim of a study demands such quantities to be calculated.

Conservation of energy states that rate of energy that stored in a control volume $\left(\dfrac{dE_{C.V.}}{dt}\right)$ must equal the rates at which energy enters (E_{in}) and leaves (E_{out}) the control volume, plus the rate of energy generated within the control volume (E_{gen}):

$$\frac{dE_{C.V.}}{dt} = \dot{E}_{in} - \dot{E}_{out} + \dot{E}_{gen} \tag{8.1}$$

Using an approach used to develop Eq. (4.40) with further consideration of generation of heat, the net increase of the energy within a building zone or control volume ($\dot{E}_{C.V.}$) based on the first law of the thermodynamics can be written as below (see Figure 8.2):

$$\frac{dE_{C.V.}}{dt} = \dot{Q}_{C.V.} - \dot{W}_{C.V.} + \dot{Q}_{gen}$$
$$+ \sum_{i=1}^{n} \dot{m}_{in-i}\left(h_{in-i} + \frac{1}{2}\vec{V}^2_{in-i} + gz_{in-i}\right)$$
$$- \sum_{j=1}^{m} \dot{m}_{out-j}\left(h_{out-j} + \frac{1}{2}\vec{V}^2_{out-j} + gz_{out-j}\right) \tag{8.2}$$

Note that the conditions to reach Eq. (8.2) is that the control volume is stationary relative to the coordinate frame, and a well-mixed condition can be assumed to ensure that the state of mass is uniform within the control volume at any time step.

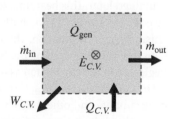

Figure 8.2 Energy balance in a zone.

Furthermore, we need to ensure that conservation of mass or continuity equation is always fulfilled in a control volume. This states that the rate at which mass enters a zone ($\dot{m}_{in}$) is equal to the rate at which mass leaves the zone ($\dot{m}_{out}$) plus the accumulation of mass within the zone $\left(\dfrac{dm_{C.V.}}{dt}\right)$ (see Eq. (4.32)):

$$\frac{dm_{C.V.}}{dt} = \sum \dot{m}_{in} - \sum \dot{m}_{out} \tag{8.3}$$

8.3 Governing Equations at Zones

At this stage, we can consider a series of assumptions to reach a more simplified form of the equation to demonstrate the energy flow within a zone. In general, we can summarize the major assumptions in a zone as:

- Changes in openings' velocity are negligible ($\sum \dot{m}_{in}\vec{V}_{in}^2 \approx \sum \dot{m}_{out}\vec{V}_{out}^2$) - As moist air is an incompressible ideal gas and areas of openings would be in the same range, we expect the velocities entering and leaving a zone to be very close to each other.
- Changes in openings' elevation are negligible ($\sum \dot{m}_{in}z_{in} \approx \sum \dot{m}_{out}z_{out}$) – this is a very valid assumption as heights of openings in a zone have normally a few centimetres difference.
- Moist air in zones is following the ideal gas behaviour ($h = C_p T$ and $u = C_v T$).
- There is no shaft work in the system – zone boundaries will not change, and the thermodynamic process can be assumed as an isochoric one.

8.4 Energy Balance Equation

As it was described in Section 8.1.2, the conservation law of energy implies that in any snapshot of time, we can define an energy balance in any defined zone. Now, applying the mentioned assumptions in Section 8.1.3, we can reduce Eq. (8.2) to the below general governing energy balance equation:

$$\frac{dE_{C.V.}}{dt} = \dot{Q}_{C.V.} + q_{gen} + \sum_{i=1}^{n} \dot{m}_{in-i}(h_{in-i}) - \sum_{j=1}^{m} \dot{m}_{out-j}\left(h_{out-j}\right) \tag{8.4}$$

Here, $\dot{Q}_{C.V.}$ represents heat rate transferred through solid boundaries of the control volume (walls, ceiling, floor, closed doors, and windows) as shown in Figure 8.3. As this will be explained later, this heat transfer rate is due to a difference between the zone (T_z) and surface temperatures (T_{si}) via both convection ($q_{conv.}$) and internal long-wave radiation mechanism (q_{LWi}). In addition, there is a heat transfer rate (q_L) associated with the difference between the moisture of the surfaces (e.g. W_{si}) and zone (W_z). The later term is often neglected in the calculation as its value is mainly insignificant in comparison to other terms.

On the other hand, q_{gen} is related to the rate of heat gained from internal loads generated by occupants and other bio-sources, appliances, and heaters as sensible or latent heat. The transmitted short-wave radiations (q_{SWt}) also will be accounted as a source of heat in a zone.

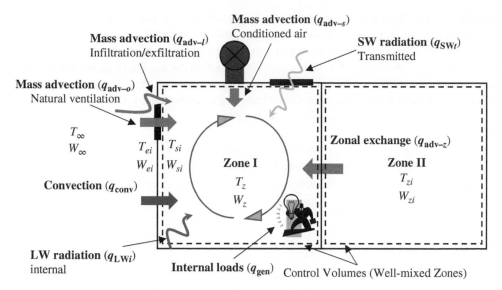

Figure 8.3 Heat transfer mechanisms in a zone.

Finally, $\sum_{i=1}^{n} \dot{m}_{\text{in}-i}(h_{\text{in}-i})$ and $\sum_{j=1}^{m} \dot{m}_{\text{out}-j}(h_{\text{out}-j})$ are related to the change in the energy of the control volume due to the moist air advection from:

1) External openings, which include unintentional infiltration/exfiltration (air leakage) throughout cracks and gaps ($q_{\text{adv}-l}$) in addition to deemed natural ventilation throughout windows and doors ($q_{\text{adv}-o}$).
2) Conditioned air from heating or cooling mechanical systems ($q_{\text{adv}-s}$).
3) Zonal exchange ($q_{\text{adv}-z}$) or internal exchanges of moist air between zones either defined in a same room or within other spaces.

Knowing that $\dfrac{dE_{C.V.}}{dt}$ represents the time derivative of energy in the control volume or the heat storage (sensible and latent) in the zone, Eq. (8.4) can be further rewritten as below:

$$\rho c_{\text{p}} V \frac{dT_z}{dt} = \sum_{i=1}^{N_s} \dot{Q}_i + \sum_{i=1}^{N_{\text{surfaces}}} h_{si\_i} A_i (T_{si\_i} - T_z)$$
$$+ \sum_{i=1}^{N_o} \dot{m}_{o\_i} c_{\text{p}}(T_\infty - T_z) + \dot{m}_l c_{\text{p}}(T_\infty - T_z) \tag{8.5}$$
$$+ \sum_{i=1}^{N_z} \dot{m}_{z\_i} c_{\text{p}}(T_{z\_i} - T_z) + \sum_{i=1}^{N_s} \dot{m}_{\text{sys}} c_{\text{p}}(T_{\text{sys}} - T_z)$$

where $\dot{m}_{o-i}$, $\dot{m}_{z\_i}$, $\dot{m}_{\text{sys}}$, and $\dot{m}_l$ are the mass flow rate from N_o openings, N_z connected zones, N_s diffusers, and cracks/gaps, respectively. Moreover, T_{z_i} and T_{sys} are the temperature of connected zones, and diffusers, respectively.

Note that the temperature of other zones and surfaces are unknown variables in Eq. (8.5) in most of the scenarios, which makes analytical solution of this equation difficult to be obtained.

In the simplest scenario, however, Eq. (8.5) can be assumed as a first-order ODE and therefore standard solution can be expected [1]. The simplified equation form and its solution are not the aim of this chapter as it is common to replace each component of Eq. (8.5) with complex equations and correlation, transforming the equation to a complex form and thus numerical methods such as finite difference are required to be employed. In the proceeding sections, we will explain implicit method to resolve Eq. (8.5), but before that we use the nodal analogy to first sketch a visual demonstration of a building energy system, and then to establish a logical connectivity between its elements together.

Example 8.1

Assume an air tightened office room with dimensions of 4 m × 4 m × 2.9 m with insulated walls and without any opening at an initial temperature of 21°C. There is one occupant who generates heat of 60 W in the office. Simplify Eq. (8.5) for this problem and solve the ODE equation. What is the necessary mass flow of 17°C dry air from a single diffuser to keep the room temperature around 21°C.

Solution

Assumptions:

- There is no heat exchange through the walls: $\sum_{i=1}^{N_{surfaces}} h_{si\_i} A_i (T_{si\_i} - T_z) \approx 0$
- There is no opening in the room: $\sum_{i=1}^{N_o} \dot{m}_{o\_i} c_p (T_\infty - T_z) \approx 0$
- The room is air tightened: $\dot{m}_1 c_p (T_\infty - T_z) \approx 0$
- The office will be assumed as one single zone: $\sum_{i=1}^{N_z} \dot{m}_{z\_i} c_p (T_{z\_i} - T_z) \approx 0$
- We assume the fluid to be dry air with $c_p \approx 1$ kJ/kg K and $\rho = 1.2$ kg/m^3
- We can assume that the mechanical ventilation is initially off: $\sum_{i=1}^{N_s} \dot{m}_{sys} c_p (T_{sys} - T_z) \approx 0$

The governing equation becomes:

$$55,680 \frac{dT_z}{dt} = 60 \Longrightarrow T_z^{(t)} = 0.0011t + c$$

$$T_z^{(0)} = 21 = c \Longrightarrow T_z^{(t)} = 0.0011t + 21$$

A controller to compensate about $60W$ is then necessary to establish a balance in the room:

$$\dot{m}_{sys} c_p (T_{sys} - T_z) + 60 = 0 \Longrightarrow \dot{m}_{sys} = \frac{0.06}{(4 - 0.0011t)} \text{ kg/s}$$

So, for each second:

$$\dot{m}_{sys} = 0.015 \text{ kg/s}$$

8.5 Nodal Analogy of the Governing Equation

As it can be seen in Figure 8.4, an example of a two-zone room with one opening and one mechanical system installed in the zone-2 is shown with the associated thermal resistances, heat sources, and heat storages. Let us assume that the short-wave and long-wave radiations are absent within the zones, the air does not contain any moisture, and solid materials and surfaces do not have any heat storage capacities.

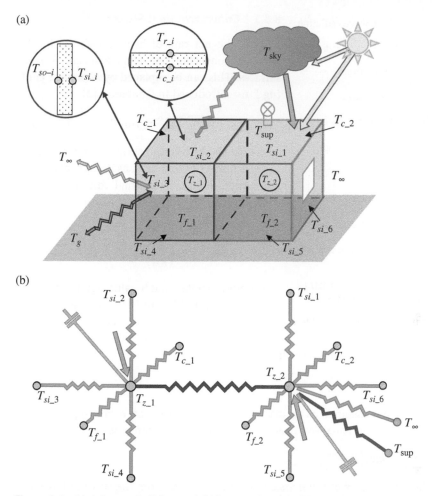

Figure 8.4 (a) A 2-zone building and (b) its associated nodal system.

Moreover, we assume that there is a heat release due to occupants and appliances, and mass is only transferred through one window while the building is air-tightened or mass advection via infiltrations is negligible. Now, before forming the nodal system of this case, we can define a colour code to represent the heat resistances, sources, and capacitors as illustrated in Figure 8.5. Each element represents a related heat transfer mechanism in the nodal system with a specific way to be calculated, which will be explained in the following sections. Now, we use the nodal analysis analogy adapted for thermal resistance introduced in Chapter 7 to develop simplified model of energetic flow in zones. Hence, we can form the nodal system of the zones as depicted in Figure 8.4b for now. The system has an advection exchange between two zones. There is a convection exchange between the zones and all interior surfaces. Exterior surfaces are exposed to the environment (external convection is taken to the account), and heat is conducted through the walls and ceiling. Short-wave and long-wave radiations only act on the external surfaces.

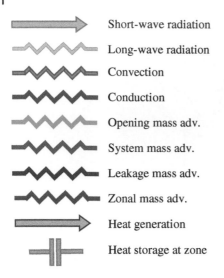

Short-wave radiation

Long-wave radiation

Convection

Conduction

Opening mass adv.

System mass adv.

Leakage mass adv.

Zonal mass adv.

Heat generation

Heat storage at zone

Figure 8.5. Colour code graph of thermal resistances, sources, and capacitors.

8.5.1 Convective Heat Fluxes

$q_{conv.}$ is the net rate of heat transferred by the means of convection between a zone and its solid surfaces. This can be replaced by **Newton's cooling Law** introduced in Chapter 6 [2]:

$$q_{conv} = \sum_{i=1}^{N_{surfaces}} h_{si\_i} A_i (T_{si\_i} - T_z) \qquad (8.6)$$

where T_{si_i} is the internal surface temperatures of a wall, ceiling, or floor. Also, as it was explained in the earlier chapter, a thermal resistance of the convection can be represented as below:

$$q_{conv} = hA\Delta T$$
$$\implies R_{conv} = \left(\frac{1}{hA}\right) \qquad (8.7)$$

It should be noted that h-value significantly varies due to the condition and position of a surface as several approximations are utilized to provide their values as explained in Chapter 7. Hence, h is a time step variant parameter, which should be calculated at each time step and for each surface. For further clarifications see Chapter 9.

8.5.2 Advective Heat Fluxes

Since air in the zone is assumed as an ideal gas, the mass advection terms (q_{adv}) through infiltration, openings, and zonal exchange in the zone can be written as:

$$q_{adv} = \sum_{i=1}^{n} \dot{m} c_p (T_i - T_z) \qquad (8.8)$$

where, T_i is the temperature of the mass advected to the zone from the surrounding environment, supplied air at the diffuser of the mechanical ventilation system, or other zones. Similar to the conduction and convection terms, the advection term can be rewritten as the format of thermal resistance between the place of the opening and zone:

$$q_{adv}\left(\frac{1}{\dot{m}_{o\_i} c_p}\right) = \Delta T$$
$$\implies R_{adv-o} = \frac{1}{\dot{m}_{o\_i} c_p} \qquad (8.9)$$

So, based on the energy exchange between a zone and connected neighbouring zones, we can write resistance for the zonal exchange:

$$R_{adv-z} = \frac{1}{\dot{m}_{z\_i} c_p} \qquad (8.10)$$

Eventually, we can obtain this term for a mechanical system:

$$R_{adv-s} = \frac{1}{\dot{m}_{sys} c_p} \qquad (8.11)$$

And, the leakage:

$$R_{\mathrm{adv}-l} = \frac{1}{\dot{m}_l c_p} \tag{8.12}$$

In all the advection resistance equations, the mass flow rate is essential to be known. This parameter, however, is another time-variant variable that should be calculated from continuity equations as explained in the following sections to be inserted in Eq. (8.5).

8.5.3 Heat Generation Fluxes

q_{gen} is accounted as an applied heat source into a zonal node when the storage term is considered in the moist air (see Eq. (8.5)).

8.5.4 Short-wave Radiative Fluxes in Zone

Sun can be assumed as a black body as explained in Chapter 6. This leads to the incident solar radiation (short-wave) to be assumed as a constant value on a building surface (at a specific time step) as depicted in Figure 8.3.

In the electrical nodal analogy, thus, this value can be approximated with a source heat generation to be applied either on the exterior node surfaces or interior node surfaces after passing through openings. If openings are closed with transparent materials such as glass, then a part of the short-wave incident will be reflected ($q_{\mathrm{SW}r-i}$) on the glass node, a part will be absorbed ($q_{\mathrm{SW}a-i}$) on the glass node, and the rest will be transmitted to the zone ($q_{\mathrm{SW}t}$) to be applied to interior surfaces' node or nodes.

Let us focus on interior spaces or zones as depicted in Figure 8.6. We find that the short-wave radiation may penetrate through openings or being transmitted through glasses ($q_{\mathrm{SW}t}$), and impact one or multiple surfaces. A part will be absorbed by each internal surface ($q_{\mathrm{SW}a-i}$) and a part will be reflected to other surfaces ($q_{\mathrm{SW}r-i}$):

$$q_{\mathrm{SW}a-i} = A_1(1-\rho_s)q_{\mathrm{SW}t} = A_1\alpha_s q_{\mathrm{SW}t} \tag{8.13}$$

where, A_1 is the area of the first impacted surface and α_s is the absorptivity of the interior surfaces against the short-wave radiation. Note that the behaviour of surfaces against short-wave radiation are barely diffuse and grey. On the other hand, the value of $C_j = A_1\alpha_s q_{\mathrm{SW}t}$ can be considered as a constant value on a surface in specific time. If the reflection is neglected, the value will be only considered on the impacted surfaces and zero on the rest of them.

Since zones can be assumed as enclosed spaces with an acceptable degree of accuracy, the reflection happens for multiple times (with the same radiative properties for all surfaces) until all the radiation is fully absorbed by the internal surfaces. Therefore, if we trace the reflected ray for multiple times, we can derive:

$$q_{\mathrm{SW}a-i}{}^{(n)} = \left(\prod_n A_n\right)\alpha_s{}^n q_{\mathrm{SW}t} \tag{8.14}$$

where $q_{\mathrm{SW}a-i}{}^{(n)}$ represents the absorption after the nth reflection over the nth surface.

Note that the number of reflections can be approximated since:

$$q_{\mathrm{SW}a-i} = \sum_n q_{\mathrm{SW}a-i}{}^{(n)} \tag{8.15}$$

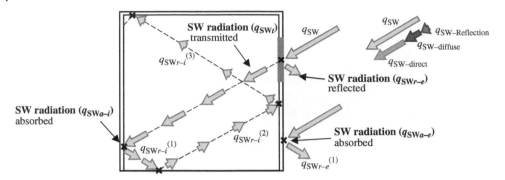

Figure 8.6 Short-wave radiation incident at exterior and interior surfaces of a zone.

Note that the q_{SW} needs to be initially calculated from available models such as those introduced in Chapter 7. While models are robust in the prediction of direct and diffuse components, the reflection components from other surrounding surfaces require their exact geometries and positions. In addition, the calculation load of reflection from surrounding surfaces will intensively increase due to ray tracing necessary for single or multiple reflection from the surfaces. The ray tracing approach follows a similar concept as what is introduced for internal ray tracing model as shown in Figure 8.6.

Example 8.2

Solar radiation at 2:00 pm in London (51.51N, −0.13W) is impacting a glass window as demonstrated in Figure 8.E2 and penetrate through a room with brick surface materials. Calculate solar incident using ASHRAE clear sky model (see Chapter 7 [3]) at 28, June, and trace solar rays until it reaches below 1 W/m^2 on a surface. The window has dimensions of 0.5 m × 0.5 m.

Solution

Glass and brick properties can be assumed as $\tau_g = 0.90$ and $\alpha_s = 0.63$.
At 14:00 pm of June 2021, the solar altitude in London is:

$$\beta = 59.75^{\circ}, A = 1093 \text{ W/m}^2; B = 0.185$$

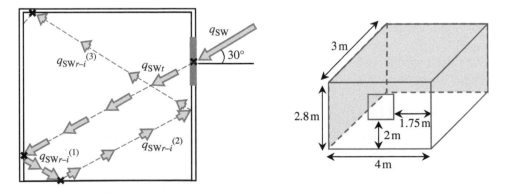

Figure 8.E2 Solar radiation in a room.

Hence:

$$I_{dn} = \frac{1093}{\exp\left(0.185/\sin\left(54.02\right)\right)} = 882.3\,\text{W/m}^2$$

And the transmitted short-wave radiation is:

$$q_{\text{SW}t} = 0.90 \times 882.3 = 794.06\,\text{W/m}^2$$

We assume a similar area of incident as the window's area:

$$q_{\text{SW}a-1} = A_1\alpha_s q_{\text{SW}t} = 125.1\,\text{W/m}^2$$

$$q_{\text{SW}a-2} = 19.7\,\text{W/m}^2$$

$$q_{\text{SW}a-3} = 3.1\,\text{W/m}^2$$

$$q_{\text{SW}a-4} = 0.5\,\text{W/m}^2$$

Now, let us assume the case study of Figure 8.4 that the transmitted short-wave radiation (direct, diffuse, and reflected) only impact floor-2 and surface-5. Here, we neglect the reflected short-wave radiation on the internal surfaces. As shown in Figure 8.7, the new nodal system includes constant values (in a specific time) on nodes T_{f_2} and T_{si_5}.

8.5.5 Long-wave Radiative Fluxes in Zone

In the previous chapter, we have defined the concept of grey surfaces in which the surfaces of an enclosure (here zone) are independent of wavelength. The surfaces have been also assumed to have radiative characteristics (reflectivity, absorptivity, and transmissivity) independent from direction or being diffuse. Therefore, for a nonparticipating medium

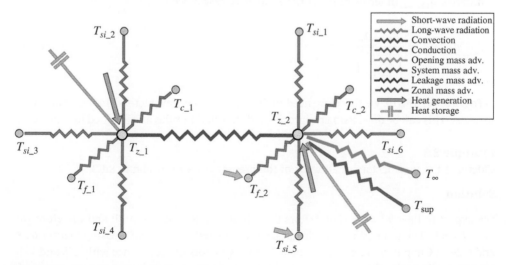

Figure 8.7 Two-zone nodal system with internal short-wave radiation when reflection is neglected.

LW Radiation ($q_{\mathrm{LW-i_n}}$)

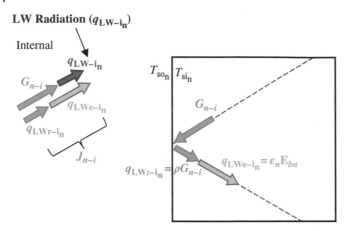

Figure 8.8 Long-wave radiation exchanges at exterior and interior surfaces of a zone.

(the moist air) and opaque surfaces, we could conclude that emissivity is equal to absorptivity against the long-wave radiation ($\varepsilon = \alpha$).

Long-wave irradiation in a zone can occur due to difference in the zones' surface temperatures, heaters, occupants, appliances, etc. the proportion of this radiative flux is relatively smaller than other contributors in the energetic balance of a zone, thus it is mainly neglected in the calculations and developed tools. Here, we provide the theory behind and the conceptual nodal system of integration of long-wave radiations if necessary.

If we assume G_{n-i} to be irradiation on a particular surface of n in a zone. As shown in Figure 8.8, a part of this incident radiation will be absorbed by the surface (αG_{n-i}), and a part will be reflected (ρG_{n-i}). The surface also emits radiation ($\varepsilon_n E_{bn}$) to all the surrounding surfaces in the zone. Thus, the difference between the radiosity (J_{n-i}) and irradiation will indicate the internal long-wave radiative heat flux ($q_{\mathrm{LW-i_n}}$). And, similar to Eq. (6.101), we can derive $q_{\mathrm{LW-i_n}}$ in terms of the surface radiative resistance ($R_m = \dfrac{1-\varepsilon_n}{\varepsilon_n A_n}$):

$$q_{\mathrm{LW}-i_n} = \frac{E_{bn} - J_n}{\left(\dfrac{1-\varepsilon_n}{\varepsilon_n A_n}\right)} \tag{8.16}$$

To solve Eq. (8.16), we therefore need to first identify the view factors amongst pair of all surfaces. This can be done using reciprocity and summation rules as explained in Chapter 6.

Example 8.3

Calculate the view factors of all walls of the presented room in Figure 8.E3.

Solution

We expect to have $n^2 = 36$ view factors though no surface can see itself and therefore we have $F_{ii} = 0$. The dimension of each surface is calculated in Table 8.E3. Using Figures 6.28 and 6.29 in Chapter 6, we calculate the below view factors in accordance with X/L and Y/L for perpendicular surfaces:

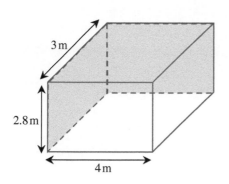
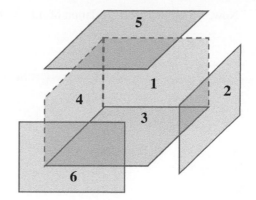

Figure 8.E3 Case study room.

Table 8.E3 Dimensions and view factors of different surfaces of the case study room.

ij	Y/X	Z/X	F_{ij}
12	1.43	1.07	0.1584
13	0.70	0.75	0.2284
14	1.43	1.07	0.1584
15	0.70	0.75	0.2284
23	0.93	1.33	0.2258
25	0.93	1.33	0.2258
26	1.07	1.43	0.2111
34	1.33	0.93	0.1581
36	0.75	0.70	0.2132
45	0.93	1.33	0.2258
46	1.07	1.43	0.2111
56	0.75	0.70	0.2132

Also, for the parallel surfaces in accordance with X/L and Y/L:

ij	X/L	Y/L	F_{ij}
16 = 61	1.33	0.93	0.2265
24 = 42	0.75	0.70	0.1260
35 = 53	1.43	1.07	0.2575

Now, we use reciprocity relation of $A_iF_{ij} = A_jF_{ji}$ to find the rest of view factors:

ij	F_{ij}
21	0.2112
31	0.2132
32	0.1581
41	0.2112
43	0.2259
51	0.2132
52	0.1581
54	0.1581
62	0.1583
63	0.2284
64	0.1583
65	0.2284

If we use the summation rule ($\sum_j F_{ij} = 1$) to check our calculations, we find:

$\sum_j F_{1j}$	$\sum_j F_{2j}$	$\sum_j F_{3j}$	$\sum_j F_{4j}$	$\sum_j F_{5j}$	$\sum_j F_{6j}$
1.000	1.000	1.000	1.000	1.000	1.000

The next step in integration of radiative resistances to our previous nodal network is to form another network amongst all pair of surfaces in a zone with N surfaces as depicted in Figure 8.9. For this purpose, and as explained in Chapter 6, first, it is required that we define some extra nodes, known as radiosity nodes, to be associated to each surface (J_{si_n}). Note that we name the surface nodes to have a temperature of $T^*_{si\_n}$ as it represents that the radiative heat flux needs to be calculated from this temperature ($E_{bn} = \sigma T^*_{si\_n}{}^4$) and not to directly be used from T_{si_n}. On the other hand, as we will explore in Section 8.1.5.6, to calculate conduction from such nodes (q_{cond_n}), we need to use T_{si_n}. This is the reason we show the value as $T^*_{si\_n}$ to be used for both radiative and conductive heat flux calculations.

The radiative heat rate between any node of n and its radiosity nodes ($q_{\text{LW}-i_n}$) represent heat passing a resistance of $\dfrac{1-\varepsilon_n}{\varepsilon_n A_n}$. The radiation then will be transferred to the rest of surfaces ($N-1$) facing their resistances of $\dfrac{1}{F_{nk}A_n}$, known as geometrical resistance, to reach their radiosity nodes:

$$q_{\text{LW}-i_n} = \sum_{k=1}^{N-1} \frac{J_n - J_k}{\left(\dfrac{1}{F_{nk}A_n}\right)} \tag{8.17}$$

It is evident that the radiative heat rates received from all $N-1$ surfaces into a particular radiosity node will be added and transferred to its associated surface node. This implies that we need to form $N^2 - N$ pairs of geometrical resistances in a zone with N surfaces.

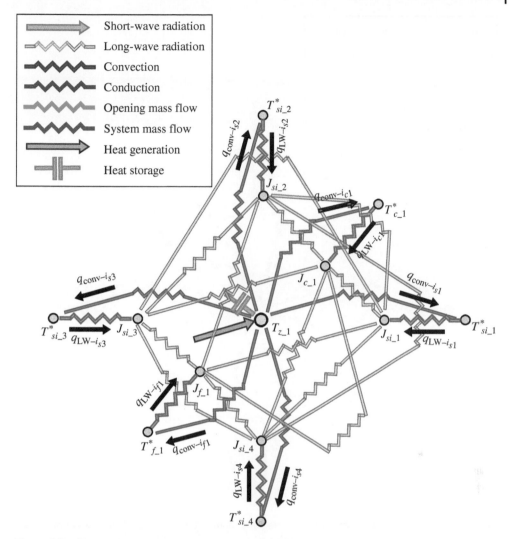

Figure 8.9 Geometrical resistance in a single zone room.

If we consider a surface to have a nonzero view factor to itself, which is barely a case in common buildings cases, then the total numbers become $N^2 - N + N = N^2$. Now, let us consider the case of Example 8.3. As shown in Figure 8.9, we have $6^2 - 6 = 30$ geometrical resistances for such a zone.

At this stage, we write the energy balance on the surface nodes while adding the effect of other contributors. For example, let us assume that the short-wave radiation is directly impacting the floor and surface-1 (the reflection is neglected). As it is shown in Figure 8.10, we can simply add short-wave radiation to the nodal system.

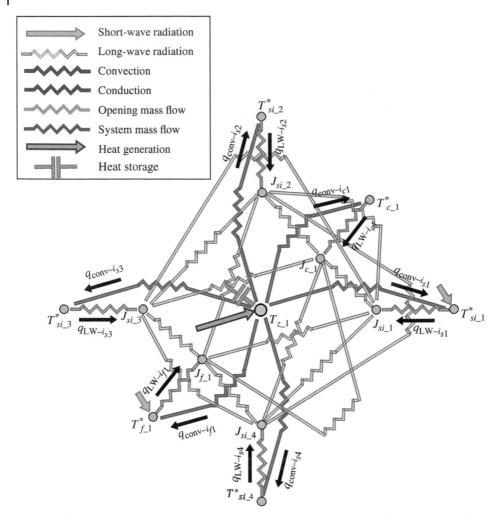

Figure 8.10 Geometrical resistance in a single zone room in the presence of short-wave radiation.

Example 8.4

Assume the surfaces of the room in Example 8.3 with the shown temperatures in Figure 8.E4. Calculate the long-wave radiation between surfaces if they are assumed to be diffuse and grey. The temperature and emissivity of each surface is given in Figure 8.E4.

Solution

We can first calculate E_b (Eq. (6.86)), radiosity (Eq. (8.17)), and surface radiative resistance $(R_m = \dfrac{1 - \varepsilon_n}{\varepsilon_n A_n})$ for each wall:

Surface number	Temp. (K)	Area (m²)	Emissivity
1	305	11.2	0.8
2	301	8.4	0.8
3	295	12.0	0.9
4	310	8.4	0.8
5	330	12.0	0.9
6	315	11.2	0.8

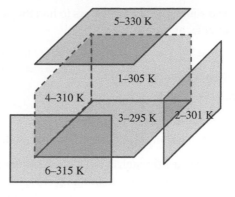

Figure 8.E4 Temperature and emissivity of the surfaces of the case study room.

Surface #	E_b (W/m²)	J (W/m²)	R_m	q_n (W/m²)
1	490.7	499.4	0.022	−390.9
2	465.4	479.5	0.030	−471.3
3	429.4	441.6	0.009	−1316.6
4	523.6	524.9	0.030	−41.7
5	672.4	654.8	0.009	1900.8
6	558.2	551.1	0.022	319.6

The rate of radiation transfer to the radiosity node is q_n (Eq. (8.16)), which should be equal to summation of all radiative exchanges towards the other surfaces:

$$q_n = \sum_{k=1}^{5} q_{LW-i_{n \to k}} = \sum_{k=1}^{5} \frac{J_n - J_k}{\left(\dfrac{1}{F_{nk}A_n}\right)}$$

We use view factors found in Example 8.3 to calculate $q_{LW-in \to k}$:

$q_{LW-i_{n \to k}}$ (W/m²)	k = 1	k = 2	k = 3	k = 4	k = 5
n = 1	35.4	147.8	−45.2	−397.6	−131.2
n = 2	−35.4	71.8	−48.1	−332.6	−127.1
n = 3	−147.8	−71.8	−158.0	−658.8	−280.2
n = 4	45.2	48.1	158.0	−246.5	−46.5
n = 5	397.6	332.6	658.8	246.5	265.3
n = 6	131.2	127.1	280.2	46.5	−265.3

We can sum up each row's values to find the net rate of radiation transfer at each surface. For example, at surface-1:

$$q_{\text{LW}-i_1} = \sum_{k=1}^{5} \frac{J_n - J_k}{\left(\dfrac{1}{F_{nk}A_n}\right)} = 35.4 + 147.8 - 45.2 - 397.6 - 131.2 = -390.9 \text{ W/m}^2$$

Which is equal to $q_1 = -390.9 \text{ W/m}^2$.

8.5.6 Conduction Heat Fluxes Through Solid Surfaces

The heat will pass through a wall with the diffusion mechanism (see Chapter 6). Due to the practicality considerations in building energy simulations to minimize the computational cost, walls are mainly assumed to be one-dimensional solid objects. This assumption is not very far from the reality as a main part of a wall in a building is behaving as a 1-D medium and only corners, location of windows, and thermal bridges have more complex and 3D heat transfer behaviours.

In the nodal system, the convective heat exchange occurs between the well-mixed moist air of a zone and the solid boundaries such as walls, ceilings, and floors of the control volume (T_{si}). Due to the conservation of energy as it is shown in Figure 8.11, the convective heat exchange will add up to radiative fluxes and then will pass through the solid surfaces via conduction mechanism. So, if we assume a control volume around the internal surfaces, we can derive:

$$q_{\text{conv}-i} + q_{\text{gen}-i} - q_{\text{rad}-i} - q_{\text{cond}} = 0 \tag{8.18}$$

where $q_{\text{gen}-i}$ is the generated heat within the wall nodes (e.g. heating pipes, electrical wires, etc.). Note that in most of the scenarios, it is assumed that there is no heat generation in a wall, or its value is negligible. So, we can obtain:

$$q_{\text{conv}-i} - q_{\text{rad}-i} - q_{\text{cond}} = 0 \tag{8.19}$$

The energy balance on the internal control volume implies that we should allocate one internal node to each wall in the nodal system and derive Eq. (8.19) for each node. The heat conduction process can be calculated by **Fourier's Law** introduced in Chapter 6 when the solid walls are presumed to have no heat storage (thermal mass):

$$q_{\text{cond}} = -kA\frac{\partial T}{\partial x} = kA\frac{T_{si} - T_{so}}{L}$$
$$\implies R_{\text{cond}} = \frac{L}{kA} \tag{8.20}$$

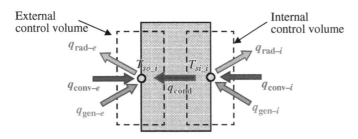

External control volume Internal control volume

$q_{\text{rad}-e}$ $q_{\text{rad}-i}$

T_{so} T_{si}

$q_{\text{conv}-e}$ q_{cond} $q_{\text{conv}-i}$

$q_{\text{gen}-e}$ $q_{\text{gen}-i}$

Figure 8.11 Heat balance in interior and exterior nodes of solid boundaries.

where $R_{cond} = \frac{L}{kA}$ is the conduction resistance and T_{so} represents the external surface temperature (e.g. see Figures 8.4 and 8.11).

Determination of q_{rad} at the internal surfaces was explained in the previous sections. Nonetheless, finding short-wave (q_{SW-e}) and long-wave (q_{LW-e}) radiations between the exterior surfaces and environment is different from enclosed spaces. The energy balance equation on the exterior surfaces, thus, can be derived as:

$$q_{conv-e} + q_{gen-e} - q_{rad-e} + q_{cond} = 0 \tag{8.21}$$

8.5.7 Short-wave Radiative Fluxes on Exterior Surfaces

As explained in Chapter 7, the short-wave radiation consists of three main parts of direct ($q_{SW-direct}$) and diffuse ($q_{SW-diffuse}$) radiations in addition to the reflected solar rays from surrounding buildings and infrastructures ($q_{SW-reflection}$). As shown in Figure 8.6, a part of q_{SW} is reflected back to the environment (ρq_{SW}). If the surface is transparent, then a part (q_{SWt}) is transmitted into the zones as discussed in Section 8.1.5.4 otherwise $q_{SWt} = 0$ and the remaining part will be absorbed on the surface (q_{SWa-e}). The amount of q_{SWa-e} on a specific surface can considerably change due to its material properties and in respect to the building orientation and inclination. If we assume the previous example of the single zone (see Figure 8.10) with internal short-wave and long-wave radiations, Figure 8.12 demonstrates the short-wave radiation absorbed on the exterior surfaces.

8.5.8 Long-wave Radiative Fluxes on Exterior Surfaces

In Chapter 7, we have learned the radiative contributors on the earth surface. While consideration of all of these contributors into the simulations are cumbersome and would drastically increase the computational costs, it is wise to only focus on the major players and neglect the rest. In building simulation studies, the long-wave radiative exchanges between a building surface and its surrounding environment are mainly limited to surface to sky, surface to ground, surface to the bulk surrounding air, and eventually surface to other buildings' surfaces.

8.5.8.1 Surface to Surface Long-wave Radiative Fluxes

In this sense, we should again be fully aware of the geometrical feature of the environment around a building to be able to calculate view factors if surface to surface calculations are targeted. Not all the available tools are equipped with such options as geometrical preparation and calculation in addition to the computational load are challenging issues. Thus, models and tools are broadly neglecting surface to surface calculation or at best only are focused on few buildings around the simulated building. In terms of theory if we have a number of surfaces (buildings, trees, other objects, etc.) around the investigated building, we need to first obtain all the related view factors. Then, we need to scrutinize to which extend the enclosed space concept can be applied to those surfaces. Consideration of reflections for both short-wave and long-wave radiative fluxes is another issue that can increase the complexities of a model.

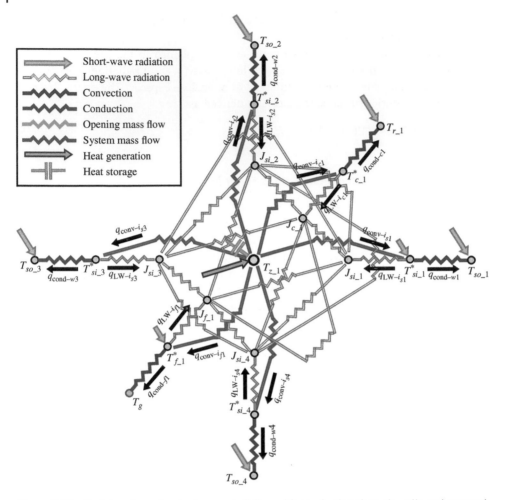

Figure 8.12 Nodal system of a single zone building with conduction through walls and external short-wave radiation on interior and exterior surfaces.

Example 8.5

Assume two buildings as illustrated in Figure 8.E5. Calculate the long-wave radiative exchange between these two buildings constructed with brick materials.

Solution

From literature, the view factors between such geometries can be assumed as:

$$F_{ij} = \frac{(\alpha\beta)^2}{\pi} \text{ if } \alpha = \frac{a}{c} < 0.2 \text{ and } \beta = \frac{b}{c}$$

Emissivity of brick can be found as $\varepsilon_{br} = 0.93$.

Thus, we can calculate $\alpha = 0.19$, $\beta = 1.15$, and $F_{ij} = 0.016$.

The radiative exchange between two parallel surfaces of 1 and 2 then can be calculated as:

$$q_{LW-12} = \sigma \varepsilon_{br} A_1 F_{12} T_1^4$$

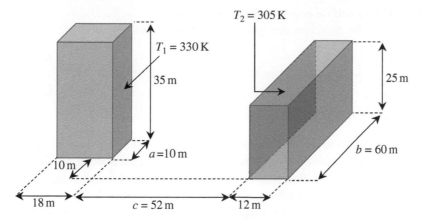

Figure 8.E5 Dimensions of two parallel buildings in an urban setting.

$$q_{LW-21} = \sigma \varepsilon_{br} A_2 F_{21} T_2^4$$

From reciprocity, we have $A_1 F_{12} = A_2 F_{21}$. Thus, the net exchange of radiative heat between two surfaces would be:

$$q_{LW-net} = q_{LW-12} - q_{LW-21} = \sigma \varepsilon_{br} A_1 F_{12} \left(T_1^4 - T_2^4 \right)$$

Replacing the values gives:

$$q_{LW-net} = 927.2 \, W$$

8.5.8.2 Surface to Environment Long-wave Radiative Fluxes

Radiative fluxes between a building surface and surrounding environment, including ground, sky, and bulk air can be very complex in the nature. The complexities, however, in the modelling is simplified using the concept of radiative heat transfer in two-surface enclosures (see Chapter 6 – Figure 6.31).

Each of the main contributors in the surrounding environment can be thus considered in a mutual radiative exchange with the building surface. As the surface of these contributors are relatively very large $\left(\dfrac{A_s}{A_{sky}} \approx \dfrac{A_s}{A_{ground}} \approx \dfrac{A_s}{A_{air}} \approx 0 \right)$, we can conclude that the radiative resistance of the environment $\left(R_{re} = \dfrac{1 - \varepsilon_e}{\varepsilon_e A_e} \right)$ to be effectively equal to zero. This is the reason that such surfaces can be assumed as the blackbody ($\varepsilon_e = 1$). On the other hand, we can assume that all the radiation leaving the building surface must reach to the environmental surface. Thus, we have $F_{so-sky} = F_{so-g} = F_{so-air} = 1$. Now, refer to the Eq. (6.106):

$$q_{LW-e} = \frac{\sigma \left(T_{so}^4 - T_e^4 \right)}{\dfrac{1 - \varepsilon}{\varepsilon A_s} + \dfrac{1}{A_s F_{so-e}} + \dfrac{1 - \varepsilon_e}{\varepsilon_e A_e}}$$

$$\Longrightarrow q_{LW-e} = \frac{\sigma \left(T_{so}^4 - T_e^4 \right)}{\dfrac{1 - \varepsilon}{\varepsilon A_s} + \dfrac{1}{A_s}} = \sigma A_s \varepsilon \left(T_{so}^4 - T_e^4 \right) \tag{8.22}$$

where, A_s is the building's surface and ε is the surface emissivity.

Hence, we can replace Eq. (8.22) for sky, ground, and the surrounding air:

$$q_{\text{LW}-\text{sky}} = \sigma A_s \varepsilon \left(T_{so}^4 - T_{\text{sky}}^4 \right) \tag{8.23}$$

$$q_{\text{LW}-\text{g}} = \sigma A_s \varepsilon \left(T_{so}^4 - T_{\text{g}}^4 \right) \tag{8.24}$$

$$q_{\text{LW}-\text{air}} = \sigma A_s \varepsilon \left(T_{so}^4 - T_{\infty}^4 \right) \tag{8.25}$$

To embed Eqs. (8.23)–(8.25) in the nodal system, we can again use the same concept as internal long-wave radiation network. This means that we can define a node in which the radiative fluxes are constructed by Eqs. (8.23)–(8.25) and the convective and conductive fluxes are formed as described in Sections 8.1.5.1 and 8.1.5.6, respectively. Unlike internal long-wave radiation fluxes, external ones are significant in an energetic balance of a building energy simulation and barely can be neglected. Therefore, it is common to linearize nonlinear power-4 radiative fluxes to make them to be easily connected to the network of Figure 8.12.

Example 8.6

Calculate the radiative fluxes between surfaces of building-1 during a typical winter day and a summer night with the given climatic data as shown in Figure 8.E6.

	Winter temperature (K)	Summer temperature (K)
Surface-1	278	295
Roof	281	300
Sky	263	285
Ground	276	298
Air	275	310

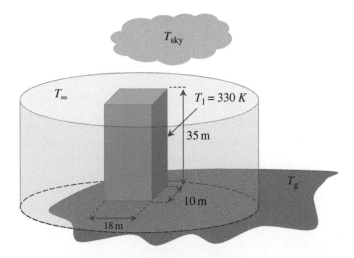

Figure 8.E6 Flux exchanges between an isolated building and its surrounding environment

Solution

First, we assume that the long-wave radiative exchanges of all four lateral surfaces are similarly denoted as surface-1. We assume a constant emissivity of 0.9 for all the surfaces in all radiative exchanges. View factors between surface-1 and environment in addition the roof and environment can be obtained from Eqs. (7.70) to (7.73):

	Surface-1	Roof
Area (m^2)	630	180
Φ	90	0
Cos (β)	0.00	1.00
β	0.71	1.00
ε	0.90	0.90
F_{sky}	0.35	0.71
F_g	0.50	0.00
F_∞	0.15	0.00

In the next step, we calculate a typical winter evening radiative exchanges of this example when the air temperature is colder than surfaces that might be slightly warmer due to solar radiation:

	Surface-1	Roof
qLW_{sky} (W/m^2)	21.4	52.3
qLW_g (W/m^2)	4.3	0.0
qLW_∞ (W/m^2)	1.9	0.0

Thus, the surfaces are significantly emitting towards the sky if the sky is clear while a small fraction is emitted to the surrounding ground and air. In a typical summer morning of this example, when the surfaces are cooler than the air, we can expect radiative fluxes towards the surface from the surrounding ground and air while the roof still emits towards the sky.

	Surface-1	Roof
qLW_{sky} (W/m^2)	17.6	54.2
qLW_g (W/m^2)	−8.0	0.0
qLW_∞ (W/m^2)	−12.4	0.0

8.5.8.3 Linearization of Long-wave Radiative fluxes

We assume that a radiative exchange (q_{LW-e}) between an external surface (T_{so}) and environmental source (T_e) in a general form of:

$$q_{LW-e} = KA\left(T_{so}^4 - T_e^4\right) \tag{8.26}$$

Note that K can include all the coefficients such as Boltzmann's constant number, emissivity, etc. Here, we can apply two methods to linearize Eq. (8.26). For the first one, we rewrite the equation as:

$$q_{LW-e} = KA(T_{so} - T_e)(T_{so} + T_e)(T_{so}^2 + T_e^2) \tag{8.27}$$

An assumption under environmental condition, which states that the variation between the temperatures is not relatively significant ($-20 \leq T_e, T_{so} \leq 50$) leads to another assumption that there is a representative mean temperature (T_m) that is in the same order of magnitude with temperatures of T_e and T_{so}. One can therefore approximate the below expression:

$$(T_{so} + T_e)(T_{so}^2 + T_e^2) \approx 2T_m(2T_m^2) \approx 4T_m^3 \tag{8.28}$$

Hence, the long-wave radiative resistance can be expressed as:

$$q_{LW-e}\left(\frac{1}{4KAT_m^3}\right) = \Delta T$$
$$\implies R_{LW-e} = \left(\frac{1}{4KAT_m^3}\right) \tag{8.29}$$

Another assumption to linearize Eq. (8.26) is to estimate the terms of Eq. (8.27) from the previous time step of a specific time in the dynamic simulation as this process will be described in Chapter 9. The embedded radiative resistance to the nodal network of the previous example of a single zone building is shown in Figure 8.13.

If the internal long-wave radiative fluxes are neglected, then Figure 8.14 represents the nodal system of a single zone. A two-zone case of Figure 8.4 is also shown in Figure 8.14 while the external nodes are taken into the consideration. In this case, the right-hand side zone has a mechanical ventilation system in addition to an opening.

8.5.9 Thermal Mass

Thermal mass is related to the capacity of materials to store heat. Furniture and walls in specific significantly store heat in a building.

8.5.9.1 Thermal Mass in Solid Surfaces

From a heat transfer prospect, solid surfaces in a building are providing a thermal shield against the climatic variations. In other words, the thermal mass cause ***decrement factor*** and ***time lag***, which are crucial parameters in design and selection of materials as will be discussed in Chapter 9 (see Example 9.9).

Now, let us assume to take the thermal mass effect in walls, ceilings, and floors, then a heat storage element should be added to each of the wall nodes. The energy balance Eq. (8.21) of the wall nodes then can be written as below:

$$\rho c \frac{\partial T}{\partial t} = q_{conv-i} + q_{gen-i} + q_{rad-i} - q_{cond} \tag{8.30}$$

where, ρ is the density and c is the specific heat of the wall materials. The transient term $\left(\frac{\partial T}{\partial t}\right)$ represents the change in the stored heat in the materials during the time. For the

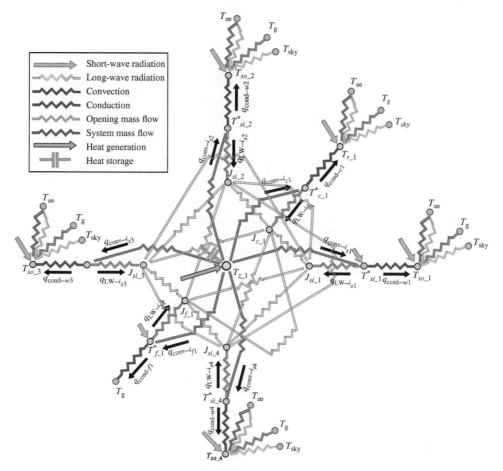

Figure 8.13 Nodal system of a single zone building in addition to external long-wave radiations with the environment.

external walls when we also have short and long-wave radiations, we can derive the equation as follows:

$$\rho c \frac{\partial T}{\partial t} = q_{conv-e} + q_{gen-e} + q_{rad-e} + q_{cond} \tag{8.31}$$

Note that in many cases, we assume the generated heat to be neglected. To include the thermal mass into the nodal system, we can assume the wall with a single control volume as depicted in Figure 8.15.

We can also split a wall to two control volumes with a half width between internal and external nodes to increase the model accuracy. This therefore leads to a more suitable assumption to consider thermal mass in walls, which is assigning an extra interior node to each wall as illustrated in Figure 8.16. This process is shown in Figure 8.16 for the similar 1-zone building presented in the previous sections.

(a)

(b)

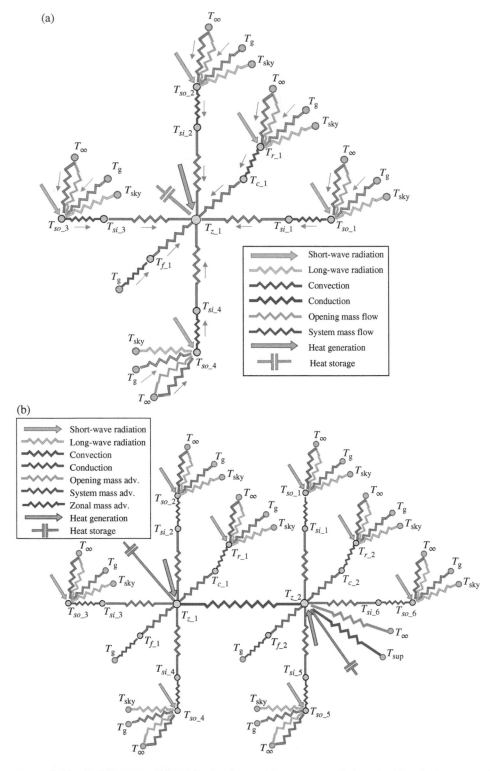

Figure 8.14 Nodal system with neglecting internal long-wave radiations for (a) a single zone and (b) a two-zone building.

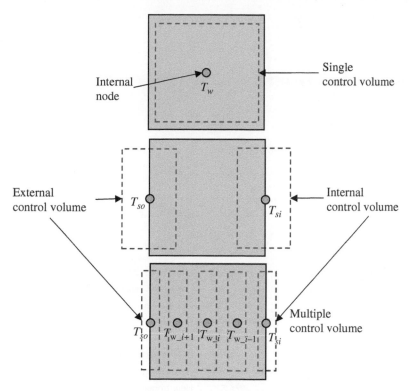

Figure 8.15 Various ways to split a wall to single or multiple control volumes.

8.5.9.2 Multi-layer Walls

It is expected that we consider more nodes than internal and external nodes within a wall on some occasion. The example is when we have a multilayer wall or the temperature distribution within a layer is important to be resolved. Let us assume that for node i in Figure 8.15, the heat conductive flux enters from node i-1 as $q_{\text{cond}\_(i-1)\rightarrow i}$ and the one that leaves to node $i + 1$ as $q_{\text{cond}\_i \rightarrow (i+1)}$. We can show the change in the conductive fluxes in a node in a differential form as:

$$q_{\text{cond}\_(i-1)\rightarrow i} - q_{\text{cond}\_i\rightarrow(i+1)} = \frac{\partial q_{\text{cond}}}{\partial x} \tag{8.32}$$

We can further write the energy balance equation for the layers within the wall (e.g. T_i) as:

$$\rho c \frac{\partial T_{w\_i}}{\partial t} = \frac{\partial q_{\text{cond}}}{\partial x} + q_{\text{gen}} \tag{8.33}$$

When $\dfrac{\partial q_{\text{cond}}}{\partial x}$ can be represented by Fourier's law and this equation is similar to heat diffusion equation presented in Chapter 6:

$$\rho c \frac{\partial T_{w\_i}}{\partial t} = \frac{\partial}{\partial x}\left(k\frac{\partial T_{w\_i}}{\partial x}\right) + q_{\text{gen}} \tag{8.34}$$

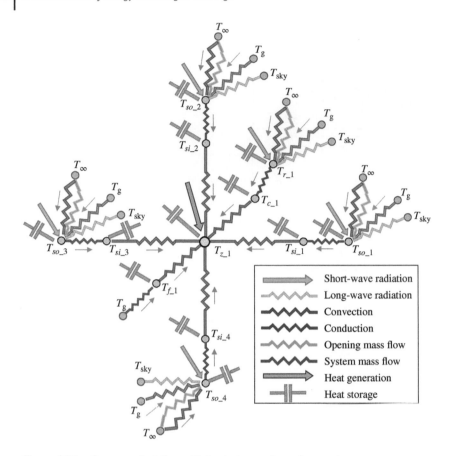

Figure 8.16 One-zone building with heat storage in surface nodes.

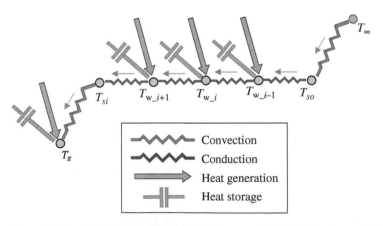

Figure 8.17 The nodal system of a multi-layer wall with thermal mass at internal nodes.

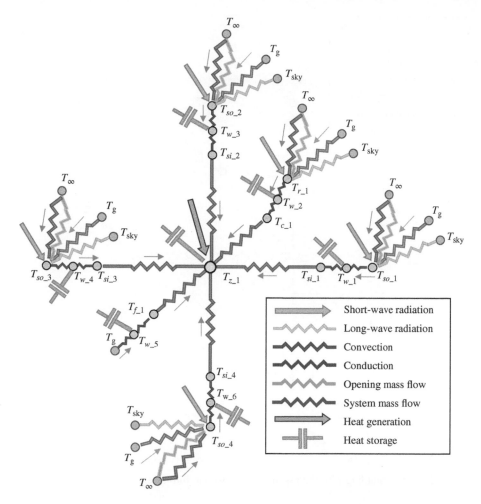

Figure 8.18 One-zone building with thermal mass only in internal wall nodes.

Now, we can allocate a thermal capacitor to each internal node at this approach in taking the thermal mass into the account. Figure 8.17 shows the nodal system of a multi-layer wall with consideration of the thermal mass only at internal nodes.

If we apply the thermal mass to the previous case study of the single zone, we can show the new nodal system as illustrated in Figure 8.18.

8.5.9.3 Thermal Mass in Furniture

Furniture can store a significant amount of heat in zones. Evidently, the thermal mass is related to their volume and material properties. In the nodal system, furniture thermal mass can be shown as an extra node (or multiple nodes) with an unknown temperature (T_{fn}), which is impacted be the zone temperature via convection mechanism. As shown in

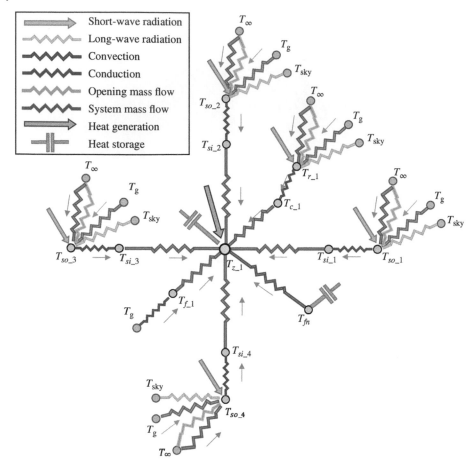

Figure 8.19 One-zone building with thermal mass in walls and furniture.

Figure 8.19, it is crucial that the volume of the furniture (V_{fn}) and its properties (ρ_{fn} and c_{fn}) to be represented with bulk values:

$$\rho_{fn} c_{fn} V_{fn} \frac{\partial T_{fn}}{\partial t} = h A_{fn}(T_{fn} - T_{z\_1}) \tag{8.35}$$

where h is the convective coefficient around a specific furniture, which can be roughly approximated in simulations and A_{fn} denotes the effective area of furniture being exposed to a zone.

Example 8.7

Assume a typical room of $5\ m \times 4\ m \times 3\ m$ with furniture from approximately $1\ m^3$ wood and $1\ m^3$ textile. If the walls and ceiling are from 0.3m concrete brick, 0.08m insulation, and 0.04 m plasterboard, compare all thermal masses for each single $^\circ C$ increase of temperature. The floor's material is from 0.2 m concrete.

Solution

The volume of each material, its specific heat capacity and density are provided in Table 8.E7.

Table 8.E7 Materials properties in a typical room.

Material	Density (kg/m³)	C_p (kJ/kg K)	Thickness (m)	Volume (m³)	Thermal mass (kJ)	Percentile (%)	Percentile wt (%)
Brick	2000	0.840	0.30	22.2	37296	85.8	*
Insulation	20	1.030	0.08	5.9	122	0.3	*
Plasterboard	700	1.000	0.04	3.0	2072	4.8	51.2
Concrete	2300	0.880	0.20	1.0	2024	4.7	*
Wood	700	1.760	*	1.0	1232	2.8	30.4
Textile	250	1.340	*	2.0	670	1.5	16.6
Air	1.2	1.005	*	60.0	72	0.2	1.8

As is evident, thermal mass of wall materials is massively dominant in a typical room. For example, bricks contain about 86% of the thermal mass (see percentile in Table 8.E7). The key point here is that the conductivity is another crucial factor when heat is stored in a solid material. In other words, insulations applied to the modern buildings' walls to a large extend isolate an indoor and outdoor of atypical room. The response time of an indoor climate, thus, would be different to outdoor climate as shown in the next chapter. In this case and with some level of approximation, we can exclude walls and floor's materials to be much involved in the thermal mass storage. This significantly changes the role of furniture in the energy balance of a room. For example, 3.3% share of furniture from wood and textile materials will increase to about 47% (see percentile wt in Table 8.E7) in this simplistic case. Obviously, less, or more furniture can even further impact on these numbers. Likewise, the thickness of construction materials is crucial. This underpins one of the main reasons of development of building energy simulation tools to help designers to find practical values for these parameters, taking several factors into the considerations.

8.6 Walls, Windows, and Thermal Bridges

Walls and windows can be further modelled in the nodal systems with a more detailed condition where multiple nodes are allocated instead of one single node. If a wall has a substantially different characteristics in thermal resistance in terms of materials and components, then it can be modelled as a parallel system as shown in Figure 8.20. This can be the place of thermal bridges, windows, and doors. If the layers are not considered to significantly store heat, then Figure 8.20a is a representative of such wall. This is also a valid assumption for glasses and doors in most of the scenarios even if the thermal storage is taken into the account. The latter thermal nodal system is shown in Figure 8.20b. When the thermal storage is not considered, the resistances can be simplified as series in Eq. (7.8) and parallel rules in Eq. (7.12).

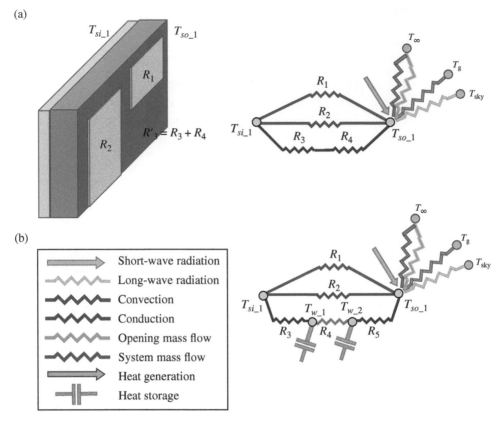

(a)

(b)

Figure 8.20 Thermal nodal network of a multilayer and multi component wall (a) without thermal storage and (b) with thermal storage.

Example 8.8

In Example 7.6 of Chapter 7, wall-2 has a surface temperature of $12°C$, the window surface temperature is $11°C$ and the external temperature is $10°C$ (Figure 8.E8.1). Calculate the total convective heat transfer resistance of wall-2 with the below correlations (see Chapter 7):

1) Simple-combined.
2) TARP
3) MoWiTT
4) DOE-2
5) Adaptive

Solution

We can split wall-2 to four areas as shown in Figure 8.E8.2 (two areas are similar numbered as '2') and then to calculate U-value of four areas being in a parallel connection to each other in accordance to Eq. (7.12) $\left(U = \dfrac{1}{R_{total}} = \sum_{k=1}^{n} \dfrac{1}{R_k} \right)$.

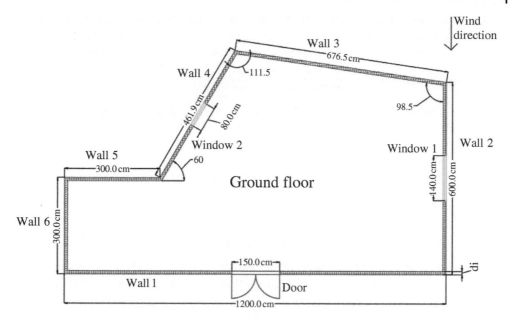

Ground floor top view

Figure 8.E8.1 Layout of the case study building.

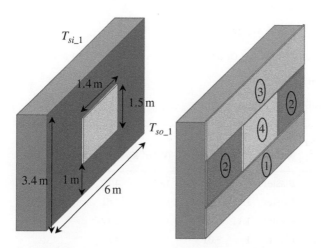

Figure 8.E8.2 Splitting of the wall of the case study building.

1) **Simple-combined:** according to the terrain description of meteorological station, we can write:

$$z_{\mathrm{met}} = 10\,\mathrm{m};\ \delta_{\mathrm{met}} = 270\,\mathrm{m};\ \alpha_{\mathrm{met}} = 0.14;\ V_{\mathrm{met}} = 3.6\,\mathrm{m/s}$$

According to the wall geometry and building location:

$$z_1 = \frac{1}{2} = 0.5\,\mathrm{m};\ \delta = 460\,\mathrm{m};\ \alpha = 0.33$$

As the wall's outer layer material is smooth plasterboard, the material roughness coefficient from ASHRAE Handbook could be found. Hence, the convective heat coefficients for the areas are (the area weighted average is used to calculate the overall value):

Area #	V_z (m/s)	A (m²)	Material	Material roughness coefficient			h (W/m² K)	U-value (W/m² K)
				D	E	F		
1	0.60	6.00	Plaster	10.22	3.1	0	12.08	12.71
2	0.91	3.45	Plaster	10.22	3.1	0	13.04	
3	1.08	5.4	Plaster	10.22	3.1	0	13.57	
4	0.91	2.10	Glass	8.23	3.33	−0.036	11.23	

2) **TARP Algorithm:** Wall-2 is not a leeward or windward surface ($W_f = \dfrac{0.5 + 1}{2} = 0.75$), therefore, Wall-2 material is just a smooth plasterboard ($R_f = 1.11$) while its perimeter $P = 14$ m and area are $A = 6$ m², respectively. Hence, the forced component should be:

$$h_f = 2.357 W_f R_f \left(\frac{PV_z}{A}\right)^{1/2} = 2.537 \times 0.75 \times 1.11 \times \left(\frac{14 \times 0.6}{6}\right)^{1/2} = 2.5\,\text{W/m}^2\,\text{K}$$

Wal-2 is a vertical wall, therefore, for the natural convection part, we can derive:

$$\Delta T = T_\infty - T_{so_1} = 10 - 12 = -2\,^\circ\text{C}$$

$$h_n = 1.31|\Delta T|^{\frac{1}{3}} = 1.31 \times |-2|^{\frac{1}{3}} = 1.65\,\text{W/m}^2\,\text{K}$$

The total convection heat coefficients ($h_f + h_n$) of areas from TARP Algorithm are:

$$h_1 = 2.49 + 1.65 = 4.14\,\text{W/m}^2\,\text{K}$$

With a similar process, we can obtain (for glass $\Delta T = T_\infty - T_{wd} = -1\,^\circ\text{C}$, $W_f = 0.8$):

Area #	Length (m)	width (m)	Perimeter (m)	Area (m²)	Material	R_f	h_f (W/m² K)	h_n (W/m² K)	h (W/m² K)	U-value (W/m² K)
1	6	1	14	6	Plaster	1.11	2.47	1.65	4.12	4.57
2	2.3	1.5	7.6	3.45	Plaster	1.11	2.96	1.65	4.61	
3	6	0.9	13.8	5.4	Plaster	1.11	3.47	1.65	5.12	
4	1.4	1.5	5.8	2.1	Glass	1	3.01	1.31	4.32	

MoWiTT Algorithm: Wall-2 is not leeward or windward surface; therefore, we can find the value with averaging from the existing values:

$$C_t = \frac{0.84 + 0.84}{2} = 0.84\,\frac{W}{m^2 K^{\frac{4}{3}}}$$

$$a = \frac{3.55 + 3.26}{2} = 3.405\,\text{W/m}^2\,\text{K(m/s)}^b$$

$$b = \frac{0.89 + 0.617}{2} = 0.7535$$

The temperature difference for the wall-2 is $\Delta T = 10 - 12 = -2\,^{\circ}C$.
Hence:

$$h_1 = \sqrt{\left[C_t(\Delta T)^{\frac{1}{3}}\right]^2 + \left[aV_z^b\right]^2} = \sqrt{\left[0.84 \times (-2)^{\frac{1}{3}}\right]^2 + \left[3.405 \times 0.6^{0.7535}\right]^2} = 2.55\,\frac{W}{m^2}K$$

Similarly ($C_t = 0.84$, $a = 3.405$, and $b = 0.7535$):

Area #	h (W/m² K)	Average h (W/m² K)
1	2.55	3.21
2	3.34	
3	3.76	
4	3.28	

3) **DOE-2 Model:** with a similar process as the previous section, we have (for glass $\Delta T = 1\,^{\circ}C$, $a = 3.405$, and $b = 0.7535$):

Area #	h_n (W/m² K)	h_f (W/m² K)	h (W/m² K)	Average h (W/m² K)
1	1.65	2.32	2.85	3.45
2	1.65	3.17	3.58	
3	1.65	3.61	3.97	
4	1.31	3.17	3.43	

4) **Adaptive Convection Algorithm:** the forced convection coefficient is calculated from McAdams as follows:

$$h_{1f} = 5.7 + 3.8V_Z = 5.7 + 3.8 \times 0.6^2 = 7.98\ W/m^2K$$

And, according to Alamdari–Hammond vertical wall method, the natural component is:

$$h_{1n} = \left\{ \left[1.5\left(\frac{|\Delta T|}{H}\right)^{\frac{1}{4}}\right]^6 + \left[1.23|\Delta T|^{\frac{1}{3}}\right]^6 \right\}^{1/6} = \left\{ \left[1.5\left(\frac{|-2|}{3.4}\right)^{\frac{1}{4}}\right]^6 + \left[1.23|-2|^{\frac{1}{3}}\right]^6 \right\}^{1/6}$$
$$= 1.63\ W/m^2K$$

Hence, the total heat convection coefficient for area-1 is $h_1 = h_{1f} + h_{1n} = 9.62\ W/m^2K$. In a similar manner:

Wall ID	h_f (W/m² K)	h_n (W/m² K)	h (W/m² K)	Average h (W/m² K)
1	7.98	1.63	9.62	10.58
2	9.15	1.63	10.78	
3	9.80	1.63	11.43	
4	9.15	1.32	10.47	

8.7 Mass Balance Equation

Mass conservation should be applied to both air and moisture in a defined zone. Considering Eq. (8.3), we can write a set of equations for the entering and leaving masses to the zone:

$$\frac{dm_{da}}{dt} = \sum \dot{m}_{da-in} - \sum \dot{m}_{da-out} \tag{8.36}$$

$$\frac{dm_v}{dt} = \sum \dot{m}_{v-in} - \sum \dot{m}_{v-out} + \sum \dot{m}_{ce} \tag{8.37}$$

where m_{da} and m_v are the mass of dry air and vapour, respectively. $\dot{m}_{ce}$ is the released (condensation) or gained (evaporation) vapour by appliances (e.g. washing, cooking, etc.) and bio sources (e.g. occupants, plants, animals, etc.).

First, we start with a simplistic condition where a 1-zone building has two openings on the opposite walls as demonstrated in Figure 8.21. In Chapter 3, we have explained how wind and stack effects induce a pressure difference between buildings' opening and force the air to circulate within a building. In the absence of these pressure differences, we expect a pressure balance between indoor and outdoor spaces and thus a zero-mass flow within the zone. We can expect the advection related to natural ventilation and infiltration to be zero in this case.

Therefore, here we define a gauge pressure of p_{da-o}, which drives mass flow into the zone. First, we assume that the air is completely dry:

$$\frac{d\rho_{da-i}V}{dt} = \rho_\infty A_1 u_1 - \rho_{da-i} A_2 u_2 = \rho_\infty Q_1 - \rho_{da-i} Q_2$$

$$\implies \frac{V}{RT_i}\frac{dP_{da-i}}{dt} = \frac{P_\infty}{RT_\infty}Q_1 - \frac{P_{da-i}}{RT_i}Q_2 \tag{8.38}$$

This is a first-order non-linear ODE, which cannot be solved with the standard solution method as velocities are unknown variables. Here, we can use Bernoulli equation as described in Chapter 3 between a point on the windward opening and middle of the zone where the velocity can be assumed to be about zero, which implies velocity is fully transformed to the pressure:

$$Q_1 = C_{d_1} A_1 \sqrt{\frac{2(P_\infty - P_{da-i})}{\rho_\infty}} \tag{8.39}$$

Figure 8.21 Mass balance in a single zone building.

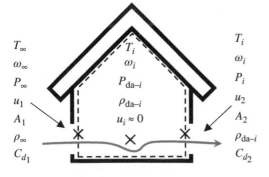

Similarly, we can derive Bernoulli equation between the zone and leeward opening. This means that the pressure will be fully transported to the velocity:

$$Q_2 = C_{d_2} A_2 \sqrt{\frac{2P_{da-i}}{\rho_{da-i}}} \tag{8.40}$$

Now, Eqs. (8.38), (8.39), and (8.40) should be solved to find variables, including u_1, u_2, and p_{da-i}.

Example 8.9

Simplify Eq. (8.38) in a typical atmospheric condition for a single zone room similar to Figure 8.21.

Solution

As we have slight change in density of air in atmospheric condition:

$$\frac{dP_{da-i}}{dt} = \frac{RT_i}{\rho_\infty V} \left(A_1 C_{d_1} \sqrt{2(P_\infty - P_{da-i})} - A_2 C_{d_2} \sqrt{2P_{da-i}} \right)$$

And, if there is not a significant storage or change in indoor pressure ($\frac{dP_{da-i}}{dt} \cong 0$), then first we can define:

$$\frac{A_1 C_{d_1}}{A_2 C_{d_2}} = k$$

Then, Eq. (8.38) can be written as:

$$\sqrt{2P_{da-i}} = k\sqrt{2(P_\infty - P_{da-i})}$$
$$\Longrightarrow P_{da-i} = \frac{k^2 P_\infty}{(1 + k^2)}$$

Now, we can find different cases for change in the area and discharge coeffect:

Area	Discharge coefficient	$\dfrac{k^2}{(1 + k^2)}$
$A_1 = A_2$	$C_{d_1} = 0.70$	0.50
	$C_{d_2} = 0.70$	
$A_1 = A_2$	$C_{d_1} = 0.63$	0.45
	$C_{d_2} = 0.70$	
$A_1 = A_2$	$C_{d_1} = 0.77$	0.55
	$C_{d_2} = 0.70$	
$A_1 = 2A_2$	$C_{d_1} = 0.70$	0.8
	$C_{d_2} = 0.70$	
$A_1 = 0.5A_2$	$C_{d_1} = 0.70$	0.2
	$C_{d_2} = 0.70$	

The finding implies that the indoor pressure is leaning towards each of windward and leeward pressures if the resistance of the associated entrance is lower. We discuss the resistance ($\Delta p = R_f Q$) in the next section.

8.7.1 Airflow Network Model

As stated in earlier section, mass flow plays a significant role in energetic balance of a buildings through advective heat fluxes from opening, infiltration, HVAC systems, and zonal exchanges. The advective resistances were also introduced while in all these terms, mass flows should be given as known variables to be able to form the nodal network system. A simplistic approach to determine these values is to explicitly input airflow rates from empirical models, which use environmental and geometrical characteristics to estimate mass flow rates. These models encounter with significant discrepancies in some scenarios where the application does not fully match with the assumptions made on their development. Besides, the conservation of mass needs to be satisfied with modifications in the simulations. Another approach is to directly employ conservation of mass law on each zone and construct another nodal system, known as airflow network (AFN).

To explain AFN, let us assume a multi-zone building as demonstrated in Figure 8.22a. Figure 8.22b also shows the thermal nodal network of this building where internal walls are assumed with only one node for the sake of simplicity. Furthermore, internal long-wave and short-wave radiative exchanges are neglected.

AFN approach assumes that air flows from one node to another while it establishes a relationship between airflow and pressure for each zone corresponding to discrete airflow passages such as doorways, construction cracks, ducts, and fans. Unlike the temperature differences in thermal nodal system, the driving force to push mass (airflow) in AFN is the pressure difference between zones.

In the AFN, we define boundary nodes as shown with orange crosses where pressures are known values to the system similar to the thermal nodal network system. Pressures and airflows at boundaries can be obtained as Cp values or other methods as explained in Chapter 3. Then, we allocate variable airflow nodes (shown with grey crosses) to each zone. Now, let us assume that mass is advected from the windward environment node (ww) with a pressure of p_{ww} to zone-1, crosses all zones one by one, and leaves to the leeward node (lw) with a pressure of p_{lw}. The advective heat fluxes are shown in Figure 8.22b. The key point here is to derive a mass balance equation at each node to form the new AFN system as depicted in Figure 8.23.

This AFN model is relatively simple as there is only one air corridor in the simulation while other advection mechanisms are assumed to be absent. This leads to derive the mass balance equation at each unknown node to obtain another set of equations represented as another linearized system of equations. Therefore, once we calculate mass flows from AFN at a time step, then we can proceed to find temperatures (and humidity) with its associated thermal nodal system at the same time step introduced in earlier sections. We can repeat the process in the next time step and so on. In other words, based on the mass flow calculated for each linkage between two nodes, the model calculates nodes' temperatures and humidity. Using these nodes' temperatures and humidity, the sensible and latent loads are resolved for each zone as well as the new pressures at this time step.

Now, we can assume that the AFN contains other openings between zones, which can potentially induce the advective fluxes among them. The AFN model of such system can be shown in Figure 8.24a with transparent lines connecting all potential zonal exchanges together. So, if an opening between two zone (e.g. internal doors) is opened in this AFN model, then the related transparent brown line should turn to a solid brown lines. The

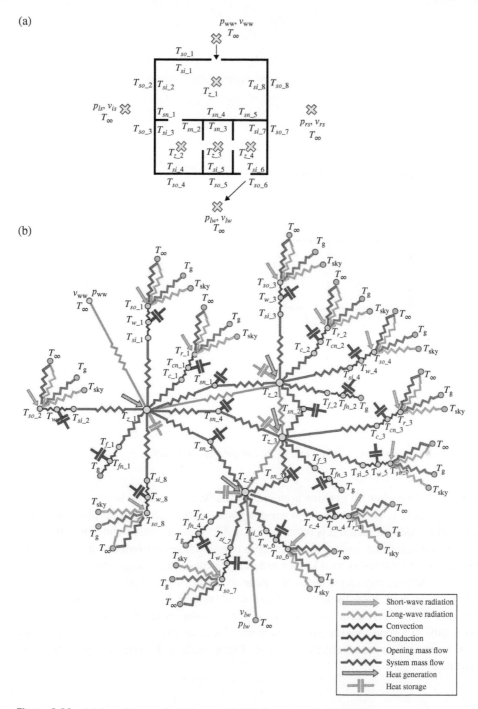

Figure 8.22 (a) A multi-zone building and its (b) thermal nodal network.

(a)

(b)

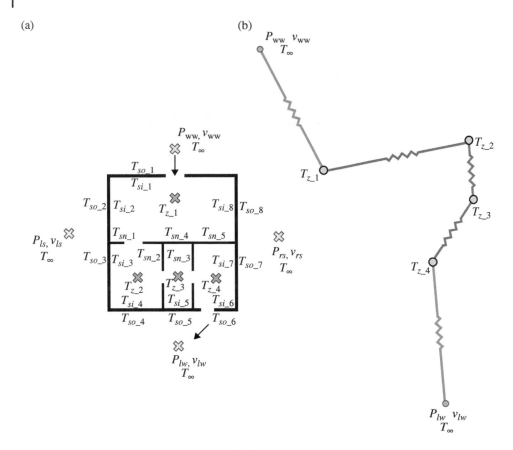

Figure 8.23 (a) A multi-zone building and its (b) airflow nodal network.

inverse case is true when an opening is closed. Similarly, an opening connected to an environmental node (e.g. external door and window) is shown by a transparent blue line if it is closed, and a solid blue line if it is opened. Figure 8.24b uses a similar concept with green lines to represent infiltration or exfiltration advective exchanges between a zone and its connected environmental nodes with known pressure distributions (orange nodes). So, if infiltration is aimed to be considered between a zone and its environmental node with a known orange pressure node, then a solid green line should be used. Eventually, we can assume a transparent violet line as a mechanical supply system assigned to each zone (see Figure 8.24b). If a zone benefits from a mechanical system, the line turns to a solid violet one to represent the mechanical advective fluxes with a known fan pressure and temperature.

8.7.2 Mass Flow Resistance

We have learned that the pressure difference between two zones causes the flow between their associated nodes in an AFN system. Similar to the thermal nodal system, but this time for pressure, we can replace pressure as voltage and airflow as intensity:

(a)

(b)

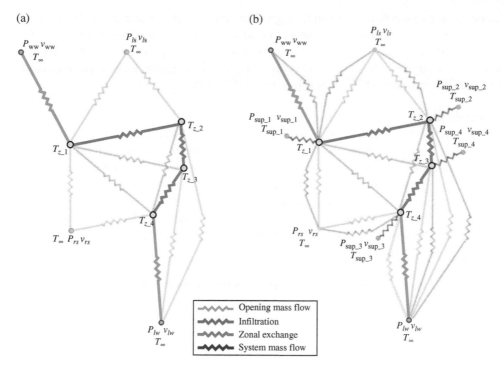

Figure 8.24 Current and potential network of (a) zonal exchanges and (b) infiltrations

$$\Delta P \propto \Delta V \text{ and } Q \propto I$$
$$\implies \Delta P = R_f Q \tag{8.41}$$

where, R_f is the resistance of the component, which causes a pressure drop against the advective mass flux. For example, we found R_f as minor and major losses for pipes and ducts of mechanical ventilation systems in Chapter 3 while many of these values are found from experimental studies. We have also addressed the pressure drop related to openings in Chapter 3 as:

$$Q = C_d A \sqrt{\frac{2\Delta P}{\rho}} \tag{8.42}$$

Hence, as it can be seen, R_f is not a linear variable and can changes in accordance with the geomtry of the opening as a function of C_d while the density is not changing in incompressible flows assumed in many building-related problems.

The resistance of building components against infiltration or exfiltration cannot be easily determined on many occasions. Nonetheless, due to the importance of the pressure drop related to infiltration and exfiltration, several empirical models have been developed as some of these commonly applied models are introduced in the next section. Most infiltration models are based on an empirical relationship between the flow and the pressure difference across a crack or opening in the building shell:

$$\dot{m}_i = C\sqrt{\rho_n}(\Delta P)^n \tag{8.43}$$

where $\dot{m}_i$ is the mass flow rate of air through element i from node n to node m. C is the flow coefficient, ρ is the air density of node n or m, ΔP denotes the total pressure loss across the element ($\Delta p = p_n - p_m$), and n is the flow exponent, which is supposed to alter between 0.5 and 1.0. For large openings expressed in Eq. (8.43), the flow exponent has values very close to 0.5 as stated before while it has values near 0.67 for small cracks.

Example 8.10

Consider a two-floor building model as shown in Figure 8.E10.1. The aim is to estimate the airflow rate through the openings. It is assumed that all windows are open while the ground and first floors are separated and can be considered as separate thermal zones. The value of the surface wind pressure and discharge coefficients for each opening related to the reference air velocity at the reference height are given for a wind angle of 30°. The reference velocity equals to 8 m/s and air temperature is 25° ($\rho = 1.18$ kg/m³). All windows have 1.5 m heights and the door has a 2.4 m height (layouts units are in centimetre).

Solution

Two AFN networks are developed for each floor as there is no link between them. Here, for the ground floor, we have four nodes and three connections. Pressure at each boundary node can be determined using their C_p values (Figure 8.E10.2):

$$P_1 = Cp_1 \frac{\rho U_{\text{ref}}^2}{2} = -18.88 \text{ Pa}$$

$$P_2 = Cp_2 \frac{\rho U_{\text{ref}}^2}{2} = 11.33 \text{ Pa}$$

$$P_3 = Cp_3 \frac{\rho U_{\text{ref}}^2}{2} = 15.10 \text{ Pa}$$

Knowing that the pressure of node-3 is negative, and the direction of the flow is from other nodes, we need to derive the mass flow rate as:

$$\dot{m}_{1z\_1} = -\rho C_{d1} A_1 \sqrt{\frac{2(P_{z\_1} - P_1)}{\rho}} = -1.29\sqrt{P_{z\_1} + 18.88}$$

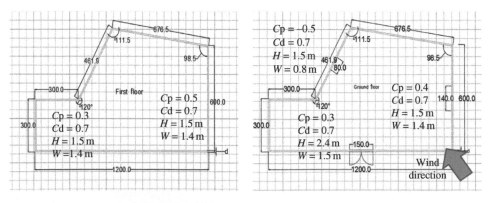

Figure 8.E10.1 Layout of ground and first floors.

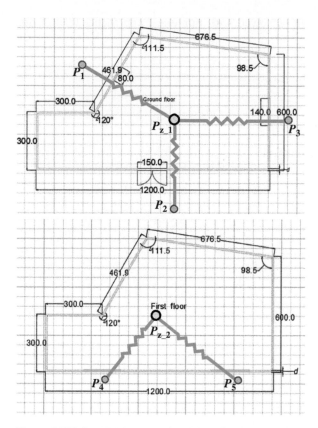

Figure 8.E10.2 AFN for ground and first floors.

The mass flow rates of other connections are calculated as below:

$$\dot{m}_{2z\_1} = \rho C_{d2} A_2 \sqrt{\frac{2(P_2 - P_{z\_1})}{\rho}} = 3.87\sqrt{11.328 - P_{z\_1}}$$

$$\dot{m}_{3z\_1} = \rho C_{d3} A_3 \sqrt{\frac{2(P_3 - P_{z\_1})}{\rho}} = 2.26\sqrt{15.104 - P_{z\_1}}$$

Then, we perform mass balance equation at the zone node using the mass balance equation:

$$\dot{m}_{1z\_1} + \dot{m}_{2z\_1} + \dot{m}_{3z\_1} = 0$$

Thus, we can obtain unknown mass flow rates and the zone pressure using the above equations:

$$-1.29\sqrt{P_{z\_1} + 18.88} + 3.87\sqrt{11.33 - P_4} + 2.26\sqrt{15.10 - P_4} = 0$$

The above non-linear equation can be solved with numerical approaches to find the values as below:

$$P_{z\_1} = 10.93 \, \text{Pa}$$

$$\dot{m}_{1z\_1} = -7.04 \, \text{kg/s}$$

$$\dot{m}_{2z\_1} = 2.43 \, \text{kg/s}$$

$$\dot{m}_{3z\_1} = 4.61 \, \text{kg/s}$$

Similarly, for the first floor, we can derive:

$$P_4 = Cp_4 \frac{\rho U_{\text{ref}}^2}{2} = 18.88 \, \text{Pa}$$

$$P_5 = Cp_5 \frac{\rho U_{\text{ref}}^2}{2} = 11.33 \, \text{Pa}$$

And, the mass balance equation at the zone node:

$$\dot{m}_{4z\_2} + \dot{m}_{5z\_2} = 0$$

And, for the mass flow rate equations:

$$\dot{m}_{4z\_2} = \rho C_{d4} A_4 \sqrt{\frac{2(P_4 - P_{z\_2})}{\rho}} = 2.26\sqrt{18.88 - P_{z\_2}}$$

$$\dot{m}_{5z\_2} = -\rho C_{d5} A_5 \sqrt{\frac{2(P_{z\_2} - P_5)}{\rho}} = -2.26\sqrt{P_{z\_2} - 11.33}$$

Thus, we can obtain the three unknown values as:

$$P_{z\_2} = 15.10 \, \text{Pa}$$

$$\dot{m}_{4z\_2} = 4.38 \, \text{kg/s} = -\dot{m}_{5z\_2}$$

8.7.3 Infiltration Mass Flow

As it is stated before, the pressure difference between a zone and the surrounding environment is the key driver in infiltration or exfiltration advective mass flow. Thus, several models have been developed on this basis and only in respect to the pressure gradient. Some of the commonly used models in building energy simulation tools include design flow rate model, effective leakage area model, and flow coefficient model, which are described in the next sections.

8.7.4 Design Flow Rate Model

This model establishes a correlation between a zone and its surrounding environment [4]:

$$\dot{m}_1 = I_d F_s \left(A + B|T_{\text{zone}} - T_\infty| + CV_z + DV_z^2 \right) \tag{8.44}$$

where, F_s denotes the schedule of fan or other devices, which pressurize or depressurize a zone. V_z is the local wind speed. A, B, and C are the correlation coefficients, which can be

Table 8.1 Coefficients of design flow rate model for different methods [4].

	A	B	C	D
Default values	1	0	0	0
BLAST	0.606	0.03636	0.1177	0
DOE-2	0	0	0.224	0

Source: Based on Coblenz and Achenbach [4].

found based on different approximations as presented in Table 8.1. DOE-2 and BLAST models are defined in Chapter 7. Eventually, I_d is a correction coefficient, the design zone infiltration airflow, which is defined for a component (for example a window) based on the below expression:

$$I_d = 10 A_{eff} \left(\frac{\Delta p}{50}\right)^{0.66} \tag{8.45}$$

where, A_{eff} is a building envelope's effective area.

8.7.5 Effective Leakage Area Model

Effective leakage area and its enhanced form, flow coefficient model, are benefiting from the combination of airflow rate related to stack (Q_s) and wind-driven (Q_w) effects:

$$\dot{m}_l = F_s \left(Q_s^2 + Q_w^2\right)^{0.5} \tag{8.46}$$

This model uses the effective air leakage area measured at 4 Pa using methods such as whole-building pressurization test [5]:

$$\dot{m}_l = F_s \frac{A_L}{1000} \left(C_s \Delta T + C_w V_z^2\right)^{0.5} \tag{8.47}$$

where, A_L is the effective air leakage area in cm^2. C_s and C_w are the stack and wind coefficients, respectively. These coefficients are provided in Tables 8.2–8.4.

8.7.6 Flow Coefficient Model

The flow coefficient model is proposed by [6], which is an enhanced version of effective leakage area model presented as the below expression:

$$\dot{m}_l = F_s \left(\left(cC_s \Delta T^n\right)^2 + \left(cC_w \{sV_z\}^{2n}\right)^2\right)^{0.5} \tag{8.48}$$

Table 8.2 Effective leakage area model stack coefficient C_s (American Society of Heating and Air-Conditioning Engineers)-Table 8.4.

	House height (stories)		
Stack coefficient	One	Two	Three
	0.000145	0.000290	0.000435

Source: Modified from ASHRAE [3]/ASHRAE.

Table 8.3 Local shelter classes (American Society of Heating and Air-Conditioning Engineers) – Table 8.5.

Shelter class	Description
1	No obstructions or local shielding
2	Typical shelter for an isolated rural house
3	Typical shelter caused by other buildings across street from building under study
4	Typical shelter for urban buildings on larger lots where sheltering obstacles are more than one building height away
5	Typical shelter produced by buildings or other structures immediately adjacent (closer than one house height: e.g. neighbouring houses on same side of street, trees, bushes)

Source: Modified from ASHRAE [3]/ASHRAE.

Table 8.4 Effective leakage area model wind coefficient C_w (American Society of Heating and Air-Conditioning Engineers) – Table 8.6.

	House height (stories)		
Shelter class	One	Two	Three
1	0.000319	0.000420	0.000494
2	0.000246	0.000325	0.000382
3	0.000174	0.000231	0.000271
4	0.000104	0.000137	0.000161
5	0.000032	0.000042	0.000049

Source: Modified from ASHRAE [3]/ASHRAE.

Table 8.5 Enhanced model wind speed multiplier G (American Society of Heating and Air-Conditioning Engineers) – Table 8.7.

	House height (stories)		
	One	Two	Three
Wind speed multiplier G	0.48	0.59	0.67

Source: Modified from ASHRAE [3]/ASHRAE.

where, c and s are the flow and shelter coefficients, respectively. These coefficients should be referred to Tables 8.5–8.7. Furthermore, n is the pressure coefficient and commonly selected as 0.67.

Table 8.6 Enhanced model stack and wind coefficients (American Society of Heating and Air-Conditioning Engineers) – Table 8.8.

	One story		Two story		Three story	
	No flue	With flue	No flue	With flue	No flue	With flue
C_s	0.054	0.069	0.078	0.089	0.098	0.107
C_w for basement/slab	0.156	0.142	0.17	0.156	0.17	0.167
C_w for crawl space	0.128	0.128	0.142	0.142	0.151	0.154

Source: Modified from ASHRAE [3]/ASHRAE.

Table 8.7 Enhanced model shelter factor s (American Society of Heating and Air-Conditioning Engineers) – Table 8.9.

Shelter class	No flue	One story with flue	Two story with flue	Three story with flue
1	1	1.1	1.07	1.06
2	0.9	1.02	0.98	0.97
3	0.7	0.86	0.81	0.79
4	0.5	0.7	0.64	0.61
5	0.3	0.54	0.47	0.43

Source: Modified from ASHRAE [3]/ASHRAE.

Example 8.11

A slab-on-grade two-storey house located in a city centre in a residential area with several similar houses in the vicinity. The house has an effective air leakage area of 527 cm^2 and a volume of 255 m^3. In addition, it is electrically heated and has no flue. The pressure difference between indoor and outdoor spaces is 4 Pa while their temperatures are $T_z = 18\,^\circ$C and $T_\infty = 10\,^\circ$C, respectively. It is assumed that the wind velocity is 6 m/s around the building and $F_s = 0.75$. Estimate the infiltration of this house using different models.

There is window located on the north wall with dimension of 2 m $\times$ 1.5 m.

Solution

1) Design flow rate model

Using Eqs. (8.44) and (8.45) with $\Delta T = T_z - T_\infty = 8\,^\circ$C:

$$I_d = 10 \times 527 \times 10^{-4} \times \left(\frac{4}{50}\right)^{0.66} = 0.0995 \text{ m}^3/\text{s m}^2$$

With the coefficient information in Table 8.1:
Default:

$$\dot{m}_1 = 0.0995 \times 0.75 \times (1 + 0 \times |18 - 10| + 0 \times 6 + 0 \times 36) = 0.0746 \text{ m}^3/\text{s}$$

BLAST:

$$\dot{m}_l = 0.0995 \times 0.75 \times [0.606 + 0.03636 \times (18 - 10) + 0.1177 \times 6 + 0] = 0.1196 \, \text{m}^3/\text{s}$$

DOE-2:

$$\dot{m}_l = 0.0995 \times 0.75 \times [0 + 0 + 0.224 \times 6] = 0.1003 \, \text{m}^3/\text{s}$$

Model	Infiltration (m³/s)
Default	0.0746
BLAST	0.1196
DOE-2	0.1003

2) Effective leakage area model

As the house is a two-storey building, we can find $C_s = 0.000290$ from Table 8.2. Moreover, the house is located on a residential area that can place it in class-3 of Table 8.3. Thus, we can find $C_w = 0.000231$ from Table 8.4.

Now, from Eq. (8.47), knowing that $A_L = 527 \, \text{cm}^2$:

$$\dot{m}_l = 0.75 \times \frac{527}{1000} \times \left(0.00029 \times (18 - 10) + 0.000231 \times \left(6^2\right)\right)^{0.5} = 0.0408 \, \text{m}^3/\text{s}$$

3) Flow coefficient model

As the building has no flue and its shelter class is 3, so from Table 8.5, we have $s = 0.7$. For a slab-on-grade two-storey house with a flue, Table 8.6 gives $C_s = 0.078$ and $C_w = 0.170$. The house has a flow coefficient of $c = 0.051 \, \text{m}^3/(\text{s/Pa}^n)$ and a pressure exponent of $n = 0.67$; this corresponds to effective leakage area of 527 cm² at 4 Pa:

$$\dot{m}_l = 0.75 \times \left[\left(0.051 \times 0.078 \times \{18 - 10\}^{0.67}\right)^2 + \left(0.051 \times 0.170 \times \{0.7 \times 6\}^{2 \times 0.67}\right)^2\right]^{0.5}$$
$$= 0.0461 \, \text{m}^3/\text{s}$$

8.7.7 Moist Air

Once the pressure is identified, vapour pressure can be driven with the below similar equation:

$$\frac{d(\omega_i m_{da-i})}{dt} = \sum \omega_\infty \dot{m}_{da-in} - \sum \omega_i \dot{m}_{v-out} + \sum \dot{m}_{ce}$$
$$\Longrightarrow \frac{V}{RT_i} \frac{dP_{da-i}}{dt} = \sum_n \frac{\omega_\infty P_\infty}{RT_\infty} Q_n - \sum_m \frac{\omega_i P_{da-i}}{RT_i} Q_m + \sum \dot{m}_{ce} \tag{8.49}$$

Note that both Eq. (8.38) for dry air and Eq. (8.49) have an interlink together ($\omega = \frac{m_v}{m_{da}}$) as it was discussed in Chapter 4.

Now, if we assume moist air as an ideal gas to reach the mass balance equation as a dependent parameter to the pressure:

$$\frac{(1 + \omega_i)V}{RT_{out}} \frac{dP_{da-i}}{dt} = \sum (1 + \omega_o) \frac{P_{da-o}}{RT_o} \dot{V} - \sum (1 + \omega_i) \frac{P_{da-i}}{RT_i} \dot{V} + \sum \dot{m}_{ce}$$

(8.50)

Note that p_{da-o}, ω_o and T_o are known variables for the mass balance equation.

The equations (8.49) and (8.50) can be only solved in very simplified conditions. Inversely, on many occasions straight forward solutions cannot be achieved due to the complexity, nonlinearity, and time-variance of their terms. Hence, numerical solutions are utilized to solve them in which known values from one will be fed to another one to be able to proceed. For example, indoor temperature (T_i) can be first obtained from Eq. (8.38) to be used in Eq. (8.50). On the other hand, the calculated indoor pressure (p_{da-i}) from Eq. (8.50) will be employed to calculate mass flow rates and thus it will be inserted to the advection terms in Eq. (8.38). The iterative process is further explained in the next chapter.

References

1 Robinson, J.C. (2004). *An Introduction to Ordinary Differential Equations*. Cambridge: Cambridge University Press.

2 Incropera, F.P. (2006). *Fundamentals of Heat and Mass Transfer*. Wiley.

3 ASHRAE (2017). *American Society of Heating R and Air-Conditioning Engineers I 2017 ASHRAE® Handbook - Fundamentals (I-P Edition)*. American Society of Heating, Refrigerating and Air-Conditioning Engineers, Inc. (ASHRAE).

4 Coblenz, C.W. and Achenbach, P.R. (1963). Field measurement of ten electrically-heated houses. *ASHRAE Transactions*. 358–365.

5 Max, H.S. and David, T.G. (1980). Infiltration-pressurization correlation: simplified physical modeling. *ASHRAE Transactions* 86 (2): 778–807.

6 Iain, S.W. and David, J.W. (1998). Field validation of algebraic equations for stack and wind driven air infiltration calculations. *HVAC&R Research* 4: 119–139.

9

Dynamic Energy Modelling in Buildings

9 Physics of an Energy Balance Problem in Buildings

In the previous chapter, we have applied the conservation laws of physics on the buildings' control volume, utilized the concept of electrical nodal network, and constructed the simplified energy model of buildings [1, 2]. Before starting to implement numerical techniques to solve the simplified models, we have to ensure that we are representing the physics of a particular problem with enough details without diverging from the objectives of a simulation. Such departures can arise when the modelling of a problem is not compatible with what is aimed. Let us assume a simple model of an isolated room with zero mass exchange with outdoor climate. Furthermore, we presume that the celling and all walls are single-layer bricks. Figure 9.1a demonstrates the sketch of the thermal diagram of this problem. In this model, it is rationale to neglect internal long-wave radiation to simplify the nodal system similar to Figure 8.18 of Chapter 8. Now, let us just replace the brick walls with thin glasses to have a greenhouse. One may use the same nodal model as Figure 9.1a by only changing wall materials and probably deleting the thermal mass elements (capacitors). Nonetheless, we know even from common sense that the indoor climates of a greenhouse and a room are very different as greenhouses are much impacted by the short-wave radiative fluxes. So, the question may arise that, what would make a massive difference in the outcome of this model, if the same model is applied for both physics. Figure 9.1b shows the nodal network of the mentioned greenhouse. Here, the storage in glasses is negligible while, due to their low thermal resistance (higher conductivities and less thicknesses), we can assume that there is no temperature gradient in the glasses similar to the lumped-heat-capacity method. The most pivotal difference of the greenhouse with the brick model is its transmissivity against the solar radiation. While a considerable fraction of solar rays is transmitted to the greenhouse and impact the surfaces (here on the floor). A part of this solar incident will be encountered with multiple reflections on the surfaces. A small fraction may leave the zone though the main fraction again will be reflected back to the surface and absorbed. Now, the heated surfaces (here the floor) will emit long-wave radiations to other surfaces. The key point is the low transmissivity of glasses against the long-wave radiative fluxes, which traps the radiation within the zone. Thus, the zone can be assumed as an enclosed space with a high absorptivity and emissivity of glasses, which emit back the

Computational Fluid Dynamics and Energy Modelling in Buildings: Fundamentals and Applications, First Edition. Parham A. Mirzaei.

(a)   (b)

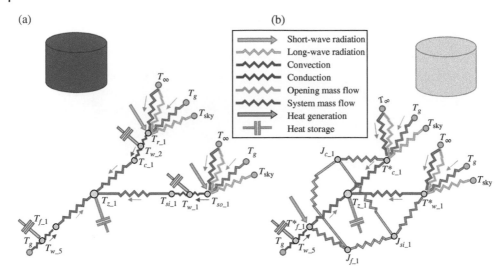

Figure 9.1 Nodal system of a cylindrical (a) brick house and (b) a greenhouse.

radiation to the other surfaces and zone air. The convection heat transfer also promotes the heat removal from all surfaces within the zone and results in the rise of the zone air temperature as expected in a greenhouse. The schematic of this process is shown in Figure 9.1b where we should account for the solar incident on the floor as well as long-wave radiation exchanges amongst all surfaces. Therefore, the nodal system of the latter problem is considerably different when the wall materials are replaced with glass. Such alterations should be carefully considered in accordance with the modelling of a problem before establishment of mathematical representation of the model.

Now, let us bring another example of cavity walls and double-glazing systems as depicted in Figure 9.2. Both technologies are widely used to minimize the heat loss due to conduction in buildings. Although, both systems have two layers with an air cavity in the middle, the underlying physics is fundamentally different. Cavity walls are built with layers of bricks where the thermal storage cannot be neglected. Meanwhile the conduction within the air can be assumed to be negligible while a natural air circulation can be expected within the cavity. As expected, many correlations are developed for the calculation of the associated convective coefficient in respect to the buoyancy-driven flow. Moreover, there are many occasions that the surface temperature of a cavity wall to be considerably different as one of the walls is exposed to the indoor space and one to the surrounding environment. This can initiate a long-wave radiation exchange between walls, which cannot be ignored. The thermal nodal system is shown in Figure 9.2a.

Now, we can analyze a double-glazing window. First, we do not expect that glass layers to store heat due to their small volume and heat capacity. While the buoyancy-driven air circulation can be expected in the cavity as illustrated in Figure 9.2b, long-wave radiation can be neglected as the temperature difference of the glass layers is normally not significant. Inversely, the transmitted short-wave radiation through the first glass is partially transmitted again through the zones and partially reflected to the cavity. The reflected fraction is then again partially transmitted to the environment and partially reflected to the cavity.

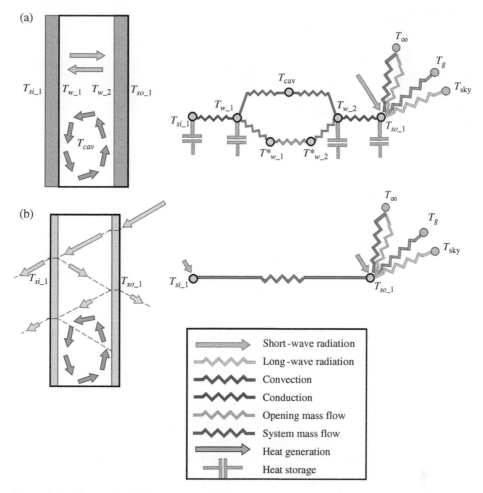

(a)

(b)

Short-wave radiation
Long-wave radiation
Convection
Conduction
Opening mass flow
System mass flow
Heat generation
Heat storage

Figure 9.2 Thermal nodal network of a (a) cavity wall and (b) double-glazing window.

This process continues in multiple times. To show the thermal nodal system of this process, although the calculation requires complex modelling methods, one can add the amount of absorbed short-wave radiation at each glass as source terms as shown in Figure 9.2b.

Example 9.1

Masonry ventilated cavity walls have air bricks to promote air circulation in the cavity. Develop a simplified model of such walls and sketch the thermal nodal system.

Solution

First, we can assume that the holes on the air brick are small enough to avoid the amount of short-wave radiation penetrating through the cavity. Each masonry wall has a high level of thermal mass, so we can allocate a capacitor to each wall as depicted in Figure 9.E1. Unlike a non-ventilated cavity wall, the air movement is due to the pressure difference between two

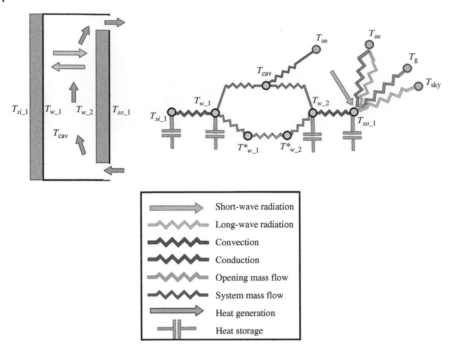

Figure 9.E1 Nodal system of the case study masonry ventilated cavity walls.

openings caused by stack effect and wind. This implies that we should consider a mass advection exchange between environment temperature (T_∞) and cavity temperature (T_{cav}). Yet, the radiative and convective exchanges between two masonry walls remain essential and thus the related connections should be considered as shown in the nodal system in Figure 9.E1.

9.1 Mathematical Representation of Buildings with Integrated Nodal System

We have seen how to form nodal network for a building after allocation of zones on it. On the next step, we must establish the mathematical models as described in the previous chapter. Hence, we apply the energy balance equation on each node with unknown variables to form the same number of equations equal to the number of variables.

In summary, we have to identify similar node types in a typical nodal system. This encompasses zonal nodes, internal wall nodes, external wall nodes, interior wall nodes, and radiosity nodes. An example case of these typical nodes is shown in Figure 9.3. In general, we should follow three steps for dynamic simulation of the mathematical model as follows:

1) Forming the nodal network to represent the physics of a problem.
2) Identification of node types and deriving equations for each node.
3) Forming the system of linearized equations and solving the matrix.

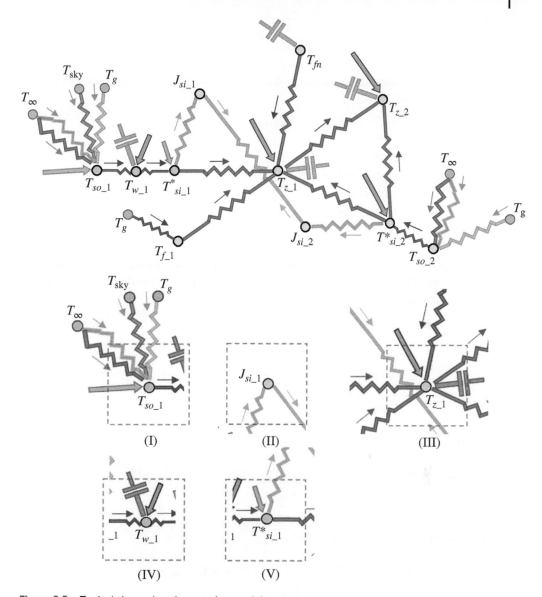

Figure 9.3 Typical thermal node types in a nodal system.

Example 9.2

In the example of Figure 9.3 identify the types of nodes and derive their equations.

Solution

The above example has 11 unknown nodes. The type and category of each node can be summarized as:

Category	Node type	Related nodes
1	(I) External surface	T_{so_1}
2	(I) External surface	T_{so_2}
3	(II) Radiosity	J_{si_1}, J_{si_2}
4	(III) Zone	T_{z_1}
5	(III) Zone	T_{z_2}
6	(IV) Internal wall	T_{w_1}
7	(V) Internal surface	T_{si_1}
8	(V) Internal surface	T_{si_2}
9	(V) Internal surface	T_{f_1}
10	(VI) Furniture	T_{fn}

This implies that we need to derive 10 different categories of equations although we only have six types of the nodes in this nodal system. Now, we refer to the definition of terms in Chapter 8 and derive each category equations as:

(I) External surface nodes (Eq. (8.31)):

Category-1 ($i = $ 'so_1' and $j = $ 'w_1'):

$$\rho_i c_i \frac{\partial T_i}{\partial t} = h_i A_i (T_\infty - T_i) + q_{SW} + \sigma A_i \varepsilon_i (T_\infty^4 - T_i^4) + \sigma A_i \varepsilon_i (T_{sky}^4 - T_i^4)$$
$$+ \sigma A_i \varepsilon_i (T_g^4 - T_i^4) - \frac{k_i A_i}{l_i} (T_i - T_j)$$

Category-2 ($i = $ 'so_2' and $j = $ 'si_2'):

$$\rho_i c_i \frac{\partial T_i}{\partial t} = h_i A_i (T_\infty - T_i) + \sigma A_i \varepsilon_i (T_\infty^4 - T_i^4) + \sigma A_i \varepsilon_i (T_{sky}^4 - T_i^4)$$
$$+ \sigma A_i \varepsilon_i (T_g^4 - T_i^4) - \frac{k_i A_i}{l_i} (T_i - T_j)$$

The linearized form is (Eq. (8.26)):

$$\rho_i c_i \frac{\partial T_i}{\partial t} = h_i A_i (T_\infty - T_i) + A_i K'(T_\infty - T_i) + A_i K''(T_{sky} - T_i)$$
$$+ A_i K'''(T_g - T_i) - \frac{k_i A_i}{l_i} (T_i - T_j)$$

where $K' = 4\sigma \varepsilon_i T_{m-\infty}^3$, $K'' = 4\sigma \varepsilon_i T_{m-sky}^3$, and $K''' = 4\sigma \varepsilon_i T_{m-g}^3$. $T_{m-\infty}$, T_{m-sky}, and T_{m-g} are the representative mean temperature between the surface and air, sky, and ground, respectively.

(II) Radiosity (Eqs. 8.16 and 8.17):

Category-3 ($i = $ 'si_1' and 'si_2', $j = $ 'si_2' and 'si_1'):

$$\frac{\sigma T_{si\_i}^4 - J_i}{\left(\frac{1-\varepsilon_i}{\varepsilon_i A_i}\right)} = \frac{J_i - J_j}{\left(\frac{1}{F_{ij} A_i}\right)}$$

(III) Zone (Eq. (8.5)):

Category-4 ($i = $ 'z_1', $j = $ 'z_2'):

$$\rho_a c_p V_i \frac{dT_i}{dt} = \dot{Q}_i + \sum_{k=1}^{si\_1, si\_2, f\_1} h_k A_k (T_k - T_i) + h A_{fn} (T_{fn} - T_i) + \dot{m}_i c_p (T_i - T_j)$$

Category-5 ($i = $ 'z_2', $j = $ 'z_1'):

$$\rho_a c_p V_i \frac{dT_i}{dt} = \dot{Q}_i + \sum_{k=1}^{si\_2} h_k A_k (T_k - T_z) + \dot{m}_i c_p (T_i - T_j)$$

(IV) Internal wall (Eq. (8.34)):

Category-6 ($i = $ 'w_1'):

$$\rho_i c_i \frac{\partial T_i}{\partial t} = \frac{\partial}{\partial x} \left(k_i \frac{\partial T_i}{\partial x} \right)$$

(V) Internal surface (Eq. (8.30)):

$$\rho_i c_i \frac{\partial T_i}{\partial t} = q_{\text{conv}-i} + q_{\text{gen}-i} - q_{\text{rad}-i} - q_{\text{cond}}$$

Category-7 ($i = $ 'si_1', $j = $ 'w_1', and $k = $ 'z_1'):

$$\rho_i c_i \frac{\partial T_i}{\partial t} = \frac{k_i A_i}{l_i} (T_j - T_i) - h_i A_i (T_i - T_k) - \frac{\sigma T_i^4 - J_i}{\left(\dfrac{1 - \varepsilon_i}{\varepsilon_i A_i} \right)} + q_{SWa-i}$$

Category-8 ($i = $ 'si_2', $j = $ 'so_2', $k = $ 'z_1' and 'z_2'):

$$\rho_i c_i \frac{\partial T_i}{\partial t} = \frac{k_i A_i}{l_i} (T_j - T_i) - \sum_{k=1}^{z\_1, z\_2} h_i A_i (T_i - T_k) - \frac{\sigma T_i^4 - J_i}{\left(\dfrac{1 - \varepsilon_i}{\varepsilon_i A_i} \right)}$$

Category-9 ($i = $ 'f_1', $j = $ 'so_2', $k = $ 'z_1'):

$$\rho_i c_i \frac{\partial T_i}{\partial t} = \frac{k_i A_i}{l_i} (T_g - T_i) - h_i A_i (T_i - T_k)$$

(VI) Furniture (Eq. (8.35)):

Category-10 ($i = $ 'fn', $j = $ 'z_1'):

$$\rho_i c_i V_i \frac{\partial T_i}{\partial t} = h_i A_i (T_{fn} - T_j)$$

For forming the system of linearized equations [1], we need to apply linearization when it is necessary (mainly for the long-wave radiation terms as described in Chapter 8) to be able to obtain a system of linear equations. Other numerical techniques such as trial-and-error can be applied to solve such equation but reaching to a system of linear equations is preferable as it is more practical and computationally economical for building energy simulation purposes. When a system of linear equations is formed for a specific time step, then a matrix related to it will be formed and then will be solved. If the processes reoccur for the next time steps, then we can have system information for a period which is called dynamic simulation of equations. In the next section, different approaches to transform the time-derivative $\left(\dfrac{\partial T_i}{\partial t}\right)$ appeared in the nodal equations will be discussed.

9.2 Numerical Solution Method for Nodal System

To solve equations found for each node with finite difference method, we can use either **explicit** or **implicit** methods used for obtaining numerical solutions of time-dependent ordinary differential equations (ODEs) and partial differential equations (PDEs). We explain these methods in further details in Chapter 10 when we introduce finite volume method, but in brief, explicit methods calculate a system of equations using the previous time step $(Y(t) = f(Y(t - \Delta t)))$ while implicit methods apply both the current state of the system and the previous one to solve a system of equations $(Y(t) = g(Y(t - \Delta t), Y(t)))$.

Nonetheless, the explicit approach demands small time steps in many circumstances, which place an unwanted restriction on the building energy calculations for as it is deemed to be conducted for a long period of time such as seasonal or annual periods. On the other hand, implicit methods do not undergo such limitations and thus are preferable in building energy calculations. As we will scrutinize in the future chapters, implicit methods are demanding extra computational loads to resolve the obtained matrices with iterative methods, and on some occasions are harder to implement. Nonetheless, and unlike computational fluid dynamics (CFD) matrices, the resultant matrices from systems of linear equations in building energy systems are relatively small and the utilized direct or iterative methods are not placing substantial extra computational load for the calculations. Hence, in this chapter we focus on the implicit solution of the governing equation on zones, radiosity nodes, and solid surfaces. The explicit method is, however, explained in detail and used for some related applications in the future chapters.

9.2.1 Zone Equations

The implicit method can be utilized to resolve the zone governing Eq. (8.5) using first-order backward difference discretization for the time derivative of the zone-i's temperature:

$$\rho c_p V \frac{T_{z\_i}^{(t)} - T_{z\_i}^{(t - \Delta t)}}{\Delta t} = \left(\sum_{j=1}^{N_s} \dot{Q}_j + \sum_{k=1}^{N_{\text{surfaces}}} h_{si\_k} A_i (T_{si\_k} - T_{z\_i}) \right.$$
$$+ \sum_{n=1}^{N_{\text{windows/doors}}} \dot{m}_{o\_n} c_p (T_\infty - T_{z\_i})$$
$$\left. + \sum_{p=1}^{N_{\text{zones}}} \dot{m}_{z\_p} c_p (T_{z\_p} - T_{z\_i}) + \dot{m}_{\text{sys}} c_p (T_{\text{sys}} - T_{z\_i}) \right)^{(t)}$$

$$(9.1)$$

where (t) represents the time of a variable or constant value, and Δt is the time step. This implies that (t) represents the current time step and $(t - \Delta t)$ refers to the previous one.

Looking at Eq. (9.1), the right-hand side parameters can be obtained within the current time step. It should be noted that many parameters are required to be predefined for the system based on the zone geometry (e.g. V and surface areas) and air properties (e.g. ρ and c_p). As it can be seen, in addition to zone temperature (T_{z_i}), many variables, including T_{si_i} and T_{z_i}, are also unknown in this equation and therefore, various expressions should be defined for each of them to be able to form a system of linear equations with equal number of variables and equations. On the other hand, mass flow rates ($\dot{m}$), convective coefficients (h_{si_i}), and outdoor climate condition (T_∞) should be specified in equation (9.1) before solving it. The treatments of latter parameters are discussed in detail in the previous chapters. The inputs summary is also discussed again in the following sections.

9.2.2 Solid Material Equations

The implicit method can be applied to the interior wall nodes using first-order backward difference discretization for the time derivative of the temperature. Let us assume a multi-layer wall as shown in Figure. 8.15, and then apply Kirchhoff's first current law to Node-i:

$$\rho_i c_i V \frac{\left[T_{w\_i}{}^{(t)} - T_{w\_i}{}^{(t-\Delta t)} \right]}{\Delta t} = \left[\frac{k_{i-1} A_{i-1}}{\Delta x_{i-1}} \left(T_{w\_i-1}^{(t)} - T_{w\_i}^{(t)} \right) \right]$$
$$- \left[\frac{k_{i+1} A_{i+1}}{\Delta x_{i+1}} \left(T_{w\_i}^{(t)} - T_{w\_i+1}^{(t)} \right) \right] + q_{\text{gen}}{}^{(t)} \tag{9.2}$$

where Δx is the spatial distance between two wall nodes. Here, $q_{\text{gen}}{}^{(t)}$ is a known value in the equation. If $k_{i-1} = k_i = k_{i+1} = \dots = k$, $\Delta x_{i-1} = \Delta x_i = \Delta x_{i+1} = \dots = \Delta x$, $\rho_{i-1} = \rho_i = \rho_{i+1} = \dots = \rho$, $c_{i-1} = c_i = c_{i+1} = \dots = c$, and $A_{i-1} = A_i = A_{i+1} = \dots = A$, then we have:

$$\left[\left[\frac{\rho c V}{\Delta t} \right] + \left(\frac{2kA}{\Delta x} \right) \right] T_{w\_i}^{(t)} + \left[\frac{-kA}{\Delta x} \right] T_{w\_i-1}^{(t)} + \left[\frac{-kA}{\Delta x} \right] T_{w\_i+1}^{(t)} = \left[\frac{\rho c V}{\Delta t} \right] T_{w\_i}{}^{(t-\Delta t)} + q_{\text{gen}}{}^{(t)} \tag{9.3}$$

From before, we know that $Fo = \dfrac{k\Delta t}{\rho c \Delta x^2}$ is Fourier number, so we can further simplify Eq. (9.3) as:

$$- Fo\, T_{w_i-1}^{(t)} + (1 + 2Fo)\, T_{w\_i}^{(t)} - Fo\, T_{w\_i+1}^{(t)} = T_{w\_i}{}^{(t-\Delta t)} + \frac{q_{\text{gen}}{}^{(t)}}{\left[\dfrac{\rho c V}{\Delta t} \right]} \tag{9.4}$$

A similar approach can be applied over the internal and external nodes as well as radiosity nodes.

Example 9.3

An isolated room with zero mass exchange with outdoor climate is assumed with a given dimension as shown in Figure 9.E3.1. The floor, celling, and three walls are insulated and heat transfer only occurs from one of the walls. Sketch the thermal diagram of this room and develop the energy balance equations at all nodes (neglect the heat storages in the walls) and find the zone temperature if a person who generates 60W of heat resides in the room. Assume $\Delta t = 15$ min and:

For air: $\rho = 1.2$ kg/m^3 and $c_p = 1.2$ J/kg K

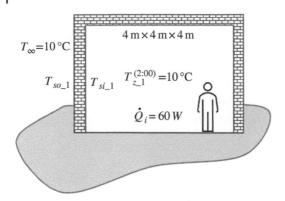

Figure 9.E3.1 Isolated room with zero mass exchange with outdoor climate.

All internal and external: $h_i = 5$ W/m² K

For brick walls: $l = 0.3$ m, $k = 6\dfrac{W}{m\,K}$, $\rho = 2000$ kg/m³ and $c_p = 840$ J/kg K

Solution

First, we represent the room with only one zone. There is no mass flow in the zone. Also, we can assume that there is no storage effect of wall. Since the heat transfer through conduction in three of walls, ceiling and floor is zero due to the insulation, we can delete their connection in the graphs. Thus, if we write the Kirchhoff's first current law in nodes T_{si_2} to T_{si_4}, T_{f_1}, and T_{c_1}, when there is no generation and storage heat (see Figure 9.E3.2):

$$q_{conv-i\_2} = q_{cond\_2}$$

$$\Rightarrow \frac{1}{R_{conv-i\_2}}(T_{si\_2} - T_{z\_1}) = 0$$

$$\Rightarrow T_{si\_2} = T_{z\_1}$$

Similarly:

$$T_{si\_2} = T_{si\_3} = T_{si\_4} = T_{f\_1} = T_{c\_1} = T_{z\_1}$$

This means that the number of unknown temperatures in the system can be reduced from 12 to 7 as T_{z_1} can be replaced to represent all five other nodes (Figure 9.E3.2). We can now write the Kirchhoff's first current law for nodes T_{so_2} to T_{so_4}, and T_{r_1}:

$$q_{conv-e\_2} = q_{cond\_2}$$

$$\Rightarrow \frac{1}{R_{conv-e\_2}}(T_\infty - T_{so\_2}) = 0$$

$$\Rightarrow T_{so\_2} = T_\infty$$

This implies that we already obtained the value for this node and other exterior ones (see Figure 9.E3.3), which reduces the number of unknown temperatures in the system from 7 to 3 (i.e. T_{z_1}, T_{si_1}, and T_{so_1}):

$$T_{so\_2} = T_{so\_3} = T_{so\_4} = T_{r\_1} = T_\infty$$

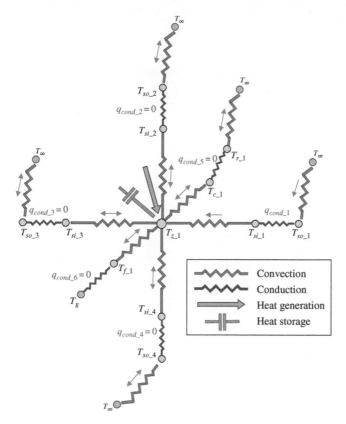

Figure 9.E3.2 Simplified nodal system of the case study isolated room.

Therefore, the nodes category and types can be written as (see Figure 9.E3.4):

Category	Node type	Related nodes
1	(I) Zone	T_{z_1}
2	(II) Internal surface	T_{si_1}
3	(III) External surface	T_{so_1}

(I) Zone: we can write the heat flux entering the zone from T_{si_1} with convection heat transfer, and simplify Eq. (9.1) as:

Category-1:

$$\rho c_p V \frac{T_{z\_1}^{(t)} - T_{z\_1}^{(t-\Delta t)}}{\Delta t} = \left(\dot{Q}_i + \frac{(T_{si\_1} - T_{z\_1})}{R_{conv-i\_1}} \right)^{(t)}$$

Or:

$$(1 + a)T_{z\_1}^{(t)} - aT_{si\_1}^{(t)} = b + T_{z\_1}^{(t-\Delta t)}$$

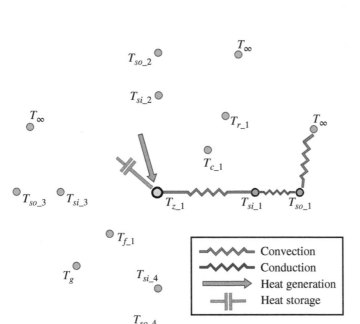

Figure 9.E3.3 Further simplification of the nodal system due to zero heat transfer in some walls, floor and ceiling.

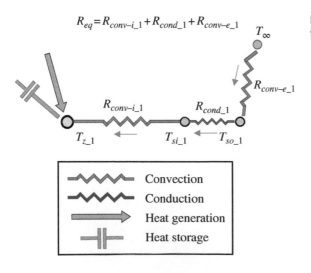

Figure 9.E3.4 Final nodal system of the isolated room.

where $a = \dfrac{\Delta t}{\rho c_p V R_{\text{conv}-i\_1}{}^{(t)}}$ and $b = \dfrac{\dot{Q}_i{}^{(t)} \Delta t}{\rho c_p V}$.

(II) Internal surface:

Development of Kirchhoff's first current law at internal wall node gives:

$$q_{\text{conv}-i\_1}{}^{(t)} = q_{\text{cond}\_1}{}^{(t)}$$

$$\Rightarrow \frac{1}{R_{\text{conv}-i\_1}{}^{(t)}} (T_{si\_1} - T_{z\_1})^{(t)} = \frac{1}{R_{\text{cond}\_1}{}^{(t)}} (T_{so\_1} - T_{si\_1})^{(t)}$$

Category-2:

$$T_{z\_1}^{(t)} - (1 + c) T_{si\_1}{}^{(t)} + c T_{so\_1}{}^{(t)} = 0$$

where $c = \left(\dfrac{R_{\text{conv}-i\_1}}{R_{\text{cond}\_1}} \right)^{(t)}$.

(III) External surface:

And, at external wall node:

$$q_{\text{conv}-e\_1}{}^{(t)} = q_{\text{cond}\_1}{}^{(t)}$$

$$\Rightarrow \frac{1}{R_{\text{conv}\_e-1}{}^{(t)}} (T_\infty - T_{so-1})^{(t)} = \frac{1}{R_{\text{cond}\_1}{}^{(t)}} (T_{so-1} - T_{si-1})^{(t)}$$

Category-3:

$$-d T_{si\_1}{}^{(t)} + (1 + d) T_{so\_1}{}^{(t)} = T_\infty^{(t)}$$

where $d = \left(\dfrac{R_{\text{conv}-e\_1}}{R_{\text{cond}\_1}} \right)^{(t)}$.

Thus, we could form three equations for each time step against three unknown variables of T_{z_1}, T_{si_1}, and T_{so_1}:

$$\begin{bmatrix} 1+a & -a & 0 \\ 1 & -1-c & c \\ 0 & -d & 1+d \end{bmatrix} \left\{ \begin{array}{c} T_{z\_1}^{(t)} \\ T_{si\_1}{}^{(t)} \\ T_{so\_1}{}^{(t)} \end{array} \right\} = \left\{ \begin{array}{c} b + T_{z\_1}^{(t-\Delta t)} \\ 0 \\ T_\infty^{(t)} \end{array} \right\}$$

We can now solve the system of linear equation for multiple time steps as follows (Figure 9.E3.5):

As it can be seen in Figure 9.E3.5, the heat loss from the wall converges to 60 W equal to the heat generated from the occupant. Thus, the room temperature would be almost stable at $13.4\,°C$ after 39 iterations when outdoor temperature is fixed to $10\,°C$. Of course, many conditions and parameters will change during time steps causing the system to respond more dynamically.

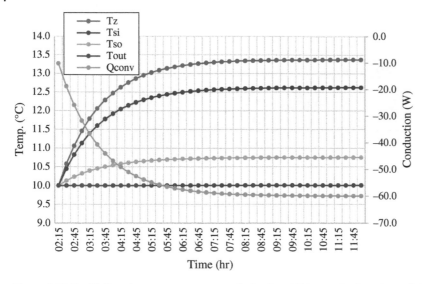

Figure 9.E3.5 Wall and zone temperature variation in addition to the heat loss of the room through time.

Example 9.4

Consider Example 9.3 while this time do not neglect the thermal storage in the wall. Find the temperature of the isolated room (Figure 9.E3.1).

Solution

We allocate one node wall, and thus we can now define four categories of nodes within four types as below (see Figure 9.E4.1):

Category	Node type	Related nodes
1	(I) Zone	T_{z_1}
2	(II) Internal surface	T_{si_1}
3	(III) Wall	T_{w_1}
4	(IV) External surface	T_{so_1}

(I) Zone: this type is similar as the previous example.

Category-1:

$$(1 + a)T_{z\_1}^{(t)} - aT_{si\_1}^{(t)} = b + T_{z\_1}^{(t-\Delta t)}$$

where $a = \dfrac{\Delta t}{\rho c_p V R_{\text{conv}-i\_1}^{(t)}}$ and $b = \dfrac{\dot{Q}_i^{(t)}\Delta t}{\rho c_p V}$.

(II) Internal surface is similar to the previous example:

$$q_{\text{cond\_1}}^{(t)} = q_{\text{conv}-i\_1}^{(t)}$$

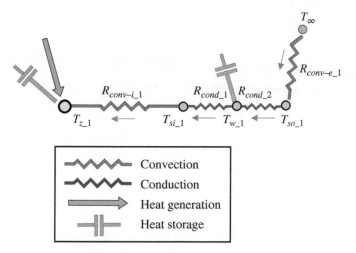

Figure 9.E4.1 Nodal system of the isolated room with the consideration of thermal mass in walls.

Category-2:

$$T_{z\_1}^{(t)} - (1 + c)T_{si\_1}^{(t)} + cT_{w\_1}^{(t)} = 0$$

where $c = \left(\dfrac{R_{\text{conv}-i\_1}}{R_{\text{cond}\_1}}\right)^{(t)}$.

(III) Wall (when $R_{\text{cond}\_1}^{(t)} = R_{\text{cond}\_2}^{(t)} = R_{\text{cond}} = \dfrac{0.5l}{KA}$):

$$\rho_{w\_1} c_{w\_1} V_{w\_1} \frac{\partial T_{w\_1}}{\partial t} = \frac{\partial}{\partial x}\left(q_{\text{cond}\_2}^{(t)} - q_{\text{cond}\_1}^{(t)}\right)$$

$$= \frac{1}{0.5lR_{\text{cond}}}\left(T_{so\_1}^{(t)} - 2T_{w\_1}^{(t)} + T_{si\_1}^{(t)}\right)$$

$$\Rightarrow \rho_{w\_1} c_{w\_1} V_{w\_1} \frac{T_{w\_1}^{(t)} - T_{w\_1}^{(t-\Delta t)}}{\Delta t} = \frac{1}{0.5lR_{\text{cond}}}\left(T_{so\_1}^{(t)} - 2T_{w\_1}^{(t)} + T_{si\_1}^{(t)}\right)$$

Category-3:

$$-dT_{si\_1}^{(t)} + (2d + 1)T_{w\_1}^{(t)} - dT_{so\_1}^{(t)} = T_{w\_1}^{(t-\Delta t)}$$

where $d = \dfrac{\Delta t}{\rho_{w\_1} c_{w\_1} V_{w\_1} 0.5lR_{\text{cond}}}$.

(IV) External surface is also same as before:

$$q_{\text{cond}\_2}^{(t)} = q_{\text{conv}-e\_1}^{(t)}$$

Category-4:

$$-eT_{w\_1}^{(t)} + (1 + e)T_{so\_1}^{(t)} = T_{\infty}^{(t)}$$

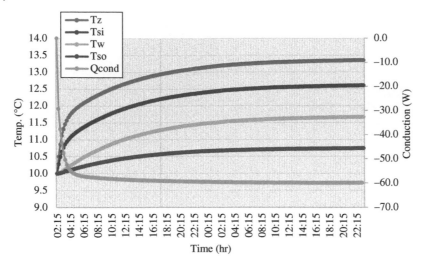

Figure 9.E4.2 Wall and zone temperature variation in addition to the heat loss of the room through time.

where $e = \left(\dfrac{R_{\text{conv}-e\_1}}{R_{\text{cond}\_2}}\right)^{(t)}$.

So, the matrix of system of linear equations can be formed as:

$$
\begin{bmatrix}
1+a & -a & 0 & 0 \\
1 & -1-c & c & 0 \\
0 & -d & 2d+1 & -d \\
0 & 0 & -e & 1+e
\end{bmatrix}
\begin{Bmatrix}
T_{z\_1}^{(t)} \\
T_{si\_1}^{(t)} \\
T_{w\_1}^{(t)} \\
T_{so\_1}^{(t)}
\end{Bmatrix}
=
\begin{Bmatrix}
b + T_{z\_1}^{(t-\Delta t)} \\
0 \\
T_{w\_1}^{(t-\Delta t)} \\
T_{\infty}^{(t)}
\end{Bmatrix}
$$

And, the solution is presented in Figure 9.E4.2. As it is expected, the thermal storage effect of the wall acts as a barrier between indoor and outdoor spaces and thus the equilibrium temperature of $13.4\,^\circ$C can be obtained after 181 iterations. Note that this case is not realistic due to constant outdoor temperature.

Example 9.5

Now assume that the floor of the room in Example 9.3 is insulated and all other walls are assumed to have one layer of brick (0.3 m) with thermal conductivity of 0.6 (W/mK). Convective heat transfer coefficients of both indoor and outdoor surfaces are assumed to be $h_{\text{out}} = h_{\text{in}} = 5$ W/m²K. If the room has an initial temperature of $10\,^\circ$C at 2:00 pm with an occupant who generates heat with a rate of 60 W, calculate the zone temperature with a $\Delta t = 15$ minutes (neglect the heat storages in the walls). What would be the required heater to warm up the room temperature to a comfortable range.

Solution

If we are not interested to find temperatures of internal and external wall nodes (while there is no heat storage in walls), and only the zone temperature is deemed, then we can consider

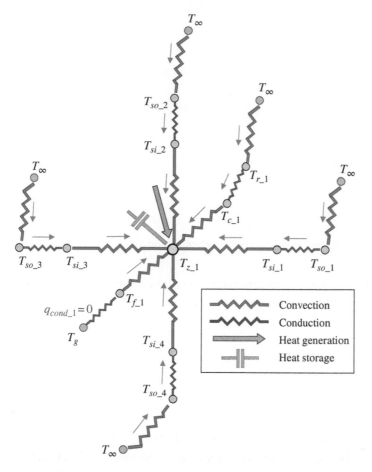

Figure 9.E5.1 Nodal system of the case study isolated room.

the heat transfer from the surrounding environment to the zone to occur with an equivalent resistance as shown in Figure 9.E5.1. We can directly calculate the convective heat transfer between the zone and environment using U-value (see Figure 9.E5.2):

$$\rho c_p V \frac{dT_z}{dt} = \dot{Q}_i + \frac{1}{R_{eq}}(T_\infty - T_z)$$

Thus, the governing equation becomes:

$$\rho c_p V \frac{T_z^{(t)} - T_z^{(t-\Delta t)}}{\Delta t} = \left(\dot{Q}_{occ} + \sum_{i=1}^{5} UA_i(T_{out} - T_z) \right)^{(t)}$$

Here, we can calculate U-value as:

$$R - value = \frac{1}{h_{in}} + \frac{l}{k} + \frac{1}{h_{out}} = 0.2 + 0.5 + 0.2 = 0.9 \ (\text{m}^2\text{K/W}) \implies U = 1.11 \ (\text{W/m}^2\text{K})$$

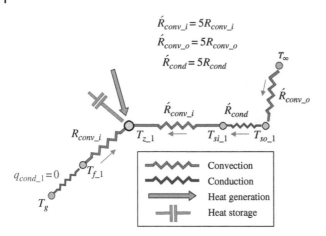

Figure 9.E5.2 Simplified nodal system of the case study isolated room.

All the surface areas and heat transfer coefficients are equal, knowing that the heat transfer from the ground is zero and $C_z = \rho c_p V$, we can write:

$$T_z^{(t)}\left(1 + \frac{\Delta t}{C_z}(5UA)\right) = \frac{\Delta t}{C_z}\dot{Q}_{occ}^{(t)} + \frac{\Delta t}{C_z}(5UA)T_\infty + T_z^{(t-\Delta t)}$$

$$\Rightarrow T_z^{(t)} = \frac{\dfrac{\Delta t}{C_z}\dot{Q}_{occ}^{(t)} + \dfrac{\Delta t}{C_z}(5UA)T_\infty + T_z^{(t-\Delta t)}}{\left(1 + \dfrac{\Delta t}{C_z}(5UA)\right)}$$

As it can be seen in the final equation, the initial temperature of $T_z^{(t-\Delta t)}$ at 2:00 is applied to find $T_z^{(t)}$ at 2:15. Similarly, the found $T_z^{(t)}$ at 2:15 would be the new $T_z^{(t-\Delta t)}$ at 2:15 to calculate the new $T_z^{(t)}$ at 2:30. The obtained zone temperatures are shown and plotted in Figure 9.E5.3:

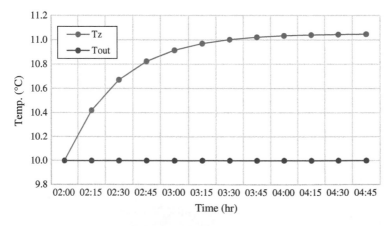

Figure 9.E5.3 Zone temperature variation through time.

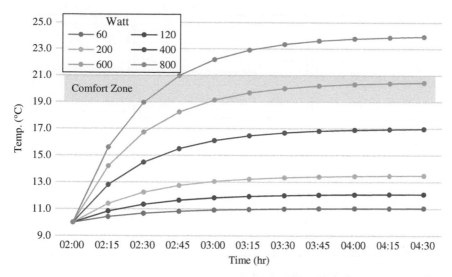

Figure 9.E5.4 Zone temperature variation through time using different heater wattages.

The room temperature reaches to a balanced temperature of 11°C (when outdoor temperature is fixed). The simple developed building energy simulation tool can be used to investigate how the room temperature can be set to about 19 − 21°C as a comfortable range. Figure 9.E5.4 shows heater wattage used to warm up the room. For example, a heater below 540 W (600 W in addition to a 60 W occupant) cannot result in a comfortable temperature after four iterations or 1 hour. If a 740 W heater (800 in addition to a 60 W occupant) is used, then the comfortable range can be achieved in two iterations or after 30 minutes.

Note that in many scenarios finding an equivalent resistance is not a straightforward procedure due to the complexity of the model. Therefore, most of the energy simulation tools use direct method of deriving equations on each node as it was explained in the earlier sections.

9.3 Inputs

The first step, in development of an energy simulation tool is to provide required inputs for the system. There are three type of inputs for each case, including user-inputs, non-time-variant, and time-variant. In the following section, we will explain these types of inputs in more details.

9.3.1 User-inputs

There are parameters that need to be adjusted in accordance with the nature of a simulation by a user as seen in Figure 9.4. The very evident example of such parameters is the time step (Δt). It should be noted that if the options to alter such parameters are not given to users in some tools, still we should account them to belong to this type of inputs. Hence, in respect to the available options, it is essential that a user plan an initial strategy to justify the selection of these parameters. Number of nodes in each wall and number of zones, which form the level of simplification of a nodal system are amongst the decisions that a user needs to

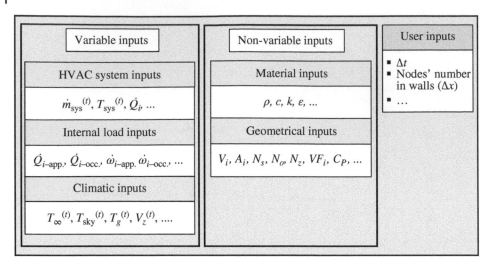

Figure 9.4 Inputs in a building energy simulation model.

make with necessary justifications to eventually develop the thermal nodal system of the simulation.

9.3.2 Non-time Variant Inputs

Non-time-variant inputs are mainly related to building fabric and materials if we assume that they have not undergone to intensive temperature and pressure change during simulations. Geometrical parameters are another family of non-time-variant inputs, which encompass parameters such as building layout, orientation, and location.

The nodal system requires multiple non-time-variant inputs during an iteration as demonstrated in Figure 9.4. Material properties, including solid surfaces' conductive characteristics such as conductivity and density in addition to radiative characteristics, including absorptivity, emissivity, reflectivity, transmissivity can be assumed to be constant values during the simulation. It should be noted that physics of a problem is stating if this is a valid assumption otherwise the time-variations of these parameters should be considered. Examples are phase change materials, which demonstrate variable conductive characteristics, or thermochromic glasses, which can change their radiative characteristics from transparent to translucent range due to a change in their temperatures.

Geometrical parameters are non-variant inputs to the nodal system. While surfaces areas and space volumes are contributing to heat transfer mechanisms, the building orientation is crucial in radiative exchanges. Such parameters should be initially synthesized to be used as constant inputs during the simulation time. Openings' condition is amongst few parameters that can be assumed as either a variable or non-variable input, depending on the coding strategy.

9.3.3 Time-Variant Inputs

Variable inputs dynamically alter during time and their non-time-variant assumptions may cause considerable discrepancies into the calculations. Examples are climatic information or heating, ventilation, and air conditioning (HVAC) systems.

9.3.3.1 HVAC System

HVAC system is diverse in terms of functionality and technology. Although these characteristics are not the concern of this book, their output parameters as the advected mass ($\dot{m}_{sys}$), temperature (T_{sys}), relative humidity (RH_{sys}), and generated heat ($\dot{Q}_i$) into a zone are of paramount importance (see Figure 9.4). The developed tools, however, might grant this option to the users to further design and control various parameters of HVAC systems. In this case, another energy balance set of equations shall be developed for the applied systems while a connection between the HVAC system and zone should be established. Yet all new sets of equations can be simultaneously solved together or shall be coupled together and shall be solved with an embedded algorithm as illustrated in Figure 9.5.

9.3.3.2 Internal Load Model

This load is referred to any generated heat within a control volume aside from the HVAC systems. Electrical appliances and occupants are the obvious examples. The generated heat from appliance can normally be extracted from their performance catalogues. Nonetheless, their utilization demands a schedule by occupants, which can be either represented by a simple on-and-off schedule similar to Figure 9.6, or it can be predicted with complex occupant behavioural models.

Occupants themselves also generate considerable heat as investigated in the previous examples. Occupants' heat generation in various activity level and conditions are also widely addressed in literature and can

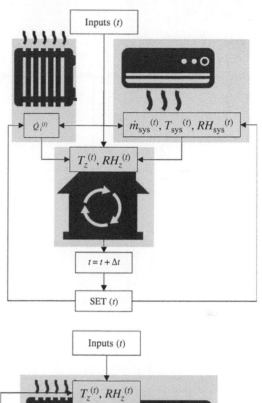

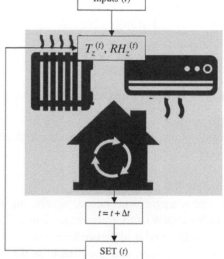

Figure 9.5 HVAC treatment in building energy simulation models.

be extracted from handbooks and guidelines such as ASHREARE [3] and CIBSE standards [4]. It is, however, very crucial to assign the schedule to each of the internal load sources. This task is not always a very straight forward procedure since the occupant behaviour is often very complex to be modelled. This topic includes many stochastic and psychological

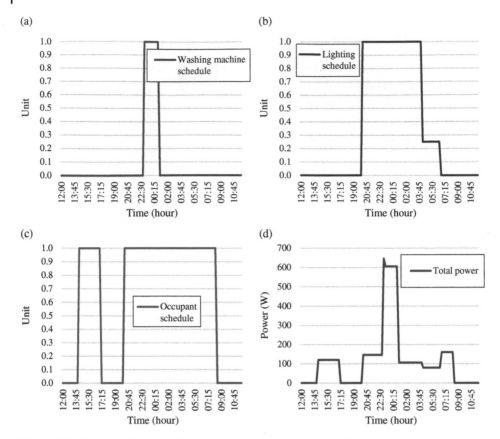

Figure 9.6 Various schedules of appliances and occupants in a typical zone.

considerations [5, 6] and many theories and understandings are proposed around this subject.

Example 9.6

Assume a single zone house, which contains five lightbulbs (12W each), one washing machine (500W) and a person as depicted in Figure 9.E6.1. The functioning schedule of them is also similar to Figure 9.6 where lightbulbs are on for 5 hours, the washing machine works for 2 hours, and the person works for 9 hours and sleeps 8 hours. The generated heat due to an average adult people activity level (with the area of 1.8 m^2) is provided in Figure 9.6 [ASHRAE Standard 55-2010].

Calculate the zone temperature using the below climatic data of Figure 9.E6.2 if there is no heater in the zone.

Solution

The variation of all generated heats due to washing machine, lightbulbs, and occupant's activity can be shown in Figure 9.6d. The room has a similar condition to Example 9.4. So, the matrix of coefficients can be found as:

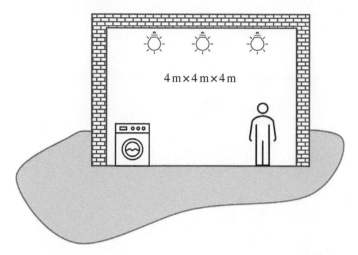

Figure 9.E6.1 Single zone house with 5 lightbulbs, one washing machine and a person.

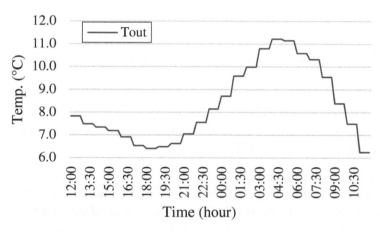

Figure 9.E6.2 Historical climatic data.

$$\begin{bmatrix} 1+a & -a & 0 & 0 \\ 1 & -1-c & c & 0 \\ 0 & -d & 2d+1 & -d \\ 0 & 0 & -e & 1+e \end{bmatrix} \left\{ \begin{array}{c} T^{(t)}_{z\_1} \\ T_{si\_1}{}^{(t)} \\ T^{(t)}_{w\_1} \\ T_{so\_1}{}^{(t)} \end{array} \right\} = \left\{ \begin{array}{c} b + T^{(t-\Delta t)}_{z\_1} \\ 0 \\ T^{(t-\Delta t)}_{w\_1} \\ T^{(t)}_{\infty} \end{array} \right\}$$

The heat generation is embedded in term-b:

$$b = \frac{\dot{Q_i}^{(t)} \Delta t}{\rho c_p V}$$

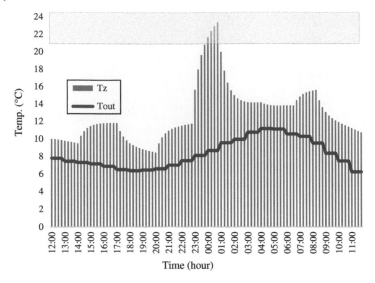

Figure 9.E6.3 Zone temperature variation through time.

Hence, we can find the zone temperature as shown in Figure 9.E6.3. Note that there are many time steps that the zone temperature is below the comfort range as the heater is not installed in the zone.

Example 9.7
Assume that we install a heater in the zone of Example 9.6 to warm up the zone as shown in Figure 9.E7.1. Calculate the zone temperature if:

1) We use a constant value of 320 W for the heater.
2) We use a simple on-and-off controller to generate a heat equal to the total loss from the walls and the ceiling (the only loss in this problem) minus the amount of the generated heat in the zone at a time step.

Solution

1) Figure 9.E7.2 demonstrates the zone temperature when a constant 320 W heat is added

   to the term-b in $b = \dfrac{\dot{Q}_i^{(t)} \Delta t}{\rho c_p V}$. As it can be seen, yet, there are many time steps that the

   zone temperature is beyond the comfort zone and probably a door or window should be opened to impose another term of loss to balance the zone temperature. This is a common case in many old heating systems without an advanced controller.

2) Now, we use a simple control strategy of heating when the zone temperature is below

   $T_z^{(t)} \leq T_{\text{SET}} = 21\,^\circ\text{C}$. Thus, we can first define the required heating power ($\dot{Q}_i^{(*)}$) during

Figure 9.E7.1 Single house zone of Example 9.6 with an additional heater.

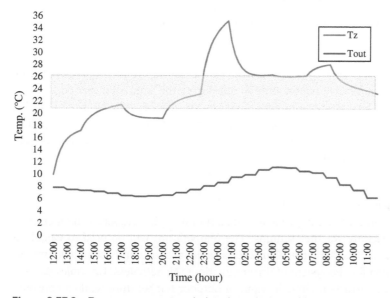

Figure 9.E7.2 Zone temperature variation through time with a constant heat flux.

the period of Δt until reaching the time-step $t + \Delta t$. $\dot{Q}_i^{(*)}$ can be calculated as the loss from surfaces minus the generated heat in the zone as:

$$\rho c_p V \frac{\left(T_{\text{SET}} - T_z^{(t)}\right)}{\Delta t} = \frac{-kA\left(T_z^{(t-\Delta t)} - T_{si}^{(t-\Delta t)}\right)}{l} + \dot{Q}_i^{(*)}$$

$$\Longrightarrow \dot{Q}_i^{(*)} = \rho c_p V \frac{\left(T_{\text{SET}} - T_z^{(t)}\right)}{\Delta t} + \frac{kA\left(T_z^{(t-\Delta t)} - T_{si}^{(t-\Delta t)}\right)}{l}$$

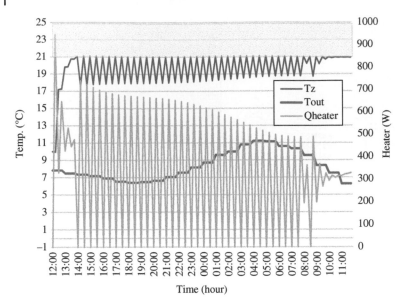

Figure 9.E7.3 Zone temperature variation through time with a variable heat flux.

Figure 9.E7.3 depicts the zone temperature, which is fluctuating around T_{SET} with a broad range of acceptable comfortable temperature found during the simulation period.

The used wattage to heat the zone in both scenarios are around 30.5 KW. Note that the utilized control strategy is a simplistic one though in reality advanced controller is applied to control indoor environments. The topic of control is a broad area in building science, which is beyond the scope of this book.

9.3.3.3 Climatic Input

Climatic inputs used in a zone energy balance equation include outdoor air temperature extracted from the vicinity weather stations. The temperature will be used to calculate the convection and mass transfer terms. Climatic variables are constantly changing, imposing a dynamic solution for the system of linear equations. Nonetheless, Examples 9.3–9.7 were solved using a non-variant climatic input of outdoor temperature while a constant convective coefficient was reflecting on an invariant air velocity around the investigated building. Expectedly, we obtained unrealistic times for change of zone temperatures. To rectify this, we need to input our models with dynamic climatic data. On the other hand, the climatic data are not necessarily available around a target building. Therefore, approximation discussed in Chapters 2 and 7 can be used to approximate these values from nearby weather stations. Data are mainly recorded in a resolution that might be larger than the selected time step in a model. For example, while hourly data are recorded in an airport's weather station, we are looking for a 15-minute time step. Hence, to approximate the variation of the climatic inputs during a calculation time step, we can use various techniques to replicate the required data. Examples are staircase fixed and linear interpolation schemes as

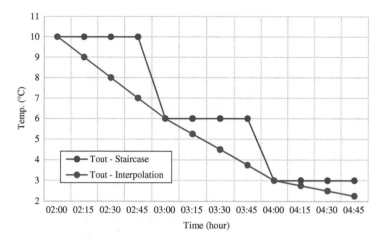

Figure 9.7 Climatic input effect to the zone temperature.

shown in Figure 9.7. The staircase fixed scheme is widely adapted in the energy building simulation tools.

Example 9.8

Solve Example 9.5 with the constant, staircase fixed, and linear interpolation on the obtained hourly outdoor temperature values (Figure 9.E8.1).

Solution

Since $\Delta t = 15$ min, and the given data are hourly basis, then each one-hour input can be assumed to be fixed or varied with a piecewise interpolation as shown in Figure 9.E8.2.

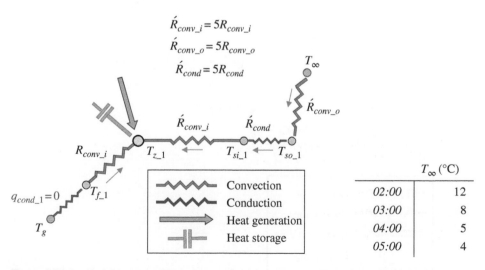

Figure 9.E8.1 Nodal system of the case study isolated room with historical climatic data.

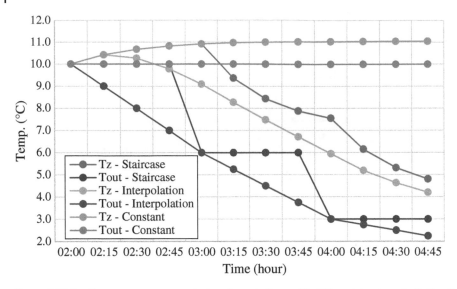

Figure 9.E8.2 Zone temperature variation through time with different treatment of climatic data.

The results of the zone temperature for the used outdoor temperature with the implicit method are also shown in the picture. While constant temperature provides exaggerated unrealistic zone temperatures, the trend of the calculated temperatures for piecewise and staircase inputs shows about 17% difference.

Example 9.9

Resolve Examples 9.3 and 9.4 with the below climatic data with and without thermal storage in a wall. Discuss the thermal storage effect of the wall (Figures 9.E9.1–9.E9.3).

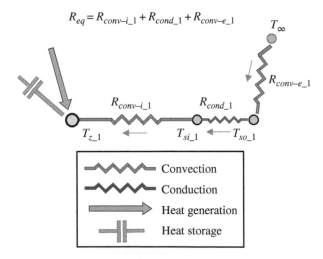

Figure 9.E9.1 Nodal system of the case study isolated room without thermal mass.

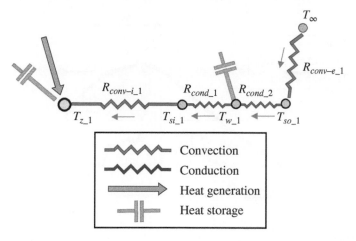

Figure 9.E9.2 Nodal system of the case study isolated room with thermal mass.

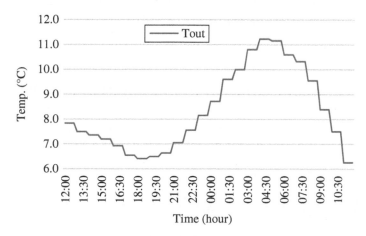

Figure 9.E9.3 Historical climatic data in the location of the case study isolated room.

Solution

The solution of both scenarios is depicted in Figure 9.E9.4. As expected, the wall is represented as a barrier against outdoor air temperature fluctuations.

Consideration of heat storage is very important in dynamic modelling of buildings and design of suitable fabrics for a household. The concept of thermal mass was well-understood from a long time ago in buildings. Traditional measures of time-lag (φ) and decrement factor have been used to address thermal mass. As it can be seen in the figure, time-lag represents a phase shift in the response of a zone to outdoor climatic fluctuation. Here, φ can be observed to be about 4 h and 15 min or 17 iterations for a model without a thermal mass consideration. Nonetheless, the thermal mass shrinks this delay in a better representation of the reality to 1 h and 30 min or about 6 iterations.

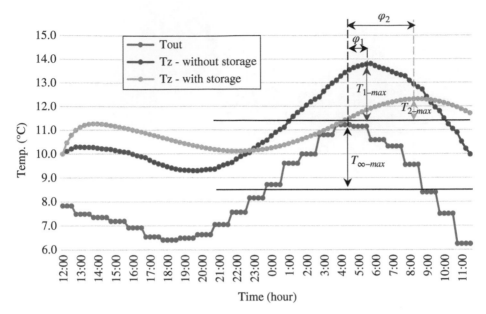

Figure 9.E9.4 Time-lag (φ) and decrement factor.

On the other hands, the decrement factor is a measure to demonstrate how amplitude of a cyclical climatic curve is squeezed due to the thermal mass effect and is defined as $\mu = \dfrac{T_{max}}{T_{\infty - max}}$. Thus, from the figure we can calculate μ for both scenarios as:

$$\mu_1 = \frac{T_{1-max}}{T_{\infty - max}} = \frac{13.8 - 11.5}{11.2 - 8.7} = 0.91$$

$$\mu_2 = \frac{T_{2-max}}{T_{\infty - max}} = \frac{12.3 - 11.5}{11.2 - 8.7} = 0.32$$

While the modelled wall without thermal mass ($\mu_1 = 0.91$) is almost 91% responsive to the climatic change, when the thermal storage is added, this number drops to 32%, almost one third of the initial value ($\mu_2 = 0.32$). This explains how a wall with a large thermal mass is a barrier against the climatic fluctuations. This model is a simplistic tool that can be used to design and select suitable materials and fabrics for a wall in different climates.

9.4 Solution Strategies

As we understood in the previous sections, for a system with 'n' as the total number of zones, radiosity, and wall nodes, we need to develop 'n' energy balance equations. These 'n' equations can be placed either in implicit or explicit solvers to obtain temperatures (and mass) for all nodes at a specific time step.

As shown in Figure 9.8, we can simultaneously solve all the developed energy balance equations in each node with the formation of the coefficient matrix of A and by using direct or iterative method to solve matrix equation of ($AX = B$). Another strategy is to avoid a large

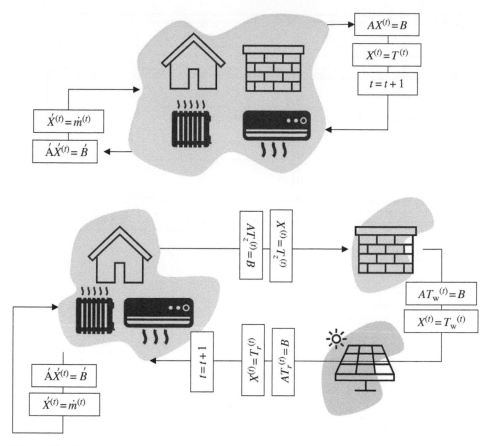

Figure 9.8 Strategies to solve system of linear equations.

matrix, to reduce the complexity of a model, and to develop a modular, flexible, and user-friendly tool is to separate different elements from each other and transfer data between them in a fixed time step. For example, it is possible to separate walls from zone and solve each part in an independent stage. The discrepancy of the latter method in loosing accuracy is not significant related to other source of errors such as temporal and spatial discretization errors (truncation error – see Chapter 10). While this process is often utilized with different strategies in different tools, we introduce two simple algorithms to establish this connection as demonstrated in Figure 9.8.

9.4.1 Integrated Solution

As we explained in Chapter 8, both thermal and mass nodal networks require inputs from each other in a specific time step. This means that the data should be processed in a solution algorithm to connect thermal and mass nodal networks in a dynamic sense. As shown in Figure 9.9, a typical integrated implicit algorithm is demonstrated as initial values are first calculated and then are used in Mass Flow Modeler to determine mass flow rates. These values are then sent to Zone and Wall Energy Balance Modeler to form an integrated matrix

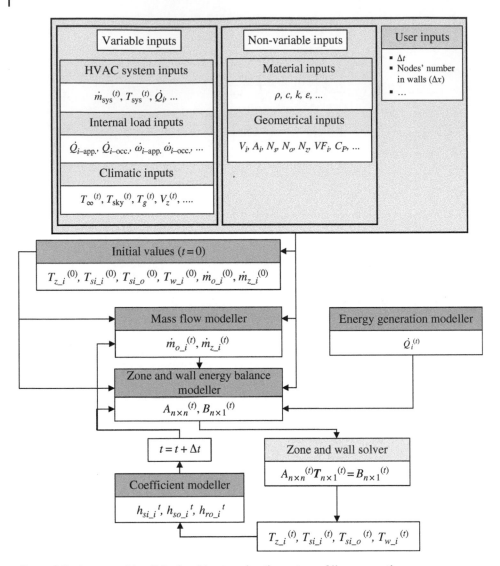

Figure 9.9 Integrated implicit algorithm to solve the system of linear equations.

and to resolve all the thermal values together. This is a very similar approach as used in the previous examples. The internal heat sources are calculated in Energy Generation Modeler and are also sent to the integrated solver module. At the end of an iteration, h-values at Coefficient Modeler are updated to again be used in mass flow and temperature calculations of the next iteration. The whole process can be looped when marching throughout time.

Example 9.10

A wall with three layers of concrete, insulation, and plasterboard is shown in Figure 9.E10.1. The materials' properties are given in the below table. The room's initial temperature is

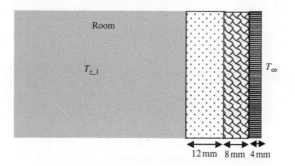

Material		Thickness mm	Conductivity W/(m.K)	Density kg/m³	Specific Heat Capacity J/(kg.K)	Resistance m² K/W	Total thickness mm
Type_1	Reinforced concrete	12	2.3	2300	1000	0.0435	
	Insulation	8	0.025	20	1030	3.256	24
	Plasterboard	4	0.21	700	1000	0.0595	

Figure 9.E10.1 Historical climatic data.

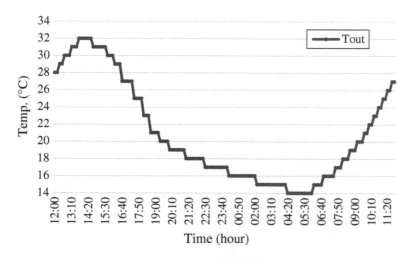

Figure 9.E10.2 A single zone building with three-layer walls.

$T_{z\_1} = 18\,^{\circ}\text{C}$, all wall nodes are at $20\,^{\circ}\text{C}$, and the internal and external convective coefficients are $h_{si\_1} = 3\ \text{W/m}^2$ and $h_{so\_1} = 6\ \text{W/m}^2$, respectively. Find the zone temperature using an integrated implicit method for a 5-node wall. There is no heat generation within the wall and the summertime outdoor temperature is as shown in Figure 9.E10.2.

Solution

The thermal diagram with five wall nodes, including internal, external, and zone nodes, is shown in Figures 9.E10.3 and 9.E10.4 ($\Delta x_1 = \Delta x_2 = \Delta x_3 = l = 0.004$ m).

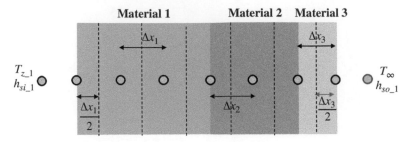

Figure 9.E10.3 Schematic of the three-layer wall.

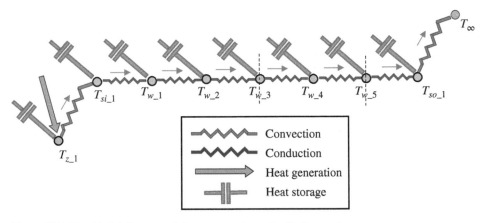

Figure 9.E10.4 Nodal diagram of the case study wall with five nodes.

We can again define four types of nodes as described before:

Category	Node type	Related nodes
1	(I) Internal surface	T_{si_1}
2	(II) Wall	$T_{w\_1}, T_{w\_2}, T_{w\_4}$
3	(II) Wall	T_{w_3}, T_{w_5}
4	(III) External surface	T_{so_1}
5	(IV) Zone	T_{z_1}

(I) Internal surface:

Category-1:

The convective heat flux from the room to the internal node is equivalent to the conduction flux that penetrates through T_{w_1}:

$$\rho_c c_c \frac{V}{2} \frac{T_{si\_1}{}^{(t)} - T_{si\_1}{}^{(t-\Delta t)}}{\Delta t} = h_{si\_1} A \left(T_{z\_1}{}^{(t)} - T_{si\_1}{}^{(t)} \right) - \frac{k_c \left(T_{si\_1}{}^{(t)} - T_{w\_1}{}^{(t)} \right)}{l}$$

which can be simplified as below:

$$\left(1 + \frac{2h_{si\_1}\Delta t}{\rho_c c_c l} + \frac{2k_c \Delta t}{\rho_c c_c l^2}\right) T_{si\_1}{}^{(t)} - \frac{2k_c \Delta t}{\rho_c c_c l^2} T_{w\_1}{}^{(t)} - \frac{2h_{si\_1}\Delta t}{\rho_c c_c l} T_{z\_1}{}^{(t)}$$

$$= T_{si\_1}{}^{(t-\Delta t)} \Longrightarrow -aT_{z\_1}{}^{(t)} + (1 + a + Fo_1) T_{si\_1}{}^{(t)} - Fo_1 T_{w\_1}{}^{(t)} = T_{si\_1}{}^{(t-\Delta t)}$$

where $a = \dfrac{2h_{si\_1}\Delta t}{\rho_c c_c l}$ and $Fo_1 = \dfrac{2k_c \Delta t}{\rho_c c_c l^2}$.

(II) Wall:

We use the implicit methods with the first-order backward discretization scheme for temporal first derivative where $q_{\text{gen}} = 0$. Similar to Eq. (9.4), we can derive the general equation as follows:

$$-FoT_{w\_i-1}^{(t)} + (1 + 2Fo)T_{w\_i}^{(t)} - FoT_{w\_i+1}^{(t)} = T_{w\_i}{}^{(t-\Delta t)}$$

Category-2:

For node-w_1 and node-w_2 with two different materials, we can derive:

$$(1 + 2Fo_1)T_{w\_1}{}^{(t)} - Fo_1\left(T_{si\_1}{}^{(t)} + T_{w\_2}{}^{(t)}\right) = T_{w\_1}{}^{(t-\Delta t)}$$

$$(1 + 2Fo_1)T_{w\_2}{}^{(t)} - Fo_1\left(T_{w\_1}{}^{(t)} + T_{w\_3}{}^{(t)}\right) = T_{w\_2}{}^{(t-\Delta t)}$$

Node-w_4 can be treated in a similar manner:

$$(1 + 2Fo_3)T_{w\_4}{}^{(t)} - Fo_3\left(T_{w\_3}{}^{(t)} + T_{w\ 5}{}^{(t)}\right) = T_{w\_4}{}^{(t-\Delta t)}$$

where $Fo_3 = \dfrac{k_i \Delta t}{\rho_i c_i l^2}$.

Category-3:

For node-w_3, there are two difference layers, thus the density and specific heat will be averaged at the node centre:

$$[1 + (k_c + k_i)Fo_2]T_{w\_3}{}^{(t)} - Fo_2\left(k_i T_{w\_4}{}^{(t)} + k_c T_{w\_2}{}^{(t)}\right) = T_{w\_3}{}^{(t-\Delta t)}$$

where $Fo_2 = \dfrac{4\Delta t}{(\rho_c + \rho_i)(c_c + c_i)l^2}$.

Similarly, node-w_5 can be treated in the same manner as node-3:

$$[1 + (k_i + k_p)Fo_4]T_{w\_5}{}^{(t)} - Fo_4\left(k_p T_{so\_1}{}^{(t)} + k_i T_{w\_4}{}^{(t)}\right) = T_{w\_5}{}^{(t-\Delta t)}$$

where $Fo_4 = \dfrac{4\Delta t}{\left(\rho_i + \rho_p\right)(c_i + c_p)l^2}$.

(III) External surface:

Category-4:

In the external surface, the heat flux enters from node-w_5 to the surface node and will leave through the outdoor node throughout a convection mechanism:

$$c_p \frac{V}{2} \frac{T_{so\_1}{}^{(t)} - T_{so\_1}{}^{(t-\Delta t)}}{\Delta t} = \frac{k_p A \left(T_{w\_5}{}^{(t)} - T_{so\_1}{}^{(t)} \right)}{l} - h_{so\_1} A \left(T_{so\_1}{}^{(t)} - T_\infty{}^{(t)} \right)$$

which can be further simplified as follows:

$$T_{so\_1}{}^{(t-\Delta t)} = -\frac{2\Delta t k_p}{\rho_p c_p l^2} T_{w\_5}{}^{(t)} + \left(1 + \frac{2\Delta t h_{so\_1}}{\rho_p c_p l} + \frac{2\Delta t k_p}{\rho_p c_p l^2} \right) T_{so\_1}{}^{(t)} - \frac{2\Delta t h_{so\_1}}{\rho_p c_p l} T_\infty{}^{(t)}$$

$$\Longrightarrow -Fo_5 T_{w_5}{}^{(t)} + (1 + b + Fo_5) T_{so_1}{}^{(t)} = T_{so_1}{}^{(t-\Delta t)} + b T_\infty{}^{(t)}$$

where $b = \dfrac{2\Delta t h_{so\_1}}{\rho_p c_p l}$ and $Fo_5 = \dfrac{2\Delta t k_p}{\rho_p c_p l^2}$.

(IV) Zone:

This type is similar to the previous examples using Eq. (9.1).

Category-5:

$$\rho c V \frac{T_{z\_1}^{(t)} - T_{z\_1}^{(t-\Delta t)}}{\Delta t} = \dot{Q}_{gen}{}^{(t)} - h_{si\_1} A \left(T^{(t)}{}_{z\_1} - T^{(t)}{}_{si\_1} \right)$$

We can further simplify the equation as:

$$(\rho c V + \Delta t h_{si\_1} A) T_{z\_1}^{(t)} - (\Delta t h_{si\_1} A) T^{(t)}{}_{si\_1} = \Delta t \dot{Q}_{gen}{}^{(t)} + (\rho c V) T_{z\_1}^{(t-\Delta t)}$$

Or we can write it in the form of:

$$c T_{z\_1}^{(t)} + d T^{(t)}{}_{si\_1} = \Delta t \dot{Q}_{gen}{}^{(t)} + (c + d) T_{z\_1}^{(t-\Delta t)}$$

where $c = \rho c V + \Delta t h_{si\_1} A$ and $d = -\Delta t h_{si\_1} A$.

And now, we can form $A_{8 \times 8}$ and $B_{8 \times 1}$ matrices, substitute the matrix with specified values, and solve the matrix by a direct method:

$$\begin{bmatrix} c & d & 0 & 0 & 0 & 0 & 0 & 0 \\ -a & 1+a+Fo_1 & -Fo_1 & 0 & 0 & 0 & 0 & 0 \\ 0 & -Fo_1 & 1+2Fo_1 & -Fo_1 & 0 & 0 & 0 & 0 \\ 0 & 0 & -Fo_1 & 1+2Fo_1 & -Fo_1 & 0 & 0 & 0 \\ 0 & 0 & 0 & -k_c Fo_2 & 1+(k_c+k_i)Fo_2 & -k_i Fo_2 & 0 & 0 \\ 0 & 0 & 0 & 0 & -Fo_3 & 1+2Fo_3 & -Fo_3 & 0 \\ 0 & 0 & 0 & 0 & 0 & -k_i Fo_4 & 1+(k_i+k_p)Fo_4 & -k_p Fo_4 \\ 0 & 0 & 0 & 0 & 0 & 0 & -Fo_5 & 1+b+Fo_5 \end{bmatrix} \begin{bmatrix} T_{z\_1}{}^{(t)} \\ T_{si\_1}{}^{(t)} \\ T_{w\_1}{}^{(t)} \\ T_{w\_2}{}^{(t)} \\ T_{w\_3}{}^{(t)} \\ T_{w\_4}{}^{(t)} \\ T_{w\_5}{}^{(t)} \\ T_{so\_1}{}^{(t)} \end{bmatrix}$$

$$= \begin{bmatrix} \Delta t \dot{Q}_{gen}{}^{(t)} + (c+d) T_{z\_1}^{(t-\Delta t)} \\ T_{si\_1}{}^{(t-\Delta t)} \\ T_{w\_1}{}^{(t-\Delta t)} \\ T_{w\_2}{}^{(t-\Delta t)} \\ T_{w\_3}{}^{(t-\Delta t)} \\ T_{w\_4}{}^{(t-\Delta t)} \\ T_{w\_5}{}^{(t-\Delta t)} \\ T_{so\_1}{}^{(t-\Delta t)} + b T_\infty{}^{(t)} \end{bmatrix}$$

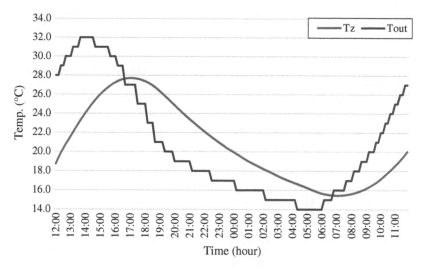

Figure 9.E10.5 Zone temperature variation through time.

We calculate wall temperature based on a selected time step, noting that the implicit scheme is unconditionally stable. The initial temperatures of the wall and zone are given as:

$$T_{si\_1}{}^{(0)} = T_{w\_1}{}^{(0)} = T_{w\_2}{}^{(0)} = T_{w\_3}{}^{(0)} = T_{w\_4}{}^{(0)} = T_{w\_5}{}^{(0)} = T_{so\_1}{}^{(0)} = 18\degree C$$

Also, we know the generation in the zone, so:

$$\dot{Q}_{gen}{}^{(\Delta t)} = 0$$

Thus, we can replace these values in the right-hand side of the above matrix for calculation of the first set of temperatures as it can be seen in Figure 9.E10.5.

9.4.2 Coupled Solution

Another solution algorithm can be written in a coupled approach to reduce the complexity of the system while equations of zones and walls are separately solved in a specific time step. The benefit of this method is in a considerable reduction of the computational cost related to the large dimension of the formed matrices and a better modularity of the simulation process.

As it can be seen in Figure 9.10, after mass flow rates are calculated in Mass flow Modeler, they are transferred to Zone Energy Balance Modeler to separately resolve the zone temperatures. The information at a same time step is then fed to Wall Energy Balance Modeler to find its values. Similar as the integrated algorithm, the coefficients are updated afterwards to be used in the next time step.

We should note that the proposed integrated and coupled algorithms shown in Figures 9.9 and 9.10 are just simplistic workflows that demonstrate how thermal and mass flow nodal networks can be linked together. Obviously, more efficient and complex algorithms are used in the developed tools. Moreover, radiosity nodes, HVAC systems, and other elements are neglected in these algorithms while in reality, they should be embedded carefully in such algorithms.

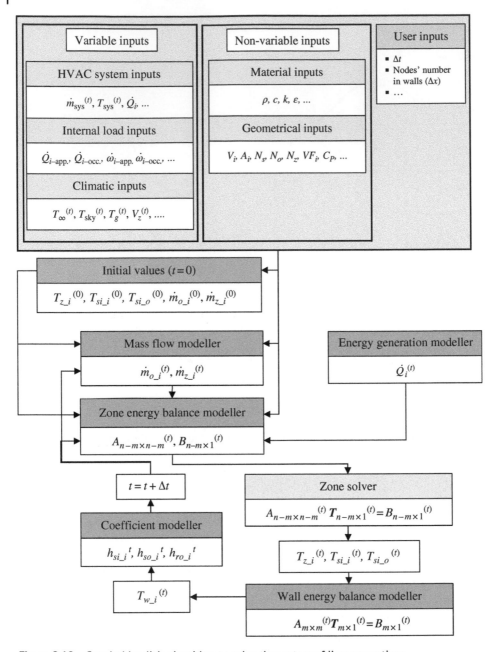

Figure 9.10 Coupled implicit algorithm to solve the system of linear equations.

Example 9.11

Consider a coupled approach when zone temperate is calculated in a specific time step as $T_z = 18°C$. Calculate the conduction through a multi-layer wall using five nodes. Assume that there is no heat generation inside the wall and the storage effect is neglected. Outdoor temperature is $T_{out} = 8°C$ while the heat convection coefficients are $h_{si\_1} = 3$ W/m² and $h_{so\_1} = 6$ W/m² for internal and external walls, respectively (Figure 9.E11.1).

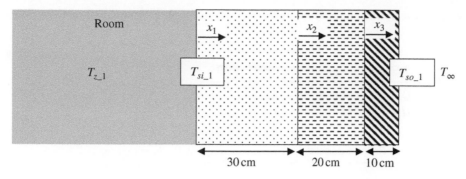

	Material	Thickness mm	Conductivity W/(m.K)	Density kg/m³	Specific Heat Capacity J/(kg.K)	Resistance m²K/W
Wall	Reinforced concrete	300	2.3	2300	1000	0.0435
	Insulation	200	0.025	20	1030	3.256
	Plasterboard	100	0.21	700	1000	0.0595

Figure 9.E11.1 A single zone building with three-layer walls.

Solution

In this case, the wall can be assumed to be in a steady-state condition. The thermal nodal diagram of the wall can also be shown in Figure 9.E11.2:

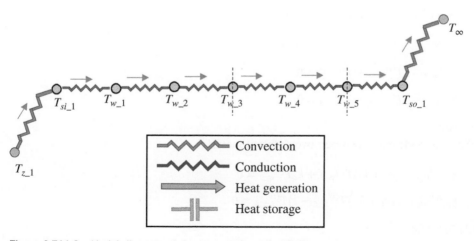

Figure 9.E11.2 Nodal diagram of the case study wall with five nodes.

We also have similar node types and categories as the last example while there is no zone node:

Category	Node type	Related nodes
1	(I) Internal surface	T_{si_1}
2	(II) Wall	$T_{w\_1}, T_{w\_2}, T_{w\_4}$
3	(II) Wall	T_{w_3}, T_{w_5}
4	(III) External surface	T_{so_1}

(I) Internal surface:

For a steady-state condition, we have:

$$q_{convi} = q_{cond\_(si\_1) \to 1}$$

Category-1:

$$h_{si\_1}\left(T^{(t)}_{z\_1} - T^{(t)}_{si\_1}\right) = \frac{k_c\left(T^{(t)}_{si\_1} - T^{(t)}_{w\_1}\right)}{l}$$

$$\Rightarrow \quad k_c T^{(t)}_{w\_1} - (h_{si_1} l + k_c) T^{(t)}_{si\_1} + h_{si\_1} l T^{(t)}_{z\_1} = 0$$

(I) Wall:

$$q_{cond\_(si\_1) \to 1} = q_{cond\_1 \to 2}$$

Category-2:

For the internal nodes in a single material, we can derive:

$$k_c\left(\frac{T^{(t)}_{si\_1} - T^{(t)}_{w\_1}}{l} - \frac{T^{(t)}_{w\_1} - T^{(t)}_{w\_2}}{l}\right) = 0$$

$$\Longrightarrow T^{(t)}_{w\_2} - 2T^{(t)}_{w\_1} + T^{(t)}_{si\_1} = 0$$

Similarly:

$$T^{(t)}_{w\_3} - 2T^{(t)}_{w\_2} + T^{(t)}_{w\_1} = 0$$

$$T^{(t)}_{w\_5} - 2T^{(t)}_{w\_4} + T^{(t)}_{w\_3} = 0$$

Category-3:

When the node exists between two different materials:

$$k_i T^{(t)}_{w\_4} - (k_c + k_i) T^{(t)}_{w\_3} + k_c T^{(t)}_{w\_2} = 0$$

$$k_p T^{(t)}_{so\_1} - (k_i + k_p) T^{(t)}_{w\_5} + k_i T^{(t)}_{w\_4} = 0$$

(III) External surface:

$$q_{cond\_5 \to so\_1} = q_{convo}$$

Category-4:

$$\frac{k_p\left(T^{(t)}_{w\_5} - T^{(t)}_{so\_1}\right)}{l} - h_{so\_1}\left(T^{(t)}_{so\_1} - T^{(t)}_{\infty}\right) = 0$$

$$\Longrightarrow h_{so\_1}l T^{(t)}_{\infty} - \left(h_{so\_1}l + k_p\right)T^{(t)}_{so\_1} + k_p T^{(t)}_{w\_5} = 0$$

Therefore, we can form A and B matrices as:

$$\begin{bmatrix} -(h_{si\_1}l + k_c) & k_c & 0 & 0 & 0 & 0 & 0 \\ 1 & -2 & 1 & 0 & 0 & 0 & 0 \\ 0 & 1 & -2 & 1 & 0 & 0 & 0 \\ 0 & 0 & k_i & -(k_c + k_i) & k_i & 0 & 0 \\ 0 & 0 & 0 & 1 & -2 & 1 & 0 \\ 0 & 0 & 0 & 0 & k_i & -(k_i + k_p) & k_p \\ 0 & 0 & 0 & 0 & 0 & k_p & -(h_{so\_1}l + k_p) \end{bmatrix} \begin{bmatrix} T^{(t)}_{si\_1} \\ T^{(t)}_{w\_1} \\ T^{(t)}_{w\_2} \\ T^{(t)}_{w\_3} \\ T^{(t)}_{w\_4} \\ T^{(t)}_{w\_5} \\ T^{(t)}_{so\_1} \end{bmatrix}$$

$$= \begin{bmatrix} -h_{si\_1}l T^{(t)}_{z\_1} \\ 0 \\ 0 \\ 0 \\ 0 \\ 0 \\ -h_{so\_1}l T^{(t)}_{\infty} \end{bmatrix}$$

Substitute the parameters with specific values and applying a direct method, the result is as follows:

Temperature (°C)		Analytical solution	FDM solution
Node	0	17.633970	17.633966
	1	17.586226	17.586223
	2	17.538482	17.538479
	3	17.490738	17.490735
	4	13.098333	13.098329
	5	8.705926	8.705922
	6	8.183014	8.183017

We can also find the analytical solution as:

Reinforced concrete layer $T(x_1) = -0.48x_1 + 17.63$

Insulation layer $T(x_2) = -43.92x_2 + 17.50$

Plasterboard layer $T(x_3) = -5.23x_3 + 8.71$

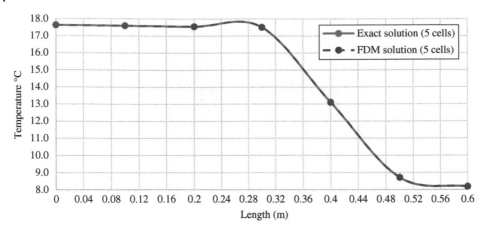

Figure 9.E11.3 Zone temperature variation through time found by exact solution and simulation.

Thus, the error between the simulation and the exact solution can be calculated as $\text{Error} = \sum_{i=1}^{n} \frac{|T_{i-\text{exact}} - T_{i-\text{FDM}}|}{T_{i-\text{exact}}}$, where n is the number of nodes. Thus, the calculated difference between the exact and numerical values of the temperature distribution can be plotted in Figure 9.E.11.3.

Example 9.12

Assume the wall model of Example 9.11. Now, use a coupled algorithm to find the zone temperature.

Solution

Unlike Example 9.11, the zone node is an unknown value to the wall system of linear equation and the zone node is separately solved.

Category	Node type	Related nodes
1	(I) Internal surface	T_{si_1}
2	(II) Wall	$T_{w\_1}, T_{w\_2}, T_{w\_4}$
3	(II) Wall	T_{w_3}, T_{w_5}
4	(III) External surface	T_{so_1}

(I) Internal surface:
Category-1:

$$\left(1 + \frac{2h_{si\_1}\Delta t}{\rho_c c_c l} + \frac{2k_c \Delta t}{\rho_c c_c l^2}\right) T_{si\_1}{}^{(t)} - \frac{2k_c \Delta t}{\rho_c c_c l^2} T_{w\_1}{}^{(t)} - \frac{2h_{si_1}\Delta t}{\rho_c c_c l} T_{z\_1}{}^{(t)}$$

$$= T_{si\_1}{}^{(t-\Delta t)} \Longrightarrow + (1 + a + Fo_1) T_{si\_1}{}^{(t)} - Fo_1 T_{w\_1}{}^{(t)} = T_{si\_1}{}^{(t-\Delta t)} + a T_{z\_1}{}^{(t)}$$

where $a = \dfrac{2h_{si\_1}\Delta t}{\rho_c c_c l}$ and $Fo_1 = \dfrac{2k_c \Delta t}{\rho_c c_c l^2}$.

Categories 2–4 remains similar as last example.

$$\begin{bmatrix} 1 + a + Fo_1 & -Fo_1 & 0 & 0 & 0 & 0 & 0 \\ -Fo_1 & 1 + 2Fo_1 & -Fo_1 & 0 & 0 & 0 & 0 \\ 0 & -Fo_1 & 1 + 2Fo_1 & -Fo_1 & 0 & 0 & 0 \\ 0 & 0 & -k_cFo_2 & 1 + (k_c + k_i)Fo_2 & -k_iFo_2 & 0 & 0 \\ 0 & 0 & 0 & -Fo_3 & 1 + 2Fo_3 & -Fo_3 & 0 \\ 0 & 0 & 0 & 0 & -k_iFo_4 & 1 + (k_i + k_p)Fo_4 & -k_pFo_4 \\ 0 & 0 & 0 & 0 & 0 & -Fo_5 & 1 + c + Fo_5 \end{bmatrix}$$

$$\begin{Bmatrix} T_{si\_1}^{(t)} \\ T_{w\_1}^{(t)} \\ T_{w\_2}^{(t)} \\ T_{w\_3}^{(t)} \\ T_{w\_4}^{(t)} \\ T_{w\_5}^{(t)} \\ T_{so\_1}^{(t)} \end{Bmatrix} = \begin{Bmatrix} T_{si\_1}^{(t-\Delta t)} + aT_{z\_1}^{(t)} \\ T_{w\_1}^{(t-\Delta t)} \\ T_{w\_2}^{(t-\Delta t)} \\ T_{w\_3}^{(t-\Delta t)} \\ T_{w\_4}^{(t-\Delta t)} \\ T_{w\_5}^{(t-\Delta t)} \\ T_{so\_1}^{(t-\Delta t)} + bT_{\infty}^{(t)} \end{Bmatrix}$$

The zone is also solved separately while the surface temperature ($T^{(t)}_{si\_1}$) is coupled to the zone:

$$cT_{z\_1}^{(t)} + dT^{(t)}_{si\_1} = \Delta t \dot{Q}_{gen}^{(t)} + (c + d)T_{z\_1}^{(t-\Delta t)}$$

$$\Longrightarrow T_{z\_1}^{(t)} = \frac{\Delta t \dot{Q}_{gen}^{(t)} + (c + d)T_{z\_1}^{(t-\Delta t)} - dT^{(t)}_{si\_1}}{c}$$

We can now plot the coupled zone temperature and compare it with the integrated one represented in Example 9.11 (Figure 9.E12).

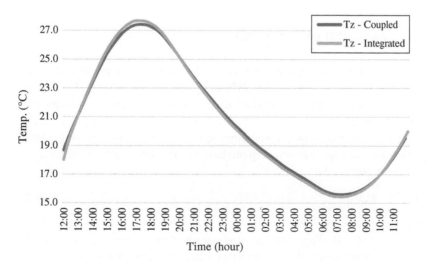

Figure 9.E12 Zone temperature variation through time using coupled and integrated solutions.

The average discrepancy between two algorithm is about 1.4% for $\Delta t = 600$ s, which can be substantially reduced when smaller time steps are employed as shown in the below table:

Δt	Discrepancy
1200	2.224
600	1.440
300	0.706
100	0.166
10	0.002

9.4.3 Nodes Connectivity

When matrices are formed to obtain temperatures, mass flows, or moistures of a formed system of linear equations, we need to ensure about the correct connectivity between nodes. Consider an arbitrary system, which resulted in a thermal nodal network as sketched in Figure 9.11. The number of unknowns in this system is 8, thus we expect to form $A_{8 \times 8}$ and $B_{8 \times 1}$ matrices as shown in Figure 9.11. For example, as it can be seen, zone-1 has four connectivity with zone-2, internal walls-1, internal wall-2, and floor-1, resulting row-1, which is related to the equation of zone-1, to contains five elements of $a_1^{(t)}$ to $a_5^{(t)}$. Note that one of the elements is associated with the connectivity of zone-1 to itself. Similarly, zone-2 has only two connectivity, and in addition with a self-connectivity, is presented with three elements of $f_1^{(t)}$ to $f_3^{(t)}$ in row-6. Row-4 and row-8 show the external wall equations, only connected with one node, and thus contain two elements. Note each non-zero element in this matrix represents that there is a connectivity between those nodes/zones. Obviously, it can be assumed that each node/zone is connected to itself and thus a non-zero element is also expected.

9.5 Temporal Variation of Parameters

The convection heat transfer coefficient as it was shown before can be approximated with various imperial and semi-imperial approaches. While the climatic data such as V_z is essential to update the forced convective part, a local ΔT should be updated for the calculation of natural convection part in each time step $(\Delta T^{(t)} = T_\infty^{(t)} - T_{so}^{(t)})$. Nonetheless, to ensure a less complex form for equation (8.5) as the governing equation, we can assume with a high level of accuracy that $\Delta T^{(t)}$ at a specific time step can be calculated from surface information from the previous time step $t - \Delta t$:

$$\Delta T^{(t)} = T_\infty^{(t)} - T_{so}^{(t - \Delta t)} \tag{9.5}$$

Another approach is introduced in Figures 9.9 and 9.10 as CHTCs $(h_n^{(t)})$ are required to be updated in each iteration after finding all temperatures at a particular time step (t). Thus, for finding temperatures at zones and walls in time step $t + \Delta t$, we again use the obtained $h_n^{(t)}$ at time step t. Such assumption for other parameters can be used on some occasions when less complexities are expected to be implemented into the simulations.

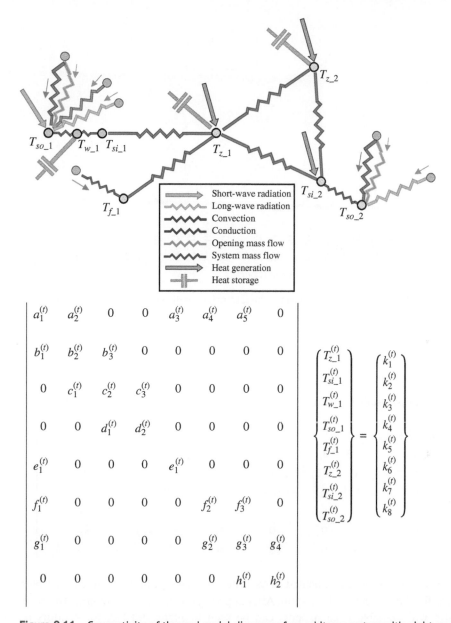

$$
\begin{vmatrix}
a_1^{(t)} & a_2^{(t)} & 0 & 0 & a_3^{(t)} & a_4^{(t)} & a_5^{(t)} & 0 \\
b_1^{(t)} & b_2^{(t)} & b_3^{(t)} & 0 & 0 & 0 & 0 & 0 \\
0 & c_1^{(t)} & c_2^{(t)} & c_3^{(t)} & 0 & 0 & 0 & 0 \\
0 & 0 & d_1^{(t)} & d_2^{(t)} & 0 & 0 & 0 & 0 \\
e_1^{(t)} & 0 & 0 & 0 & e_1^{(t)} & 0 & 0 & 0 \\
f_1^{(t)} & 0 & 0 & 0 & 0 & f_2^{(t)} & f_3^{(t)} & 0 \\
g_1^{(t)} & 0 & 0 & 0 & 0 & g_2^{(t)} & g_3^{(t)} & g_4^{(t)} \\
0 & 0 & 0 & 0 & 0 & 0 & h_1^{(t)} & h_2^{(t)}
\end{vmatrix}
\begin{Bmatrix}
T_{z\_1}^{(t)} \\
T_{si\_1}^{(t)} \\
T_{w\_1}^{(t)} \\
T_{so\_1}^{(t)} \\
T_{f\_1}^{(t)} \\
T_{z\_2}^{(t)} \\
T_{si\_2}^{(t)} \\
T_{so\_2}^{(t)}
\end{Bmatrix}
=
\begin{Bmatrix}
k_1^{(t)} \\
k_2^{(t)} \\
k_3^{(t)} \\
k_4^{(t)} \\
k_5^{(t)} \\
k_6^{(t)} \\
k_7^{(t)} \\
k_8^{(t)}
\end{Bmatrix}
$$

Figure 9.11 Connectivity of thermal nodal diagram of an arbitrary system with eight nodes.

Example 9.13

Assume the integrated algorithm of **Example 9.11**. If the internal and external convective heat transfer coefficients due to climatic and zonal temperature variations are found as shown in Figure 9.E13.1, then calculate the zone temperature.

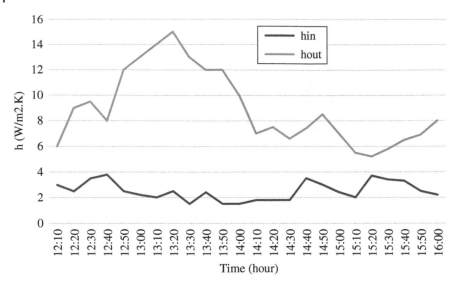

Figure 9.E13.1 Internal and external convective heat transfer coefficients due to climatic and zonal temperature variations.

Solution

Alteration in the convective heat transfer coefficients will imply changes in the A matrix:

$$
\begin{bmatrix}
-(h_{si\_1}l + k_c) & k_c & 0 & 0 & 0 & 0 & 0 \\
1 & -2 & 1 & 0 & 0 & 0 & 0 \\
0 & 1 & -2 & 1 & 0 & 0 & 0 \\
0 & 0 & k_i & -(k_c + k_i) & k_i & 0 & 0 \\
0 & 0 & 0 & 1 & -2 & 1 & 0 \\
0 & 0 & 0 & 0 & k_i & -(k_i + k_p) & k_p \\
0 & 0 & 0 & 0 & 0 & k_p & -(h_{so\_1}l + k_p)
\end{bmatrix}
$$

Thus, the A matrix should be updated in each iteration ($A^{(n)}$) unlike previous examples when A was fixed in the entire simulation. After applying the mentioned changes, the zone temperature with variable h coefficients can be seen in Figure 9.E13.2 for 4 hours of simulation:

The average in the discrepancy between two approaches is about 1% with some points to have about 2% difference. Such difference might seem insignificant for a simple case and few hours of simulations; however, this discrepancy will be propagated for longer period of simulations due to the variation in the convective heat transfer coefficients.

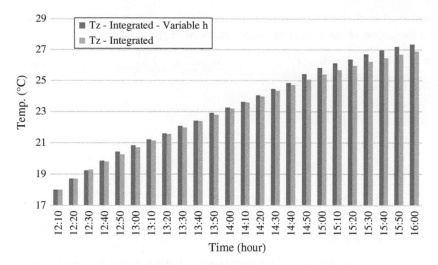

Figure 9.E13.2 Zone temprature variation through time with constant and variable convective heat transfer coeffients.

9.6 Linearization of the Radiation

As it was explained before, long-wave radiation equation is not initially in a linear form (see Figure 9.12):

$$q''_{\mathrm{LWR}} + q_{\mathrm{conv\_o}} = q_{\mathrm{cond}}$$

$$\Longrightarrow \varepsilon_g \sigma F_g \dot{A} \left(T_g^4 - T_{so\_1}^4 \right) + \varepsilon_{\mathrm{sky}} \sigma F_{\mathrm{sky}} A \left(T_{\mathrm{sky}}^4 - T_{so\_1}^4 \right) + \varepsilon_\infty \sigma F_\infty A \left(T_\infty^4 - T_{so\_1}^4 \right)$$

$$+ hA(T_\infty - T_{so\_1}) = \frac{kA}{l}(T_{so\_1} - T_{w\_1})$$

$$(9.6)$$

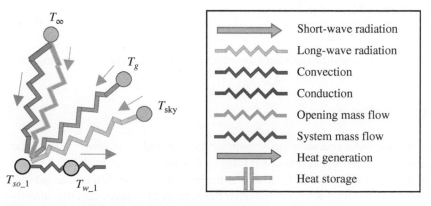

Figure 9.12 Heat balance on the external surface of a building.

As it was explained in the previous chapter, the long-wave radiation terms shall be linearized for the formation of the system of linear equations of the solid walls:

$$q''_{LWR} = h_{r\_g}\left(T_g - T_{so}\right) + h_{r\_sky}\left(T_{sky} - T_{so}\right) + h_{r\_\infty}\left(T_\infty - T_{so}\right) \tag{9.7}$$

The radiative coefficients were defined as:

$$h_{r\_g} = \frac{\varepsilon\sigma F_g\left(T_g^4 - T_{so}^4\right)}{T_g - T_{so}} \tag{9.8}$$

$$h_{r\_sky} = \frac{\varepsilon\sigma F_{sky}\left(T_{sky}^4 - T_{so}^4\right)}{T_{sky} - T_{so}} \tag{9.9}$$

$$h_{r\_\infty} = \frac{\varepsilon\sigma F_\infty\left(T_\infty^4 - T_{so}^4\right)}{T_\infty - T_{so}} \tag{9.10}$$

In one of the feasible approaches, we can calculate the radiative heat flux coefficients based on the known values of the surface and the associated radiation flux. The assumption here is based on an insignificant variation between the surface and ground, air and sky temperatures during the time step of the calculations. Therefore, we can rewrite the long-wave radiation as below:

$$q''_{LWR} = h_{r\_gnd}^{(i+1)}\left(T_{gnd}^{(i+1)} - T_{so}^{(i+1)}\right) + h_{r\_sky}^{(i+1)}\left(T_{sky}^{(i+1)} - T_{so}^{(i+1)}\right) + h_{r\_air}^{(i+1)}\left(T_{air}^{(i+1)} - T_{so}^{(i+1)}\right) \tag{9.11}$$

Thus, the new radiative coefficients can be written as:

$$h_{r\_gnd}^{(i+1)} = \frac{\varepsilon\sigma F_{gnd}\left(T_{so}^{(i)4} - T_{gnd}^{(i)\,4}\right)}{T_{so}^{(i)} - T_{gnd}^{(i)}} \tag{9.12}$$

$$h_{r\_sky}^{(i+1)} = \frac{\varepsilon\sigma F_{sky}\left(T_{so}^{(i)4} - T_{sky}^{(i)\,4}\right)}{T_{so}^{(i)} - T_{sky}^{(i)}} \tag{9.13}$$

$$h_{r\_air}^{(i+1)} = \frac{\varepsilon\sigma F_{air}\left(T_{so}^{(i)4} - T_{air}^{(i)\,4}\right)}{T_{so}^{(i)} - T_{air}^{(i)}} \tag{9.14}$$

Therefore, all h_r values will be calculated based on the previous time step information.

9.7 Mass Imbalance

There are two models widely employed to account for the required mass advection terms for the thermal network systems. The first one is forming the mass flow network based on the conservation of mass as described in Chapter 8 as the network can be then embedded into any integrated or coupled algorithm. On the other hand, there are approaches that directly calculate mass advection due to HVAC systems, infiltrations, and openings independently from other elements of the system. Examples are design flow rate as introduced in Chapter 8 and the below model that estimates mass flow through an opening using variables such as wind speed, thermal stack effect, and the area of the opening.

Example 9.14

The Wind and Stack with Open Area Model developed by ASHRAE [7] has two components, including one related to the mass advection driven by wind:

$$Q_w = C_w A F_{schedule} V_z$$

And, due to the stack effect:

$$Q_s = C_d A F_{schedule} \sqrt{2g \Delta H_{NPL}(|T_z - T_\infty|/T_z)}$$

where C_w is the effectiveness of opening calculated and is set to be 0.5–0.6 for perpendicular winds and 0.25–0.35 for diagonal winds. ΔH_{NPL} is the height from midpoint of the lower opening to the natural pressure level. $F_{schedule}$ is defined as the open area fraction variation in time. C_d is the discharge coefficient of the opening defined as $C_d = 0.4 + 0.0045|T_z - T_\infty|$.

The total ventilation rate then can be calculated by the quadrature sum of the wind and stack air flow components:

$$Q_{total} = \sqrt{Q_w^2 + Q_s^2}$$

Now, calculate the ventilation rate for the below sketched opening as illustrated in Figure 9.E14 (it is assumed that $F_{schedule} = 0.75$).

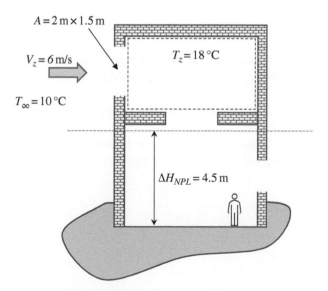

$A = 2\,m \times 1.5\,m$

$V_z = 6\,m/s$

$T_\infty = 10\,°C$

$T_z = 18\,°C$

$\Delta H_{NPL} = 4.5\,m$

Figure 9.E14 A naturally ventilated building case study.

Solution

The opening has an area of $2 \times 1.5 = 3 \text{ m}^2$. Here, the wind is perpendicular, and we assume $C_w = 0.55$. The discharge coefficient for opening can also be found as:

$$C_d = 0.4 + 0.0045 \times |18 - 10| = 0.44$$

Therefore, the airflow rates can be calculated as:

$$Q_w = 0.55 \times 3 \times 0.75 \times 6 = 7.43 \text{ m}^3/\text{s}$$
$$Q_s = 0.436 \times 3 \times 0.75 \times \sqrt{2 \times 9.81 \times 20 \times (18 - 10)/18} = 6.15 \text{ m}^3/\text{s}$$

The total ventilation rate calculated by this model is:

$$Q_{\text{total}} = \sqrt{7.43^2 + 6.15^2} = 9.64 \text{ m}^3/\text{s}$$

As it can be seen from Example 9.14, the calculated mass flow of the upper opening is independent from other mass flows either entering or leaving the zone. In other words, the model can feed the mass flow to the coupled and integrated algorithm without solving another mass flow network. In such cases, as the conservation of mass is not necessarily complied, we need to provide a dummy value to the zone to maintain a mass balance in each time step.

References

1 Dongarra, J.J. et al. (1998). *Numerical Linear Algebra for High Performance Computers*. Society for Industrial and Applied Mathematics.

2 Incropera, F.P. (2006). *Fundamentals of Heat and Mass Transfer*. Wiley.

3 N/A (2013). *ASHRAE Handbook - Fundamentals (SI Edition)*. American Society of Heating, Refrigerating and Air-Conditioning Engineers, Inc. (ASHRAE).

4 Guide, A. (2006). *Environmental Design*. Chartered Institute of Building Services Engineers (CIBSE).

5 Hong, T. et al. (2018). *Occupant behavior models: a critical review of implementation and representation approaches in building performance simulation programs*. Building Simulation 11 (1): 1–14.

6 Balvedi, B.F., Ghisi, E., and Lamberts, R. (2018). *A review of occupant behaviour in residential buildings*. Energy and Buildings 174: 495–505.

7 ASHRAE (2019). *ASHRAE Handbook of Fundamentals*, vol. Chapter 16. Atlanta, GA: American Society of Heating, Refrigerating and Air Conditioning Engineers.

10

Fundamental of Computational Fluid Dynamics – A Finite Volume Approach

10   What Is CFD

As it was stated earlier, analytical, experimental, and numerical techniques are amongst the approaches to study fluid dynamics. As the governing equation of flows, known as Navier–Stokes's equation, was introduced in earlier chapters, analytical solutions appeared to be only feasible in a limited scenarios where the fluid flow was simplified in terms of complexities and was idealized to a common standard sort of problems. Nonetheless, the demand for comprehension of fluid flow, specifically related to a diverse type of industrial, and scientific applications, has obliged stakeholders to employ different experimental studies, benefiting from various types of observational techniques. The extensive cost of the measurements and the required manpower to perform them in addition to the monotonous and time-consuming procedure involved in the experimental studies have easily justified the popularity of numerical techniques in fluid dynamics. Very soon computational fluid dynamics (CFD) was emerged and matured to respond to model complex and multi-scale phenomena such as car industry, aerospace, weather forecast, medical sciences, built environment, etc. The exponential improvement in the ability of high-performance computers, clusters, and super-computers further fuelled the popularity of CFD to be adapted as a cost-effective, time-efficient, and reliable technique.

10.1   Steps in CFD

In general, the governing equations as stated in Chapter 2, represent the conservation of mass, momentum, and energy in a fluid. Thus, the molecular behaviours of the fluid are not discussed while macroscopic behaviours are represented in these set of partial differential equations (PDEs). These PDEs will then be transformed into integral forms (as here finite volume method [FVM] is used) derived in a discrete space represented within finite time steps and control volumes. To perform this procedure, we need to follow three major steps in a CFD simulation, including pre-processing, solution, and post-processing steps. Each step is further explained in the following sections.

Computational Fluid Dynamics and Energy Modelling in Buildings: Fundamentals and Applications,
First Edition. Parham A. Mirzaei.
© 2023 John Wiley & Sons Ltd. Published 2023 by John Wiley & Sons Ltd.

10.1.1 Preprocessing

10.1.1.1 Understanding the Physics of a Problem

As stated in the energy balance modelling in Chapter 9, underlying physics stands again as the main starting point in the CFD modelling. Gravitational forces might not be a significant factor in a pipe modelling while it is crucial in a buoyant model flow in a natural ventilation scenario. Coriolis effects are essential in atmospheric and meso-scale CFD modelling but having minor impacts on airflow in microclimate CFD models. Such examples are numerous while a user starts to develop a model, especially nowadays that commercial packages allow the users to literally tick the related boxes to activate models and parameters in a tool while not necessarily being an expert in the field. Hence, valid assumptions and simplifications should always be justified by a CFD user using the related guidelines, developed literature, expert opinions, etc.

10.1.1.2 Geometry and Domain Creation

The computational domain is the spatial discrete space where the Navier–Stokes's equations are used to calculate the airflow. The size, dimension, and of details of geometrical elements is essential in the meshing process as is described in the next section.

10.1.1.3 Mesh Generation

A mesh is a cluster of non-overlapped control volumes, cumulatively creating a domain in which the Navier–Stokes's equations are discretized. As this will be explained in detail in the following sections, a mesh can have diverse forms and types while its treatment is one of the most challenging and sometimes boresome steps in the pre-processing stage. Since thousands and millions of cells are created in a typical mesh, common algorithms are implemented to proceed automatically with the mesh generation while some are developed to adapt to a finer resolution when quantities' gradients are high. Nonetheless, algorithms encounter limitations on many occasions where a specific type of mesh should be generated, which demands users' experience in controlling the embedded algorithms. Therefore, using available automated mesh generators are not necessarily a solution in many scenarios, especially in the related built environment problems where computational cost and accuracy are unilaterally crucial.

10.1.1.4 Assigning Boundary and Initial Conditions

The Navier–Stokes equations are nonlinear, coupled PDEs that are very sensitive to the boundary and initial conditions. As this will be described in the following sections, different simplifications result in different classification of conservation equations, which respond differently to distinct types of boundary conditions. While most of the tools provide all sort of possible boundaries and the equations can be plausibly converged with different choices, a justified and suitable assignment of boundary conditions are equally vital in the preprocessing step. Understanding the physics, following the associated guidelines, reading the state-of-the-art of the subject area, asking expert opinion, etc. are recommended approaches in the selection of boundary conditions.

10.1.1.5 Definition of Solid and Fluid Materials' Properties

Common material properties are available as charts and tables while most of the time they are embedded in bespoke and commercial tools. Nonetheless, properties' variation due to temperature and pressure changes are sometimes key in the simulations, which should be taken into the account.

10.1.2 Solution

The product of this step is transforming the Navier–Stokes's equations to system of linear algebraic equations. From a mathematical point of view, approximation of the variables and discretization of PDEs can be conducted using different techniques. If the integral form of the equations is discretized, the method is called FVM while discretization of differential form can result into finite difference method (FDM). Other methods such as finite elements method (FEM), which is also based on integral form of the equations, or spectral methods are using different techniques in solving PDEs, which is beyond the scope of this book.

In general, and as described comprehensively in Chapter 11, once the Navier–Stokes's equations are integrated over the generated cells, the obtained system of linear algebraic equations can be solved with direct or iterative methods. Each approach demonstrates advantages and disadvantages that needs to be decided by a user in accordance with many arisen considerations as will be discussed later.

10.1.3 Post-processing

After a system of linear algebraic equations is solved over all cells, post-processing is necessary to translate the crude matrices into understandable information and illustrative results. Examples are velocity vectors in buildings or temperature contours of stratified air in a room. Plots and analysis of quantities in various simulation sensors, including point, lines, planes, and volumes, as will be defined in the next chapter are essential on many occasions for assessment of the results. Commercial tools are widely improved to offer illustrative figure and visual features of flow in a simulation. Nonetheless, users need awareness about the methods that the post-processing is conducted. A common challenging example is about averaging of quantities in cells where mass-weighted or area-weighted can technically result in different answers when flow characteristics and conditions are different. Again, analysing results of a CFD simulation needs attention to multiple factors that can be found in the related literature, guidelines, expert opinions, etc.

10.2 Classification of Conservation Equations

To understand the extent to which a PDE is behaving is crucial when numerical methods are selected to solve a system of equations. In particular, conservation equations are classified in the categories of elliptic, parabolic, and hyperbolic as they can fit to the general form of the below linear second-order PDE with two independent variables:

$$a\frac{\partial^2 u}{\partial x^2} + b\frac{\partial^2 u}{\partial x \partial y} + c\frac{\partial^2 u}{\partial y^2} + d\frac{\partial u}{\partial x} + e\frac{\partial u}{\partial y} + fu = h \tag{10.1}$$

Forming the roots equation of the local characteristic equations as a function of x and y will classify the type of the PDE [1]. Thus, when discriminant of the PDE is resolved as $b^2 - 4ac < 0$, $b^2 - 4ac = 0$, or $b^2 - 4ac > 0$, then the equation would be elliptic, parabolic, or hyperbolic, respectively. The classification of a second-order PDE with N independent variables can be also found in the reference books such as [1, 2], which is beyond the scope of this book.

Example 10.1

Find the classification type of the 2D-steady and 1D-unsteady heat diffusion equations with no heat generation and constant values at boundaries (Dirichlet condition).

$$2D - \text{Steady} : \frac{\partial^2 \phi}{\partial x^2} + \frac{\partial^2 \phi}{\partial y^2} = 0$$

$$1D - \text{Unsteady} : \frac{\partial^2 \phi}{\partial x^2} = \frac{1}{\alpha} \frac{\partial \phi}{\partial t}$$

Solution

Refer to Eq. (10.1), for the steady case, we have $a = c = 1$ and $b = 0$, thus we can derive $b^2 - 4ac = -4 < 0$. This implies that the PDE is elliptic.

For the unsteady case, $a = 1$ and $b = c = 0$, thus $b^2 - 4ac = 0$ and hence the PDE is parabolic.

In general, basic differences between various types of PDEs is related to their characteristics, the so-called ***domain of dependence*** (DoD) and ***domain of influence*** (DoI).

As shown in Figure 10.1, DoD of a specific point $P = (x_0, t_0)$ within a space (here, 2D space of x–t coloured as blue) bounded with specific boundaries (here by lines $x = x_0 - at_0$ and $x = x_0 + at_0$) can be defined as all the points on the bounded region, which are required

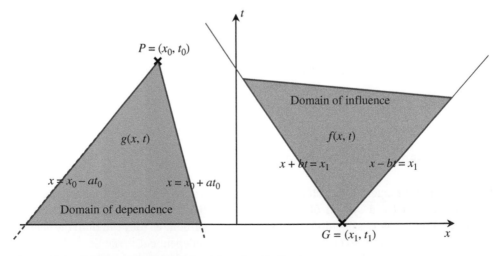

Figure 10.1 Domain of dependence and domain of influence.

to be taken to the account to uniquely resolve the solution at point P. In other words, the solution of $g(x_0, t_0)$ depends on all the values of $g(x, t)$ in the blue bounded region.

DoI of a specific point $G = (x_1, t_1)$ is also defined as the cluster of points at a constrained region (here, 2D space of $x-t$ coloured as green) bounded with specific boundaries (here by lines $x + bt = x_1$ and $x - bt = x_1$) in which their solution is altered when a change in the solution occurs at G.

DoD in elliptic equations is a curve or surface completely enclosing the point P as illustrated in Figure 10.2a. For example, in a 2D-steady heat diffusion equation of Example 10.1, the temperature of an arbitrary point (P_1, P_2, or P_3) in a domain is changing, when the temperature is altered any other point of the domain (here t can be replaced by the second spatial dimension). DoD in parabolic equations is determined by the intersection of the so-called characteristic curves with the boundary lines. These curves also form parts of DoI. At a marching problem, thus, DoD is changing throughout the time as it can be seen in Figure 10.2b shown by DoD1, DoD2, and DoD3. Eventually, the DoD of a hyperbolic marching problem is shown in Figure 10.2c. DoD for a point $P_1 = (x_1, t_1)$ is bounded by boundary conditions and characteristic curves. This implies that the solution at this point is influenced by DoD due to a propagation speed (shown by a in Figure 10.1). On the other hand, any change at point $P_1 = (x_1, t_1)$ will impact on its DoI, bounded again within the characteristic curves. Note that the propagation speed can be assumed to be infinite in elliptic and parabolic equations.

Classification of fluid flow in equilibrium and marching problems can appear due to the desired simplification of Navier–Stokes's equations in accordance with the nature of a problem. These classifications can enable us to relate to general format of the equations and their expected solutions compatible with the introduced behaviour of them.

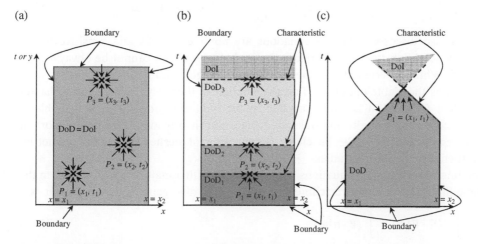

Figure 10.2 Domains of dependence and influence for (a) elliptic, (b) parabolic, and (c) hyperbolic PDEs.

10.3   Difference of Finite Difference and Finite Volume

FDM has been developed earlier than FVM back to years way before emerging of computers. The research around development, implementation, and application of both methods is massive and beyond the scope of this book to be discussed. Here, we only present the main difference of these methods and refer the interested readers to some references such as [3–5].

In general, FDM methods are mainly applied over structured grids though they can be applied on unstructured ones. FVM, nonetheless, are applied over cells, which gives flexibilities in generation of more complex grids, especially on imposing boundary conditions. Interpolation methods should be applied to approximate values on the exchange location of quantities, which are cells' surfaces. Discretization over cells using integral form of Navier–Stokes's equations automatically satisfies conservation laws while this is not necessary the case in FDM where discretization schemes should be cautiously treated to ensure the conservation of quantities. FVM are challenging to be discretized for higher-order schemes while FDM has a straightforward procedure in obtaining such schemes though mainly on structured grids.

10.4   Integral Form of the Conservation Equations

In Chapter 2, the integral form of the Navier–Stokes's equations was derived for the continuum space when the pressure gradient term is not considered. So, the continuity equation is introduced as:

$$\frac{\partial(\rho\phi)}{\partial t} + \nabla \cdot (\rho\phi\boldsymbol{u}) = \nabla \cdot (\Gamma\nabla\phi) + S_\phi \tag{10.2}$$

Also, the momentum equations become:

$$\int_{CV} \frac{\partial(\rho\phi)}{\partial t} dV + \int_{CV} \nabla \cdot \left(\rho\phi\vec{u}\right) dV = \int_{CV} \nabla \cdot (\Gamma\nabla\phi) dV + \int_{CV} S_\phi dV \tag{10.3}$$

It was also discussed that NS equations are nonlinear PDEs and sensitive to the boundary condition type and therefore the analytical solution cannot be always easily obtained. While experimental studies and a wide range of numerical techniques were discussed as the viable options to find the solution of these equations, in this chapter we explain FVM. FVM is widely applied in the built environment related problems and developed tools due to multiple advantages that it provides. The most important reason can be named as a flexible choice of generation of meshes in complex geometries of built environment case studies.

The integral form of NS equations is used in FVM with a major transformation in the volume integral terms to surface integrals using Gauss divergence theorem known as:

$$\int_{CV} (\nabla \cdot \mathbf{F}) dV = \int_A (\mathbf{F} \cdot n) dA \tag{10.4}$$

Applying Eq. (10.4) over Eq. (10.3), we reach to an integral form for advection and diffusion terms to be restated in surface integral forms:

$$\frac{\partial}{\partial t}\left(\int_{CV}\rho\phi dV\right) + \int_A \boldsymbol{n}\cdot\left(\rho\phi\vec{\boldsymbol{u}}\right)dA = \int_A \boldsymbol{n}\cdot(\Gamma\nabla\phi)dA + \int_{CV}S_\phi dV \tag{10.5}$$

We deal with other terms later as we keep them as volume integrals for now. In this equation, the net flux entering a control volume is the summation of flux integrals over the surfaces (two faces for a one-dimensional mesh, four faces for a two-dimensional mesh, and six faces for a three-dimensional mesh). We also have:

$$\int_A f dA = \sum_i \int_{A_i} f dA \tag{10.6}$$

Thus, Eq. (10.5) can be rewritten as:

$$\frac{\partial}{\partial t}\left(\int_{CV}\rho\phi dV\right) + \sum_i\int_{A_i} \boldsymbol{n}\cdot\left(\rho\phi\vec{\boldsymbol{u}}\right)dA = \sum_i\int_{A_i} \boldsymbol{n}\cdot(\Gamma\nabla\phi)dA + \int_{CV}S_\phi dV \tag{10.7}$$

So, with a proper control volumes allocation, we can develop the discrete form of the integrals. The discretization, as it is discussed in the following sections, occurs in control volumes that subdivide the continuous space.

10.5 Grid (Mesh)

Regardless of the technique that will be used to solve the NS equation, the first step is always to generate a mesh or grid to represent the continuous space with discrete elements called as control volumes. To simplify a certain problem, the physics can be reduced on many occasions from three-dimensions (3D) to two-dimensions (2D) and even on some instances to one-dimension (1D). A crossflow in a room, for example, can be assumed to some extend as a 2D problem while a fully developed pipe flow can be even simulated as a 1D problem. Hence, a grid can be 1D, 2D, or 3D in accordance with the nature of the problem. The general characteristics of cells in multiple dimensions are shown in Figure 10.3.

In general, a cell is defined as an imaginary control volume into which domain is discretized. A cell has a centre where material properties can be assigned to it in a cell-centred grid and thus nowhere else is assumed to contain material in such case. Note that a grid can be defined to be vertex-centred when the material properties are assigned to the vertex (grid point) and therefore nowhere else is assumed to contain material. As shown in Figure 10.3, each cell has several faces, varying in accordance with the dimension of a grid. A group of nodes, faces, and cells form a grid or mesh. Meshes or grids are generally classified to structured, unstructured and hybrid types.

In the structured meshes, there is a regular connectivity between a vertex and its neighbouring vertices (Figure 10.4a). All the cells can be arranged in a systematic way and hence be stored in files based on their regular connectivity with the neighbouring cells. This will make the calculation and communication between the modules of a code in a faster and more organized manner. The drawback of structured meshing is related to the preparation time and the selected meshing strategy, which is sometimes very sophisticated and even impractical in some scenarios.

Unstructured meshes provides as illustrated in Figure 10.4b more flexible options to discretize the geometry, resulting in a faster and easier mesh generation. As a major

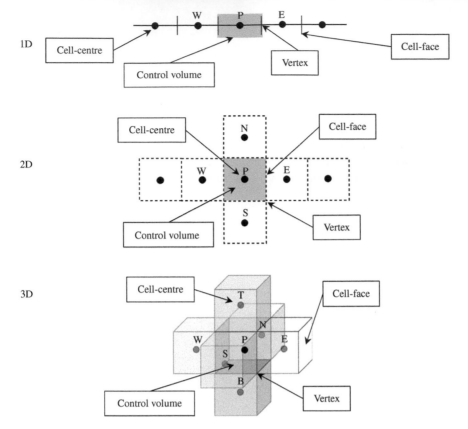

Figure 10.3 1D, 2D and 3D cells.

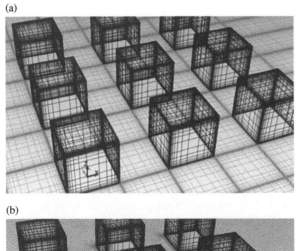

Figure 10.4 (a) Structured, and (b) unstructured meshes.

(a)

(b)

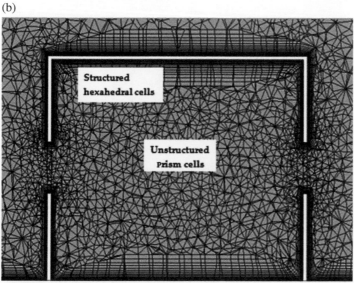

Figure 10.5 Hybrid cells with (a) denser cells around the investigated object. (b) Structured hexahedral cells on boundary layer. *Source:* (a) Adapted from Zhang et al. [6]. (b) Adapted from Shirzadi et al. [7].

disadvantage, these meshes require a substantially larger storage spaces and CPU overhead since the connectivity with the neighbouring cells is irregular and needs to be explicitly defined. Such limitation is potentially exacerbated in 2D and 3D domains.

To benefit from advantage of each method, as shown in Figure 10.5a, hybrid meshes are widely adapted in CFD simulations. The vicinity of objects of interests with complex

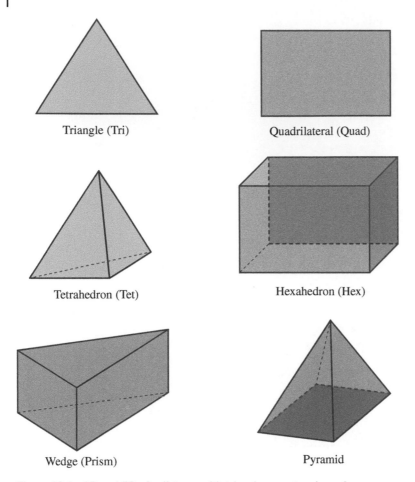

Triangle (Tri)

Quadrilateral (Quad)

Tetrahedron (Tet)

Hexahedron (Hex)

Wedge (Prism)

Pyramid

Figure 10.6 2D and 3D of cell types with triangle or rectangle surfaces.

geometries or locations with higher gradients can be meshed with unstructured cells to provide enough cells to capture the flow characteristics at those locations. On the other hand, the regions far from the complex geometries and high gradients can be filled with structure and larger cells, which are reducing the cell numbers. Inversely, when a boundary layer is required to be fully resolved (see Figure 10.5b), structured cells are preferred where aside from the solid boundaries, unstructured and larger cells can be used.

As it can be observed in Figure 10.6, the most common cell types are kwon as 'tri' standing for triangle and 'quad' standing for quadrilateral in 2D spaces. For 3D meshing, other types of cells such as 'tet', standing for tetrahedral with four faces, pyramid, and wedge with five faces, 'hex', standing for hexahedron with six faces are preferred as demonstrated in the figure. In general, hexahedron meshes are the most preferable ones in built environment simulations as shown in Chapter 12 when a high accuracy solution is required, especially adjacent to walls where a high-resolution mesh is required to resolve the boundary layer effect.

Figure 10.7 Mesh density around a simulation object (here a building). *Source:* Adapted from Aydin and Mirzaei [8]].

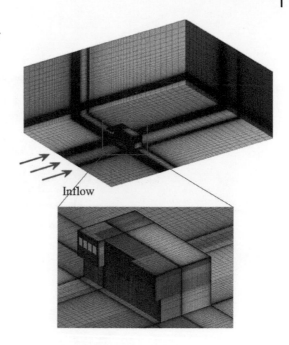

Inflow

10.5.1 Mesh Quality

In an ideal mesh, all the cells should be identical and equilateral with small dimensions, but this is often impossible due to the limitations related to the computational power, embedded meshing algorithm, and geometry's complexities. To reduce the number of cells in a mesh, therefore, in many applications, a higher number of cells are applied to the regions where the fluid flow solutions are more crucial to be found (see Figure 10.7a). Inversely, coarser cells are applied in regions with less favourable solutions. Even with aiming to implement identical cells in a mesh, sometimes the geometry of the problems imposes the employment of non-identical and non-equilateral cells as illustrated in Figure 10.7b.

These barriers can highly impact on the quality of a mesh, implying that generation of a high-quality mesh will increase the accuracy of the CFD results in addition to an elevation in the speed of convergence. These points will be further explained in the following sections. The main point here remains the definition of a high-quality mesh. Many criteria can be defined to characterize a mesh quality, but there are some common measures introduced to identify the quality of a mesh as examples include skewness, smoothness, and aspect ratio.

10.5.1.1 Skewness

Skewness demonstrates how cells in a mesh are deviated from ideal geometries to minimize the impact of cell geometries on the solution when the NS equations are applied on them. As shown in Figure 10.8, an equilateral triangle in a 2D mesh is known as the ideal geometry

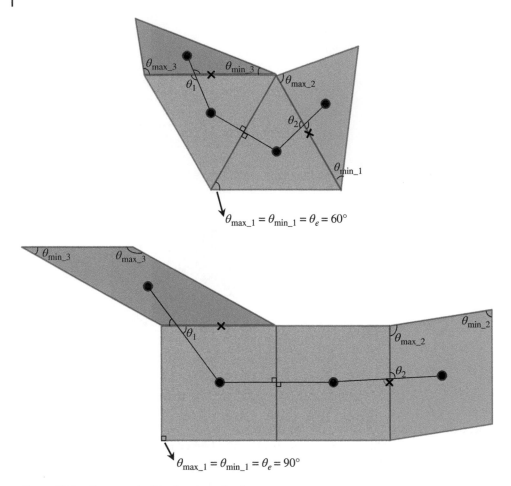

Figure 10.8 Skewness in 2D tri and quad cells.

while a square represent an ideal quad cell. When a cell is skewed, its connection with the neighbouring cells is deviating from being perpendicular, and thus the approximations on the interface will start to not be ideal at the centre of the surface. When the skewness is very high, the associated calculations can cause accuracy and convergence issues. Thus, the below equation can simply measure the deviation of a cell from the ideal forms:

$$\text{Skewness} = \text{Max}\left[\frac{\theta_{\max} - \theta_e}{180 - \theta_e}, \frac{\theta_e - \theta_{\min}}{\theta_e}\right] \tag{10.8}$$

where $\theta_{\max}$ and $\theta_{\min}$, denote the largest and smallest angles of the geometry, and θ_e is the angle for an equiangular face, which is 90° for quad and 60° for tri cells.

Acceptability of the skewness highly depends on the application. In general, Table 10.1 can be a good guideline to control the skewness of the generated cells in a mesh [9].

Table 10.1 Cell quality against various skewness values.

Skewness	0.98–1.00	0.95–0.97	0.80–0.95	0.50–0.80	0.25–0.50	0–0.25
Cell quality	Inacceptable	Low	Poor	Acceptable	Good	High

Figure 10.9 Smoothness of 2D quad cells in a structured grid.

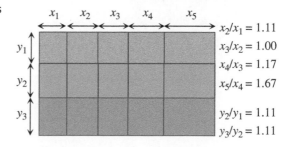

10.5.1.2 Smoothness

Volume and shape of a cell is very essential in flux transfer of a quantity within a mesh. The quantity transfer should occur smoothly in the discrete space similar to a continuum space where it is filled with attached materials. Therefore, it is essential for a generated mesh to be smoothly structured with suitable cells, meaning that sudden alterations in the geometry are not favourable on many occasions. Figure 10.9 illustrates that the length ratio of two subsequent cells is defined as the smoothness ratio. Many guidelines suggest keeping this ratio below 1.2 while with a closer ration to 1, a better result is expected. For example, a rectification of the ratio x_5/x_4 to a lower value can result in a high-quality grid in Figure 10.9.

10.5.1.3 Aspect Ratio

The aspect ratio of a cell is another criterion to evaluate quality of a grid. One of the common definitions of the aspect ratio is the ratio of the longest edge to the shortest one. Examples of cells with the unity aspect ratios are $Cell_2$ and $Cell_4$ in Figure 10.10 for both tri and quad cells, respectively. For $Cell_1$, the largest edge of $Cell_1$ is l_2 and the shortest is l_1. Thus, the aspect ratio can be calculated as $l_2/l_1 = 1.45$. For $Cell_3$, the aspect ratio is $l_2/l_4 = 1.41$. In another definition, this ratio is considered amongst longest and shortest normal vectors amongst all edges for triangle cells. This can be calculated as $s_3/s_2 = 2.55$ for $Cell_1$.

In an ideal situation, the aspect ratio value should be close to one where equilateral cells are generated. Again, to preserve the number of a mesh in a practical range for controlling the computational expenses, it is not always feasible to generate equilateral cells. Nonetheless, it is crucial to measure aspect ratio of cells similar to other criteria in order to ensure an accurate and smooth convergence. In general, decomposition and re-meshing strategies should be considered to improve the quality of a mesh.

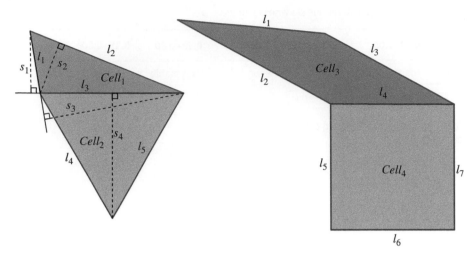

Figure 10.10   Aspect ratio in 2D tri and quad cells.

Example 10.2

Provide different grid types for the below buildings layout and discuss the mesh quality (layouts units are in centimetre) (Figure 10.E2.1).

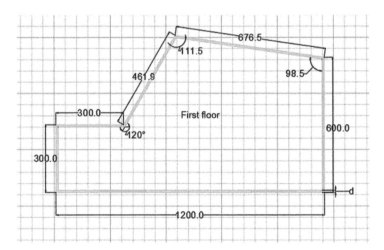

Figure 10.E2.1   Layout of the building

Solution

The below strategies are used to mesh the layout (Figure 10.E2.2):

(a)

(b)

(c)

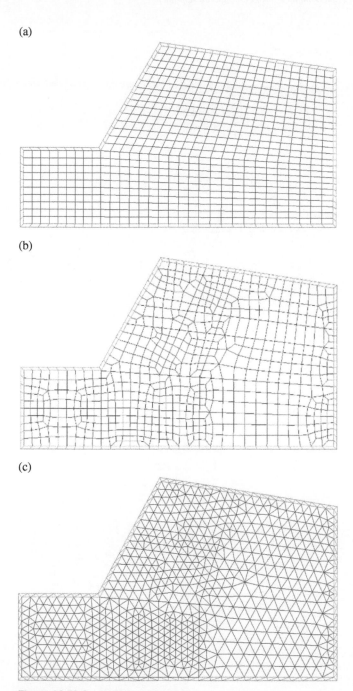

Figure 10.E2.2 Various grid types

(d)

(e)

(f)

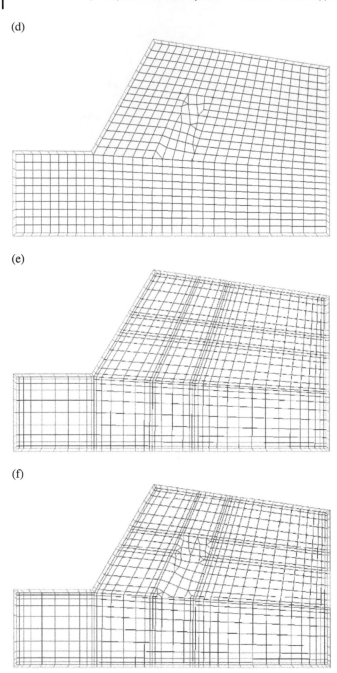

Figure 10.E2.2 (Continued)

The details are provide in the tables:

Mesh description		Cell number	Min. orthogonality	Max skewness	Max AR	Smoothness
				Quality		
No boundary layer	(a) Structured Hex	698	7.071E−01	0.2923	5.435	—
	(b) Unstructured-Hex	723	6.024E−01	0.3976	5.047	—
	(c) Unstructured-Tet	1472	3.508E−01	0.5105	7.473	—
	(d) Unstructured-hybrid	756	6.976E−01	0.2922	4.725	—
With boundary layer	(f) Structured Hex	1106	7.091E−01	0.2909	6.729	1.20
	(e) Unstructured-hybrid	1098	6.808E−01	0.2909	6.730	1.20

10.6 Diffusion Equation

The simplest form of the Navier–Stokes's equations can be represented as the 1D diffusion equation when unsteady and convection accelerations (see Eq. 10.7) are absent and there is no pressure gradient in the equation:

$$\int_{CV} \nabla \cdot (\Gamma \nabla \phi) dV + \int_{CV} S_\phi dV = 0 \tag{10.9}$$

Green Theorem $\int_{CV} (\nabla \cdot \mathbf{F}) dV = \int_A (\mathbf{F} \cdot n) dA$

As it was stated, we apply Green theorem to the divergence part and thus we obtain ($\Gamma \nabla \phi = \mathbf{F}$):

$$\int_{CV} \nabla \cdot (\Gamma \nabla \phi) dV + \int_{CV} S_\phi dV = \int_A (\Gamma \nabla \phi) \cdot n \, dA + \int_{CV} S_\phi dV = 0 \tag{10.10}$$

Now, we apply this equation to a one-dimensional grid as shown in Figure 10.11. The integral form of the $\Gamma \nabla \phi$ over the eastern (e) and western (w) faces of Cell-P can be written as below:

$$\int_A (\Gamma \nabla \phi) \cdot n \, dA + \int_{CV} S_\phi dV = (\Gamma A \nabla \phi)_e - (\Gamma A \nabla \phi)_w + \overline{S} \Delta V = 0 \tag{10.11}$$

Note that here the source/sink term (S_ϕ) assumed with a constant value of $\overline{S}$ at the centre of the Cell-P:

$$\int_{CV} S_\phi dV = \overline{S} \Delta V \tag{10.12}$$

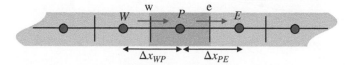

Figure 10.11 One dimensional grid with a central Cell-P and neighbouring western cell of W and eastern cellof E

It can be also defined in any other shapes, such as a linear function, in accordance with the nature of a problem. Now, to continue, we need to find the values of $\Gamma A \nabla \phi$ at w and e faces. Since, the quantities of ϕ, such as velocity, pressure, or temperature, are only defined at the cell centres (e.g. W, P, and E) in the defined 1D discrete space, then we need to interpolate them in w and e faces. Therefore, here we use linear interpolation for Γ, which is a material property of the discrete space; for example, it represents conductivity in diffusion of heat or viscosity in diffusion of momentum:

$$\Gamma_w = \frac{\Gamma_W + \Gamma_P}{2} \tag{10.13}$$

$$\Gamma_e = \frac{\Gamma_P + \Gamma_E}{2} \tag{10.14}$$

The values of Γ are linearly estimated from the available values from the vicinity cell centres. We can also use the central difference method (discussed in the following sections) to define the values of $\nabla\phi$ at w and e faces using available data from the vicinity cell centres:

$$(\nabla\phi)_e = \left(\frac{\phi_E - \phi_P}{\Delta x_{PE}}\right) \tag{10.15}$$

$$(\nabla\phi)_w = \left(\frac{\phi_P - \phi_W}{\Delta x_{WP}}\right) \tag{10.16}$$

Then, we can rewrite the Eq. (10.11) as:

$$(\Gamma A \nabla \phi)_e - (\Gamma A \nabla \phi)_w + \bar{S}\Delta V = 0$$
$$\Longrightarrow \Gamma_e A_e \left(\frac{\phi_E - \phi_P}{\Delta x_{PE}}\right) - \Gamma_w A_w \left(\frac{\phi_P - \phi_W}{\Delta x_{WP}}\right) + \bar{S}\Delta V = 0 \tag{10.17}$$

To simplify Eq. (10.17), we can further assume that $\bar{S}\Delta V$ is changing with the below linear equation as depicted in Figure 10.12:

$$\bar{S}\Delta V = S_u + S_p \phi_P \tag{10.18}$$

Here, S_u and S_p are constant values. It should be noted that higher order approximations can be also fitted to the source/sink term, if results with higher levels of accuracy are deemed. Now, with the combination of Eqs. (10.17) and (10.18), and after rearrangement of the expression at a cell centre:

$$\Gamma_e A_e \left(\frac{\phi_E - \phi_P}{\Delta x_{PE}}\right) - \Gamma_w A_w \left(\frac{\phi_P - \phi_W}{\Delta x_{WP}}\right) + S_u + S_p \phi_P = 0$$
$$\Longrightarrow \left(\frac{\Gamma_w}{\Delta x_{WP}}A_w + \frac{\Gamma_e}{\Delta x_{PE}}A_e - S_p\right)\phi_P = \left(\frac{\Gamma_w}{\Delta x_{WP}}A_w\right)\phi_W + \left(\frac{\Gamma_e}{\Delta x_{PE}}A_e\right)\phi_E + S_u \tag{10.19}$$

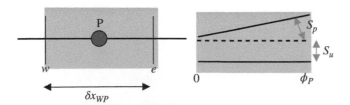

Figure 10.12 Linear variation of source/sink in a control volume.

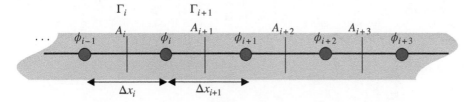

Figure 10.13 Internal cells for a one-dimensional grid.

It should be mentioned that the terms before ϕ_P, ϕ_W, and ϕ_E are known values in the above equation and thus we have:

$$a_P\phi_P = a_W\phi_W + a_E\phi_E + S_u \tag{10.20}$$

where $a_W = \dfrac{\Gamma_w}{\Delta x_{WP}}A_w$, $a_E = \dfrac{\Gamma_e}{\Delta x_{PE}}A_e$, and $a_P = a_W + a_E - S_P$. For simplifying the equation, we denote $\dfrac{\Gamma_w}{\Delta x_{WP}} = D_w$ and $\dfrac{\Gamma_e}{\Delta x_{PE}} = D_e$.

The above equation can be derived for any internal cell of a 1D grid, implying that the number of internal equations will be $n-2$ where n is the number of cells and $1 < i < n$. As shown in Figure 10.13, Eq. (10.20) can be applied over all cells as:

$$\int_A f dA = \sum_i \int_{A_i} f dA \tag{10.21}$$

Thus:

$$\int_A (\Gamma\nabla\phi)\cdot n\, dA + \int_{CV} S_\phi dV = \sum_i \int_{A_i} (\Gamma\nabla\phi)\cdot n\, dA + \int_{CV} S_\phi dV = 0$$

$$\Longrightarrow \sum_i \Gamma_{i-1} A_{i-1}\left(\frac{\phi_i - \phi_{i+1}}{\Delta x_i}\right) - \Gamma_i A_i\left(\frac{\phi_i - \phi_{i+1}}{\Delta x_{i+1}}\right) + S_u + S_i\phi_i = 0 \tag{10.22}$$

It should be noted that cells $i = 1$ and $i = n$ are boundary cells and should be treated separately. Hence, with the treatment of cells $i = 1$ and $i = n$, two more equations will be obtained. This results in having total equations of n while the total unknown (ϕ_i) are n, which will lead us to form a system of linear equations. The system of linear equation will form a matrix of $n \times n$ for the quantity ϕ, which then can be solved with various direct or iterative methods depending on the size of the matrix as this will be discussed in detail in Chapter 11.

10.7 Boundary Treatment

When the amount of the quantity ϕ is known at a boundary (ϕ_0), the condition will be called *Dirichlet boundary condition*. As it can be seen in Figure 10.14, a boundary node-1 can be treated as below:

$$\Gamma_e A_e\left(\frac{\phi_2 - \phi_1}{\Delta x_1}\right) - \Gamma_w A_w\left(\frac{\phi_1 - \phi_0}{\dfrac{\Delta x_1}{2}}\right) + S_u + S_p\phi_1 = 0 \tag{10.23}$$

$$\Longrightarrow \left(2\frac{\Gamma_w}{\Delta x_0}A_w + \frac{\Gamma_e}{\Delta x_1}A_e - S_p\right)\phi_1 = \left(2\frac{\Gamma_w}{\Delta x_0}A_w\right)\phi_0 + \left(\frac{\Gamma_e}{\Delta x_1}A_e\right)\phi_2 + S_u$$

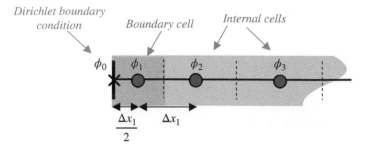

Figure 10.14 Dirichlet boundary condition at the left-side cell of a 1D grid.

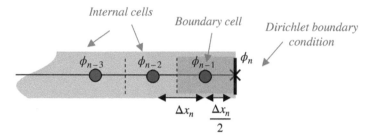

Figure 10.15 Dirichlet boundary condition at the right-side cell of a 1D grid.

Thus, the equation can be simplified to:

$$\left(2D_{\mathrm{w}}A_{\mathrm{w}} + D_{\mathrm{e}}A_{\mathrm{e}} - S_p\right)\phi_1 = D_{\mathrm{e}}A_{\mathrm{e}}\phi_2 + \left[\left(2D_{\mathrm{w}}A_{\mathrm{w}}\phi_0\right) + S_u\right] \tag{10.24}$$

Note if there is not a source/sink in the diffusion equation, then S_p and S_u will be zero. A similar approach can be followed to obtain the boundary equation for the right-side boundary cell as illustrated in Figure 10.15:

$$\Gamma_{\mathrm{e}}A_{\mathrm{e}}\left(\frac{\phi_n - \phi_{n-1}}{\frac{\Delta x_n}{2}}\right) - \Gamma_{\mathrm{w}}A_{\mathrm{w}}\left(\frac{\phi_{n-1} - \phi_{n-2}}{\Delta x_n}\right) + S_u + S_p\phi_{n-1} = 0$$

$$\Longrightarrow \left(D_{\mathrm{w}}A_{\mathrm{w}} + 2D_{\mathrm{e}}A_{\mathrm{e}} - S_p\right)\phi_{n-1} = D_{\mathrm{w}}A_{\mathrm{w}}\phi_{n-2} + \left[\left(2D_{\mathrm{e}}A_{\mathrm{e}}\phi_n\right) + S_u\right] \tag{10.25}$$

Now, we can form the coefficient table of all equations of the diffusion equation as illustrated in Table 10.2.

Table 10.2 Coefficient of diffusion equation for a 1D grid.

Cell	a_W	a_E	S_p	S_u	$a_P = a_E + a_W - S_p$
1	0	$D_{\mathrm{e}}A_{\mathrm{e}}$	$-2D_{\mathrm{w}}A_{\mathrm{w}} + S_p$	$2D_{\mathrm{w}}A_{\mathrm{w}}\phi_0 + S_u$	$D_{\mathrm{e}}A_{\mathrm{e}} + 2D_{\mathrm{w}}A_{\mathrm{w}} - S_p$
2, 3, $n-2$	$D_{\mathrm{w}}A_{\mathrm{w}}$	$D_{\mathrm{e}}A_{\mathrm{e}}$	S_p	S_u	$D_{\mathrm{e}}A_{\mathrm{e}} + D_{\mathrm{w}}A_{\mathrm{w}} - S_p$
$n-1$	$D_{\mathrm{w}}A_{\mathrm{w}}$	0	$-2D_{\mathrm{e}}A_{\mathrm{e}} + S_p$	$2D_{\mathrm{e}}A_{\mathrm{e}}\phi_n + S_u$	$D_{\mathrm{w}}A_{\mathrm{w}} + 2D_{\mathrm{e}}A_{\mathrm{e}} - S_p$

Example 10.3

A building wall is assumed with one 20 cm brick layer as shown in Figure 10.E3. The materials' properties are assumed to be homogenous as given in the table. The heat conduction through wall is also assumed to follow a one-dimensional and steady-state behaviour. The interior and exterior wall surface temperatures are 20 and 35 $^\circ$C, respectively. Assume a 1D mesh with equal cell intervals of 5 cm. Discretize the diffusion equation and find the temperature on each cell using the FVM. Compare your 1D numerical and analytical results, and calculate the average and local errors.

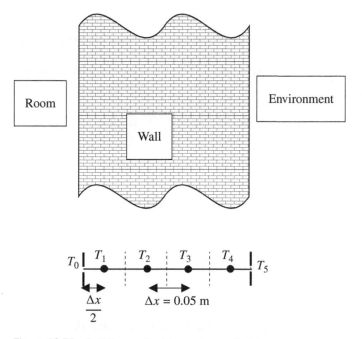

Figure 10.E3 Building wall with one layer of brick.

Material	Brick
Thermal conductivity (W/m K)	0.8

Solution

Due to homogeneity, and as intervals are equal to $\Delta x = 0.05$ cm, we can derive the below equations for internal cells of T_2 to T_3 using Eq. (10.22):

$$\left(\frac{k}{\Delta x} + \frac{k}{\Delta x}\right) T_2 = \frac{k}{\Delta x} T_1 + \frac{k}{\Delta x} T_3$$

$$\left(\frac{k}{\Delta x} + \frac{k}{\Delta x}\right) T_3 = \frac{k}{\Delta x} T_2 + \frac{k}{\Delta x} T_4$$

For boundary cells:

$$\left(\frac{k}{\Delta x} + \frac{k}{\frac{\Delta x}{2}}\right) T_1 = \frac{k}{\frac{\Delta x}{2}} T_0 + \frac{k}{\Delta x} T_2$$

$$\left(\frac{k}{\Delta x} + \frac{k}{\frac{\Delta x}{2}}\right) T_4 = \frac{k}{\Delta x} T_3 + \frac{k}{\frac{\Delta x}{2}} T_5$$

Or, assuming $a_W = D_w A_w$, $a_E = D_e A_e$, and $D_w = D_e = \dfrac{k}{\Delta x}$ for internal cells, $\dfrac{D_w}{2} = D_e = \dfrac{k}{\Delta x}$ for boundary Cell-0 and $D_w = \dfrac{D_e}{2} = \dfrac{k}{\Delta x}$ for Cell-4, we can form a similar result as the coefficient table:

Cell	a_W	a_E	S_P	S_u	a_P
1	0	$\dfrac{kA}{\Delta x}$	$-\dfrac{2kA}{\Delta x}$	$\dfrac{2kAT_0}{\Delta x}$	$\dfrac{3kA}{\Delta x}$
2	$\dfrac{kA}{\Delta x}$	$\dfrac{kA}{\Delta x}$	0	0	$\dfrac{2kA}{\Delta x}$
3	$\dfrac{kA}{\Delta x}$	$\dfrac{kA}{\Delta x}$	0	0	$\dfrac{2kA}{\Delta x}$
4	$\dfrac{kA}{\Delta x}$	0	$-\dfrac{2kA}{\Delta x}$	$-\dfrac{2kAT_5}{\Delta x}$	$\dfrac{3kA}{\Delta x}$

We can now rearrange the coefficients in a matrix form as we can divide all the terms by $\dfrac{kA}{\Delta x}$:

$$\begin{bmatrix} 3 & -1 & 0 & 0 \\ -1 & 2 & -1 & 0 \\ 0 & -1 & 2 & -1 \\ 0 & 0 & -1 & 3 \end{bmatrix} \begin{bmatrix} T_1 \\ T_2 \\ T_3 \\ T_4 \end{bmatrix} = \begin{bmatrix} 2T_0 \\ 0 \\ 0 \\ 2T_5 \end{bmatrix} = \begin{bmatrix} 40 \\ 0 \\ 0 \\ 70 \end{bmatrix}$$

Hence, we can use the inverse method to find the solution while it becomes same as the exact solution because it is independent of cell size (Δx) for the steady diffusion equation. The exact solution is:

$$T(x) = \frac{x}{0.2}(T_2 - T_1) + T_1$$

And, the results are as follows:

Cell	Temperature (°C)
0	20
1	21.875
2	25.625
3	29.375
4	33.125
5	35

10.7.1 Neumann Boundary Type I

Another form of boundary treatment is **Neumann boundary condition** where gradient of a quantity ϕ ($\nabla\phi$) is given on the boundary (type I). The gradient can be directly known as q'' as shown in Figure 10.16.

For the Cell-1, the diffusion equation can be written as:

$$\Gamma_e A_e \left(\frac{\phi_2 - \phi_1}{\Delta x_1}\right) - \Gamma_w A_w \left(\frac{\phi_1 - \phi_0}{\frac{\Delta x_1}{2}}\right) + S_p \phi_1 + S_u = 0$$

$$\implies \left(\frac{2\Gamma_w}{\Delta x_1} A_w + \frac{\Gamma_e}{\Delta x_1} A_e - S_p\right)\phi_1 = \left(\frac{2\Gamma_w}{\Delta x_1} A_w\right)\phi_0 + \left(\frac{\Gamma_e}{\Delta x_1} A_e\right)\phi_2 + S_u \tag{10.26}$$

Cell-face-0 can be found as:

$$-\Gamma_e A_e \left(\frac{\phi_1 - \phi_0}{\frac{\Delta x_1}{2}}\right) = q_0'' A_w$$

$$\implies \frac{2\Gamma_e A_e}{\Delta x_1} \phi_0 = \frac{2\Gamma_e A_e}{\Delta x_1} \phi_1 + q_0'' A_w \tag{10.27}$$

For Cell-n-1, we can find (Figure 10.17):

$$\Gamma_e A_e \left(\frac{\phi_n - \phi_{n-1}}{\frac{\Delta x_1}{2}}\right) - \Gamma_w A_w \left(\frac{\phi_{n-1} - \phi_{n-2}}{\Delta x_n}\right) + S_p \phi_{n-1} + S_u = 0$$

$$\implies \left(\frac{\Gamma_w}{\Delta x_n} A_w + \frac{2\Gamma_e}{\Delta x_n} A_e - S_p\right)\phi_{n-1} = \left(\frac{\Gamma_w}{\Delta x_n} A_w\right)\phi_{n-1} + \left(\frac{2\Gamma_e}{\Delta x_n} A_e\right)\phi_n + S_u \tag{10.28}$$

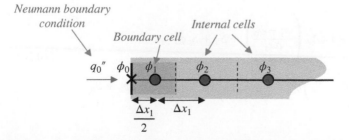

Figure 10.16 Neumann constant flux at the left-side cell of a 1D grid.

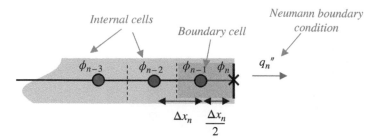

Figure 10.17 Neumann constant flux at the right-side cell of a 1D grid.

Table 10.3 Boundary cell and cell-face coefficients (type I) of diffusion equation for a 1D grid.

Cell	a_W	a_E	S_P	S_u	a_P
0	0	$2D_eA_e$	0	$q_0^{''}A_w$	$2D_eA_e$
1	$2D_wA_w$	D_eA_e	S_p	S_u	$2D_wA_w + D_eA_e - S_p$
n-1	D_wA_w	$2D_eA_e$	S_p	S_u	$D_wA_w + 2D_eA_e - S_p$
n	$2D_wA_w$	0	0	$-q_n^{''}A_e$	$2D_wA_w$

And for Cell-face-n (Figure 10.17):

$$-\Gamma_wA_w\left(\frac{\phi_n - \phi_{n-1}}{\frac{\Delta x_n}{2}}\right) = q_n^{''}A_e$$

$$\Longrightarrow \frac{2\Gamma_wA_w}{\Delta x_n}\phi_n = \frac{2\Gamma_wA_w}{\Delta x_n}\phi_{n-1} - q_n^{''}A_e$$

(10.29)

Therefore, when the heat flux is given, we can form the coefficient table for the boundary cells as shown in Table 10.3.

10.7.2 Neumann Boundary Type II

Another type of Neumann boundary condition (type II) is when the gradient is implemented to be calculated from Newton cooling law as demonstrated in Figure 10.18. This boundary type requires convective heat transfer coefficient (h) and environmental temperature. For the boundary Cell-1, we can derive:

$$\Gamma_eA_e\left(\frac{\phi_2 - \phi_1}{\Delta x_1}\right) - \Gamma_wA_w\left(\frac{\phi_1 - \phi_0}{\frac{\Delta x_1}{2}}\right) + S_u + S_p\phi_1 = 0$$

$$\Longrightarrow \left(\frac{2\Gamma_w}{\Delta x_1}A_w + \frac{\Gamma_e}{\Delta x_1}A_e - S_p\right)\phi_1 = \left(\frac{2\Gamma_w}{\Delta x_1}A_w\right)\phi_0 + \left(\frac{\Gamma_e}{\Delta x_1}A_e\right)\phi_2 + S_u$$

(10.30)

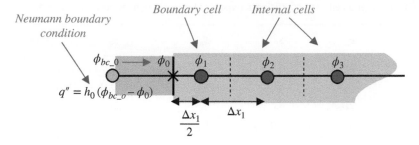

Figure 10.18 Neumann convective flux at the left-side cell of a 1D grid.

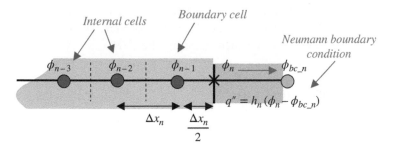

Figure 10.19 Neumann convective flux at the right-side cell of a 1D grid.

And, we can derive the equation on Cell-face-0 as illustrated in Figure 10.18:

$$
h_0 A_w \left(\phi_{bc\_0} - \phi_0 \right) = -\Gamma_e A_e \left(\frac{\phi_1 - \phi_0}{\frac{\Delta x_1}{2}} \right)
$$

$$
\Longrightarrow \left(\frac{2\Gamma_e}{\Delta x_1} A_e + h_0 A_w \right) \phi_0 = \frac{2\Gamma_e}{\Delta x_1} A_e \phi_1 + h_0 A_w \phi_{bc\_0}
$$

(10.31)

For the boundary Cell-n (Figure 10.19):

$$
\Gamma_e A_e \left(\frac{\phi_n - \phi_{n-1}}{\frac{\Delta x_n}{2}} \right) - \Gamma_w A_w \left(\frac{\phi_{n-1} - \phi_{n-2}}{\Delta x_n} \right) + S_u + S_p \phi_{n-1} = 0
$$

$$
\Longrightarrow \left(\frac{\Gamma_w}{\Delta x_n} A_w + \frac{2\Gamma_e}{\Delta x_n} A_e - S_p \right) \phi_{n-1} = \left(\frac{\Gamma_w}{\Delta x_n} A_w \right) \phi_{n-2} + \frac{2\Gamma_e}{\Delta x_n} A_e \phi_n + S_u
$$

(10.32)

And, for the right Cell-face-n:

$$
h_n A_e \left(\phi_n - \phi_{bc\_n} \right) = -\Gamma_w A_w \left(\frac{\phi_n - \phi_{n-1}}{\frac{\Delta x_n}{2}} \right)
$$

$$
\Longrightarrow \left(\frac{2\Gamma_w}{\Delta x_n} A_w + h_n A_e \right) \phi_n = \frac{2\Gamma_w}{\Delta x_n} A_w \phi_{n-1} + h_n A_e \phi_{bc\_n}
$$

(10.33)

In conclusion, when environmental conditions are known, we can rearrange the coefficients in Table 10.4.

Table 10.4 Boundary cell and cell-face coefficients (type II) of diffusion equation for a 1D grid.

Cell	a_W	a_E	S_p	S_u	a_P
0	0	$2D_e A_e$	$-h_0 A_w$	$h_0 A_w \phi_{bc\_0}$	$2D_e A_e + h_0 A_w$
1	$2D_w A_w$	$D_e A_e$	S_p	S_u	$D_e A_e + 2D_w A_w - S_p$
$n-1$	$D_w A_w$	$2D_e A_e$	S_p	S_u	$2D_e A_e + D_w A_w - S_p$
n	$2D_w A_w$	0	$-h_n A_e$	$h_n A_e \phi_{bc\_n}$	$2D_w A_w + h_n A_e$

Example 10.4

For the building wall of Example 10.3, the room temperature and h-value are and 25 °C and 5 W/m^2 K, respectively. These numbers are 35°C and 10 W/m^2 K for the environment around the building. Assume a 1D mesh with equal cell intervals of 5 cm. Discretize the diffusion equation and find the temperature on each cell using the FVM.

Solution

We use Neumann boundary condition type II, and thus for the Cell-face-0 and boundary Cell-1:

$$\left(h_0 + \frac{2k}{\Delta x} \right) T_0 = \frac{2k}{\Delta x} T_1 + h_0 T_{bc\_0}$$

$$\frac{3k}{\Delta x} T_1 = \frac{k}{\Delta x} T_2 + \frac{2k}{\Delta x} T_0$$

With a similar approach, for Cell-face-5 and boundary Cell-4:

$$\frac{3k}{\Delta x} T_4 = \frac{2k}{\Delta x} T_5 + \frac{k}{\Delta x} T_3$$

$$\left(h_n + \frac{2k}{\Delta x} \right) T_5 = \frac{2k}{\Delta x} T_4 + h_n T_{bc\_n}$$

This implies that:

Cell	a_W	a_E	S_p	S_u	a_P
0	0	$\dfrac{2kA}{\Delta x}$	$-h_0 A$	$h_{0A} T_{bc\_0}$	$\dfrac{2kA}{\Delta x} + h_0 A$
1	$\dfrac{2kA}{\Delta x}$	$\dfrac{kA}{\Delta x}$	0	0	$\dfrac{3kA}{\Delta x}$
2	$\dfrac{kA}{\Delta x}$	$\dfrac{kA}{\Delta x}$	0	0	$\dfrac{2kA}{\Delta x}$
3	$\dfrac{kA}{\Delta x}$	$\dfrac{kA}{\Delta x}$	0	0	$\dfrac{2kA}{\Delta x}$
4	$\dfrac{kA}{\Delta x}$	$\dfrac{2kA}{\Delta x}$	0	0	$\dfrac{3kA}{\Delta x}$
5	$\dfrac{2kA}{\Delta x}$	0	$-h_i A$	$h_i A T_{bc\_in}$	$\dfrac{2kA}{\Delta x} + h_i A$

And, the matrix of system of linear equations become:

$$
\begin{bmatrix}
58 & -48 & 0 & 0 & 0 & 0 \\
-48 & 72 & -24 & 0 & 0 & 0 \\
0 & -16 & 32 & -16 & 0 & 0 \\
0 & 0 & -16 & 32 & -16 & 0 \\
0 & 0 & 0 & -24 & 72 & -48 \\
0 & 0 & 0 & 0 & -48 & 53
\end{bmatrix}
\begin{bmatrix}
T_0 \\ T_1 \\ T_2 \\ T_3 \\ T_4 \\ T_5
\end{bmatrix}
=
\begin{bmatrix}
350 \\ 0 \\ 0 \\ 0 \\ 0 \\ 125
\end{bmatrix}
$$

So, the results can be found as:

Cell/cell-face	Temperature ($^\circ$C)
0	32.86
1	32.41
2	31.52
3	30.63
4	29.73
5	29.29

10.8 Expansion to Higher Dimensions

We follow the same approach to expand the discretization to a 2D space. For this purpose, consider a two-dimensional structured grid with quadrilateral cells (quad) as shown in Figure 10.20. Again, we must apply the simplified transport equation over the 2D spaces:

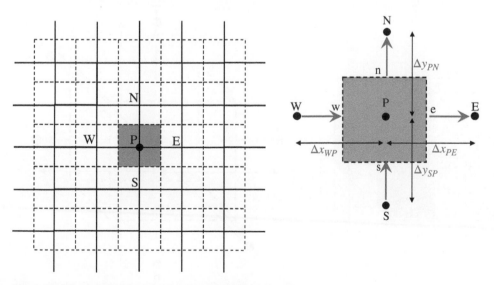

Figure 10.20 A 2D grid with structured quadrilateral cells.

$$\int_{CV} \nabla \cdot (\Gamma \nabla \phi) dV + \int_{CV} S_\phi dV = 0 \tag{10.34}$$

As it can be seen in Figure 10.20, now we can derive the fluxes that enter or leave the 2D control volume of P from the eastern (E), western (W), northern (N), and southern (S) neighbour cells as below:

$$\int_A (\Gamma \nabla \phi) \cdot \boldsymbol{n}\, dA + \int_{CV} S_\phi dV = 0$$

$$\Longrightarrow [(\Gamma A \nabla \phi)_e - (\Gamma A \nabla \phi)_w] + [(\Gamma A \nabla \phi)_n - (\Gamma A \nabla \phi)_s] + \overline{S} \Delta V = 0$$

$$\Longrightarrow \Gamma_e A_e \left(\frac{\phi_E - \phi_P}{\Delta x_{PE}} \right) - \Gamma_w A_w \left(\frac{\phi_P - \phi_W}{\Delta x_{WP}} \right) + \Gamma_n A_n \left(\frac{\phi_N - \phi_P}{\Delta y_{PN}} \right) - \Gamma_s A_s \left(\frac{\phi_P - \phi_S}{\Delta y_{PS}} \right) + S_u + S_P \phi_P = 0$$

$$\tag{10.35}$$

Note again that $\overline{S} \Delta V$ term is estimated by a linear change ($S_u + S_P \phi_P$) in the control volume. By rearranging Eq. (10.35), we can reach to:

$$\left(\frac{\Gamma_w A_w}{\Delta x_{WP}} + \frac{\Gamma_e A_e}{\Delta x_{PE}} + \frac{\Gamma_s A_s}{\Delta y_{PS}} + \frac{\Gamma_n A_n}{\Delta y_{PN}} - S_P \right) \phi_P = \left(\frac{\Gamma_w A_w}{\Delta x_{WP}} \right) \phi_W$$

$$+ \left(\frac{\Gamma_e A_e}{\Delta x_{PE}} \right) \phi_E + \left(\frac{\Gamma_s A_s}{\Delta y_{PS}} \right) \phi_S + \left(\frac{\Gamma_n A_n}{\Delta y_{PN}} \right) \phi_N + S_u = 0 \tag{10.36}$$

Similar to the Eq. (10.20), we can then rewrite the above equation as follows:

$$a_P \phi_P = a_W \phi_W + a_E \phi_E + a_S \phi_S + a_N \phi_N + S_u \tag{10.37}$$

We can similarly apply the above methodology into a three-dimensional domain using a hexahedral (Hex) structured mesh. As it can be seen, there are two more cells at top (T) and bottom (B) of the central Cell-P included in the balance of fluxes from the neighbouring cells to the control volume as it can be seen in Figure 10.21.

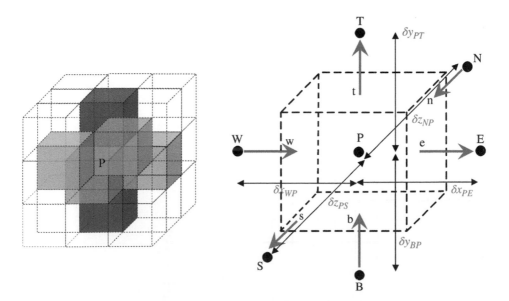

Figure 10.21 A 3D grid with structured hexahedral cells.

With a similar approach as 1D and 2D, we can obtain the below expression for a 3D structured Hex mesh:

$$a_P \phi_P = a_W \phi_W + a_E \phi_E + a_S \phi_S + a_N \phi_N + a_B \phi_B + a_T \phi_T + S_u \tag{10.38}$$

The labels e, w, n, s, t, and b stand respectively for east, west, north, south, top, and bottom surfaces of the control volume (P). Thus, in a general term, the diffusion equation can be written as:

$$a_P \phi_P = \sum a_{nb} \phi_{nb} + S_u \tag{10.39}$$

where, a_{nb} represent the neighbouring cells.

Note that a similar procedure can be followed for unstructured grids as well as non-standard cells (other polyhedrons).

Example 10.5

A (heat sink) Microchip needs intensive cooling in a very packed area to avoid serious damage because of an increase in the temperature. As depicted in Figure 10.E5.1, the thermal conduction module (TCM) technique is utilized for cooling of multichip modules. The TCM consists of a multilayer ceramic substrate where the chips are soldered in an aluminium piston that is in direct contact with the chip, and a spring that connects the aluminium piston with a cold plate that is cooled down by passing water through small rectangular channels. The heat conduction takes place from the chip to the aluminium piston as the spring presses the piston, promoting direct contact. Then, the heat goes thorough the piston until it reaches the spring that is connected to the cold plate. The main interest is to obtain the steady-state temperature distribution within the cold plate, which has symmetry and can be assumed as the below 2D scenario.

Also, the water condition and the related material properties can be defined as below:

Aluminium thermal conductivity: $k = 230$ W/mK

Heat flux: $q_s'' = 10^4$ W/m^2

Convection heat transfer coefficient: $h = 300$ W/m^{2}K

Temperature of the water: $T_\infty = 20°$C

Generate a structured 2D mesh for the solid cold plate (the water and air will not be modelled) assuming $\Delta x = 2.5$ mm and $\Delta y = 4$ mm (aspect ratio $\frac{\Delta y}{\Delta x} = 1.6$). Derive the steady heat conduction equation for all cells using central difference method. Use direct method and find the temperature distribution. The processor should not run with a temperature higher than 75°C.

Solution

Figure 10.E5.2 shows only one unit of the cold plate channel as we can assume that there are a series of similar plates at both sides of this unit where a symmetrical boundary is applied. The plate is also insulated on the top.

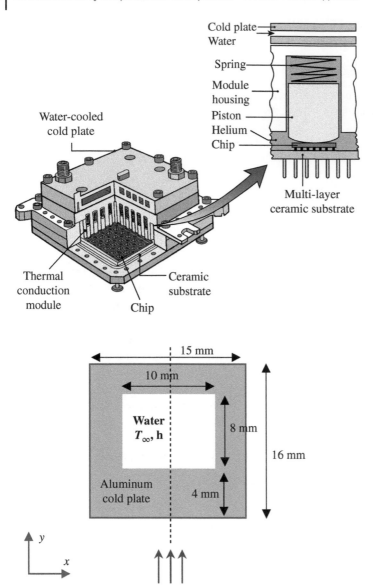

Figure 10.E5.1 Schematic of a microchip with an intensive cooling system. Left (*Source:* [10]– figure 5).

For Neumann boundary type I, the equation and table of coefficient can be found as below:

$$q_s'' A_s = -kA_n \left(\frac{T_{15} - T_8}{\frac{\Delta y}{2}} \right) \Longrightarrow 2D_n A_n T_{15} = 2D_n A_n T_8 + q_s'' A_s$$

Figure 10.E5.2 Simplified mesh for the microchip.

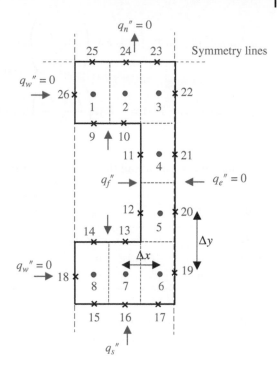

However, since we have $q_w'' = q_e'' = q_n'' = 0$, we can skip deriving equations for cell-faces 18, 19, 20, 21, 22, 23, 24, 25, and 26. For example, we can derive the equations for cell-face-26:

$$q_w A_w = -kA_e \left(\frac{T_1 - T_{26}}{\frac{\Delta x}{2}} \right) \Longrightarrow T_{26} = T_1$$

And, for Neumann boundary type II applied over cell-face-9:

$$q_f'' = hA_s(T_\infty - T_9) = -kA_n \left(\frac{T_1 - T_9}{\frac{\Delta y}{2}} \right) \Longrightarrow (2D_n A_n + hA_s)T_9 = 2D_n A_n T_1 + hA_s T_\infty$$

Cell	a_W	a_E	a_N	a_S	S_P	S_u	a_P
18, 26	0	$2D_e A_e$	0	0	0	$q_w'' A_w$	$2D_e A_e$
23, 24, 25	0	0	0	$2D_s A_s$	0	$-q_n'' A_n$	$2D_s A_s$
19, 20, 21, 22	$2D_w A_w$	0	0	0	0	$-q_e'' A_e$	$2D_w A_w$
15, 16, 17	0	0	$2D_n A_n$	0	0	$q_s'' A_s$	$2D_n A_n$

$$q_f'' = hA_w(T_\infty - T_{11}) = -kA_e\left(\frac{T_4 - T_{11}}{\frac{\Delta x}{2}}\right) \implies (2D_eA_e + hA_w)T_{11} = 2D_eA_eT_4 + hA_wT_\infty$$

$$q_f'' = hA_n(T_\infty - T_{14}) = -kA_s\left(\frac{T_8 - T_{14}}{\frac{\Delta y}{2}}\right) \implies (2D_sA_s + hA_n)T_{14} = 2D_sA_sT_8 + hA_nT_\infty$$

Again, we can replace all the coefficients in the table:

Cell	a_W	a_E	a_N	a_S	S_P	S_u	a_P
9, 10	0	0	$2D_nA_n$	0	$-hA_s$	hA_sT_∞	$2D_nA_n + hA_s$
11, 12	0	$2D_eA_e$	0	0	$-hA_w$	hA_wT_∞	$2D_eA_e + hA_w$
13, 14	0	0	0	$2D_sA_s$	$-hA_n$	hA_nT_∞	$2D_sA_s + hA_n$

All cells are boundary types and there is no source and sink term. So, for boundary Cell-1 where it is exposed to three boundary nodes, we can derive:

$$k\Delta y\left(\frac{T_2 - T_1}{\Delta x}\right) - k\Delta x\left(\frac{T_1 - T_9}{\frac{\Delta y}{2}}\right) = 0$$

$$\implies (D_eA_e + 2D_sA_s)T_1 = D_eA_eT_2 + 2D_sA_sT_9$$

where $A_w = A_e = \Delta y$ and $A_n = A_s = \Delta x$. Also, $D_w = D_e = \dfrac{k}{\Delta x}$ and $D_n = D_s = \dfrac{k}{\Delta y}$.

The same approach will be applied to Cell-8, which is surrounded by three boundary nodes and in the same condition. Similarly, we can find the below coefficients for other boundary cells:

Cell	a_W	a_E	a_N	a_S	S_P	S_u	a_P
1	0	D_eA_e	0	$2D_sA_s$	0	0	$D_eA_e + 2D_sA_s$
2,7	D_wA_w	D_eA_e	0	$2D_sA_s$	0	0	$D_wA_w + D_eA_e + 2D_sA_s$
3	D_wA_w	0	0	D_sA_s	0	0	$D_wA_w + D_sA_s$
4,5	$2D_wA_w$	0	D_nA_n	D_sA_s	0	0	$2D_wA_w + D_nA_n + D_sA_s$
6	D_wA_w	0	D_nA_n	$2D_sA_s$	0	0	$D_wA_w + D_nA_n + 2D_sA_s$
8	0	D_eA_e	$2D_nA_n$	$2D_sA_s$	0	0	$D_eA_e + 2D_nA_n + 2D_sA_s$

Now, we can replace the values into the system of linear equation:

A																	B
1	2	3	4	5	6	7	8	9	10	11	12	13	14	15	16	17	B
655.5	-368	0	0	0	0	0	0	-287.5	0	0	0	0	0	0	0	0	0
-368	1023.5	-368	0	0	0	0	0	0	-288	0	0	0	0	0	0	0	0
0	-368	511.8	-144	0	0	0	0	0	0	0	0	0	0	0	0	0	0
0	0	-144	1024	-143.8	0	0	0	0	0	-736	0	0	0	0	0	0	0
0	0	0	-143.8	1024	-143.8	0	0	0	0	0	-736	0	0	0	0	0	0
0	0	0	0	-143.8	799.25	-368	0	0	0	0	0	-287.5	0	0	0	0	0
0	0	0	0	0	-368	1311	-368	0	0	0	0	0	-287.5	0	0	0	0
0	0	0	0	0	0	-368	943	0	0	0	0	0	0	-287.5	0	0	0
-287.5	0	0	0	0	0	0	0	295	0	0	0	0	0	0	-287.5	0	150
0	-287.5	0	0	0	0	0	0	0	295	0	0	0	0	0	0	-287.5	150
0	0	0	-736	0	0	0	0	0	0	748	0	0	0	0	0	0	240
0	0	0	0	-736	0	0	0	0	0	0	748	0	0	0	0	0	240
0	0	0	0	0	-287.5	0	0	0	0	0	0	295	0	0	0	0	150
0	0	0	0	0	0	-287.5	0	0	0	0	0	0	295	0	0	0	150
0	0	0	0	0	0	0	-287.5	0	0	0	0	0	0	287.5	0	0	250
0	0	0	0	0	0	0	0	-287.5	0	0	0	0	0	0	287.5	0	250
0	0	0	0	0	0	0	0	0	-287.5	0	0	0	0	0	0	287.5	250

Finally, the cells' temperatures can be found as below:

Cell/node	Temperature ($^\circ$C)
1	30.7
2	30.9
3	31.4
4	32.5
5	34.6
6	37.9
7	38.5
8	38.8
9	30.4
10	30.6
11	32.3
12	34.3
13	38.0
14	38.3
15	39.7
16	39.4
17	38.8
Average	34.4

10.9 Discretization Methods

As we have seen in the previous section, the central difference method with second-order of magnitude in the accuracy was used to define the values of $\nabla \phi$ at w and e cell-faces using available data from the neighbouring cell centres. Other discretization forms can be also used to define such values. For examples, the first-order forward difference of the first-derivate in a uniform one-dimensional grid at the eastern face can be shown in Figure 10.22 and written as below:

$$(\nabla \phi)_e = \left(\frac{\phi_E - \phi_e}{\Delta x_{PE}} \right)$$

(10.40)

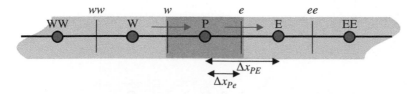

Figure 10.22   First-order forward difference of first-derivate in a uniform one-dimensional grid.

Or, the second-order backward difference of the first-derivate for a uniform grid can be derived as:

$$(\nabla\phi)_e = \left(\frac{1/2\phi_w - 2\phi_P + 3/2\phi_e}{\Delta x_{Pe}}\right) \tag{10.41}$$

Similarly, the fourth-order central difference for a uniform grid can be written:

$$(\nabla\phi)_e = \left(\frac{1/12\phi_w - 2/3\phi_P + 2/3\phi_E - 1/12\phi_{ee}}{\Delta x_{Pe}}\right) \tag{10.42}$$

Note that these approximations require the values at faces 'ee', 'e', 'w', 'ww', etc., which can be themselves approximated with various methods. Example is the linear approximation as below:

$$\phi_w = \frac{\phi_W + \phi_P}{2} \tag{10.43}$$

$$\phi_e = \frac{\phi_P + \phi_E}{2} \tag{10.44}$$

Example 10.6

Use Eq. (10.41) as the backward difference scheme to solve Example 10.3 with the interior and exterior wall surface temperatures to be fixed at 20 and 35 °C, respectively.

Solution

First, we assume a similar area and Γ for all cells where there is no source or sink term. Using such approximation, the heat diffusion equation, for example with second-order backward difference of the first-derivate, becomes:

$$(\Gamma A \nabla\phi)_e - (\Gamma A \nabla\phi)_w + \bar{S}\Delta V = 0$$

$$\Longrightarrow \Gamma_e A_e \left(\frac{1/2\phi_w - 2\phi_P + 3/2\phi_e}{\Delta x/2}\right) - \Gamma_w A_w \left(\frac{1/2\phi_{ww} - 2\phi_W + 3/2\phi_w}{\Delta x/2}\right) = 0$$

We apply approximation of Eqs. (10.43) and (10.44):

$$\Longrightarrow \left[1/2\left(\frac{\phi_W + \phi_P}{2}\right) - 2\phi_P + 3/2\left(\frac{\phi_P + \phi_E}{2}\right)\right]$$

$$- \left[1/2\left(\frac{\phi_{WW} + \phi_W}{2}\right) - 2\phi_W + 3/2\left(\frac{\phi_W + \phi_P}{2}\right)\right] = 0$$

After further rearranging the terms:

$$\left(\frac{-7}{4}\right)\phi_P = \left(+\frac{1}{4}\right)\phi_{WW} + \left(\frac{-5}{4}\right)\phi_W + \left(\frac{-3}{4}\right)\phi_E$$

This expression can be rearranged as the below equation only for Cell-3:

$$a_P\phi_P = a_{WW}\phi_{WW} + a_W\phi_W + a_E\phi_E + S_u$$

For boundary Cell-4:

$$\left[1/2\left(\frac{T_3 + T_4}{2}\right) - 2T_4 + 3/2T_5\right] - \left[1/2\left(\frac{T_2 + T_3}{2}\right) - 2T_3 + 3/2\left(\frac{T_3 + T_4}{2}\right)\right] = 0$$

$$\Longrightarrow \left(\frac{-10}{4}\right)T_4 = \left(\frac{-3}{2}\right)T_5 + \left(\frac{-5}{4}\right)T_3 + \left(\frac{1}{4}\right)T_2$$

Boundary Cell-2:

$$\left[1/2\left(\frac{T_1 + T_2}{2}\right) - 2T_2 + 3/2\left(\frac{T_2 + T_3}{2}\right)\right] - \left[1/2T_0 - 2T_1 + 3/2\left(\frac{T_1 + T_2}{2}\right)\right] = 0$$

$$\Longrightarrow \left(\frac{-7}{4}\right)T_2 = \left(\frac{-3}{4}\right)T_3 + \left(\frac{-6}{4}\right)T_1 + \left(\frac{1}{2}\right)T_0$$

For boundary Cell-1, we need two boundary nodes to replace against 'W' and 'ww'. Nonetheless, cell-face-ww does not exist and therefore we define a mirror cell of T_{00} at this point as seen in Figure 10.E6. The expression, then, can be rewritten as:

$$\left[1/2\left(\frac{T_0 + T_1}{2}\right) - 2T_1 + 3/2\left(\frac{T_1 + T_2}{2}\right)\right] - 2\left[1/2T_{00} - 2T_0 + 3/2\left(\frac{T_0 + T_1}{2}\right)\right] = 0$$

$$\Longrightarrow \left(\frac{-10}{4}\right)T_1 = \left(\frac{-3}{4}\right)T_2 + \left(\frac{-11}{4}\right)T_0 + T_{00}$$

If we can define a linear approximation to estimate T_{00}:

$$T_{00} = \frac{3T_0 - T_1}{2}$$

Now, we replace this equation into the previous one:

$$-2T_1 = \left(\frac{-3}{4}\right)T_2 + \left(\frac{-5}{4}\right)T_0$$

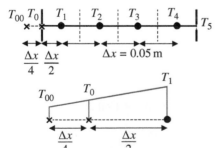

Figure 10.E6 Definition of a mirror cell of T_{00}.

The matrix of system of linear equations can be now formed as:

A				B
−2.0	0.8	0.0	0.0	−25.0
1.5	−1.8	0.8	0.0	10.0
−0.3	1.3	−1.8	0.8	0.0
0.0	−0.3	1.3	−2.5	−52.5

And, the solution gives:

Backward difference	Analytical	Error (%)
22.310	21.875	2.0
26.160	25.625	2.1
29.753	29.375	1.3
33.260	33.125	0.4
	Average	1.4

It should be again noted that even though a more accurate scheme is developed here, but more terms are included in the calculations (here ϕ_{WW}) that can potentially increase the computational cost with creating larger matrix of the coefficients for the system of the linearized equations. This can become troublesome when the size of matrices turns out to be larger. More detailed evaluation of using the higher order approximations will be discussed in the following sections and the next chapter.

10.10 Steady-State Diffusion–Convection Equation

After elimination of the transient term in the transport equation and taking the convection term into the account, we can obtain:

$$\nabla \cdot (\rho u \phi) = \nabla \cdot (\Gamma \nabla \phi) + S_\phi \tag{10.45}$$

Similar to the previous section, after integration over the control volumes and using Gauss divergence theorem, we may reach to the diffusion–convection equation:

$$\int_A n \cdot (\rho \phi u) dA = \int_A n \cdot (\Gamma \nabla \phi) dA + \int_{CV} S_\phi dV \tag{10.46}$$

Similar as before, we can again start to write the diffusion-convection equation in a one-dimensional form as follows:

$$\frac{\partial}{\partial x}(\rho u \phi) = \frac{\partial}{\partial x}\left(\Gamma \frac{\partial \phi}{\partial x}\right) + S_\phi$$

$$\Longrightarrow \int_A n \cdot (\rho \phi u) dA = \int_A n \cdot \left(\Gamma \frac{\partial \phi}{\partial x}\right) dA + \int_{CV} S_\phi dV \tag{10.47}$$

We should also note that the continuity equation in one-dimensional form can be written as:

$$\frac{\partial}{\partial x}(\rho u) = 0 \tag{10.48}$$

For reaching the numerical solution of the above 1D diffusion-convection equation, we can again apply Gauss divergence theorem to the equation:

$$\int_A n \cdot (\rho u) dA = 0 \tag{10.49}$$

For a particular control volume as shown in Figure 10.23, the above equations on the cell-faces ('w' and 'e') can be derived as below:

$$\int_A n \cdot (\rho u \phi) dA = \int_A n \cdot \left(\Gamma \frac{\partial \phi}{\partial x} \right) dA + \int_{CV} S_\phi dV$$
$$\Longrightarrow (\rho u \phi A)_e - (\rho u \phi A)_w = \left(\Gamma A \frac{\partial \phi}{\partial x} \right)_e - \left(\Gamma A \frac{\partial \phi}{\partial x} \right)_w + \overline{S} \Delta V \tag{10.50}$$

And, for the continuity equation, one can derive the below expression:

$$\int_A n \cdot (\rho u) dA = 0$$
$$\Longrightarrow (\rho u A)_e - (\rho u A)_w = 0 \tag{10.51}$$

If the flow is incompressible, which is the case in most of the related cases in the atmospheric condition inside and around buildings, the equation can be simplified to:

$$(u)_e - (u)_w = 0 \tag{10.52}$$

The areas are the same ($A_w = A_e = A$) in the considered 1D mesh of Figure 10.23. Note that in this case, the entering ('w') or leaving ('e') quantities are perpendicular to the cell-faces. Assuming $F = \rho u$, $D = \dfrac{\Gamma}{\delta x}$, using a central difference scheme for the diffusion term, and a linear change for $\overline{S} \Delta V$, the transport equation can be further simplified as:

$$F_e A_e \phi_e - F_w A_w \phi_w = D_e A_e (\phi_E - \phi_P) - D_w A_w (\phi_P - \phi_W) + (S_u + S_P \phi_P) \tag{10.53}$$

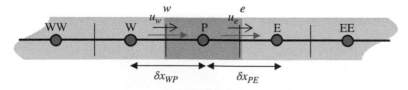

Figure 10.23 Convective and diffusive fluxes in a 1D grid.

Table 10.5 Non-boundary cells coefficients of diffusion–convection equation in a 1D grid.

Cell	a_W	a_E	S_P	S_u	$a_P = a_E + a_W - S_P$
Non-boundary cells	$\left(D_w + \dfrac{F_w}{2}\right)A_w$	$\left(D_e - \dfrac{F_e}{2}\right)A_e$	S_P	S_u	$\left(D_w + \dfrac{F_w}{2}\right)A_w + \left(D_e - \dfrac{F_e}{2}\right)$ $A_e + (F_e A_e - F_w A_w) - S_P$

As stated before in Eqs. (10.43) and (10.44), ϕ_e and ϕ_w can be approximated with an arithmetic averaging of the values at the neighbouring cell centres. So, we can rearrange Eq. (10.53) as:

$$F_e A_e (\phi_P + \phi_E)/2 - F_w A_w (\phi_W + \phi_P)/2$$
$$= D_e A_e (\phi_E - \phi_P) - D_w A_w (\phi_P - \phi_W) + (S_u + S_P \phi_P)$$
$$\Longrightarrow \left[\left(D_w + \frac{F_w}{2}\right)A_w + \left(D_e - \frac{F_e}{2}\right)A_e + (F_e A_e - F_w A_w) - S_P \right]\phi_P = \left(D_w + \frac{F_w}{2}\right)A_w \phi_W$$
$$+ \left(D_e - \frac{F_e}{2}\right)A_e \phi_E + S_u$$

$$(10.54)$$

Equation (10.54) can be further shown with a standard expression similar to the diffusion equation as $(a_P \phi_P = a_W \phi_W + a_E \phi_E + S_u)$ demonstrated in Table 10.5.

We can see later that other approximation rather than the arithmetic averaging is feasible at cell-faces. For treatment of the boundary conditions, the same approach is applied to the diffusion term as explained in diffusion equation section. Additional boundaries, however, are required to be assigned as velocities, which cause advection of the quantity (ϕ) across the discrete space. This is the place that the conservation of mass is taken into the account to resolve the velocity within the grid using its defined value on the boundaries. Thus, from the continuity Eq. (10.52), we can derive:

$$F = F_0 = F_1 = \cdots = F_n \tag{10.55}$$

And, the convection–diffusion equation at Cell-1 yields (Figure 10.24):

$$F_e A_e (\phi_1 + \phi_2)/2 - F_w A_w \phi_0 = D_e A_e (\phi_2 - \phi_1) - 2D_w A_w (\phi_1 - \phi_0) + (S_u + S_P \phi_1)$$
$$\Longrightarrow \left[2D_w A_w + \left(D_e - \frac{F_e}{2}\right)A_e + F_e A_e - S_P \right]\phi_1 = \left(D_e - \frac{F_e}{2}\right)A_e \phi_2 + [(2D_w + F_w)A_w \phi_0 + S_u]$$

$$(10.56)$$

Similarly, for the boundary Cell-n, we can obtain the below expression (Figure 10.25):

$$F_e A_e \phi_n - F_w A_w (\phi_{n-2} + \phi_{n-1})/2$$
$$= 2D_e A_e (\phi_n - \phi_{n-1}) - D_w A_w (\phi_{n-1} - \phi_{n-2}) + (S_u + S_P \phi_{n-1})$$
$$\Longrightarrow \left[\left(D_w + \frac{F_w}{2}\right)A_w + 2D_e A_e - F_w A_w - S_P \right]\phi_{n-1}$$
$$= \left(D_w + \frac{F_w}{2}\right)A_w \phi_{n-2} + [(2D_e - F_e)A_e \phi_n + S_u]$$

$$(10.57)$$

And, the coefficient table can be written in Table 10.6.

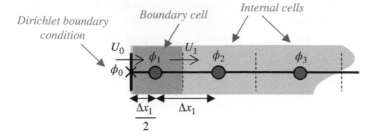

Figure 10.24 Left-side boundary treatment of diffusion–convection equation in a 1D grid.

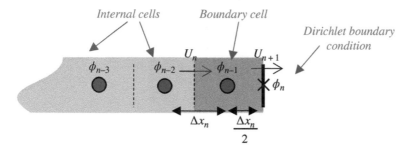

Figure 10.25 Right-side boundary treatment of diffusion–convection equation in a 1D grid.

Table 10.6 Dirichlet boundary cell coefficients of diffusion equation for a 1D grid.

Cell	a_W	a_E	S_P	S_u	$a_P = a_E + a_W - S_P$
1	0	$\left(D_e - \dfrac{F_e}{2}\right)A_e$	$-2D_w A_w - F_e A_e + S_p$	$(2D_w + F_w) A_w \phi_0 + S_u$	$\left(D_e - \dfrac{F_e}{2}\right)A_e + 2D_w A_w + F_e A_e - S_p$
n -1	$\left(D_w + \dfrac{F_w}{2}\right)A_w$	0	$-2D_e A_e + F_w A_w + S_p$	$(2D_e - F_e) A_e \phi_n + S_u$	$\left(D_w + \dfrac{F_w}{2}\right)A_w + 2D_e A_e - F_w A_w - S_p$

Example 10.7

The quantity ϕ is transported in a system with a velocity of 2 m/s obeying a steady-state 1D advection–diffusion behaviour (Figure 10.E7.1). The diffusivity and density of ϕ are 0.1 kg/m s and 1.2 kg/m³, respectively. The quantity has a value of 5 at $x = 0$ while it reduces to 1 at

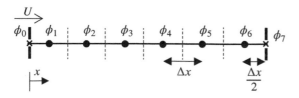

Figure 10.E7.1 1D mesh with six identical cells.

the end of the $L = 80$ cm medium. Use FVM with six cells to calculate the transportation along the medium. Compare the numerical results with the analytical solution. Investigate whether using a finer grid can improve the accuracy of the results.

Solution

The exact solution of the 1D transport equation can be found as below:

$$\frac{\phi(x) - \phi_0}{\phi_L - \phi_0} = \frac{(\exp^{Pe_{x/L}}) - 1}{(\exp^{Pe}) - 1}$$

All areas are equal and can be cancelled out in both side of the equations. Moreover, there is no source and sink term in the equations. Since $D_w = D_e = \frac{\Gamma}{\delta_x} = 0.75$ and $F_w = F_e = \rho u = 2.4$ for cells 2, 3, 4, and 5, we can directly find the coefficients from Table 10.5:

Cell	a_W	a_E	S_P	S_u	$a_P = a_E + a_W - S_P$
Non-boundary cells	1.95	−0.45	0	0	1.5

And for Dirichlet boundary conditions, we can use Table 10.6:

Cell	a_W	a_E	S_P	S_u	$a_P = a_E + a_W - S_P$
1	0	$\left(D - \dfrac{F}{2}\right)$	$-2D - F$	$(2D + F)\phi_0$	$\left(D - \dfrac{F}{2}\right) + 2D + F$
n-1	$\left(D + \dfrac{F}{2}\right)$	0	$-2D + F$	$(2D - F)\phi_n$	$\left(D + \dfrac{F}{2}\right) + 2D - F$

Now, we can find the matrix form of the system of linear equations:

			A				
1	2	3	4	5	6		B
3.45	0.45	0	0	0	0		19.5
−1.95	1.5	0.45	0	0	0		0
0	−1.95	1.5	0.45	0	0		0
0	0	−1.95	1.5	0.45	0		0
0	0	0	−1.95	1.5	0.45		0
0	0	0	0	−1.95	1.05		−0.9

The results of the numerical solution and its comparison with the analytical solution can be calculated as:

x	Central difference	Analytical	Error (%)
0	5.000	5.000	
0.067	4.999	5.000	0.0
0.200	5.007	5.000	0.1
0.333	4.971	5.000	0.6
0.467	5.128	4.999	2.6
0.600	4.447	4.967	10.5
0.733	7.401	4.192	76.5
0.8	1.000	1.000	
		Average	15.1

We can also plot the quantity over the discrete domain and see an obviously unwanted oscillation around the analytical ones (red graphs) (Figure 10.E7.2):

To improve the results and mitigate the oscillations, we may try to improve the results by increasing the number of cells in the 1D grid to 16:

								A								
1	2	3	4	5	6	7	8	9	10	11	12	13	14	15	16	B
7.2	−0.8	0	0	0	0	0	0	0	0	0	0	0	0	0	0	32
−3.2	4	−0.8	0	0	0	0	0	0	0	0	0	0	0	0	0	0
0	−3.2	4	−0.8	0	0	0	0	0	0	0	0	0	0	0	0	0
0	0	−3.2	4	−0.8	0	0	0	0	0	0	0	0	0	0	0	0
0	0	0	−3.2	4	−0.8	0	0	0	0	0	0	0	0	0	0	0
0	0	0	0	−3.2	4	−0.8	0	0	0	0	0	0	0	0	0	0
0	0	0	0	0	−3.2	4	−0.8	0	0	0	0	0	0	0	0	0
0	0	0	0	0	0	−3.2	4	−0.8	0	0	0	0	0	0	0	0
0	0	0	0	0	0	0	−3.2	4	−0.8	0	0	0	0	0	0	0
0	0	0	0	0	0	0	0	−3.2	4	−0.8	0	0	0	0	0	0
0	0	0	0	0	0	0	0	0	−3.2	4	−0.8	0	0	0	0	0
0	0	0	0	0	0	0	0	0	0	−3.2	4	−0.8	0	0	0	0
0	0	0	0	0	0	0	0	0	0	0	−3.2	4	−0.8	0	0	0
0	0	0	0	0	0	0	0	0	0	0	0	−3.2	4	−0.8	0	0
0	0	0	0	0	0	0	0	0	0	0	0	0	−3.2	4	−0.8	0
0	0	0	0	0	0	0	0	0	0	0	0	0	0	−3.2	4.8	1.6

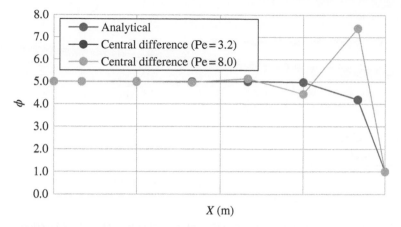

Figure 10.E7.2 Solution of steady-state 1D advection–diffusion using central difference discretization scheme with different Pe numbers.

The local and average errors are considerably decreased to a reasonable range:

x	Central difference	Analytical	Error (%)
0	5.000	5.000	
0.025	5.000	5.000	0.0
0.075	5.000	5.000	0.0
0.125	5.000	5.000	0.0
0.175	5.000	5.000	0.0
0.225	5.000	5.000	0.0
0.275	5.000	5.000	0.0
0.325	5.000	5.000	0.0
0.375	5.000	5.000	0.0
0.425	5.000	5.000	0.0
0.475	5.000	4.998	0.0
0.525	4.998	4.995	0.1
0.575	4.994	4.982	0.2
0.625	4.975	4.940	0.7
0.675	4.900	4.801	2.1
0.725	4.600	4.339	6.0
0.775	3.400	2.805	21.2
0.8	1.000	1.000	
		Average	1.9

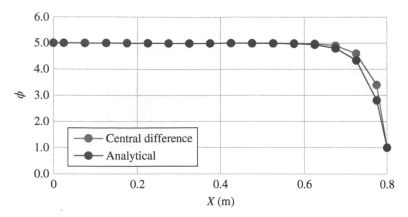

Figure 10.E7.3 Elimination of unwanted oscillations in the solution of steady-state 1D advection–diffusion using central difference discretization.

The plot of the new sets of results are as follows (Figure 10.E7.3).

To understand the nature of oscillations (often known as wiggles) and avoid them in the solution, we will later introduce Peclet number ($P_e = \dfrac{F}{D} = \dfrac{\rho u}{\Gamma / \delta_x}$), which is a ratio of advective over diffusive transport rates. This number identifies when the oscillations in diffusion-convection equations can be avoided. P_e for the central difference scheme using $\Delta x = 0.13$ m and $\Delta x = 0.05$ m can be calculated as $P_e = 3.2$ and $P_e = 1.2$, respectively. We can see later that this scheme is only stable when $P_e \leq 2$. As it can be seen in Figure 10.E7.2 if we increase $P_e = 8$, for example while increasing the velocity to 5 m/s, the oscillation increases drastically.

10.11 Other Approximation Methods

As we discussed before, material and quantity values of ϕ in the discrete space are assigned to and stored in the centre of cells in cell-centred grids and are approximated in cell-faces and vertices. In addition to the cell-centred grid, one can generate a vertex-centred mesh (Figure 10.26) in which all values of ϕ are assigned to the vertices and approximated in cell centres.

The most common approximation in cell-centred grids is the mid-point rule in which the quantity of ϕ is integrated at the middle of the faces and can be approximated as described earlier in this chapter:

$$\int_i f(\phi)dA_i = f_i(\phi)A_i + O(\Delta y^2) \tag{10.58}$$

where, i can be two, four and six cell-faces for 1D, 2D, and 3D grids, respectively. We have also demonstrated how $f_i(\phi)$ can be interpolated from the neighbouring cells. Other linear and nonlinear techniques can be also utilized to approximate the integration of the quantity on the cell-faces based on the values of the neighbouring cells, e.g. Trapezoidal method,

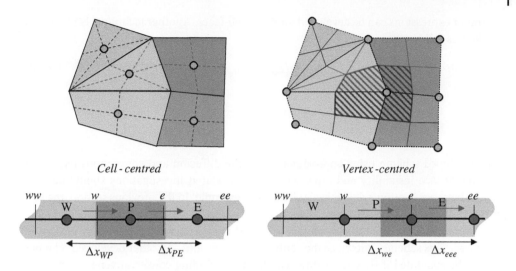

Cell-centred   *Vertex-centred*

Figure 10.26   Cell-centred and vertex-centred control volumes.

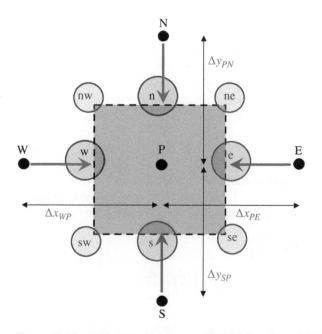

Figure 10.27   Utilization of corner vertices in the calculation of fluxes from cell-faces.

which is using corner vertices of northwest (nw) and southwest (sw) as illustrated in Figure 10.27 to calculate flux from the west cell-face:

$$\int_i f dA_i = \left(\frac{f_{ne} + f_{se}}{2} \right) A_i + O(\Delta y^2) \tag{10.59}$$

Similar expressions can be developed for other cell-faces. Another technique is Simpson's method as follows:

$$\int_i f dA_i = \left(\frac{f_{ni} + 4f_i + f_{si}}{6}\right) A_i + O(\Delta y^4) \tag{10.60}$$

10.12 Scheme Evaluation

The developed scheme using approximation for the diffusion and convections terms can potentially face instability and other issues as the related investigations should be conducted before accepting a discretized scheme. For example, in the previous section, the second-order central difference was used for the second-spatial derivative of the velocity for the diffusion term and a linear interpolation was used to approximate the unknown values at cell-faces. This scheme, known as the central difference, and any other developed schemes should be evaluated if they meet three conditions, including conservativeness, boundedness, and transportiveness [3].

10.12.1 Conservativeness

A developed scheme should allow the flux quantities to pass through cells to represent the conservation law of the physics. In other words, when quantity ϕ leaves a cell, the same amount of ϕ should enter the adjacent cells connected to the leaving face. If in an equation, such as convection–diffusion, the fluxes are defined with derivatives, then the approximation of them should be in way to satisfy the conservation of the fluxes in the cell-faces. As an example, consider that a quantity ϕ is passing through a 1D grid with the below equation:

$$\rho \frac{\partial \phi}{\partial x} = 0 \tag{10.61}$$

As shown in Figure 10.28, applying a central difference approximation for the derivative and a linear interpolation for the density, the flux leaving cell 'P' at face 'e' should be equal to the flux entering cell 'E' at face 'e'. Thus, we can derive this expression as:

$$\left(\frac{\partial \phi}{\partial x}\right)_e @P = \left(\frac{\partial \phi}{\partial x}\right)_w @E$$

$$\Longrightarrow \left(\frac{1}{2}(\rho_P + \rho_E)\right)\left(\frac{(\phi_E - \phi_P)}{\delta_x}\right) = \left(\frac{1}{2}(\rho_P + \rho_W)\right)\left(\frac{(\phi_E - \phi_P)}{\delta_x}\right) \tag{10.62}$$

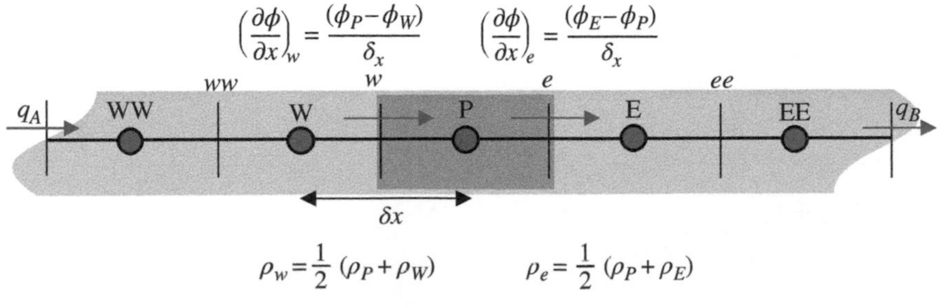

Figure 10.28 Conservation of central difference approximation in a 1D grid.

As you can see this equation is only valid when the density is not changing across the grid or $\rho_E = \rho_W$. Now, if we apply the same concept to the whole domain, and assuming that density is not changing across all the grid, the only remaining terms are boundary ones while interior terms should cancel out each other. For example, in the five-cell grid of Figure 10.28 and when $\rho_{ww} = \rho_w = \rho_e = \rho_{ee} = \rho$, we can write:

$$\left[\rho\left(\frac{(\phi_W - \phi_{WW})}{\delta_x}\right) - q_A\right] - \left[\rho\left(\frac{(\phi_P - \phi_W)}{\delta_x}\right) - \rho\left(\frac{(\phi_W - \phi_{WW})}{\delta_x}\right)\right] - \left[\rho\left(\frac{(\phi_E - \phi_P)}{\delta_x}\right) - \rho\left(\frac{(\phi_P - \phi_W)}{\delta_x}\right)\right]$$
$$- \left[\rho\left(\frac{(\phi_{EE} - \phi_E)}{\delta_x}\right) - \rho\left(\frac{(\phi_E - \phi_P)}{\delta_x}\right)\right] - \left[q_B - \rho\left(\frac{(\phi_{EE} - \phi_E)}{\delta_x}\right)\right] = q_B - q_A$$

$$(10.63)$$

As a **telescoping series**, all terms cancel out with each other except the boundary ones. One can therefore ensure that the scheme preserve the flux q_A, entering the grid and leaving it with the flux of q_B as expected from the physics of the problem in the continuous space.

10.12.2 Boundedness

Unlike the conservativeness, which represents the physics behind the law of conservation in a grid, the **boundedness** representing the mathematical characteristics of a scheme towards reaching a solution. As FVM will end up to a system of linear equations that should be solved through matrix algebra, the iterative methods are preferred in many scenarios when a grid has a large number of cells. To ensure a stability in the matrix solution, the formed matrix of coefficient should satisfy the boundedness condition. In the mathematical sense, the coefficients of the discretized equations ($a'_P \phi_P = \sum a_{nb} \phi_{nb} + S_u$) should be **diagonally dominant** to provide a **sufficient condition** to ensure a convergence to a solution:

$$\frac{\sum |a_{nb}|}{|a'_P|} \begin{cases} \leq 1 \text{ at all cells} \\ < 1 \text{ at one cell at least} \end{cases}$$

$$(10.64)$$

where $a'_P = a_P - S_P$.

As a physical explanation to this mathematical condition, the boundedness ensures that the values in the internal cells are not exceeding the values at boundary when the source/sink term does not exist, e.g. the temperatures in a 1D wire are not calculated to be 95 or 78 °C when boundary values at two ends are 80 and 89 °C – all internal values should be between the boundary values.

After the sufficient condition is satisfied in the above equation, another requirement to guarantee a boundedness condition is to have all the coefficient of the developed scheme (i.e. a'_P and a_{nb}) to be either positive or negative (with the same sign):

$$\begin{cases} \text{all } a_{nb} \text{ and } a'_P > 0 \\ \text{or} \\ \text{all } a_{nb} \text{ and } a'_P < 0 \end{cases}$$

$$(10.65)$$

Again, the physical meaning of this mathematical condition can be expressed as the fact that all neighbouring cells should be impacted similarly when a particular cell shows increase or decrease in the quantity of ϕ.

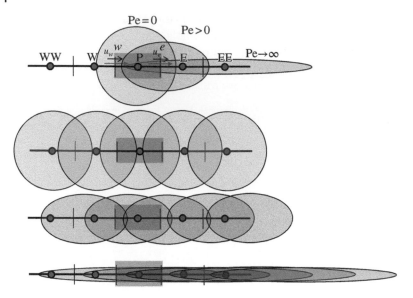

Figure 10.29 Peclet number as a ratio of convective to diffusive transport rates.

10.12.3 Transportiveness

Transportiveness in a convection–diffusion equation is presented by the ratio of convective to diffusive transport rates known as **Peclet** number:

$$P_e = \frac{F}{D} = \frac{\rho u}{\Gamma/\delta_x} \tag{10.66}$$

As it can be seen in Figure 10.29, when Pe number is large, the quantity ϕ is mainly transported by convection. And when Pe number is small, the diffusion is the main mechanism in the transport of the quantity ϕ.

When a discretization scheme is developed, it is crucial that the scheme is logically capable of representing the transport mechanism. As it can be seen in Figure 10.29, the flow direction biases the contribution of cells in the numerical solution. For example, when Pe = 0 or the diffusion is dominant, the contribution of cells P and E on the face 'e' and cells W and P on the face 'w' is similarly important, which makes a circular diffusion of the quantity ϕ around cell P. However, when Pe number increases, the contribution of cells located in the upstream side are more dominant in the transport mechanism. In this case, as it can be observed, the effect of the upstream cell W on cell P or cell P on cell E is more dominant, which forms elliptical contours of the transport of the quantity ϕ. If the Pe is very large (pure convection), then the elliptical transport equation is stretched out towards shaping a straight line when Pe $\longrightarrow \infty$.

10.13 Common Schemes

In this section, we present the commonly used schemes in the FVM, which are also widely used in built environment-related problems, and at the end provide their evaluations. It should be noted that the below schemes are not all the existing ones and many more schemes have been developed and utilized in different studies in accordance with the nature of a subject area. As discussed earlier, the main difference in these schemes is related to the approximation provided for the quantity ϕ at cell-faces using information from its related control volume and its neighbouring cells.

10.13.1 Central Difference

As explained previously, the quantities of ϕ are simply interpolated via arithmetic averaging from the values of both adjacent cells (in a one-dimensional grid) in this second-order of accuracy scheme as those expressed in Eq. (10.43) ($\phi_e = \dfrac{(\phi_P + \phi_E)}{2}$) and Eq. (10.44) ($\phi_w = \dfrac{(\phi_W + \phi_P)}{2}$). Thus, the discretized form of the transport equation with the assumption of central discretization for the diffusion term yields:

$$\left[\left(D_w - \frac{F_w}{2}\right) + \left(D_e + \frac{F_e}{2}\right) + S_P\right]\phi_P = \left(D_w + \frac{F_w}{2}\right)\phi_W + \left(D_e - \frac{F_e}{2}\right)\phi_E + S_u$$

(10.67)

The example provided in the conservativeness section represents that the central scheme satisfies this condition. In an incompressible flow, the continuity equation dictates $F_e - F_w = 0$, which implies that:

$$a_P = \left[(D_w + D_e) + S_P + \left(\frac{F_e - F_w}{2}\right)\right] \Longrightarrow a_P = [(D_w + D_e) + S_P]$$

(10.68)

Since D_w, D_e, and S_P are all positive, then a_P is a positive number. As $D_w + \dfrac{F_w}{2}$ is also a positive number, to satisfy the boundedness condition, all coefficients should be positive:

$$D_e - \frac{F_e}{2} \geq 0 \Longrightarrow F_e/D_e = P_{e_e} < 2$$

(10.69)

Furthermore, we can simply notice that the scheme does not sense the flow direction or the strength of convection relative to diffusion, and thus the transportiveness condition is not satisfied. Hence, either velocity should be very low or grid spacing should be small in this scheme as one of its major drawbacks. Interested readers can refer to many related studies such as [3].

10.13.2 First-Order Upwind

To impose the influence of flow direction, we may use first-order Upwind scheme. In this scheme, when the flow is in the positive direction ($F_e > 0$, $F_w > 0$) as it can be seen in Figure 10.30, the information from the upstream cells will be transferred to the downstream cells with the below simple assumption:

$$\phi_e = \phi_P$$

(10.70)

$$\phi_w = \phi_W$$

(10.71)

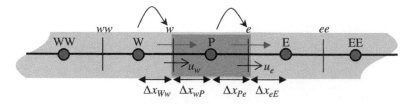

Figure 10.30   First-order Upwind in a 1D grid

Now, if we replace these terms in the convection–diffusion equation, we may reach:

$$
\begin{aligned}
F_e\phi_e - F_w\phi_w &= D_e(\phi_E - \phi_P) - D_w(\phi_P - \phi_W) + S_P\phi_P + S_u \\
\Longrightarrow F_e\phi_P - F_w\phi_W &= D_e(\phi_E - \phi_P) - D_w(\phi_P - \phi_W) + S_P\phi_P + S_u \quad (10.72) \\
\Longrightarrow (F_e + D_w + D_e - S_P)\phi_P &= D_e\phi_E + (D_w + F_w)\phi_W + S_u
\end{aligned}
$$

Or, we can rearrange Eq. (10.72) to:

$$
[(D_w + F_w) + D_e + (F_e - F_w) - S_P]\phi_P = (D_w + F_w)\phi_W + D_e\phi_E + S_u \quad (10.73)
$$

Here, the diffusion term is discretized with the central difference scheme while the quantity ϕ is updated from the upstream neighbour in each cell, e.g. at the cell 'P', we have $\phi_w = \phi_W$ and $\phi_e = \phi_P$. Similarly, when the flow is in the negative direction ($F_e < 0$, $F_w < 0$), we can reach the below expression:

$$
[D_w + D_e - F_w - S_P]\phi_P = D_w\phi_W + (D_e - F_e)\phi_E + S_u \quad (10.74)
$$

If we apply the same condition for a Dirichlet boundary at cell-face-0 ($\phi_e = \phi_1$) as illustrated in Figure 10.31:

$$
\begin{aligned}
F_e\phi_1 - F_w\phi_0 &= D_e(\phi_2 - \phi_1) - 2D_w(\phi_1 - \phi_0) + S_P\phi_1 + S_u \\
\Longrightarrow (F_e + D_e + 2D_w - S_P)\phi_1 &= D_e\phi_2 + [(F_w + 2D_w)\phi_0 + S_u]
\end{aligned} \quad (10.75)
$$

With a similar approach for cell-face-n-1 as seen in Figure 10.32, we can derive ($\phi_w = \phi_{n-2}$):

$$
\begin{aligned}
F_e\phi_{n-1} - F_w\phi_{n-2} &= 2D_e(\phi_n - \phi_{n-1}) - D_w(\phi_{n-1} - \phi_{n-2}) + S_P\phi_{n-1} + S_u \\
\Longrightarrow (2D_e + D_w + F_e - S_P)\phi_{n-1} &= (D_w + F_w)\phi_{n-2} + [2D_e\phi_n + S_u]
\end{aligned} \quad (10.76)
$$

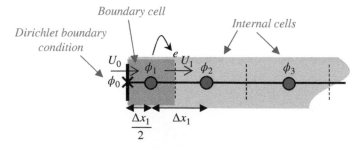

Figure 10.31   Dirichlet left-side boundary condition for first-order Upwind in a 1D grid.

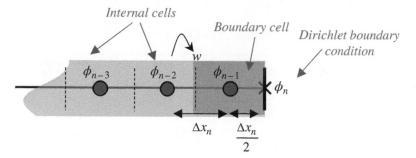

Figure 10.32 Dirichlet right-side boundary condition for first-order Upwind in a 1D grid

Table 10.7 First-order Upwind coefficients of diffusion equation for a 1D grid.

Cell	a_W	a_E	S_P	S_u	$a_P = a_E + a_W - S_P$
1	0	D_e	$-F_e - 2D_w + S_P$	$(F_w + 2D_w)\phi_0 + S_u$	$F_e + D_e + 2D_w - S_P$
2, 3, ..., $n-1$	$(D_w + F_w)$	D_e	S_P	S_u	$(D_w + F_w) + D_e + (F_e - F_w) - S_P$
n	$(D_w + F_w)$	0	$-2D_e + F_w - F_e + S_P$	$2D_e\phi_n + S_u$	$2D_e + D_w + F_e - S_P$

Eventually, we can place all the coefficients in Table 10.7.

First-order Upwind is developed to capture the flow direction and demonstrate transportiveness in the nature. Similar to the central difference scheme, if we write the telescoping series of the above equation, again we can reach to $q_B - q_A$, and therefore we can conclude the scheme is conservative. The scheme is also bounded since all coefficients are already positive (unless there is a large source term in the grid that can potentially impact the transport of ϕ from one cell to another one towards the downstream direction).

Example 10.8

Solve Example 10.7 using first-order Upwind scheme with six cells and discuss the accuracy of the results (see Figure 10.E8.1).

Figure 10.E8.1 1D mesh with six identical cells.

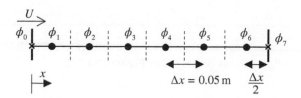

Solution

We can apply use Table 10.7 to find the coefficients as follows:

Cell	a_W	a_E	S_P	S_u	$a_P = a_E + a_W - S_P$
1	0	D	$-F - 2D$	$(F + 2D)T_0$	$F + 3D$
2, 3, 4, 5	$(D + F)$	D	0	0	$F + D$
6	$(D + F)$	0	$-2D$	$2DT_n$	$3D + F$

Now, we can replace the above table with the values to find the matrix form of the system of linear equations:

A						B
4.65	−0.75	0	0	0	0	19.5
−3.15	3.9	−0.75	0	0	0	0
0	−3.15	3.9	−0.75	0	0	0
0	0	−3.15	3.9	−0.75	0	0
0	0	0	−3.15	3.9	−0.75	0
0	0	0	0	−3.15	4.65	1.5

Finally, the results are shown in the below table. As it can be seen while there is no oscillation in the results even when $P_e = 3.2$, the error can be significantly reduced using only six cells.

x	First-order Upwind	Analytical	Error (%)
0.000	5.000	5.000	
0.067	4.999	5.000	0.0
0.200	4.996	5.000	0.1
0.333	4.980	5.000	0.4
0.467	4.913	4.999	1.7
0.600	4.634	4.967	6.7
0.733	3.462	4.192	17.4
0.800	1.000	1.000	
		Average	4.4

The error even can be reduced further to 2.2% when a finer grid with 16 cells is employed ($P_e = 1.2$), as presented in Figure 10.E8.2.

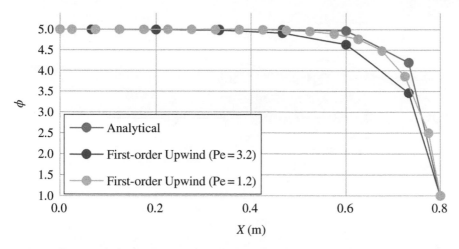

Figure 10.E8.2 Solution of steady-state 1D advection-diffusion using 1st-order Upwind discretization scheme with different Pe numbers.

10.13.3 Second-Order Upwind

This scheme is developed to benefit from the effective transportiveness of Upwind scheme while improving its accuracy to a second-order one. In this scheme, ϕ quantities at cell-faces are computed using a multidimensional linear reconstruction approach obtained through a Taylor series expansion of the cell-centred solution about the cell centroid (see Figure 10.33):

$$\phi_{\mathrm{w}} = \phi_W + \frac{\partial \phi_W}{\partial x} \Delta x \tag{10.77}$$

where $\dfrac{\partial \phi_W}{\partial x}$ can be calculated as discussed previously:

$$\phi_w = \phi_W + \frac{(\phi_W - \phi_{WW})}{2\Delta x} \Delta x$$
$$\Longrightarrow \phi_{\mathrm{w}} = \frac{3}{2}\phi_W - \frac{\phi_{WW}}{2} \tag{10.78}$$

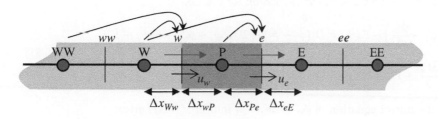

Figure 10.33 Second-order Upwind in a 1D grid

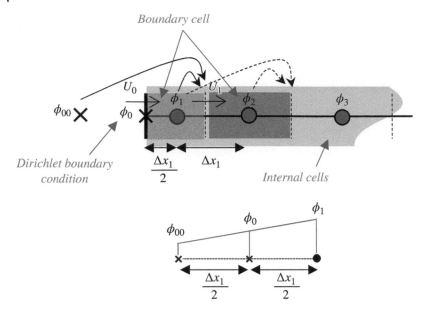

Figure 10.34 Mirror vertex for Dirichlet boundary condition of second-order Upwind in a 1D grid.

Similarly, we can write:

$$\phi_e = \phi_P + \frac{\partial \phi_P}{\partial x} \Delta x$$

$$\Longrightarrow \phi_e = \frac{3}{2}\phi_P - \frac{\phi_W}{2}$$

(10.79)

Replacing Eqs. (10.78) and (10.79) into the convection–diffusion equation gives:

$$F_e\left(\frac{3}{2}\phi_P - \frac{\phi_W}{2}\right) - F_w\left(\frac{3}{2}\phi_W - \frac{\phi_{WW}}{2}\right) = D_e(\phi_E - \phi_P) - D_w(\phi_P - \phi_W) + S_P\phi_P + S_u$$

$$\Longrightarrow \left(\frac{3}{2}F_e + D_e + D_w - S_P\right)\phi_P = \left(\frac{-F_w}{2}\right)\phi_{WW} + \left(\frac{F_e}{2} + \frac{3}{2}F_W + D_w\right)\phi_W + D_e\phi_E + S_u$$

(10.80)

For the boundary cell-faces, we again should define mirror cells. As shown in Figure 10.34, we can write the below expression for the Cell-1:

$$F_e\left(\frac{3}{2}\phi_1 - \frac{\phi_{00}}{2}\right) - F_w\phi_0 = D_e(\phi_2 - \phi_1) - 2D_w(\phi_1 - \phi_0) + S_P\phi_1 + S_u$$

$$\Longrightarrow \left(\frac{3F_e}{2} + D_e + 2D_w - S_P\right)\phi_1 = \frac{F_e}{2}\phi_{00} + (F_w + 2D_w)\phi_0 + D_e\phi_2 + S_u$$

(10.81)

Replacing the mirror equation of $\phi_{00} = 2\phi_0 - \phi_1$ into Eq. (10.81) gives:

$$(2F_e + D_e + 2D_w - S_P)\phi_1 = D_e\phi_2 + [(F_e + F_w + 2D_w)\phi_0 + S_u]$$

(10.82)

For Cell-2 (see Figure 10.34):

$$F_e\left(\frac{3}{2}\phi_2 - \frac{\phi_1}{2}\right) - F_w\left(\frac{3}{2}\phi_1 - \frac{\phi_{00}}{2}\right) = D_e(\phi_3 - \phi_2) - D_w(\phi_2 - \phi_1) + S_P\phi_2 + S_u$$

$$\Longrightarrow \left(\frac{3}{2}F_e + D_e + D_w - S_P\right)\phi_2 = \left(\frac{-F_w}{2}\right)\phi_{00} + \left(\frac{F_e}{2} + \frac{3}{2}F_W + D_w\right)\phi_1 + D_e\phi_3 + S_u$$

$$(10.83)$$

Knowing that $\phi_{00} = 2\phi_0 - \phi_1$, Eq. (10.85) becomes:

$$\left(\frac{3}{2}F_e + D_e + D_w - S_P\right)\phi_2 = \left(\frac{F_e}{2} + 2F_W + D_w\right)\phi_1 + D_e\phi_3 + [-F_w\phi_0 + S_u]$$

$$(10.84)$$

And, for Cell-n-1 with a similar approach as Cell-1, we can derive:

$$F_e\left(\frac{3}{2}\phi_{n-1} - \frac{\phi_{n-2}}{2}\right) - F_w\left(\frac{3}{2}\phi_{n-2} - \frac{\phi_{n-3}}{2}\right)$$

$$= 2D_e(\phi_n - \phi_{n-1}) - D_w(\phi_{n-1} - \phi_{n-2}) + S_P\phi_{n-1} + S_u$$

$$\Longrightarrow \left(\frac{3}{2}F_e + 2D_e + D_w - S_P\right)\phi_{n-1} = \left(\frac{-F_w}{2}\right)\phi_{n-3} + \left(\frac{F_e}{2} + \frac{3}{2}F_W + D_w\right)\phi_{n-2} + [2D_e\phi_n + S_u]$$

$$(10.85)$$

The use of more than one point in the second-order Upwind (or any other scheme) results in inclusion of more terms in the general equation and consequently more non-zero elements in the matrix of coefficient. For example, in a 1D grid, three diagonal matrices will be changed to a wider five-diagonal while in a 2D grid, it will be expanded from five to nine diagonals, and in 3D from 7 to 13. The second-order Upwind scheme similarly follows the necessary evaluation criteria as discussed before for first-order Upwind.

Example 10.9

Solve Example 10.7 using second-order Upwind scheme with six cells and discuss the accuracy of the results.

Solution:

First, we form the table of coefficient for the internal and boundary cells as follows:

Cell	a_{WW}	a_W	a_E	S_P	S_u	$a_P = a_E + a_W - S_P$
1	0	0	D	$-2F - 2D$	$(2F + 2D)\phi_0$	$2F + 3D$
2	0	$\frac{5F}{2} + D$	D	$F - D$	$-F\phi_0$	$\frac{3F}{2} + 2D$
3, 4, 5	$\frac{-F}{2}$	$2F + D$	D	0	0	$\frac{3F}{2} + 2D$
6	$\frac{-F}{2}$	$\frac{5F}{2} + D$	0	$\frac{F}{2} - 2D$	$2D\phi_n$	$\frac{3F}{2} + 3D$

Then, we replace the values into it to find the matrix form of the system of linear equations:

A						B
7.05	−0.75	0	0	0	0	31.5
−6.75	5.1	−0.75	0	0	0	−12
1.2	−5.55	5.1	−0.75	0	0	0
0	1.2	−5.55	5.1	−0.75	0	0
0	0	1.2	−5.55	5.1	−0.75	0
0	0	0	1.2	−6.75	5.85	1.5

The results, then, can be resolved as below with an average error of 3.3% when $P_e = 3.2$:

x	Second-order Upwind	Analytical	Error (%)
0.000	5.000	5.000	
0.067	5.000	5.000	0.0
0.200	5.000	5.000	0.0
0.333	5.000	5.000	0.0
0.467	5.000	4.999	0.0
0.600	5.000	4.967	0.7
0.733	5.000	4.192	19.3
0.800	1.000	1.000	
		Average	3.3

The plot of first- and second-order Upwind schemes are demonstrated in the same graph (Figure 10.E9.1).

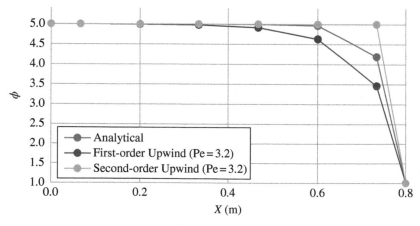

Figure 10.E9.1 Solution of steady-state 1D advection–diffusion using first- and second-order Upwind discretization schemes

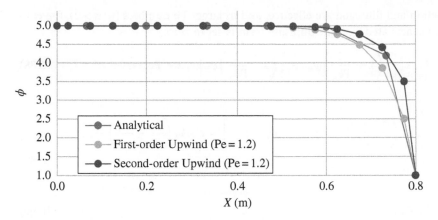

Figure 10.9.2 Solution of steady-state 1D advection–diffusion using first- and second-order Upwind discretization schemes after increasing the cell numbers.

While the overall error of second-order Upwind is lower, still the higher gradients of ϕ cannot be captured with this approach. Therefore, we can increase the number of cells to 16 to be able to better resolve the gradients at this region. The results of both schemes against $P_e = 1.2$ are plotted in the below while the average error can be reduced from 2.2% to 1.8% using second-order Upwind instead of first-order Upwind. The local error of the region with a higher gradient can be also decreased using a scheme with a higher order of accuracy (Figure 10.E9.2).

10.13.4 Power Law

The first-order accuracy power law scheme has been developed based on the exact solution of 1D transport equation. This scheme tends to set diffusion term equal to zero for $P_e > 10$ (for a west cell-face for a left-to-right velocity direction):

$$\phi_w = \phi_W$$
$$\Longrightarrow q_w = F_w A_w \phi_W \ (10 < Pe_w) \tag{10.86}$$

When Peclet number is not within this range, the diffusion term is calculated as follows:

$$q_w = F_w A_w \left[\phi_W - \frac{(1 - 0.1Pe_w)^5}{Pe_w} (\phi_P - \phi_W) \right] (0 < Pe_w < 10) \tag{10.87}$$

Once again, the scheme satisfies all the evaluation criteria.

10.13.5 Hybrid

Hybrid differencing scheme (first-order accurate) is a combination of the central difference scheme (second-order) and Upwind scheme to account for the flow direction. Thus, this scheme benefits from the advantages of each scheme in various cells of a grid. In other words, in small Pe numbers, this scheme is similar to the central difference scheme where both convection and diffusion are effective, and in larger Pe numbers ($|Pe| > 2$), the convection term is activated, and the diffusion term is set to zero:

$$q_w = F_w A_w \left[\frac{1}{2}\left(1 + \frac{2}{Pe_w}\right)\phi_W + \frac{1}{2}\left(1 - \frac{2}{Pe_w}\right)\phi_P \right] - 2 < Pe_w < 2 \qquad (10.88)$$

Or:

$$q_w = F_w A_w \left(\frac{\phi_W + \phi_P}{2}\right) - \left[\frac{F_w A_w}{Pe_w}(\phi_P - \phi_W)\right] = F_w A_w \left(\frac{\phi_W + \phi_P}{2}\right) - D_w(\phi_P - \phi_W) \qquad (10.89)$$

Also, we have:

$$q_w = F_w A_w \phi_W 2 \le Pe_w \qquad (10.90)$$

This scheme is also satisfying all the evaluation criteria.

10.13.6 QUICK

Quadratic Upwind differencing scheme – quadratic upstream interpolation for convective kinetics (QUICK) is a second-order scheme. It initiates a second-order polynomial intersects among three points of WW, W, and P for the calculation of ϕ_w (when $0 < u_e$ and $0 < u_w$):

$$f(x) = a + bx + cx^2 \qquad (10.91)$$

As seen in Figure 10.35, this implies that we can apply three values of ϕ at WW, W, and P assuming equal intervals between cells (Δx) and Cell-W to be the origin ($x = 0$):

$$\phi_{x=0} = \phi_W = a + b(0) + c(0)^2 \qquad (10.92)$$

$$\phi_{x=-\Delta x} = \phi_{WW} = a + b(-\Delta x) + c(-\Delta x)^2 \qquad (10.93)$$

$$\phi_{x=\Delta x} = \phi_P = a + b(\Delta x) + c(\Delta x)^2 \qquad (10.94)$$

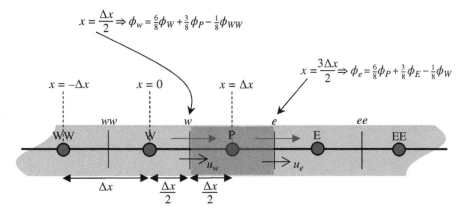

Figure 10.35 Development of QUICK scheme in a 1D grid.

Thus, by performing some basic algebra, we can reach the general form of the polynomial:

$$\phi_x = (\phi_W) + \left(\frac{\phi_P - \phi_{ww}}{2\Delta x}\right)x + \left(\frac{\phi_P - 2\phi_W + \phi_{ww}}{2\Delta x^2}\right)x^2 \tag{10.95}$$

Here, to obtain ϕ at 'e' and 'w', we should place $x = \frac{\Delta x}{2}$ into Eq. (10.97):

$$\phi_w = \frac{6}{8}\phi_W + \frac{3}{8}\phi_P - \frac{1}{8}\phi_{ww} \tag{10.96}$$

With a similar approach, we can derive the below equation for ϕ_e at $x = \frac{3\Delta x}{2}$:

$$\phi_e = \frac{6}{8}\phi_P + \frac{3}{8}\phi_E - \frac{1}{8}\phi_w \tag{10.97}$$

The general transport equation, thus, can be rearranged as:

$$\left[F_e\left(\frac{6}{8}\phi_P + \frac{3}{8}\phi_E - \frac{1}{8}\phi_w\right) - F_w\left(\frac{6}{8}\phi_W + \frac{3}{8}\phi_P - \frac{1}{8}\phi_{ww}\right)\right]$$

$$= D_e(\phi_E - \phi_P) - D_w(\phi_P - \phi_w) + S_P\phi_P + S_u$$

$$\implies \left(\frac{6}{8}F_e - \frac{3}{8}F_w + D_e + D_w - S_P\right)\phi_P =$$

$$\left(\frac{-1}{8}F_w\right)\phi_{ww} + \left(D_w + \frac{1}{8}F_e + \frac{6}{8}F_w\right)\phi_w + \left(D_e - \frac{3}{8}F_e\right)\phi_E + S_u \tag{10.98}$$

The treatment of boundary is similar as before using a mirror cell. For example, for Cell-2, we can derive the below expression as illustrated in Figure 10.34:

$$\left[F_e\left(\frac{6}{8}\phi_2 + \frac{3}{8}\phi_3 - \frac{1}{8}\phi_1\right) - F_w\left(\frac{6}{8}\phi_1 + \frac{3}{8}\phi_2 - \frac{1}{8}\phi_{00}\right)\right]$$

$$= D_e(\phi_3 - \phi_2) - D_w(\phi_2 - \phi_1) + S_P\phi_2 + S_u$$

$$\implies \left(\frac{6}{8}F_e - \frac{3}{8}F_w + D_e + D_w - S_P\right)\phi_2 = \tag{10.99}$$

$$\left(\frac{-1}{8}F_w\right)\phi_{00} + \left(D_w + \frac{1}{8}F_e + \frac{6}{8}F_w\right)\phi_1 + \left(D_e - \frac{3}{8}F_e\right)\phi_3 + S_u$$

where, $\phi_{00} = 2\phi_0 - \phi_1$:

$$\left(\frac{6}{8}F_e - \frac{3}{8}F_w + D_e + D_w - S_P\right)\phi_2 =$$

$$\left(D_w + \frac{1}{8}F_e + \frac{7}{8}F_w\right)\phi_1 + \left(D_e - \frac{3}{8}F_e\right)\phi_3 + \left[\frac{-1}{4}F_w\phi_0 + S_u\right] \tag{10.100}$$

For Cell-1 and Cell-n-1, we need to find a high-order accuracy for approximation of the values at ϕ_0 and ϕ_n. Thus, we can again fit a polynomial at boundary Cell-1 while using Cell-1, Cell-2, and Cell-face-0 into the account and Cell-P to be the origin ($x = 0$):

$$\phi_{x=0} = \phi_P = a + b(0) + c(0)^2 \tag{10.101}$$

$$\phi_{x = \frac{-\Delta x}{2}} = \phi_A = a + b\left(\frac{-\Delta x}{2}\right) + c\left(\frac{-\Delta x}{2}\right)^2 \tag{10.102}$$

$$\phi_{x = \Delta x} = \phi_E = a + b(\Delta x) + c(\Delta x)^2 \tag{10.103}$$

This leads to the below general form of:

$$\phi_x = (\phi_P) + \left(\frac{3\phi_P - 4\phi_A + \phi_E}{3\Delta x}\right)x + \left(\frac{-6\phi_P + 4\phi_A + 2\phi_E}{3\Delta x^2}\right)x^2 \tag{10.104}$$

And:

$$\frac{\partial \phi_x}{\partial x} = b + 2cx = \left(\frac{3\phi_P - 4\phi_A + \phi_E}{3\Delta x}\right) + 2\left(\frac{-6\phi_P + 4\phi_A + 2\phi_E}{3\Delta x^2}\right)x$$

$$\Longrightarrow D_w \phi_w = \left(\Gamma A \frac{\partial \phi}{\partial x}\right)_{x = \frac{-\Delta x}{2}} = \left(\frac{3\phi_P - 4\phi_A + \phi_E}{3\Delta x}\right) + \left(\frac{+6\phi_P - 4\phi_A - 2\phi_E}{3\Delta x}\right) \tag{10.105}$$

$$= \frac{9\phi_P - 8\phi_A - \phi_E}{3\Delta x}$$

Hence, for Cell-1:

$$\left[F_e\left(\frac{6}{8}\phi_1 + \frac{3}{8}\phi_2 - \frac{1}{8}\phi_{00}\right) - F_w\phi_0\right] = D_e(\phi_2 - \phi_1) - \frac{1}{3}D_w(9\phi_1 - 8\phi_0 - \phi_2) + S_P\phi_1 + S_u$$

$$\Longrightarrow \left(\frac{6}{8}F_e + D_e + \frac{9}{3}D_w - S_P\right)\phi_1 =$$

$$\left(\frac{1}{8}F_e\right)\phi_{00} + \left(D_e + \frac{1}{3}D_w - \frac{3}{8}F_e\right)\phi_2 + \left[\left(\frac{8}{3}D_w + F_w\right)\phi_0 + S_u\right] \tag{10.106}$$

Again, we apply $\phi_{00} = 2\phi_0 - \phi_1$:

$$\left(\frac{7}{8}F_e + D_e + \frac{9}{3}D_w - S_P\right)\phi_1 = \left(D_e + \frac{1}{3}D_w - \frac{3}{8}F_e\right)\phi_2 + \left[\left(\frac{8}{3}D_w + F_w + \frac{1}{4}F_e\right)\phi_0 + S_u\right] \tag{10.107}$$

For Cell-n-1:

$$\left[F_e\phi_n - F_w\left(\frac{6}{8}\phi_{n-2} + \frac{3}{8}\phi_{n-1} - \frac{1}{8}\phi_{n-3}\right)\right]$$

$$= \frac{1}{3}D_e(8\phi_n - 9\phi_{n-1} + \phi_{n-2}) - D_w(\phi_{n-1} - \phi_{n-2}) + S_P\phi_{n-1} + S_u$$

$$\Longrightarrow \left(-\frac{3}{8}F_w + \frac{9}{3}D_e + D_w - S_P\right)\phi_{n-1} =$$

$$\frac{-1}{8}F_w\phi_{n-3} + \left(\frac{6}{8}F_w + D_w + \frac{1}{3}D_e\right)\phi_{n-2} + \left[\left(\frac{8}{3}D_e - F_e\right)\phi_n + S_u\right] \tag{10.108}$$

It can be shown that while QUICK demonstrates transportiveness and it complies with conservativeness, it has a condition of $P_e \leq \frac{8}{3}$ to be bounded and thus is stable.

Example 10.10

Solve Example 10.7 using QUICK scheme with six cells and discuss the accuracy of the results.

Solution

We use Eq. (10.98) for the internal cells, and Eqs. (10.100), (10.107), and (10.108) for boundary cells of Cell-2, Cell-1, and Cell-n-1, respectively. The table of coefficients, therefore, can be found as:

Cell	a_{WW}	a_W	a_E	S_P	S_u	$a_P = a_E + a_W - S_P$
1	0	0	$\frac{-3}{8}F + \frac{4}{3}D$	$\frac{-10}{8}F - \frac{8}{3}D$	$\left(\frac{8}{3}D + \frac{5}{4}F\right)\phi_0$	$\frac{7}{8}F + 4D$
2	0	$F + D$	$\frac{-3}{8}F + D$	$\frac{-6}{8}F$	$\frac{-1}{4}F\phi_0$	$\frac{3}{8}F + 2D$
3, 4, 5	$\frac{-1}{8}F$	$\frac{7}{8}F + D$	$\frac{-3}{8}F + D$	0	0	$\frac{3}{8}F + 2D$
6	$\frac{-1}{8}F$	$\frac{6}{8}F + \frac{4}{3}D$	0	$F - \frac{8}{3}D$	$\left(\frac{8}{3}D - F\right)\phi_n$	$\frac{-3}{8}F + 4D$

Again, by replacing values into the above table, we may reach to the matrix form of the system of linear equations:

		A				*B*
5.1	−0.1	0	0	0	0	25.00
−3.15	2.4	0.15	0	0	0	−3.00
0.3	−2.85	2.4	0.15	0	0	0
0	0.3	−2.85	2.4	0.15	0	0
0	0	0.3	−2.85	2.4	0.15	0
0	0	0	0.3	−2.8	2.1	−0.40

This yields the below results with about 6.1% average in errors with only six cells:

x	QUICK	Analytical	Error (%)
0.000	5.000	5.000	
0.067	5.000	5.000	0.0
0.200	5.000	5.000	0.0
0.333	5.000	5.000	0.0
0.467	5.002	4.999	0.1
0.600	4.959	4.967	0.2
0.733	5.707	4.192	36.1
0.800	1.000	1.000	
		Average	6.1

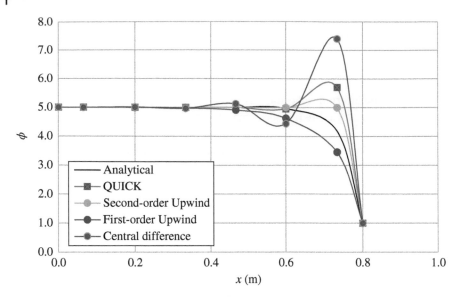

Figure 10.E10.1 Solution of steady-state 1D advection–diffusion using QUICK discretization scheme.

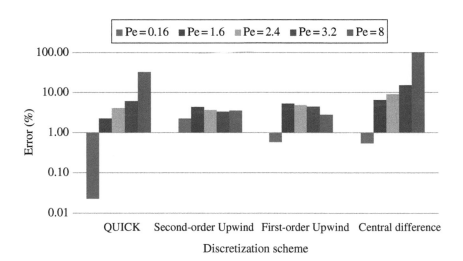

Figure 10.E10.2 Performance of various discretization schemes under different Pe numbers.

We can also plot all the schemes using six cells with $P_e = 3.2$ (Figure 10.E10.1).

If we change the velocities in all schemes to obtain a range of Peclet numbers, we can investigate the performance of each scheme as shown in Figure 10.E10.2. For small Pe numbers, the performance of QUICK scheme is incredibly high while Upwind schemes perform better in higher Pe numbers.

10.14 Unsteady Diffusion Equation

The solution of the steady form of the transport equation was discussed in detail in the previous sections. In this part, we consider the transient term in Eq. (10.7) as demonstrated as below:

$$\frac{\partial}{\partial t}(\rho\phi) + \nabla \cdot (\rho u\phi) = \nabla \cdot (\Gamma\nabla\phi) + S_\phi \tag{10.109}$$

First, following the same approach as before, we employ a spatial integration over the control volume and utilize Gauss divergence theorem to transform the integration on the control volume to the cell-faces. Then, we need to perform a temporal integration over the transport equation due to having a time-dependent term:

$$\int_{CV}\left(\int_t^{t+\Delta t}\frac{\partial}{\partial t}(\rho\phi)dt\right)dV + \int_t^{t+\Delta t}\left(\int_A n \cdot (\rho u\phi)dA\right)dt$$
$$= \int_t^{t+\Delta t}\left(\int_A n \cdot (\Gamma\nabla\phi)dA\right)dt + \int_t^{t+\Delta t}\left(\int_{CV}S_\phi dV\right)dt \tag{10.110}$$

Again, to simplify the process, we can start with the one-dimensional form while omitting the advection term:

$$\int_t^{t+\Delta t}\int_{CV}\rho\frac{\partial\phi}{\partial t}dVdt = \int_t^{t+\Delta t}\left(\int_A n \cdot \left(\Gamma\frac{\partial\phi}{\partial x}\right)dA\right)dt + \int_t^{t+\Delta t}\left(\int_{CV}S_\phi dV\right)dt \tag{10.111}$$

Using a first-order backward differencing scheme for the temporal derivative of the temperature over a control volume '*P*', we can derive the left-hand side term as:

$$\int_{CV}\left[\int_t^{t+\Delta t}\rho\frac{\partial\phi}{\partial t}dt\right]dV = \rho\left(\phi_P^{(t+\Delta t)} - \phi_P^{(t)}\right)\Delta V \tag{10.112}$$

Note that the upper index denotes the time of each quantity. Also, we can rewrite the right-hand side of the general equation as below:

$$\rho\left(\phi_P^{(t+\Delta t)} - \phi_P^{(t)}\right)\Delta V = \int_t^{t+\Delta t}\left[\left(\Gamma_e A_e\frac{\phi_E - \phi_P}{\delta x_{PE}}\right) - \left(\Gamma_w A_w\frac{\phi_P - \phi_W}{\delta x_{WP}}\right)\right]dt + \int_t^{t+\Delta t}\overline{S}\Delta V dt \tag{10.113}$$

Here, we need an assumption for the temporal variation of ϕ_P, ϕ_E, and ϕ_W. Using a **weight function** $(0 < \theta < 1)$, we can define temporal variations of the quantity as:

$$\int_t^{t+\Delta t}\phi_P dt = \left[\theta\phi_P^{(t+\Delta t)} + (1-\theta)\phi_P^{(t)}\right]\Delta t \tag{10.114}$$

$$\int_t^{t+\Delta t}\phi_E dt = \left[\theta\phi_E^{(t+\Delta t)} + (1-\theta)\phi_E^{(t)}\right]\Delta t \tag{10.115}$$

$$\int_t^{t+\Delta t}\phi_W dt = \left[\theta\phi_W^{(t+\Delta t)} + (1-\theta)\phi_W^{(t)}\right]\Delta t \tag{10.116}$$

Further rearrangements result in:

$$\rho\left(\frac{\phi_P^{(t+\Delta t)} - \phi_P^{(t)}}{\Delta t}\right)\Delta V = \theta\left[\left(\Gamma_e A_e \frac{\phi_E^{(t+\Delta t)} - \phi_P^{(t+\Delta t)}}{\delta x_{PE}}\right) - \left(\Gamma_w A_w \frac{\phi_P^{(t+\Delta t)} - \phi_W^{(t+\Delta t)}}{\delta x_{WP}}\right)\right]$$

$$+ (1-\theta)\left[\left(\Gamma_e A_e \frac{\phi_E^{(t)} - \phi_P^{(t)}}{\delta x_{PE}}\right) - \left(\Gamma_w A_e \frac{\phi_P^{(t)} - \phi_W^{(t)}}{\delta x_{WP}}\right)\right] + \bar{S}\Delta V$$

(10.117)

We can further rearrange Eq. (10.117) ($\Delta V = A\Delta x$):

$$\left[\rho\frac{\Delta V}{\Delta t} + \theta\left(\frac{\Gamma_e A_e}{\delta x_{PE}} + \frac{\Gamma_w A_w}{\delta x_{WP}}\right)\right]\phi_P^{(t+\Delta t)}$$

$$= \frac{\Gamma_e A_e}{\delta x_{PE}}\left[\theta\phi_E^{(t+\Delta t)} + (1-\theta)\phi_E^{(t)}\right] + \frac{\Gamma_w A_w}{\delta x_{WP}}\left[\theta\phi_W^{(t+\Delta t)} + (1-\theta)\phi_W^{(t)}\right]$$

$$+ \left[\rho\frac{\Delta x}{\Delta t} - (1-\theta)\frac{\Gamma_e A_e}{\delta x_{PE}} - (1-\theta)\frac{\Gamma_w A_w}{\delta x_{WP}}\right]\phi_P^{(t)} + \bar{S}\Delta V$$

$$\Rightarrow a_P\phi_P^{(t+\Delta t)} = a_E\left[\theta\phi_E^{(t+\Delta t)} + (1-\theta)\phi_E^{(t)}\right] + a_W\left[\theta\phi_W^{(t+\Delta t)} + (1-\theta)\phi_W^{(t)}\right]$$

$$+ \left[a_P' - (1-\theta)a_W - (1-\theta)a_E\right]\phi_P^{(t)} + S_u$$

(10.118)

A common example of Eq. (10.118) is the unsteady heat conduction equation where we can assume $\phi = T$ and $\Gamma = k$. Equation (10.118) also can be shown in a standard form similar to Eq. (10.39) in which $a_P' = \rho\frac{\Delta V}{\Delta t}$ and $a_P = \theta(a_W + a_E) + a_P' - S_P$. In general, when $\theta = 0$, $\phi_E^{(t)}$, $\phi_W^{(t)}$, and $\phi_P^{(t)}$ are used for the calculation of $\phi_P^{t+\Delta t}$ in a new time step as the method called **Explicit scheme**. Inversely, when $0 < \theta < 1$, quantities in the new time step ($\phi_E^{(t+\Delta t)}$ and $\phi_W^{(t+\Delta t)}$) are also used to calculate $\phi_P^{(t+\Delta t)}$; this method is called **Implicit scheme**. Furthermore, the implicit scheme is called **Crank–Nicolson** when $\theta = \frac{1}{2}$, and **Fully implicit** scheme when $\theta = 1$.

10.14.1 Explicit Scheme

For the general expression, we can assume $\theta = 0$ and thus rewrite Eq. (10.118) as ($\bar{S}\Delta V = S_P\phi_P^{(t)} + S_u$):

$$a_P\phi_P^{(t+\Delta t)} = \left[a_E\phi_W^{(t)} + a_W\phi_E^{(t)} + \left[a_P' + S_P - a_E - a_W\right]\phi_P^{(t)} + S_u\right]$$

10.119)

where, $a_P = a_P'$.

The right-hand side of Eq. (10.119) is a known value that can be fully obtained from a previous time step. To check the boundedness of the explicit scheme, the coefficient of $\phi_P^{(t)}$ should be positive, so if $S_p = 0$ and $\delta x_{PE} = \delta x_{WP} = \Delta x$ ($\Delta V = A\Delta x$):

$$\rho c\frac{\Delta V}{\Delta t} - a_W - a_E > 0$$

$$\Rightarrow \rho c\frac{\Delta V}{\Delta t} > \frac{2Ak}{\Delta x}$$

(10.120)

$$\Rightarrow \Delta t < \rho c\frac{(\Delta x)^2}{2k} \Rightarrow Fo = \frac{\alpha\Delta t}{(\Delta x)^2} < \frac{1}{2}$$

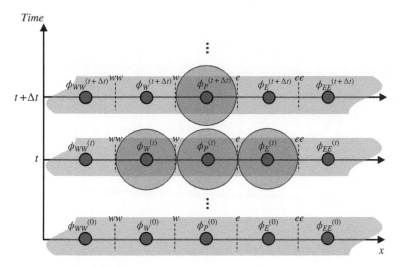

Figure 10.36 Explicit scheme for a 1D grid.

where $\dfrac{\alpha \Delta t}{(\Delta x)^2}$ is called Fourier number (Fo) and implies a constraint on the chosen time-step for the investigated scheme.

As Δx normally has a small value to provide high-resolution results, to keep Fo $< \dfrac{1}{2}$, the chosen Δt must be even one order of the magnitude smaller. Therefore, the application of the explicit scheme for long-time calculations, which are common in the airflow modelling of built environment, are computationally expensive as it requires too many time steps. In the explicit scheme, each unknown variable in a particular time step $(t + \Delta t)$ is calculated from the known variables from the previous time step (t) as shown in Figure 10.36. Thus, in this scheme, there is no need to form the matrix of system of linear equations and the unknown variable can be directly calculated at each time step.

The expansion of the unsteady diffusion equation in 2D and 3D explicit forms are similar to what was proposed in the previous sections. Note that the boundedness conditions enforce Fourier number to be below $\dfrac{1}{4}$ and $\dfrac{1}{8}$ in 2D and 3D spaces, respectively.

Example 10.11

A rectangular aluminium plate with a dimension of 0.1 m $\times$ 0.1 m is at initial homogenous temperature of $200\,^{\circ}$C as shown in Figure 10.E11.1. If the plate is exposed to $0\,^{\circ}$C Dirichlet boundaries at all sides. Calculate the temporal temperature distribution of the plate using a 2D 4×4 grid and an explicit approach. Discuss the stability of the solution (aluminium properties: $\rho = 2700$ kg/m^3, $c_p = 910$ J/kg K, $k = 205$ W/m K).

Solution

In general, we have $\Gamma = k$, $\Delta x = \Delta y$, and $D = D_e = D_w = D_n = D_s = \dfrac{k}{\Delta x}$. For internal cells of 6, 7, 10, and 11, we have $\theta = 0$ and $\overline{S} = 0$ $\left(a'_P = \rho c \dfrac{\Delta V}{\Delta t}\right.$ and $a_P = \theta(a_W + a_E + a_N + a_S) + a'_P = a'_P$):

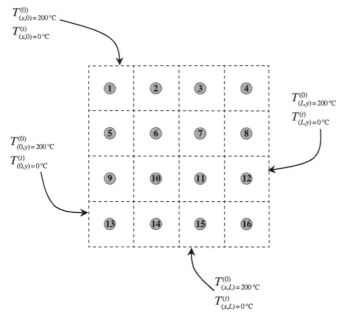

Figure 10.E11.1 Temperature distribution across a rectangular aluminium plate.

$$a'_P T_P^{(t+\Delta t)} = DAT_E^{(t)} + DAT_W^{(t)} + DAT_N^{(t)} + DAT_S^{(t)} + [a'_P - 4DA]\,T_P^{(t)}$$

For example, in Cell-6 ($Fo = D/a'_P$):

$$T_P^{(t+\Delta t)} = Fo\left(T_E^{(t)} + T_W^{(t)} + T_N^{(t)} + T_S^{(t)}\right) + [1 - 4Fo]T_P^{(t)}$$

Thus, we have:

Cell	a_W	a_E	a_N	a_S	S_P	S_u	a_P
6, 7, 10, 11	0	0	0	0	0	$Fo\left(T_E^{(t)} + T_W^{(t)} + T_N^{(t)} + T_S^{(t)}\right) + [1-4Fo]T_P^{(t)}$	1

As we could expect, all the values in the new time step of $t + \Delta t$ will be calculated directly based on the values of the previous time step (t). Now, we can derive a similar statement for all the boundary nodes, which have the same condition due to a similar Dirichlet condition. For example, for cells 2 and 3, ($T_N^{(t)} = 0$):

$$\rho c \Delta V \left(\frac{T_P^{(t+\Delta t)} - T_P^{(t)}}{\Delta t}\right) = DA\left[\left(T_E^{(t)} - T_P^{(t)}\right) - \left(T_P^{(t)} - T_W^{(t)}\right)\right] + DA\left[2\left(0 - T_P^{(t)}\right) - \left(T_P^{(t)} - T_S^{(t)}\right)\right]$$

$$\Rightarrow T_P^{(t+\Delta t)} = Fo\left(T_E^{(t)} + T_W^{(t)} + T_S^{(t)}\right) + [1 - 5Fo]T_P^{(t)}$$

With a similar approach for other cells, hence, we can form the table of coefficient as:

Cell	a_W	a_E	a_N	a_S	S_P	S_u	a_P
2, 3	0	0	0	0	0	$Fo\left(T_E^{(t)} + T_W^{(t)} + T_S^{(t)}\right) + [1 - 5Fo]T_P^{(t)}$	1
5,9	0	0	0	0	0	$Fo\left(T_E^{(t)} + T_N^{(t)} + T_S^{(t)}\right) + [1 - 5Fo]T_P^{(t)}$	1
8,12	0	0	0	0	0	$Fo\left(T_W^{(t)} + T_N^{(t)} + T_S^{(t)}\right) + [1 - 5Fo]T_P^{(t)}$	1
14,15	0	0	0	0	0	$Fo\left(T_E^{(t)} + T_W^{(t)} + T_N^{(t)}\right) + [1 - 5Fo]T_P^{(t)}$	1
1	0	0	0	0	0	$Fo\left(T_E^{(t)} + T_S^{(t)}\right) + [1 - 6Fo]T_P^{(t)}$	1
4	0	0	0	0	0	$Fo\left(T_W^{(t)} + T_S^{(t)}\right) + [1 - 6Fo]T_P^{(t)}$	1
13	0	0	0	0	0	$Fo\left(T_E^{(t)} + T_N^{(t)}\right) + [1 - 6Fo]T_P^{(t)}$	1
16	0	0	0	0	0	$Fo\left(T_W^{(t)} + T_N^{(t)}\right) + [1 - 6Fo]T_P^{(t)}$	1

Note that for corner cells of 1, 4, 13, and 16, two neighbours have zero values. Now, we can replace the values in the above table and solve each value at a new time step. While all the values are given in the problem statement, Δt is our choice to be selected. Figure 10.E11.2, thus, demonstrates the values of three typical cells (due to the boundary and initial conditions) marching through time using $\Delta t = 1$ or $Fo = 1.3$.

If we would like to increase Δt to 2.5 s ($Fo = 0.33$) and 3s ($Fo = 0.40$), then we increase Fo and the results will become unstable. T_6 results are shown in Figure 10.E11.3 as an example ($\Delta t = 0.5$ or $Fo = 0.07$ is also shown for the comparison reasons).

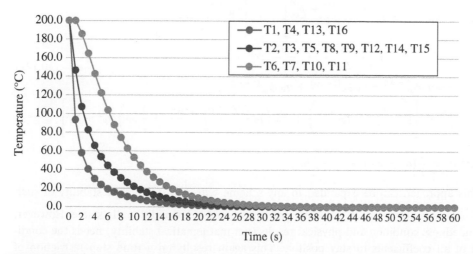

Figure 10.E11.2 Temporal temperature variations of three typical cells of the rectangular aluminium plate using an implicit scheme when $\Delta t = 1$s or $Fo = 1.3$.

As it is clear above $Fo > 0.25$, as expected for a 2D grid, the results are unstable.

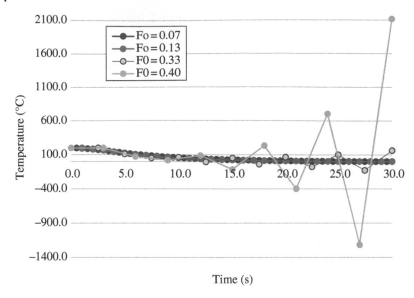

Figure 10.E11.3 Temporal temperature variations of T_6 using an implicit scheme after increasing Δt.

10.14.2 Implicit Scheme

Deriving the unsteady diffusion equation in an implicit form results in multivariable equations obtained from each cell. Refer to Eq. (10.118), we can first develop a semi-implicit approach of Crank–Nicolson by setting $\theta = 0.5$ ($\overline{S}\Delta V = \frac{1}{2}S_P\phi_P^{(t+\Delta t)} + \frac{1}{2}S_P\phi_P^{(t)} + S_u$):

$$a_P\phi_P^{(t+\Delta t)} = a_E\left[\frac{\phi_E^{(t+\Delta t)} + \phi_E^{(t)}}{2}\right] + a_W\left[\frac{\phi_W^{(t+\Delta t)} + \phi_W^{(t)}}{2}\right]$$

$$+ \left[\rho c\frac{\Delta V}{\Delta t} - \frac{a_W}{2} - \frac{a_E}{2}\right]\phi_P^{(t)} + S_u$$

$$\Longrightarrow a_P\phi_P^{(t+\Delta t)} = a_E\phi_E^{(t+\Delta t)} + a_W\phi_W^{(t+\Delta t)}$$

$$+ \left[\left(a_E\frac{\phi_E^{(t)}}{2} + a_W\frac{\phi_W^{(t)}}{2}\right) + \phi_P^{(t)}\left(\rho c\frac{\Delta V}{\Delta t} - \frac{a_E}{2} - \frac{a_W}{2} + \frac{1}{2}S_P\right)\phi_P^{(t)} + S_u\right]$$

(10.121)

where $a_P = \frac{1}{2}(a_E + a_W) + a_P' - \frac{1}{2}S_P$.

This approach can be expanded to any scheme with $\frac{1}{2} \leq \theta < 1$. The stability of such approaches is automatically satisfied, and the scheme is unconditionally stable. However, boundedness condition and physical results (not mathematical stability) needs the condition of all coefficients to stay positive. This again results on a time step restriction of $Fo = \frac{\alpha\Delta t}{(\Delta x)^2} < 1$. Choosing such implicit schemes results in flexibility to select larger Δt, which essentially can speed up the solution of an unsteady transport solution though yet not very practical for 2D and 3D scenarios with numerous cells.

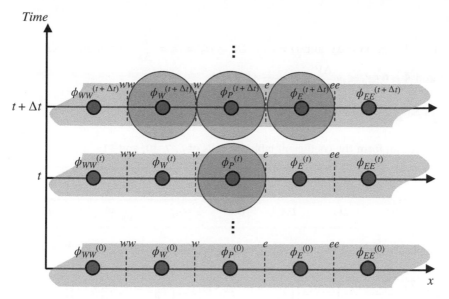

Figure 10.37 Implicit scheme for a 1D grid.

Note that each unknown variable in the implicit scheme at a time step $(t + \Delta t)$ is calculated from unknown neighbours' variables in the current time step $(t + \Delta t)$ and known variables from the previous time step (t) as depicted in Figure 10.37. Unlike to the explicit scheme, deriving the equations in each cell results in a multivariable equation in that a system of linear equations should be formed as 'n' is the number of variables or cells at each time-step. The system of linear equations can be transformed to the matrix form. In this case and when the mesh contains many cells, iterative methods are taken to the account to solve the matrix, otherwise direct methods are preferred to simultaneously obtain the variables at all cell centres. Iterative and direct methods will be discussed in the next chapter.

When $\theta = 1$, we can form the fully implicit schemes as below $\left(\overline{S} = S_P \phi_P^{(t + \Delta t)} + S_u \right)$:

$$a_P \phi_P^{(t + \Delta t)} = a_E \phi_E^{(t + \Delta t)} + a_W \phi_W^{(t + \Delta t)} + \left[a_P' \phi_P^{(t)} + S_u \right]$$

where $a_P = a_E + a_W + a_P' - S_P$.

This scheme is stable and provides realistic results. Nonetheless, the matrix size is a major issue and iterative methods are widely used to solve the resulting matrices as discussed in the next chapter.

Example 10.12
Calculate Example 10.11 with a fully implicit scheme.

Solution

Again, we have $\Gamma = k$, $\Delta x = \Delta y$, and $D = D_e = D_w = D_n = D_s = \dfrac{k}{\Delta x}$. For internal cells, we have $\theta = 1$ and $\overline{S} = 0$ ($a'_p = \rho c \dfrac{\Delta V}{\Delta t}$ and $a_P = 4D + a'_P$):

$$a_P T_P^{(t+\Delta t)} = DA T_E^{(t+\Delta t)} + DA T_W^{(t+\Delta t)} + DA T_N^{(t+\Delta t)} + DA T_S^{(t+\Delta t)} + a'_p T_P^{(t)}$$

Therefore, we can form the coefficient table for cells of 6, 7, 10, and 11:

Cell	a_W	a_E	a_N	a_S	S_P	S_u	a_P
6, 7, 10, 11	DA	DA	DA	DA	$-a'_p$	$a'_p T_P^{(t)}$	$4DA + a'_p$

For other cells, we rewrite Eq. (10.117). For example, for cells 2 and 3 ($T_N^{(t+\Delta t)} = 0$):

$$\rho c \Delta V \left(\frac{T_P^{(t+\Delta t)} - T_P^{(t)}}{\Delta t} \right) = DA \left[\left(T_E^{(t+\Delta t)} - T_P^{(t+\Delta t)} \right) - \left(T_P^{(t+\Delta t)} - T_W^{(t+\Delta t)} \right) \right]$$
$$+ DA \left[2 \left(0 - T_P^{(t+\Delta t)} \right) - \left(T_P^{(t+\Delta t)} - T_S^{(t+\Delta t)} \right) \right]$$

$$\Rightarrow \left(5DA + a'_p \right) T_P^{(t+\Delta t)} = DA \left(T_E^{(t+\Delta t)} + T_W^{(t+\Delta t)} + T_S^{(t+\Delta t)} \right) + a'_p T_P^{(t)}$$

Therefore, the table can be filled as:

Cell	a_W	a_E	a_N	a_S	S_P	S_u	a_P
2, 3	DA	DA	0	DA	$-2D - a'_p$	$a'_p T_P^{(t)}$	$5DA + a'_p$
5, 9	0	DA	DA	DA	$-2D - a'_p$	$a'_p T_P^{(t)}$	$5DA + a'_p$
8, 12	DA	0	DA	DA	$-2D - a'_p$	$a'_p T_P^{(t)}$	$5DA + a'_p$
14, 15	DA	DA	DA	0	$-2D - a'_p$	$a'_p T_P^{(t)}$	$5DA + a'_p$
1	0	DA	0	DA	$-4D - a'_p$	$a'_p T_P^{(t)}$	$6DA + a'_p$
4	DA	0	0	DA	$-4D - a'_p$	$a'_p T_P^{(t)}$	$6DA + a'_p$
13	0	DA	DA	0	$-4D - a'_p$	$a'_p T_P^{(t)}$	$6DA + a'_p$
16	DA	0	DA	0	$-4D - a'_p$	$a'_p T_P^{(t)}$	$6DA + a'_p$

The main difference of the implicit scheme with the implicit one is that at this stage, we will reach a system of linear equations. Here, in this example, we have a 16×16 matrix. Now, if for instance, we solve the matrix for a reasonable $\Delta t = 0.01$ s ($Fo = 0.01$), we can have a similar result to the explicit method as shown in Figure 10.E12.1.

If we increase $\Delta t = 2$ s ($Fo = 0.27$), the implicit scheme become unstable while the implicit scheme can resolve the results as illustrated in Figure 10.E12.2.

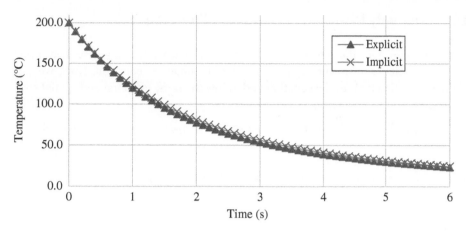

Figure 10.E12.1 Comparison of implicit and explicit schemes for Example 10.11 when Δt = 0.01 s (Fo = 0.01).

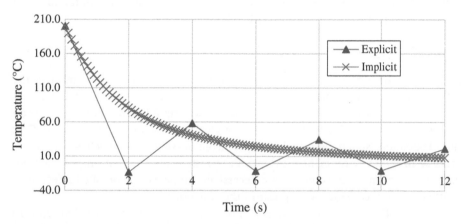

Figure 10.E12.2 Comparison of implicit and explicit schemes for Example 10.11 when Δt = 2 s (Fo = 0.27).

For a 3D space, a similar implicit approach can be followed and therefore an equation can be derived as bellow:

$$a_P\phi_P^{t+\Delta t} = a_W\phi_W^{t+\Delta t} + a_E\phi_E^{t+\Delta t} + a_S\phi_S^{t+\Delta t} + a_N\phi_N^{t+\Delta t} + a_B\phi_B^{t+\Delta t} + a_T\phi_T^{t+\Delta t} + \left[a_P'\phi_P^t + S_u\right]$$

(10.122)

where $a_P = a_E + a_W + a_N + a_S + a_B + a_T + a_P' - S_P$.

Note that we can initially divide the left-hand side in Eq. (10.117) by area (A) if the grid is assumed to be homogenous. This can result on neighbouring coefficients of a_{nb} to exclude areas, and thus one can eliminate these areas in all the equations across the unsteady sections and examples and change ΔV in a_P' to Δx.

10.15   Unsteady Diffusion–Convection Equation

In the previous section, the temporal discretization of each term was described while explicit or implicit schemes were applied to solve the obtained equation. Now, we can include the advection term to the equation and follow the same temporal and spatial strategies to discretize this new term.

$$a_P \phi_P^{t+\Delta t} = a_W \phi_W^{t+\Delta t} + a_E \phi_E^{t+\Delta t} + a_S \phi_S^{t+\Delta t} + a_N \phi_N^{t+\Delta t} + a_B \phi_B^{t+\Delta t} + a_T \phi_T^{t+\Delta t} + a_P' \phi_P^t + S_u$$

(10.123)

where $a_P = a_W + a_E + a_S + a_N + a_B + a_T + a_P' - S_p$ and $a_P' = \rho c \dfrac{\Delta V}{\Delta t}$.

10.16   Pressure–Velocity Coupling

The pressure term appears in Navier–Stokes' equations was not considered in Eq. (10.2) and thus in the utilized FVM. Despite its rather simplistic form, consideration of the pressure gradient term is not always a straightforward task as shown as:

$$a_P \phi_P^{t+\Delta t} = \left(\sum a_{nb} \phi_{nb}^{t+\Delta t} + a_P' \phi_P^t + S_u \right) - \frac{P_e - P_w}{\delta x} \Delta V$$

(10.124)

If the pressure gradient is known, its value can be assumed as a source term in the momentum equations. Nonetheless, in fluids problems, we deal with the situations that we need to take this term to the account as an unknown variable. The pressure then can be linked to the equation of state (for example an ideal gas law for the air as $p = \rho RT$) as a new set of equation. Nonetheless, in the case of incompressible flow, which is also a common practice in built environment problems, the constant pressure can impose another constraint on the momentum equation and necessitates implementation of a correct pressure as otherwise a discrepancy will appear in the equations.

Yet, the velocity field, which was used as a known value to calculate advection term (F) needs to be resolved simultaneously with other flow variables in Eq. (10.2) or its more complete format when the pressure gradient is present. This further imposes a term of uu, uv, or uw in the equations, which is in fact a nonlinear term. On the other word, F itself is not a known value as it was used in the previous sections.

These two major constraints making the solution of complex flows even more difficult if not treated. Nonetheless, while there is a comprehensive literature related to the efforts to resolve such problems, we refer the interested readers to advance books introduced in the references and will just provide the common solution employed in many commercial and bespoke tools. The solution was introduced by embedding a coupling technique between staggered grids of scalar and velocity fields. Since the proposition of this strategy multiple algorithms have been developed and further modified to facilitate the coupling technique. We intend to simply explain the integration of the pressure gradient at this section though do not cover the details of such algorithms in this book.

As shown in Figure 10.38, the proposed technique is about guessing F with a velocity, and then using it in addition to a guessed pressure field to resolve the momentum equations.

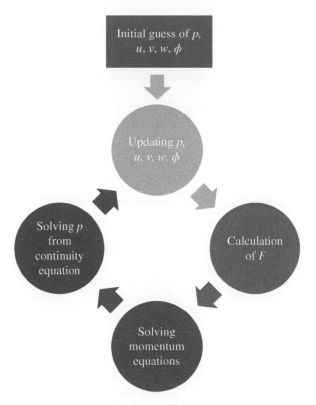

Figure 10.38 Coupling algorithm of velocity and pressure fields.

Then, the continuity will be checked with the found velocities and again will be utilized to resolve the velocity and pressure fields. Thus, the continuity equation turns to an equation of transport for the pressure. This process will be repeated until a desired convergence is obtained.

Another important technique in deployment of the coupling technique of Figure 10.38 is that the explained algorithm should be implemented in a ***staggered grid***. The idea of using such grid is that storing scalar variables such as pressure and temperature in the same cell centres as quantity ϕ can end up in misleading results. A famous example is when both pressures at W and E neighbouring cells are the same (let us say p_1) and different from Cell-P (let's say $p_2 \neq p_1$). Thus, the pressure gradient using Eq. (10.124) at the cell-faces of 'e' and 'w' can be calculated as:

$$\frac{\partial p}{\partial x} = \frac{p_e - p_w}{\delta x} = \frac{\left(\frac{p_E + p_P}{2}\right) - \left(\frac{p_P + p_W}{2}\right)}{\delta x}$$

$$\implies \frac{\partial p}{\partial x} = \frac{p_E - p_W}{2\delta x} = \frac{p_1 - p_1}{2\delta x} = 0$$

(10.125)

This, however, is an unrealistic result where there is a change of pressure between P and E as well as P and W. A 2D staggered gird is shown in Figure 10.39 as a remedy example to such potential problem. In such grid, scalar variables are stored at a grid in which the cell

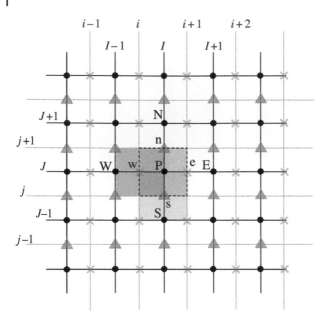

Figure 10.39 2D staggered grid.

centres are shown with dots. The coordinate of this grid is specified with capital letters 'I' and 'J'. For example, pressures are denoted as $p_E^{(m)}$, which implies pressure at Cell-E in the main grid. Now, assume that another 2D grid is created for the x-component velocities (u) with shifting the main grid to the left as seen in Figure 10.40a. Similarly, the pressure at Cell-E in this grid is denoted as $p_E^{(u)}$. The cell centres of this grid have coordinates with lowercase letter of 'i' and capital letter of 'J'. Now, we can rewrite Eq. (10.125) as:

$$\frac{\partial p^{(m)}}{\partial x} = \frac{p_e^{(m)} - p_w^{(m)}}{\delta x}$$

$$\Longrightarrow \frac{\partial p^{(m)}}{\partial x} = \frac{p_E^{(u)} - p_P^{(u)}}{\delta x} = \frac{p_2 - p_1}{\delta x} \neq 0$$

(10.126)

Thus, such shifted grid can now return nonzero values as expected. In the case of necessity to calculate y-component values in a 2D grid where the shifted cells result in undefined variables at cell-faces, interpolation can be applied as shown in Figure 10.40b:

$$\frac{\partial p^{(m)}}{\partial y} = \frac{p_n^{(m)} - p_s^{(m)}}{\delta y}$$

$$\Longrightarrow \frac{\partial p^{(m)}}{\partial y} = \frac{\left[\frac{p_{nw}^{(u)} + p_{ne}^{(u)} + p_w^{(u)} + p_e^{(u)}}{4}\right] - \left[\frac{p_w^{(u)} + p_e^{(u)} + p_{sw}^{(u)} + p_{se}^{(u)}}{4}\right]}{\delta y}$$

$$\Longrightarrow \frac{\partial p^{(m)}}{\partial y} = \frac{p_{nw}^{(u)} + p_{ne}^{(u)} - p_{sw}^{(u)} - p_{se}^{(u)}}{4\delta y}$$

(10.127)

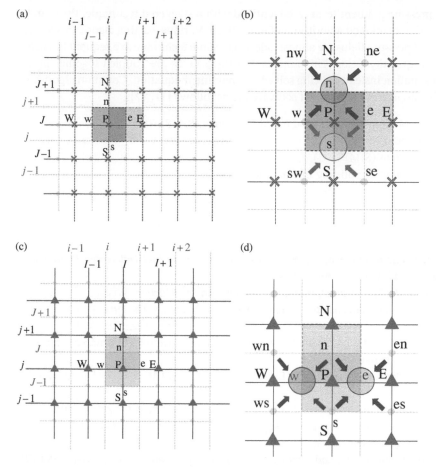

Figure 10.40 (a) x-Component (u-grid), (b) interpolation of scalar variables in u-grid, (c) y-component (v-grid), and (d) interpolation of scalar variables in v-grid.

A similar procedure can be applied to shift the main grid downwards to create a y-component grid. The pressure at Cell-E in this grid is denoted as $p_E^{(v)}$. The cell centres of this grid have coordinates with capital letter of 'I' and lowercase letter of 'j' (see Figure 10.40c). with a similar calculation, we can reach the below gradients at cell-faces:

$$\frac{\partial p^{(m)}}{\partial y} = \frac{p_P^{(V)} - p_S^{(V)}}{\delta y} \tag{10.128}$$

And (Figure 10.40d):

$$\frac{\partial p^{(m)}}{\partial x} = \frac{\left[\dfrac{p_n^{(V)} + p_{en}^{(V)} + p_s^{(V)} + p_{es}^{(V)}}{4}\right] - \left[\dfrac{p_{wn}^{(V)} + p_n^{(V)} + p_{ws}^{(V)} + p_s^{(V)}}{4}\right]}{\delta x}$$

$$\Longrightarrow \frac{\partial p^{(m)}}{\partial x} = \frac{p_{en}^{(V)} + p_{es}^{(V)} - p_{wn}^{(V)} - p_{ws}^{(V)}}{4\delta x} \tag{10.129}$$

Once the pressure gradient terms are identified with a staggered technique, they can be replaced into the mentioned algorithm of Figure 10.38. Various existing algorithms to couple velocity and pressure fields are almost following a same technique in a sequent to update the scalar and quantity fields. This implies that these fields are coupled together in the momentum equation instead of being solved as a nonlinear equation. Common algorithms being widely adopted in commercial software are as follows [11, 12]:

- Simi-Implicit Method for Pressure-Linked Equations (SIMPLE)
- SIMPLE Revised (SIMPLER)
- SIMPLE-Consistent (SIMPLEC)
- Pressure Implicit with Splitting of Operations (PISO)

References

1 Pinchover, Y. and Rubinstein, J. (2005). *An Introduction to Partial Differential Equations*. Cambridge: Cambridge University Press.
2 Strauss, W.A. (2007). *Partial Differential Equations: An Introduction*. Wiley.
3 Versteeg, H.K., *An introduction to Computational Fluid Dynamics: The Finite Volume Method*. 1995: Harlow, Essex, England; New York Longman Scientific & Technical; Wiley.
4 Tu, J., Yeoh, G.H., and Liu, C. (2007). *Computational Fluid Dynamics: A Practical Approach*. Butterworth-Heinemann.
5 Ferziger, J.H., Perić, M., and Street, R.L. (2002). *Computational Methods for Fluid Dynamics*, vol. 3. Springer.
6 Zhang, R., Gan, Y., and Mirzaei, P.A. (2020). *A new regression model to predict BIPV cell temperature for various climates using a high-resolution CFD microclimate model. Advances in Building Energy Research* 14 (4): 527–549.
7 Shirzadi, M., Tominaga, Y., and Mirzaei, P.A. (2020). *Experimental and steady-RANS CFD modelling of cross-ventilation in moderately-dense urban areas. Sustainable Cities and Society* 52: 101849.
8 Aydin, Y.C. and Mirzaei, P.A. (2017). *Wind-driven ventilation improvement with plan typology alteration: A CFD case study of traditional Turkish architecture. Building Simulation* 10 (2): 239–254.
9 Gabriel, J.D. and John, A.S. (1985). *ANSYS Engineering Analysis System User's Manual*. Houston, PA: Swanson Analysis Systems.
10 Ellsworth, M.J., et al. The evolution of water cooling for IBM large server systems: Back to the future. In 2008 11th Intersociety Conference on Thermal and Thermomechanical Phenomena in Electronic Systems. 2008.
11 Anderson, D., Tannehill, J.C., and Pletcher, R.H. (2016). *Computational Fluid Mechanics and Heat Transfer*. Taylor & Francis.
12 Patankar, S. (1980). *Numerical Heat Transfer and Fluid Flow*. London, UK: Taylor & Francis.

11

Solvers and Solution Analysis

11   Introduction

11.1   Solvers of Algebraic Equation Systems

In the previous chapter, we scrutinized the finite volume method (FVM) to transform the Navier–Stokes (NS) equations as a set of nonlinear partial differential ones to a system of linear algebraic equations. The process was performed by discretization of the NS equations discussed in Section 10.6. After discretization, a matrix, or multiple matrices of the system of linear equations, depending on the type of the simplification of NS, could be formed. Then after, solving the system of equations will provide numerical solution of the problem. Nonetheless, this stage, similar to pre- and post-processing stages, needs careful considerations as the size of linear algebraic equation systems can be extremely large, which is the case for many common engineering problems, specifically built environment related ones. The commonly used methods are based on direct and iterative approaches.

Direct approaches, as their name indicates, can provide the exact solution of a system of linear equation. On the other hand, iterative technique reaches to a solution close to the exact solution of the system of linear equation via performing a series of calculations within a loop. Evidently, the closeness of the results needs to be defined by a user based on the nature of a problem and the desired accuracy from a computational fluid dynamics (CFD) simulation. We discuss each of direct and iterative methods with their associated advantages and disadvantages in more detail in the following sections.

11.2   Direct Method

Common examples of a direct solution of system of linear algebraic equation include ***Cramer's rule in matrix inversion***, ***Gaussian elimination***, and ***one-dimensional tri-diagonal matrix algorithm (TDMA)***. If a system of equations has N unknowns (the-number of cells in a mesh), it is well understood that the number of operations to utilize such approaches is in order of magnitude of N^3. Storage of required coefficients is also a challenge when N becomes large as it is calculated to be in the order of magnitude of N^2 in many direct

Computational Fluid Dynamics and Energy Modelling in Buildings: Fundamentals and Applications, First Edition. Parham A. Mirzaei.
© 2023 John Wiley & Sons Ltd. Published 2023 by John Wiley & Sons Ltd.

methods. Aside from being computationally expensive and dependent on high storage capacities, the direct methods result in exact solutions of a system of linear equations.

11.2.1 Cramer's Rule

The most basic way to find the solution to a system of linear equations is Cramer's rule. Considering a linear system of algebraic equations as below:

$$A\phi = B \tag{11.1}$$

where the matrix A $(n \times n)$ has a non-zero determinant, and ϕ is the column vector of the variables. It can be proved that this system has a unique solution, multiplying both sides of equation by the inverse A, known as matrix A^{-1} $(n \times n)$:

$$A^{-1}A\phi = A^{-1}b$$
$$\Longrightarrow I\phi = \phi = A^{-1}B \tag{11.2}$$

where I is the identity matrix and we know that $AA^{-1} = A^{-1}A = I$.

Therefore, to find the solution vector, we need to first obtain A^{-1} with the below equation:

$$A^{-1} = \frac{1}{\det(A)} \operatorname{adj}(A) \tag{11.3}$$

where $\operatorname{adj}(A)$ is the adjugate matrix of A, which is the transpose of cofactor matrix expressed as:

$$\operatorname{adj}(A) = C^{\mathrm{T}} = \begin{bmatrix} \operatorname{cof}(A,1,1) & \operatorname{cof}(A,1,2) & \cdots & \operatorname{cof}(A,1,n) \\ \operatorname{cof}(A,2,1) & \operatorname{cof}(A,2,2) & \cdots & \operatorname{cof}(A,2,n) \\ \vdots & \vdots & & \cdots \\ \operatorname{cof}(A,n,1) & \operatorname{cof}(A,n,2) & \cdots & \operatorname{cof}(A,n,n) \end{bmatrix}^{\mathrm{T}} \tag{11.4}$$

Thus, each element of the cofactor matrix can be calculated as:

$$C = (-1)^{i+j} \det(A(i \,|\, j)) \tag{11.5}$$

where $\det(A(i \,|\, j))$ implies the determinant of A when the row and column are associated with the eliminated $a_{i,j}$ element.

Example 11.1

Assume the below linear system of algebraic equations and find the vector of solution using Cramer' rule.

$$8x_1 + 2x_2 + 2x_4 = 20$$

$$-6x_2 - x_3 + 2x_4 = -7$$

$$3x_1 - 2x_2 + 7x_3 - x_4 = 16$$

$$-x_1 + x_2 + 2x_3 - 5x_4 = -13$$

Solution

First, we form the matrix of the system of linear equations:

$$\begin{bmatrix} 8 & 2 & 0 & 2 \\ 0 & -6 & -1 & 2 \\ 3 & -2 & 7 & -1 \\ -1 & 1 & 2 & -5 \end{bmatrix} \begin{bmatrix} x_1 \\ x_2 \\ x_3 \\ x_4 \end{bmatrix} = \begin{bmatrix} 20 \\ -7 \\ 16 \\ -13 \end{bmatrix}$$

The inverse of matrix A can be found as:

$$A^{-1} =$$

0.141	0.067	−0.015	0.086
−0.008	−0.180	−0.005	−0.074
−0.071	−0.092	0.157	−0.097
−0.058	−0.086	0.065	−0.271

Thus, we can have the vector of solution as:

$$\begin{bmatrix} x_1 \\ x_2 \\ x_3 \\ x_4 \end{bmatrix} = \begin{bmatrix} 1 \\ 2 \\ 3 \\ 4 \end{bmatrix}$$

11.2.2 Gaussian Elimination

Gaussian elimination method is one of the basic direct methods, which applies an elimination process to reduce the size of a linear system of algebraic equations. It starts with transformation of Matrix A in the equation system ($A\phi = B$) to an **upper triangular matrix** (U) with multiplying its first row by $\dfrac{A_{2,1}}{A_{1,1}}$ and then subtract it from the second row, which results to $A_{2,\,1}$ becomes zero:

$$A' = \begin{bmatrix} A_{1,1} & A_{1,2} & \cdots & A_{1,n} \\ A_{2,1} - A_{1,1}\left(\dfrac{A_{2,1}}{A_{1,1}}\right) & A_{2,2} - A_{1,2}\left(\dfrac{A_{2,1}}{A_{1,1}}\right) & \cdots & A_{2,n} - A_{1,n}\left(\dfrac{A_{2,1}}{A_{1,1}}\right) \\ \vdots & \vdots & \cdots & \\ A_{n,1} & A_{n,2} & & A_{n,n} \end{bmatrix} \tag{11.6}$$

With further elimination $A_{2,1}, A_{3,1}, ...,$ and $A_{n,1}$, we make all the elements in the column below $A_{1,1}$ to turn to zero; thus, we have $A_{i,1} = 0 \, (1 < i \leq n)$. Then, we repeat the elimination procedure with the elements of $A_{3,2}, A_{4,2}, ...,$ and $A_{n,2}$ until we have $A_{i,2} = 0 \, (2 < i \leq n)$. We continue the procedure to transform all the elements below the diagonal elements to become zero and to reach an U-matrix as expressed below:

$$U = \begin{bmatrix} A_{1,1} & A_{1,2} & \cdots & A_{1,n} \\ 0 & A'_{2,2} & \cdots & A'_{2,n} \\ \vdots & \vdots & & \cdots \\ 0 & 0 & \cdots & A'_{n,n} \end{bmatrix} \text{ and } B' = \begin{Bmatrix} B'_1 \\ B'_2 \\ \vdots \\ B'_n \end{Bmatrix} \tag{11.7}$$

Obviously, the elements of the vector of the variables (B') as well as the product U-matrix differ from the original matrix of A, except for the first row. The whole process is known as the **forward elimination** until reaching to $U\phi = B'$. Now, the **back-substitution** procedure can be applied to the U-matrix to find the column vector of the variables as follows:

$$\phi_n = \frac{B'_n}{A'_{n,n}} \tag{11.8}$$

The found element of ϕ_n is then used in the next equation to find ϕ_{n-1}:

$$\phi_{n-1} = \frac{B'_{n-1} - A'_{n-1,n}\phi_n}{A'_{n-1,n-1}} \tag{11.9}$$

The back-substitution in general can be represented as:

$$\phi_i = \frac{B'_i - \sum_{j=i+1}^{n} A'_{i,j}\phi_j}{A'_{i,i}} \tag{11.10}$$

Example 11.2
Solve the system of linear equation of Example 11.1 using Gaussian elimination.

Solution

The initial matrix form is as below:

$$\begin{bmatrix} 8 & 2 & 0 & 2 \\ 0 & -6 & -1 & 2 \\ 3 & -2 & 7 & -1 \\ -1 & 1 & 2 & -5 \end{bmatrix} \begin{bmatrix} x_1 \\ x_2 \\ x_3 \\ x_4 \end{bmatrix} = \begin{bmatrix} 20 \\ -7 \\ 16 \\ -13 \end{bmatrix}$$

Dividing row-1 by 8:

$$\begin{bmatrix} 1 & 0.25 & 0 & 0.25 & | & 2.5 \\ 0 & -6 & -1 & 2 & | & -7 \\ 3 & -2 & 7 & -1 & | & 16 \\ -1 & 1 & 2 & -5 & | & -13 \end{bmatrix}$$

Now, reducing three times of row-1 from row-3, and adding row-1 to row-4 gives:

$$
\left[\begin{array}{cccc|c}
1 & 0.25 & 0 & 0.25 & 2.5 \\
0 & -6 & -1 & 2 & -7 \\
0 & -2.75 & 7 & -1.75 & 8.5 \\
0 & 1.25 & 2 & -4.75 & -10.5
\end{array}\right]
$$

After dividing row-2 by -6:

$$
\left[\begin{array}{cccc|c}
1 & 0.25 & 0 & 0.25 & 2.5 \\
0 & 1 & 0.167 & -0.333 & 1.167 \\
0 & -2.75 & 7 & -1.75 & 8.5 \\
0 & 1.25 & 2 & -4.75 & -10.5
\end{array}\right]
$$

We add 2.75 times of row-2 to row-3, and -1.25 times of it to row-4:

$$
\left[\begin{array}{cccc|c}
1 & 0.25 & 0 & 0.25 & 2.5 \\
0 & 1 & 0.167 & -0.333 & 1.167 \\
0 & 0 & 7.458 & -2.667 & 11.708 \\
0 & 0 & 1.792 & -4.333 & -11.958
\end{array}\right]
$$

Again, we divide row-3 by 7.458 and reduce 1.792 time of it from row-4:

$$
\left[\begin{array}{cccc|c}
1 & 0.25 & 0 & 0.25 & 2.5 \\
0 & 1 & 0.167 & -0.333 & 1.167 \\
0 & 0 & 1 & -0.358 & 1.570 \\
0 & 0 & 0 & -3.693 & -14.772
\end{array}\right]
$$

Now, we find:

$$
x_4 = \frac{-14.772}{-3.693} = 4
$$

And, with back-substitution procedure we find:

$$
x_3 = 1.570 - 4 \times x_4 = 3
$$

And, similarly:

$$
x_2 = 2 \text{ and } x_1 = 1
$$

11.2.3 1D TDMA

This algorithm is suitable for a range of problems when the resultant matrix of system of linear algebraic equation has many zero elements, known as a **spare matrix**. This was the case of many of the presented equations in Chapter 10, resulting from application of FVM on steady or transient diffusion and diffusion-convection equations. In general, non-zero elements found to be diagonally located for 1D, 2D, and 3D grids with structured types as shown in Figure 11.1.

TDMA is a Gaussian elimination approach in one-dimensional conditions, which is inexpensive in term of arithmetic operations, and requires a minimum amount of storage. Nonetheless, it transforms into an iterative method in multi-dimensional problems.

To explain TDMA, let us consider a tri-diagonal matrix as below:

$$
\begin{bmatrix}
b_1 & c_1 & 0 & & & & \\
& & & \cdots & & 0 & \\
a_2 & b_2 & c_2 & & & & \\
& \cdots & & \cdots & & \cdots & \\
& & a_{n-1} & b_{n-1} & c_{n-1} & \\
0 & & & \cdots & & \\
& & 0 & a_n & b_n &
\end{bmatrix}
\begin{bmatrix}
\phi_1 \\
\phi_2 \\
0 \\
\phi_{n-1} \\
\phi_n
\end{bmatrix}
=
\begin{bmatrix}
d_1 \\
d_2 \\
0 \\
d_{n-1} \\
d_n
\end{bmatrix}
\tag{11.11}
$$

Hence, the general form can be written as:

$$a_i\phi_{n-1} + b_i\phi_n + c_i\phi_{n+1} = d_i \tag{11.12}$$

Now, after applying **_forward elimination_** $(2 \leq i \leq n-1)$, we can reach to a general form of:

$$c^*_i = \frac{c_i}{b_i - a_i c^*_{i-1}} \tag{11.13}$$

$$d^*_i = \frac{d_i - a_i d^*_{i-1}}{b_i - a_i c^*_{i-1}} \tag{11.14}$$

And, we have:

$$c^*_1 = \frac{c_1}{b_1} \tag{11.15}$$

$$d^*_1 = \frac{d_1}{b_1} \tag{11.16}$$

$$d^*_n = \frac{d_n - a_n d^*_{n-1}}{b_n - a_n c^*_{n-1}} \tag{11.17}$$

At this stage, we apply the **_back-substitution_** as explained in Gaussian elimination method in a reverse order to obtain $\phi_n, \phi_{n-1}, \phi_{n-2}, ..., \phi_2$:

$$\phi_n = d^*_n \tag{11.18}$$

And, then for $i = n-1$ to $i = 1$:

$$\phi_i = d^*_i - c^*_i \phi_{i+1} \tag{11.19}$$

(a)

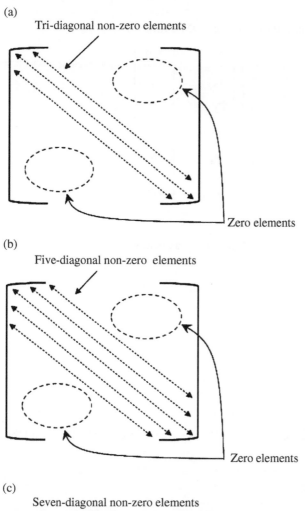

(b)

(c)

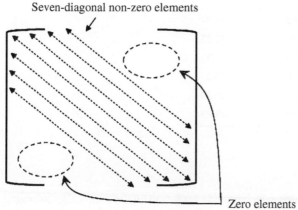

Figure 11.1 Diagonal matrices in (a) 1D, (b) 2D, and (c) 3D conditions.

Example 11.3

Solve the below tri-diagonal matrix equation using TDMA method.

$$
\begin{bmatrix}
26 & -16 & 0 & 0 & 0 & 0 \\
16 & -32 & 16 & 0 & 0 & 0 \\
0 & 16 & -32 & 16 & 0 & 0 \\
0 & 0 & 16 & -32 & 16 & 0 \\
0 & 0 & 0 & 16 & -32 & 16 \\
0 & 0 & 0 & 0 & -16 & 21
\end{bmatrix}
\begin{bmatrix}
T_0 \\ T_1 \\ T_2 \\ T_3 \\ T_4 \\ T_5
\end{bmatrix}
=
\begin{bmatrix}
350 \\ 0 \\ 0 \\ 0 \\ 0 \\ 125
\end{bmatrix}
$$

Solution

First, we find coefficients of Eq. (11.12) for all six nodes as below:

i	a_i	b_i	c_i	d_i
1	*	26	−16	350
2	16	−32	16	0
3	16	−32	16	0
4	16	−32	16	0
5	16	−32	16	0
6	−16	21	*	125

Replacing the matrices value in Eqs. (11.13) and (11.14), we can use Eq. (11.19) to find the solution as:

i	c^*_i	d^*_i	ϕ_i
1	−0.615	13.462	33.37
2	−0.722	9.722	32.35
3	−0.783	7.609	31.33
4	−0.821	6.250	30.31
5	−0.848	5.303	29.29
6	*	28.265	28.27

11.3 Iterative Method

Iterative solvers are based on algorithms, which provide a solution for the matrix after imposing an initial guess for the solution. The initial guess, thus, improves to reach the results via repetition or looping around a defined procedure. The advantage of iterative methods is on their cheaper computational cost when the system of equations is large as the number of arithmetic operations on them is in the order of magnitude of N for a $N \times N$ matrix per each iteration.

However, it should be noted that the number of iterations and thus the exact computational cost of such methods highly depends on the predefined convergence criteria, which will be discussed in the following sections. Furthermore, iterative methods are more efficient in storing non-zero coefficients in the matrices. As we will explain the disadvantages of iterative methods in details in the following sections, reaching to a satisfactory convergence is not always a straightforward process and needs careful considerations and user experience. Below, we introduce some of the most common iterative algorithms, widely being used in FVM.

11.3.1 Jacobi

We start with a simple, but commonly utilized iterative algorithms, known as **Jacobi**. The general form of the algorithm can be derived based on an expression for the diagonal coefficients, in which each row is solved in isolation, assuming other variables to be known from the former iteration as shown in Figure 11.2 with grey circles. On other words, we assume $\sum_{\substack{j=1 \\ j \neq i}}^{n} (A_{ij})\phi_j$ to be constant value in an iteration shown by C_i, and thus the general for-

mulation of Jacobi can be written as:

$$A_{ii}\phi_i = B_i - \sum_{\substack{j=1 \\ j \neq i}}^{n} (A_{ij})\phi_j$$

$$\implies A_{ii}\phi_i = B_i - C_i = D_i \tag{11.20}$$

Then, we can extract ϕ_i from this equation while k indicates the number of iteration ($i = 1, 2, ..., n$):

$$\phi_i^{(k)} = \frac{B_i}{A_{ii}} - \sum_{\substack{j=1 \\ j \neq i}}^{n} \left(\frac{A_{ij}}{A_{ii}}\right)\phi_j^{(k-1)}$$

$$\implies \phi_i^{(k)} = \frac{D_i}{A_{ii}} \tag{11.21}$$

The condition that this algorithm requires to work or be converged is that the matrix A is strictly diagonally dominant, which implies the absolute value of a diagonal term is larger than the summation of absolute values of other terms in a particular row:

$$|A_{ii}| > \sum_{j \neq i} |A_{ij}| \tag{11.22}$$

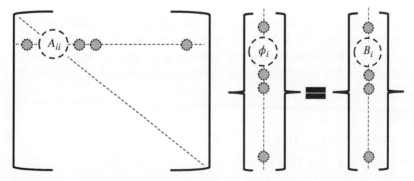

Figure 11.2 Jacobi Algorithm uses values only from the last iteration in a specific row.

Example 11.4

Solve the matrix of Example 11.1 using Jacobi method with initial guess of zero for all values.

Solution

$$
\begin{bmatrix}
8 & 2 & 0 & 2 \\
0 & -6 & -1 & 2 \\
3 & -2 & 7 & -1 \\
-1 & 1 & 2 & -5
\end{bmatrix}
\begin{bmatrix}
x_1 \\
x_2 \\
x_3 \\
x_4
\end{bmatrix}
=
\begin{bmatrix}
20 \\
-7 \\
16 \\
-13
\end{bmatrix}
$$

Based on the above matrix, the equations become:

$$x_1^{(k)} = \left(20 - 2x_2^{(k-1)} - 2x_4^{(k-1)}\right)/8$$

$$x_2^{(k)} = \left(-7 + x_3^{(k-1)} - 2x_4^{(k-1)}\right)/(-6)$$

$$x_3^{(k)} = \left(16 - 3x_1^{(k-1)} + 2x_2^{(k-1)} + x_4^{(k-1)}\right)/(7)$$

$$x_4^{(k)} = \left(-13 + x_1^{(k-1)} - x_2^{(k-1)} - 2x_3^{(k-1)}\right)/(-5)$$

With applying $x_1^{(0)} = x_2^{(0)} = x_3^{(0)} = x_4^{(0)} = 0$, we can iterate and find the values as:

0	1	2	3	4	...	28	29
0.000000	2.500000	1.558333	1.275000	1.171052		1.000000	1.000000
0.000000	1.166667	1.652381	1.929365	1.869824		2.000000	2.000000
0.000000	2.285714	1.919048	2.553912	2.774308		3.000000	3.000000
0.000000	2.600000	3.247619	3.386429	3.752438		3.999999	4.000000

First, we can define the residual, which will be discussed in detail in the following sections as:

$$\left\| R^{(k)} \right\|_1 = \sum_i \left| B_i - \sum_j^n \left(A_{ij}\right) \phi_j^{(k)} \right|$$

Or, another definition as:

$$\left\| R^{(k)} \right\|_2 = \sqrt{\sum_i \left| B_i - \sum_j^n \left(A_{ij}\right) \phi_j^{(k)} \right|^2}$$

If we plot $\left\| R^{(k)} \right\|_2$ in a logarithmic scale as below, we can see that the residuals tend to gradually decrease. We can stop the calculation once it is below a threshold residual (e.g. here 10^{-6} occurs after 28 iterations), which is shown as a red line in Figure 11.E4.

Also, we expect the matrix A to easily converge as the condition of Eq. (11.22) is satisfied:

$$i = 1 \Longrightarrow |8| > |2| + |0| + |2| = 4$$

$$i = 2 \Longrightarrow |-6| > |0| + |-1| + |2| = 3$$

$$i = 3 \Longrightarrow |7| > |3| + |-2| + |-1| = 6$$

$$i = 4 \Longrightarrow |-5| > |-1| + |1| + |2| = 4$$

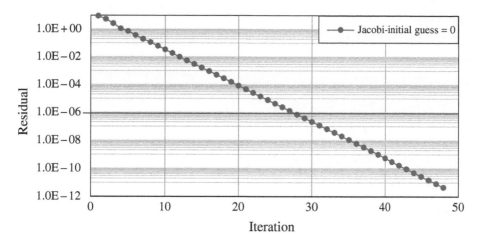

Figure 11.E4 Plot of residual of Jacobi method after multiple iterations with an initial guess of zero.

11.3.2 Gauss–Seidel

The difference between Gauss–Seidel and Jacobi algorithms, is in the way that we update $\phi_i^{\langle k \rangle}$ in the equation. Whilst in Jacobi only $\phi_j^{\langle k-1 \rangle}$ ($1 \leq j \leq n$ and $j \neq i$) values from the previous iteration are employed to calculate $\phi_i^{\langle k \rangle}$, Gauss–Seidel simultaneously utilizes values from the current iteration (shown with the grey stars in Figure 11.3), but only those have been already resolved ($\phi_j^{\langle k \rangle}$ where $1 \leq j \leq i-1$) in addition to the values from the previous iteration ($\phi_j^{\langle k-1 \rangle}$ ($i+1 \leq j \leq n$) as shown in with grey dots in Figure 11.3. Thus, the general equation can be derived as:

$$\phi_i^{\langle k \rangle} = \frac{B_i}{A_{ii}} - \sum_{j=1}^{i-1} \left(\frac{A_{ij}}{A_{ii}} \right) \phi_j^{\langle k \rangle} - \sum_{j=i+1}^{n} \left(\frac{A_{ij}}{A_{ii}} \right) \phi_j^{\langle k-1 \rangle} \tag{11.23}$$

Gauss–Seidel has typically a faster convergence, which can be examined with the below example. To ensure a certain convergence for Gauss–Seidel method, A matrix should be strictly diagonally dominant. In addition to this condition, A matrix should be positive-definite.

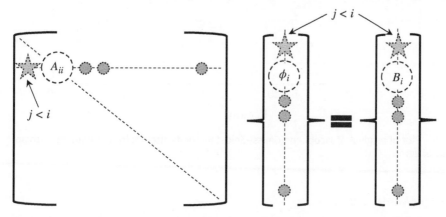

Figure 11.3 Gauss–Seidel algorithm uses values from both current and last iterations in a specific row.

Example 11.5

Solve Example 11.1 using Gauss–Seidel method with an initial guess of zero for all values.

Solution

The former equations in Gauss–Seidel become:

$$x_1^{(k)} = \left(20 - 2x_2^{(k-1)} - 2x_4^{(k-1)}\right)/8$$

$$x_2^{(k)} = \left(-7 + x_3^{(k-1)} - 2x_4^{(k-1)}\right)/(-6)$$

$$x_3^{(k)} = \left(16 - 3x_1^{(k)} + 2x_2^{(k)} + x_4^{(k-1)}\right)/(7)$$

$$x_4^{(k)} = \left(-13 + x_1^{(k)} - x_2^{(k)} - 2x_3^{(k)}\right)/(-5)$$

Again, we have $x_1^{(0)} = x_2^{(0)} = x_3^{(0)} = x_4^{(0)} = 0$, so we can find variables in other iterations as:

0	1	2	3	4	...	10	11
0.000000	2.500000	1.470238	1.093835	1.021010		1.000002	1.000000
0.000000	1.166667	1.892857	1.974235	1.994890		1.999999	2.000000
0.000000	1.547619	2.618197	2.914110	2.981210		2.999998	3.000000
0.000000	2.952381	3.731803	3.941724	3.987260		3.999999	4.000000

Hence, the convergence with a residual of 10^{-6} occurs after 10 iterations, almost more than twice faster as the Jacobi method. The plot in Figure 11.E5 compares the residual of both methods.

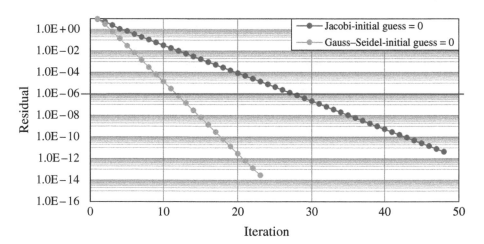

Figure 11.E5 Plot of residual of Jacobi and Gauss–Seidel methods after multiple iterations with an initial guess of zero.

11.3.3 Higher-order TDMA

We have introduced 1D TDMA as an effective direct method for three-diagonal problems in Section 11.4.3. Here, we further expand this algorithm to higher dimensions, including 2D and 3D, as an iterative solver.

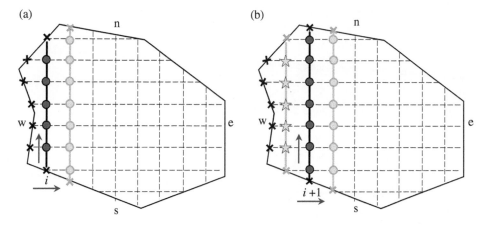

Figure 11.4 The sweeping process in 2D TDMA.

We start with a 2D discretized transport problem in a structured mesh as illustrated in Figure 11.4a. Consider the very eastern column as a 1D equation and its system of equations in the form of $A\phi = B$. If we assume the southern and northern boundary values to be fixed, this equation can be directly solved with a 1D TDMA. The only difference is associated with the terms related to the second dimension that are embedded in term B as known variables from the previous iteration (grey circles) or found from the current iteration (grey stars). In general, a 2D transport applied to a row can be written as:

$$a_P\phi_P^{\langle k \rangle} - a_S\phi_S^{\langle k \rangle} - a_N\phi_N^{\langle k \rangle} = a_W\phi_W^{\langle k \rangle} + a_E\phi_E^{\langle k \rangle} + S_u^{\langle k \rangle} \tag{11.24}$$

For the first row, the western values are known from boundary conditions and the eastern boundaries can be assumed from the previous iteration as shown by the grey circles. For the next column as seen in Figure 11.4b, the western values can be obtained from already found values of this column as illustrated by the grey stars. On the other hand, the eastern values again can be assumed from the previous iteration (grey circles). Thus, we can rewrite the Eq. (11.24) as:

$$a_P\phi_P^{\langle k \rangle} - a_S\phi_S^{\langle k \rangle} - a_N\phi_N^{\langle k \rangle} = a_W\phi_W^{\langle k \rangle} + a_E\phi_E^{\langle k-1 \rangle} + S_u^{\langle k \rangle} \tag{11.25}$$

The right-hand side values of Eq. (11.25) is a constant value. Thus, we can apply a 1D TDMA to find $\phi_P^{\langle k \rangle}$s at each column and then we move to the next right-hand side column until reaching the eastern boundary conditions. The process is called ***sweeping*** that should be iterated several times until a desired convergence is achieved. The first iteration, hence, requires an initial guess for $\phi_W^{\langle 0 \rangle}$ and $\phi_E^{\langle 0 \rangle}$ as we can define:

$$a_P\phi_P^{\langle 1 \rangle} - a_S\phi_S^{\langle 1 \rangle} - a_N\phi_N^{\langle 1 \rangle} = a_W\phi_W^{\langle 0 \rangle} + a_E\phi_E^{\langle 0 \rangle} + S_u^{\langle 0 \rangle} \tag{11.26}$$

Equation (11.26) can be solved with the direct 1D TDMA as explained before while the known boundary values at north and south boundaries of a particular column of 'i' in the structured grid are used. Once, all $\phi^{\langle 1 \rangle}$s are resolved in the ith-column, we can move to the next one

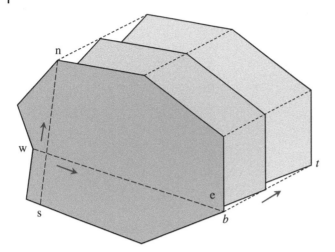

Figure 11.5 The sweeping process in 3D TDMA.

$(i + 1)$ and repeat the same procedure until all the 2D mesh is covered and hence all $\phi^{\langle 1 \rangle}$ in the mesh are solved. The process should start from either east or west side of the mesh and towards to the other side, which is known as the ***sweep direction***. Note, however, in Eq. (11.26) and depending on the sweep direction, the values of either $\phi_W^{\langle 0 \rangle}$ (west-sweep) or $\phi_E^{\langle 0 \rangle}$ (east-sweep) in $(i + 1)$-column is now obtained from the previously resolved (i)-column. The sweeping process will be continued until a desired convergence is reached in the nth iteration.

The extension of TDMA in 3D geometries is similar to the 2D one while in this case, a 3D mesh is sliced to various 2D planes as depicted in Figure 11.5. The procedure is therefore the first to resolve ϕ-values on each line, and then continues the sweeping process on each plane. After that, we can move to the next plane and repeat the procedure. Therefore, a 3D transport equation can be written as:

$$a_P \phi_P^{\langle k \rangle} - a_S \phi_S^{\langle k \rangle} - a_N \phi_N^{\langle k \rangle} = C^{\langle k \rangle} \tag{11.27}$$

where:

$$C^{\langle k \rangle} = a_W \phi_W^{\langle k \rangle} + a_E \phi_E^{\langle k-1 \rangle} + a_B \phi_B^{\langle k \rangle} + a_T \phi_T^{\langle k \rangle} + S_u^{\langle k \rangle} \tag{11.28}$$

11.4 Solution Analysis

While methods to iteratively reach the solution for the system of algebraic equations is introduced in the previous sections, the main concern remains in the plausibility of guaranteeing a convergence for a numerical solution. This task is, however, depending on the ***consistency*** and ***stability*** of a matrix of coefficient of that system, which should be always taken to the account in the numerical analysis. Thus, these concepts can later ensue to define the definition of convergence. Hence, we briefly explain consistency and stability in the following sections though fundamentals of these criteria include advanced mathematical analysis as interested readers are referred to [1–3].

11.4.1 Consistency

The consistency is a measure, which evaluates how a discretized scheme can numerically represent a partial differential equation (PDE). In an ideal condition, therefore, when time step (Δt) and spatial spacing (Δx, Δy, Δz) approaches to zero, we expect to obtain a zero-truncation error. This implies that the discretised form of the equations should transform to the partial equation when the selected temporal and spatial intervals are reaching to zero. In reality, nonetheless, a ***truncation error*** occurs as a product of this fact that the time step and spatial spacing are practically non-zero quantities. Hence, we can define the concept of consistency for a discretized partial equation in which the truncation error should hypothetically become zero or vanish if any or all of Δt, Δx, Δy, $\Delta z \rightarrow 0$.

Example 11.6

Examine the consistency of explicit forward difference scheme for time and central difference scheme for space of the 1D unsteady heat conduction equation without the source/sink term.

Solution

1D unsteady heat conduction equation can be discretized with forward difference scheme for time and central difference scheme for space:

$$\rho c \frac{\Delta x}{\Delta t} \tilde{T}_P^{t+\Delta t} = \frac{k}{\Delta x} \tilde{T}_E^t + \frac{k}{\Delta x} \tilde{T}_W^t + \left[\rho c \frac{\Delta x}{\Delta t} - \frac{2k}{\Delta x} \right] \tilde{T}_P^t \qquad (E6.1)$$

So, if $\alpha = \frac{k}{\rho c}$, we can rearrange the equation as:

$$\left(\frac{\tilde{T}_P^{t+\Delta t} - \tilde{T}_P^t}{\Delta t} \right) = \alpha \left(\frac{\tilde{T}_E^t + \tilde{T}_W^t - 2\tilde{T}_P^t}{\Delta x^2} \right) \qquad (E6.2)$$

Using Taylor series, the first derivative in (E6.2) can be found as:

$$T_P^{t+\Delta t} = T_P^t + (\Delta t) \frac{\partial T_P^t}{\partial t} + \frac{(\Delta t)^2}{2!} \frac{\partial^2 T_P^t}{\partial t^2} + \frac{(\Delta t)^3}{3!} \frac{\partial^3 T_P^t}{\partial t^3} + \dots$$

$$\implies \frac{T_P^{t+\Delta t} - T_P^t}{\Delta t} = \frac{\partial T_P^t}{\partial t} + O(\Delta t) \qquad (E6.3)$$

Similarly, for other derivatives in (E6.2):

$$T_E^t = T_P^t + (\Delta x) \frac{\partial T_P^t}{\partial x} + \frac{(\Delta x)^2}{2!} \frac{\partial^2 T_P^t}{\partial x^2} + \frac{(\Delta x)^3}{3!} \frac{\partial^3 T_P^t}{\partial x^3} + \dots \qquad (E6.4)$$

$$T_W^t = T_P^t - (\Delta x) \frac{\partial T_P^t}{\partial x} + \frac{(\Delta x)^2}{2!} \frac{\partial^2 T_P^t}{\partial x^2} - \frac{(\Delta x)^3}{3!} \frac{\partial^3 T_P^t}{\partial x^3} + \dots \qquad (E6.5)$$

Now, with adding (E6.4), and (E6.5):

$$\frac{T_E^t + T_W^t - 2T_P^t}{\Delta x^2} = \frac{\partial^2 T_P^t}{\partial x^2} + O(\Delta x^2) \qquad (E6.6)$$

We can add (E6.3) and (E6.6):

$$\left(\frac{T_P^{t+\Delta t} - T_P^t}{\Delta t}\right) + O(\Delta t) = \alpha\left(\frac{T_E^t + T_W^t - 2T_P^t}{\Delta x^2}\right) + O(\Delta x^2) \tag{E6.7}$$

Therefore, the difference of the discretized equation of 1D diffusion $(\tilde{T})$ (E6.2) from its differential equation (T_P^t) (E6.7) gives the truncation error (ε_P^t) as:

$$\varepsilon_P^t = O(\Delta t, \Delta x^2) \tag{E6.8}$$

Hence, we can observe that when Δt and $\Delta x \to 0$, then $\varepsilon_P^t \to 0$.

11.4.2 Stability

In the numerical solution of PDEs, it is of paramount importance to ensure that errors are not growing through the iterative solutions. On the other words, the solution yields a converge if errors remain bounded within a range or the scheme is **stable** or has a **stability condition**. The knowledge behind the stability condition again demands advanced mathematical knowledge (e.g. Von-Neumann stability analysis) and thus is beyond the scope of this book.

Example 11.7

Discuss the stability of 1D unsteady heat conduction equation without the source/sink term discretized in Example 11.6.

Solution

This was discussed in Chapter 10 – Section 10.14.1 where explicit method was described. In order to achieve consistency, the condition of $\Delta t < \dfrac{(\Delta x)^2}{2\alpha}$ should be satisfied, which implies that we cannot take Δt and Δx independently to zero. A similar answer can be achieved using other techniques such as Von-Neumann stability analysis. Example of unstable results were discussed in Example 10.11.

11.4.3 Grid Convergence

A PDE with a particular boundary and initial conditions can be transformed into a system of linear equations using a numerical approach. A numerical solution then can be guaranteed if consistency and stability conditions are satisfied after implementation of a discretization scheme. Likewise, an iterative solution of the equation can be obtained if that scheme is stable and consistent. If this is the case, the numerical solution can be assumed to have **grid convergence**, which implies that it approaches the exact solution.

11.4.4 Iterative Convergence and Residual

Iterative convergence is a measure to ensure a reduction in the difference between the exact and numerical solutions $(R^{\langle k \rangle})$ at each iteration $\langle k \rangle$. Hence, in an ideal condition, $R^{\langle k \rangle} \to 0$ when $k \to \infty$. In a practical aspect, we are bounded by computational and time resources to interrupt the iterations at some stages, which enforces $R^{\langle k \rangle} \neq 0$. This value, which is known as the **residual**, however, should be always kept below a specific range $(R^{\langle k \rangle} < \varepsilon)$, depending on the nature of a tackled problem. ε is also denoted as the convergence criteria.

In the mathematical terms for a system of linear equations, we can find $R^{\langle k \rangle}$ at iteration k for a specific control volume (i) as:

$$a_P \phi_P = \sum a_{nb} \phi_{nb} + S_u$$

$$\Longrightarrow R^{\langle k \rangle}{}_i = \left(\sum_{nb} a_{nb} \phi_{nb} + S_u - a_P \phi_P \right)^{\langle k \rangle} \tag{11.29}$$

where, S_u is the constant part of the source term ($S_u + S_P \phi_P$). Now, if we sum all the residuals (m) across the flow field, we can rewrite Eq. (11.29) as:

$$R^{\langle k \rangle} = \sum_{i=1}^{m} \left| \left(\sum_{nb} a_{nb} \phi_{nb} + S_u - a_P \phi_P \right)^{\langle k \rangle} \right|_i \tag{11.30}$$

The procedure is almost the same as the one introduced in Example 11.4. It is also a common practice to normalize the residual as the summation will end up to large numbers in flow field with large magnitude of quantities. The normalization procedure can occur with different methods though we define two common practices as the below equations:

$$\overline{R}^{\langle k \rangle} = \frac{\sum_{i=1}^{m} \left| \left(\sum_{nb} a_{nb} \phi_{nb} + S_u - a_P \phi_P \right)^{\langle k \rangle} \right|_i}{\overline{R}^{\langle k_0 \rangle}} \tag{11.31}$$

In this equation, the residual is normalized by its own size at the initial iterations $\langle k_0 \rangle$, which is normally inequal to the first one. Or, in another format, the residual is normalized by the absolute number of the left-hand side elements summed in all cells as represented as:

$$\overline{R}^{\langle k \rangle} = \frac{\sum_{i=1}^{m} \left| \left(\sum_{nb} a_{nb} \phi_{nb} + S_u - a_P \phi_P \right)^{\langle k \rangle} \right|_i}{\sum_{i=1}^{m} \left| (a_P \phi_P)^{\langle k \rangle} \right|_i} \tag{11.32}$$

Example 11.8

Solve matrix of Example 11.3 by Jacobi and Gauss–Seidel methods and find the residual after using the normalization defined in Eqs. (11.31) and (11.32).

$$\begin{bmatrix} 26 & -16 & 0 & 0 & 0 & 0 \\ 16 & -32 & 16 & 0 & 0 & 0 \\ 0 & 16 & -32 & 16 & 0 & 0 \\ 0 & 0 & 16 & -32 & 16 & 0 \\ 0 & 0 & 0 & 16 & -32 & 16 \\ 0 & 0 & 0 & 0 & -16 & 21 \end{bmatrix} \begin{bmatrix} T_0 \\ T_1 \\ T_2 \\ T_3 \\ T_4 \\ T_5 \end{bmatrix} = \begin{bmatrix} 350 \\ 0 \\ 0 \\ 0 \\ 0 \\ 125 \end{bmatrix}$$

Solution

The table of coefficients for the above matrix would be as follows:

a_W	a_E	a_P	S_p	S_u
0	16	26	−10	350
16	16	32	0	0
16	16	32	0	0
16	16	32	0	0
16	16	32	0	0
16	0	21	−5	125

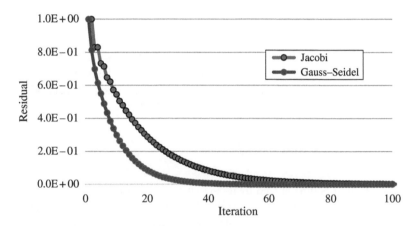

Figure 11.E8.1 Plot of residual of normalized Jacobi and Gauss–Seidel methods after multiple iterations with a zero initial guess.

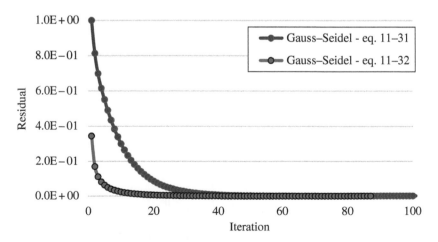

Figure 11.E8.2 Plot of residual of Gauss–Seidel method after multiple iterations with a zero initial guess using two normalization approaches.

Therefore, the residuals for Jacobi and Gauss–Seidel methods can be shown in Figure 11.E8.1.

For example, it takes 111 and 222 iterations for Gauss–Seidel and Jacobi methods to reach a residual below 10^{-6}, respectively. We can utilize Eqs. (11.31) and (11.32) to find residuals for example for Gauss–Seidel (Figure 11.E8.2).

Using Eq. (11.32), the solution converges to a residual below 10^{-6} after 88 iterations instead of 111 using Eq. (11.31).

11.4.5 Initial Guess

An initial guess is crucial to start the iteration procedure in the iterative methods. Let us assume Jacobi method for instance; the first set of solutions requires initial values to proceed to the second set of solutions until reaching a desired convergence. For example, after 28 iterations a convergence of 10^{-6} was found in Example 11.4 using a zero vector as the initial guess. Nonetheless, the rate of convergence is rigidly connected to the initial guess. The rate of convergence is extremely important when large matrices should be iteratively solved, particularly related to the built environment problems with a large number of cells. Thus, minimizing the number of iterations can save lots of calculation time. The methods to generate a good initial guess, and not a random or zero guess, again demands advanced mathematical discussions, which is an existing research topic in the CFD subject area. Methods such as multi-grid are widely investigated and are further implemented in the CFD tools. While we again refer the interested readers to some related references, we provide the below example to demonstrate how convergence can drastically be improved upon utilization of a better initial guess.

Example 11.9

Solve matrix of Example 11.3 by Gauss–Seidel method using an initial guess of 31°C as an average of the solution vector; note that we are using these values from the previous examples' results replaced as a good guess.

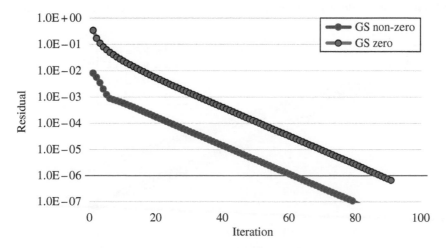

Figure 11.E9 Plot of residual of Gauss–Seidel method after multiple iterations with zero and non-zero initial guesses.

Solution

As it can be seen in Figure 11.E9, for such a small matrix, still a faster convergence of 10^{-6} can be achieved after 61 iterations compared to 87 iterations for a solution with a zero-vector guess.

11.4.6 Under-Relaxation

To prevent the solution of a system of equations from changing too quickly or inversely to impose a faster alteration, the change in dependent variables from one iteration to another can be modified by 'relaxing' them. It can be shown that the relation factor does not impact the final solution as it only adjusts the pathway to a solution; this implies that this pathway can be a better one in comparison to the initial one. Again, finding an optimum pathway is not necessarily a straightforward task, and depending on a problem, needs user experience and mathematical knowledge to be effectively controlled.

A relaxation value of $\alpha = 1.0$ uses 100% of the current dependent variable value whereas a relaxation factor of below ($0 < \alpha < 1$ is known as **under-relaxation**) or above ($1 < \alpha$ known as **over-relaxation**) unity uses α% of the previous iteration value in combination with $(1-\alpha)$% of the current value. For example, $\alpha = 0.7$ would combine 70% of the previous iteration value with 30% of the current value. The general expression for the relaxation factor can be written as:

$$\phi = (1-\alpha)\phi_{\text{old}} + \alpha\phi_{\text{new}} \tag{11.33}$$

Example 11.10

Solve matrix of Example 11.3 by Gauss–Seidel method using a zero-vector initial guess using relaxation factors of 0.3, 0.7, and 1.2.

Solution

As shown in Figure 11.E10, under-relaxations of 0.1, 0.3, and 0.7 results in the solution matrix to reach a 10^{-6} convergence after 121, 91, and 129 iterations, respectively. While an over-relaxation of 1.2 does not help the convergence to improve or change as the

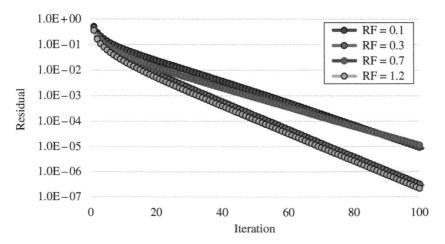

Figure 11.E10 Plot of residual of Gauss–Seidel method after multiple iterations with a zero initial guess using multiple relaxation factors.

convergence of 10^{-6} occurs after 88 iterations similar to the case of $\alpha = 1$. Note again that a smaller number does not mean an optimum pathway to solution in all the problems as a general rule, and as it was stated earlier, it is unique for each system of the equations.

11.5 Physical Uncertainty

As it was stated before, CFD is a powerful tool in design and development of a range of applications, specifically in the built environment related problems. The accuracy of a CFD simulation, thus, is of paramount importance as it can lead to miscalculations and incorrect results. Therefore, we are obliged to understand source of plausible errors and develop measures to evaluate and control them in accordance with our expectation from results of simulations. Knowing such uncertainty in CFD results helps to assess the confidence level of our simulation results.

In general, a CFD simulation can be initially generated based on an incorrect modelling of a realistic problem, known as physical uncertainty. This states that the physics of a deemed problem can be inaccurately replicated due to (i) **oversimplification**, (ii) **complexity**, or (iii) **lack of understanding** in working mechanism of a model. The inaccuracy in model replications can occur in any preprocessing, solving, or post-processing stages of a CFD simulation, for example, in geometry generation, meshing, boundary condition assignment, material property definition, etc.

Oversimplification can occur in boundary conditions assignment, isothermal or non-isothermal assumption of the flow, steady-state or transient assumption of the flow, turbulence model adaption, etc. As shown in Figure 11.6, for example, if wind assumed as a constant velocity in the atmospheric boundary layer, we may end up oversimplifying the problem in which the negligence of the wind profile with power-law or log-law (see Chapter 3, Section 3.1.2) will

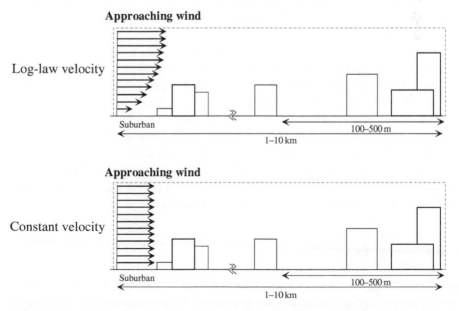

Figure 11.6 Oversimplification of the inflow boundary condition and adding a source of error into the modelling.

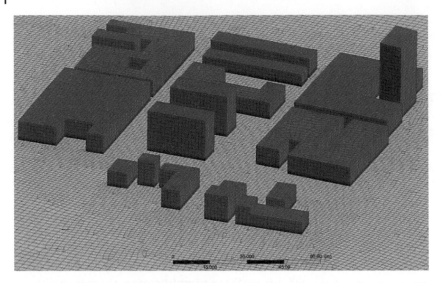

Figure 11.7 Ignoring long-wave radiation modelling in a city neighbourhood and adding a source of error due to the complexity of modelling.

introduce a significant error in the calculation of wind load, the energetic budget in buildings, pedestrian comfort, etc.

Let us assume another example of a city neighbourhood as shown in Figure 11.7. We can explore that the calculation of the long-wave radiation between each pair of surfaces can be a complex process as the calculation of view factors can be a monotonous and computationally extensive process. Here, we have 125 surfaces, so more than 15.5k view factors should be calculated. On the other hand, the benefits in the calculation of instance energy demand of the buildings might not exceed few percents. Hence, we may prefer to intentionally ignore the calculation of long-wave radiation, implying that we are adding a complexity error to the calculations.

We can provide an example of turbulence modelling as another source of error, which refers to lack of understanding of the underlying physics of a phenomenon – here turbulence transport. As it can be seen in Figure 11.8a, the experimental measurement of airflow around a cuboid bluff body reveals complex separation regions behind and on the lateral sides of the object. Nonetheless, the commonly applied standard $k-\varepsilon$ model cannot capture very well these regions as illustrated in Figure 11.8b. This is a well-known unsatisfactory prediction of reproducing the wake regions and reattachment zones over the buildings. The reason mainly describes an overestimated turbulent production in the impingement region on the frontal areas. Thus, such obligation in the use of turbulence or other models can potentially add up further errors to the simulation results.

11.6 Numerical Errors

The physical uncertainty is related to the modelling of a realistic problem, or in other words, due to the translation of physics to mathematics equations. Such translation typically is coming from the state of the art of a specific subject area to minimize these sources of errors. Let us

(a)   (b)

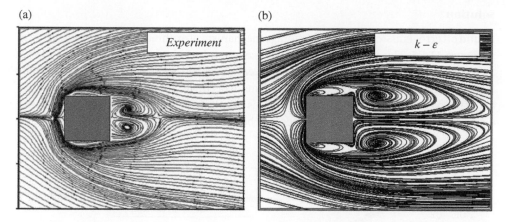

Figure 11.8 Lack in understanding of the underlying physic of turbulence can add another source of error. (a) Experiment conducted by Yoshie et al. [4], and simulations of (b) the standard $k - \varepsilon$ model [5]. *Source:* Yoshie et al. [4]/With permission of Elsevier.

assume that the latter source of errors are minimized with careful consideration of assumptions and enhancement in representation of various physics of phenomena in a particular problem. In other words, the problem is strongly interpreted into the governing mathematical equations. Even after a good representation of a realistic problem with a fair governing of mathematical equations, there are significant errors associated with the numerical methods, when they are chosen to solve the governing equations. Let us assume FVM as our chosen technique to solve a simplified format of the Navier–Stokes equations with suitable justification to represent a real-world problem without losing much of reality. Now, solution of these sets of mathematical equations with FVM (or any other numerical technique) encounters three sources of numerical errors, including (i) **roundoff**, (ii) **truncation**, and (iii) **convergence** errors. Before explaining each of these numerical errors, we should state that there are other possible errors related to user's role in coding and setting up a CFD tool, which are assumed to be not the case in this chapter. Thus, we neglect human errors though they can be a serious and significant source of error as tones of investigation is available in literature related to the techniques and procedures to minimize them.

11.6.1 Roundoff Error

Real numbers are stored and processed in computers only up to a certain level of digits in accordance with the available computational resources. For example, a single precision format stores up to seven digits while in double precision, this number is 15. Hence, the number is rounded after a defined decimal in arithmetic calculations, which impose an unwanted error to the calculations. The magnitude of this error is related to a solver algorithm and used hardware.

Example 11.11
Solve matrix of Example 11.3 by Gauss–Seidel method using a zero-vector initial guess while storing only 2, 4, and 7 decimals of the arithmetic calculations in the code.

Solution

Round off error is associated with the storage mechanism of arithmetic calculations as shown in Figure 11.E5. This implies that, for instance, the calculation with two decimals cannot reach a convergence below 10^{-4}. Similarly, it cannot tackle a 10^{-5} convergence when only four decimals are stored in the code. Storing more decimals, let us say seven decimals, will allow the convergence to easily reach 10^{-6}. This means that there is always a round off error, depending on how many digits in a code can be afforded to be stored.

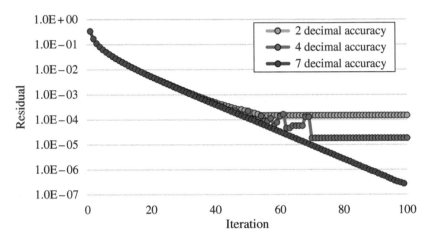

Figure 11.E11 Plot of residual of Gauss–Seidel method after multiple iterations with a zero initial guess while storing only 2, 4 and 7 decimals of the arithmetic calculations.

11.6.2 Truncation Error

In the previous chapter, we have explained the approximation method to define discrete form of derivatives (including temporal and spatial) of PDEs. Different derivatives with different order of magnitude include infinite higher-order terms that are truncated after keeping a certain number of terms (e.g., see Example 11.6). The truncation of terms in Taylor series, thus, introduce an inherent source of error into the numerical calculations associated with the PDEs, known as **discretization** or **truncation error**.

While the error should be decreased using smaller cells or time intervals, generation of fine meshes is always a challenge in CFD problems, especially in buildings and microclimate modelling where typically many cells should be utilized in a mesh. It should be noted that truncation error can be accumulated in time marching problems, and, hence, extra care should be always given to choose a suitable temporal derivative and time step.

Example 11.12
Consider the below PDE equation, which is similar to steady heat-diffusion equation with a source term:

$$\frac{\partial^2 \phi}{\partial x^2} + c = 0$$

Investigate the truncation error of the utilized second-order central difference method for the spatial derivative.

Solution

From the definition of the truncation error, we expect the error to be in the order of:

$$\text{Err} = O\left(\Delta x^2\right) \approx C\Delta x^2$$

Applying logarithmic scale on both side of the equation gives:

$$\log\left(\text{Err}\right) = \log\left(C\Delta x^2\right) = \log\left(C\right) + \log\left(\Delta x^2\right)$$
$$\implies \log\left(\text{Err}\right) = K + 2\log\left(\Delta x\right)$$

We investigate the validity of this equation in Example 11.13.

11.6.3 Iterative Convergence Error

If the iteration proceeds in a convergent solution, the discrepancy between the final solution and the current solution at an iteration k reduces. Nonetheless, the convergence error occurs when a solution has not entirely converged or k is not sufficiently large due to interruption related to lack of computational resources or time. The discrepancy is more significant in transient solutions when the solver is set to march from a specific iteration to another one even if the desired convergence criteria has not met. Such errors between the current solution and the final solution are known as the ***iterative convergence error***. Hence, it is crucial to utilize measures in a specific iteration to reduce the convergence error. ***Residual*** is a commonly accepted measure as it was explained in the preceding section.

11.7 Verification and Validation

A logical transformation of a real-world problem to a correct model implies the ***validation*** process. This needs an accurate representation of the governing equation, suitable choice of boundary and initial conditions, etc. One can, thus, relate the validation process to demonstrate the ***correct physics*** of a defined realistic problem. In other words, validation defines the uncertainty in a problem.

On the other hand, ***verification*** is an essential step in CFD simulations to ensure that an accurate representation and interpretation of the conceptual model into the numerical solution is initially occurred. Solving the differential equations in an accurate way (mathematics), correct implementation of equations into programming codes, suitable choice of the required setups, etc. should be initially verified in a CFD simulation. Thus, verification quantifies errors, or we can relate it to the ***correct mathematics*** of a defined model.

Although validation is an essential step in CFD simulations, it is not always easy to be conducted. The challenges are associated with the limitation of knowing the correct answers to compare with the obtained simulation results. In this aspect, experimental studies are essential to be generated as a mean of comparison to CFD results as this concept will be discussed in the following sections. Analytical or exact solutions are only possible for few classic scenarios, mainly when Navier–Stokes equations are undergone through significant simplifications. Such scenarios were essential in the development of CFD field though not very handy when it comes to the modern complex problems. Hence, as it was stated before, the observational methods, including in situ, field or wind-tunnel measurements, are essential in many CFD subject areas.

Example 11.13

Water flows at 20 °C between two infinite plates as shown in Figure 11.E13.1. The flow is fully developed and laminar. The pipe diameter (D) is 1 m and $\partial P/\partial x$ is constant (0.1 N/m). Now, assume a 1D mesh with five equal cells. Discretize the simplified equation in y-direction (which is similar to the diffusion equation) and find the velocity on each node using FVM. Validate your 1D numerical and analytical results and calculate the average and local errors.

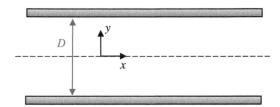

Figure 11.E13.1 Water flows between two infinite plates.

Solution

At 20 °C, water properties can be obtained as $\mu = 1.002 \times 10^{-3}$ Pa.s and $\rho = 998.237$ kg/m^3. The general form of the velocity after simplification of the Navier–Stokes equations can be shown as:

$$u(y) = \frac{y^2}{2\mu} \cdot \frac{\partial p}{\partial x} - \frac{D^2}{8\mu} \cdot \frac{\partial p}{\partial x}$$

From a FVM point of view, the simplified equation of the flow has a similar format as 1D diffusion equation when the pressure gradient is assumed as a constant number $\left(\frac{\partial p}{\partial x} = -c \right)$:

$$\mu \frac{\partial^2 u}{\partial y^2} - \frac{\partial p}{\partial x} = 0$$

Boundary conditions are velocities equal to zero ($u = 0$), which has resemblance to Dirichlet boundary of temperature in the diffusion equation. Using five similar control volumes, we can write the below general equation similar to Example 10.3 of Chapter 10 ($\Delta y = 0.2$ m and $D = 1$ m) (Figure 11.E13.2):

For internal nodes of 2, 3, and 4:

$$\left[\mu A \left(\frac{U_E - U_P}{\Delta y} \right) - \mu A \left(\frac{U_P - U_w}{\Delta y} \right) \right] - \frac{dp}{dx} A \Delta y = 0$$

Figure 11.E13.2 1D domain with five similar cells.

After rearranging:

$$\left(\frac{2\mu A}{\Delta y}\right)U_p = \frac{\mu A}{\Delta y}U_w + \frac{\mu A}{\Delta y}U_E - \frac{dp}{\Delta x}A\Delta y$$

For boundary face node-1:

$$\left[\mu A\left(\frac{U_E - U_p}{\Delta y}\right) - \mu A\left(\frac{U_p - u_A}{\frac{1}{2}\Delta y}\right)\right] - \frac{dp}{dx}A\Delta y = 0$$

$$\left(\frac{3\mu A}{\Delta y}\right)U_p = \frac{\mu A}{\Delta y}U_E + \left(\frac{2\mu A}{\Delta y}u_A - \frac{dp}{\Delta x}A\Delta y\right)$$

And, for face node-5:

$$\left[\mu A\left(\frac{u_B - U_p}{\frac{1}{2}\Delta y}\right) - \mu A\left(\frac{U_p - U_w}{\Delta y}\right)\right] - \frac{dp}{dx}A\Delta y = 0$$

$$\left(\frac{3\mu A}{\Delta y}\right)U_p = \frac{\mu A}{\Delta y}U_w + \left(\frac{2\mu A}{\Delta y}u_B - \frac{dp}{\Delta x}A\Delta y\right)$$

Therefore, we can form the coefficient table as:

Node	a_W	a_E	a_P	S_P	S_u
1	0	$\frac{\mu A}{\Delta y}$	$\frac{3\mu A}{\Delta y}$	$\frac{-2\mu A}{\Delta y}$	$-\frac{dp}{dx}A\Delta y + \frac{2\mu A}{\Delta y}u_A$
2, 3, 4	$\frac{\mu A}{\Delta y}$	$\frac{\mu A}{\Delta y}$	$\frac{2\mu A}{\Delta y}$	0	$-\frac{dp}{\Delta x}A\Delta y$
5	$\frac{\mu A}{\Delta y}$	0	$\frac{3\mu A}{\Delta y}$	$\frac{-2\mu A}{\Delta y}$	$-\frac{dp}{dx}A\Delta y + \frac{2\mu A}{\Delta y}u_B$

After replaying the variables with their values and forming the linear system of equations (A is assumed as the unit area), we use a direct method to find the solution in general aspect of $\frac{dp}{\Delta x}$:

y (m)	FVM velocity (m/s)	Analytical velocity (m/s)	Error (%)
0.04	$4.99\text{E} - 01\frac{dp}{dx}$	$4.49\text{E} - 01\frac{dp}{dx}$	11.1
0.02	$1.10\text{E} + 00\frac{dp}{dx}$	$1.05\text{E} + 00\frac{dp}{dx}$	4.8
0	$1.30\text{E} + 00\frac{dp}{dx}$	$1.25\text{E} + 00\frac{dp}{dx}$	4.0

(Continued)

y (m)	FVM velocity (m/s)	Analytical velocity (m/s)	Error (%)
−0.02	$1.10\text{E}+00\dfrac{dp}{dx}$	$1.05\text{E}+00\dfrac{dp}{dx}$	4.8
−0.04	$4.99\text{E}-01\dfrac{dp}{dx}$	$4.49\text{E}-01\dfrac{dp}{dx}$	11.1

Note that $\dfrac{dp}{\Delta x}$ is not affecting the error as it will be canceled out in the error expression. In an internal flow (see Chapter 2, Section 2.8), the flow can be assumed to be laminar when $\text{Re}_D = \dfrac{\rho u D}{\mu} < 2100$:

$$u_{\max}(y=0) < \frac{2100\mu}{\rho D} = 2.108 \times 10^{-2}\,\text{m/s}$$

This implies that $u_{\max}$ occurs at the center of the pipe $y = 0$. So, from the general analytical answer, we can write:

$$4.990 \times 10^2 (-0.0025)\frac{\partial p}{\partial x} < 2.108 \times 10^{-2}$$

$$\Rightarrow -1.690 \times 10^{-2} < \frac{\partial p}{\partial x}$$

Since the velocity should have a positive value ($0 < u$), the pressure gradient requires to be negative and the maximum value to comply with the laminar flow condition is $-1.696 \times 10^{-5} < \dfrac{\partial p}{\partial x}$. Replacing this number into the analytical and numerical expressions in the previous table, we find CFD and analytical solution of the maximum velocities of a laminar flow as:

y (m)	FVM velocity (m/s)	Analytical velocity (m/s)	Error (%)
0.04	8.43E−03	7.59E−03	11.1
0.02	1.85E−02	1.77E−02	4.8
0	2.19E−02	2.11E−02	4.0
−0.02	1.85E−02	1.77E−02	4.8
−0.04	8.43E−03	7.59E−03	11.1

Now, we investigate the validity of the equation found in Example 11.12, which is similar to the general form of this example, with increasing the number of cells (reducing Δy) for a fixed $\frac{\partial p}{\partial x}$ related to $\text{Re} = 2100$ ($u_{\max} = 0.0211$ m/s). For this purpose, we define 1D meshes with 5, 10, 20, and 40 cells, and then use the table of coefficient to find the system of equations for each of them. Hence, we can find the velocities in different meshes as shown in Figure 11.E13.3.

Now, we can calculate errors of each mesh in relation to the analytical solution and then plot log(Err) against log(Δy) as shown in Figure 11.E13.4.

As it can be seen the slope of the graph is about 1.8, which is quite close with the expected truncation error of 2.

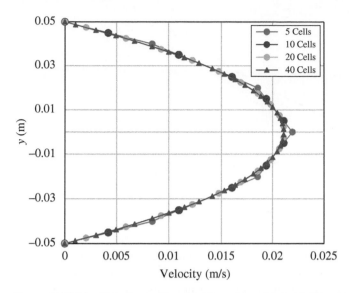

Figure 11.E13.3 Velocity profiles for 1D meshes with 5, 10, 20, and 40 cells.

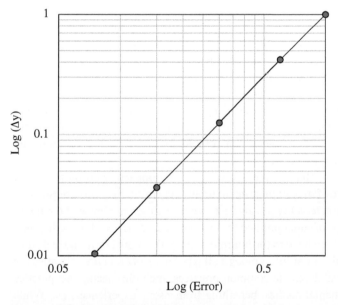

Figure 11.E13.4 Logarithmic plot of errors against change of mesh sizes.

11.8 Measures to Minimize Errors

As we explained earlier, different sources of error can potentially result in generation of inaccurate, distorted, and unreliable solutions. Minimizing the associated errors and uncertainties in a CFD simulation, therefore, demands several measures to be applied in different

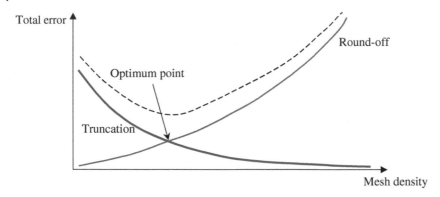

Figure 11.9 The relation between truncation and roundoff errors.

steps from preprocessing to post-processing stages. If we minimize the possibility of development an incorrect programming code as well as user mistakes in setting up a simulation, yet we expect errors and uncertainties in the results. In general, the complex nature of PDF equations, hardware limitations, and deficiencies in the numerical approximation makes the process of identification and controlling of each specific error almost impossible in a separate sense.

A common example, as shown in Figure 11.9, is the reduction of truncation errors when cell sizes are reduced. Inversely, the roundoff error expected to increase as the magnitude of numbers are reduced, and the number of arithmetic operations is increased. While the proportions are not necessarily accurate in Figure 11.9, an optimum value always is expected for the mesh size. In other words, these complex behaviours of errors, therefore, bounds the mesh density in a certain range if an optimum mesh is aimed to be generated. This means that increasing the number of cells in a mesh would not necessarily be the solution for enhancing the resolution of a CFD model as the complexity and roundoff errors of the model will increase at the same time.

Nonetheless, we should specify that the mesh size is not only impacting on these specific errors. For example, generation of smaller unstructured cells may result in a worse mesh quality, which can cause a worse convergence. As another example, different choice of boundary conditions may change the type of equations and thus the associated errors while at the same time impact on the uncertainties related to the interpretation of the model from the real word problem. Therefore, the complex interlinks amongst all types of errors and uncertainties are suggested to be verified and validated via common practices, including using best practice guidelines, validation experimental studies, benefiting from users' experiences, etc. While the verification and validation are advanced topics with many challenges on various aspects, we discuss some of the basic concepts useful for beginner users in the following sections.

11.8.1 Simulation Resource Assessment

Computational resources and time are limited, and management of these resources should take place initially in the pre-process stage. Especially, in the built environment related CFD simulations as discussed in more details in the next chapter, inspection of available

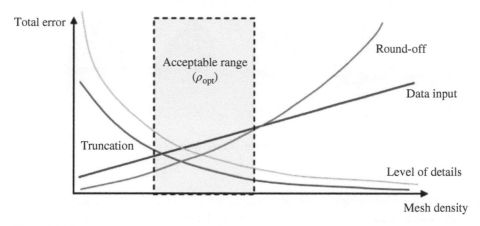

Figure 11.10 Acceptable range (ρ_{opt}) of the overall error in built environment CFD models.

hardware resources is of paramount importance to deliver a rigorous simulation practice. As stated in the previous section, one of the common steps is checking the practicality of using higher precisions for the arithmetic operations, resulting in reducing errors due to rounding off the numbers. This is mainly occurring when the number of calculations is increased due to a higher number of cells (smaller cell sizes) while arithmetic operations over each cell encounter calculations over smaller numbers. Thus, enormous number of arithmetic calculations with small numbers shall be minimized to ensure that round-off error is reduced accordingly. Also, using high-order accuracy derivative can lead to truncation error to be decreased. The optimum point of cell sizes was shown in Figure 11.9.

Now, let us investigate the mesh versus error from a system point of view. If we assume a built environment problem as a system, higher level of details (granularity) of a model can results in smaller cell sizes in average. From a system perspective, an error associated with the level of details will be declined [6] as it is increased or cell sizes are decreased; this is shown with the orange line in Figure 11.10. One can define the **mesh density** as the average number of cells in a fixed sized domain of a built environment CFD model. On the other hand, increasing the level of details demands more data to be provided as boundary, geometry, material properties, sub-model, etc. for the system. Hence, the increase in number of data introduces a linear **data input error**, which is shown by a red line in Figure 11.10.

Now, combining two concepts, we can identify an acceptable range of mesh density (ρ_{opt}) for generation of a mesh in a particular built environment CFD model. Finding this range should be recognized and might be found with trial and error investigations by users though it is not always a straightforward procedure to produce a mesh in this expected range. Therefore, we can relate the identified errors to the available resources with a preliminary assessment of the model in accordance with the computational power.

11.8.2 Geometry Simplification

Creation of a geometry is the first step of the pre-processing stage. In relation to the previous section explanation of system details, a CFD model can have various level of details (LoD). This concept is also systematically defined in the built environment 3D model as illustrated in

Figure 11.11 Level of details concept for buildings. *Source:* Biljecki et al. [7]/With permission of Elsevier.

Figure 11.11 [7]. While a higher LoD seems more attractive on many occasions, we must simultaneously make a balance between the geometry detail (flow field resolution) and mesh size (computational load). Refer to input data errors, users should consider LoD of buildings, roads, infrastructures, trees, etc. and they should further assign boundary conditions and materials properties to them, which can become an intensive procedure for larger domains.

11.8.3 Grid Sensitivity Analysis

We have introduced the CFD preparation process in the previous sections where the first step was the pre-processing stage. After preparation of the geometry, mesh generation was the action to discrete the continuous space to several finite control volumes or cells. The mesh generation process was discussed to be not a straightforward process for many complex geometries. However, there are measures, known as mesh quality indices, in hand to ensure a minimized error associated to this step. While mesh quality was explained in the previous chapter, in this section, we introduce a mesh related assessment, known as ***mesh independency*** test or ***grid sensitivity*** (GS) analysis.

The purpose of mesh independency test is to marginalize the discretization error as much as possible in respect to the affordable computational resources. Thus, one should monitor a monotonic fluctuation in the residuals of the quantities of interest. Nonetheless, it is crucial to define monotonic alterations in the GS analysis in accordance with the maximum available computational resources for a specific problem. For example, if we do not have resources to resolve the finest mesh that we can reach to a monotonic behaviour, then we must redevelop our model, rethink of simplifications in our problem, compromise in the residual, etc.

To monitor quantities of interest, we first need to generate and assign one or multiple discrete points in our geometry; here we define them as ***simulation sensors,*** to be able to extract the required values. Simulation sensors can be interpreted as nodes (points), surfaces, or volumes in our domain of study. Node-simulation sensors are demonstrated in Figure 11.12 for a simple rectangular domain example while the mesh is refined three times. The places of simulation sensors, however, remain unchanged for the comparison of the quantity in different refined domains.

Note that there are many studies to implement advanced approaches to perform GS analysis though we only define a simple approach to compare two subsequently refined meshes together using the below equation for the GS:

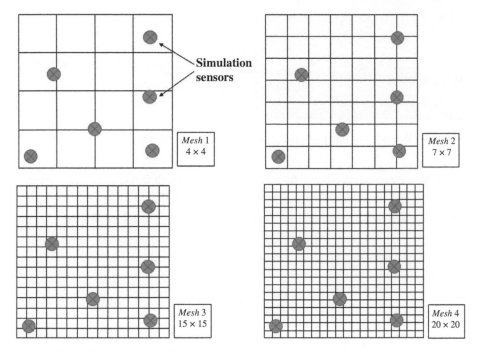

Figure 11.12 Demonstration of node-simulation sensors in refined meshes.

$$\text{GS}^{(j)} = \sum_{i=1}^{n} \left| \frac{\phi_i^{(j+1)} - \phi_i^{(j)}}{\phi_i^{(j)}} \right| \tag{11.34}$$

where n is the number of simulation sensors, and $\text{GS}^{(j)}$ is the jth comparison of quantities of ϕ between two subsequently refined grids.

Where to place simulation sensors, highly depends on the flow structure, geometry, the expected quantities, turbulence model, and many other factors. User-experience, the related state-of-the-art in a subject area, and established guidelines are some of the methods, which can help on a careful selection of them in a specific problem. Note again that one can define **plane-simulation sensors** or **volume-simulation sensors** and employ a similar calculation procedure on them.

Example 11.14

Define some node-simulation sensors in the fully developed laminar flow of Example 11.13 and perform a mesh independency test on the results after refining the initial 1D five-cell mesh for multiple times.

Solution

It is important that we ensure that the grid on the results of the flow is not significant. To monitor the flow in the laminar regime, we can first justify defining a node close to one of the pipe walls (for example $\frac{y}{D} = 0.4$) and one at the centre of it ($\frac{y}{D} = 0$) (Figure 11.E14.1).

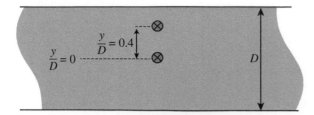

Figure 11.E14.1 Definition of node-simulation sensors in the study domain.

We start with a three-cell mesh and then refine the initial mesh three more times to generate 1D meshes with 5, 10, and 20 cells. We solve the equations similar to Example 11.13, and then extract and linearly interpolate values at $\frac{y}{D} = 0$ and $\frac{y}{D} = 0.4$ as follows:

	y/D = 0.4	y/D = 0
3-Cell	1.218E−02	2.342E−02
5-Cell	8.432E−03	2.192E−02
10-Cell	7.588E−03	2.108E−02
20-Cell	7.588E−03	2.108E−02

We can clearly observe that the values at the node-simulation sensors significantly change up to the 10-cell mesh and then it remains constant. We can also use Eq. (11.34) to calculate $GS^{(j)}$ and plot it as shown in Figure 11.E14.2.

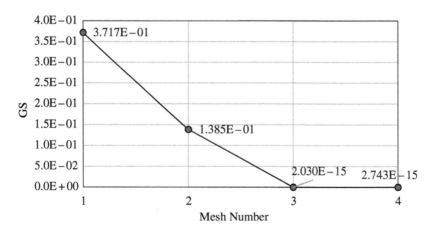

Figure 11.E14.2 Plot of grid sensitivity against mesh size.

Thus, we can make the conclusion that refining the 1D mesh after 5 cells will not significantly impact on the CFD results and thus it is a good enough size of the mesh to be used. For a very much lower error related to the mesh size, we may use 10 cells, but using 20 cells cannot be justified as there is a very small change in the results.

11.8.4 Iterative Convergence Control

In a converged solution, target quantities are expected to demonstrate a monotonic alteration behaviour. This implies that the truncation error due to reasons, for example, limited computational resources, avoiding the solution to reach to the iterative convergence should be in their minimum values. Thus, if the solution is constrained to be interrupted after a certain number of iterations, an iterative convergence would be meaningful at this point if quantities demonstrate slight variations even after more iterations are continued. One can, therefore, monitor one or multiple target quantities, for example, mass flow rate at a specific cross-section of cross-ventilation case, lift force on an airfoil, or pressure coefficient over a building surface to speculate these variations. The choice and the place of a monitored quantity again is substantially depending on a user to be well experienced, to well understand the related literature, and to properly contemplate the existing guidelines.

Example 11.15
Solve the transient Example 10.12 in Chapter 10 with Gauss–Seidel method with convergence criteria of 10^{-1} and 10^{-6}.

Solution

First, we form the matrix of coeffect as below:

	a_W	a_E	a_N	a_S	a_P	S_U
1	0	8200	0	8200	663450	122850000
2	8200	8200	0	8200	655250	122850000
3	8200	8200	0	8200	655250	122850000
4	8200	0	0	8200	663450	122850000
5	0	8200	8200	8200	655250	122850000
6	8200	8200	8200	8200	647050	122850000
7	8200	8200	8200	8200	647050	122850000
8	8200	0	8200	8200	655250	122850000
9	0	8200	8200	8200	655250	122850000
10	8200	8200	8200	8200	647050	122850000
11	8200	8200	8200	8200	647050	122850000
12	8200	0	8200	8200	655250	122850000
13	0	8200	8200	0	663450	122850000
14	8200	8200	8200	0	655250	122850000
15	8200	8200	8200	0	655250	122850000
16	8200	0	8200	0	663450	122850000

Now, we apply the Gauss–Seidel method with a $\Delta t = 0.1$ s and then we form the residuals and stop the calculations in each time step until the residual is below the defined criteria.

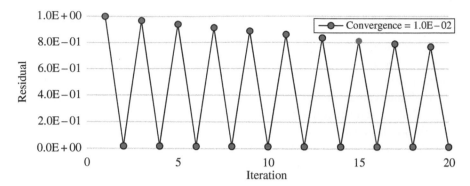

Figure 11.E15.1 Plot of residual of Gauss-Seidel method with a convergence criterion of 10^{-1} and a time-step of Δt = 0.1 s.

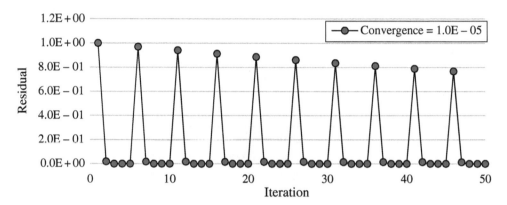

Figure 11.E15.2 Plot of residual of Gauss-Seidel method with a convergence criterion of 10^{-6} and a time-step of Δt = 0.1 s.

For 10^{-1}, the convergence will be achieved after two iterations for almost all the time steps and 20 iterations in total for 1 s of simulations as it can be seen in Figure 11.E15.1.

In the case of convergence criteria of 10^{-6}, each time step demands about five iterations (see Figure 11.E15.2).

Now, if we compare some of the obtained temperatures using a direct method in Chapter 10 and Gauss–Seidel for the convergence criteria of 10^{-6}, we can notice a very slight difference as shown in Figure 11.E15.3.

For the convergence criteria of 10^{-6}, the average difference for T_1, T_2, and T_6 is about 1.05E−09, 8.34E−10, and 1.30E−10, respectively. Also, for the convergence criteria of 10^{-1}, the average differences are 1.09E−05, 8.97E−06, and 8.89E−06, respectively. Obviously, the choice of convergence criteria is very important in the computational cost and should be justified in accordance with the requirement of a specific problem.

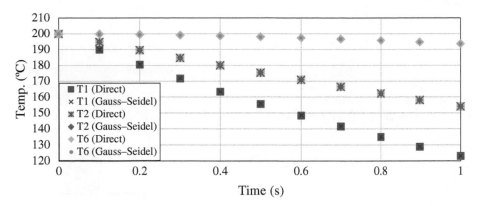

Figure 11.E15.3 Comparison of some of the temperatures using a direct method and Gauss–Seidel method with a convergence criterion of 10^{-6}.

11.8.5 Comparison with Experimental Benchmarks

As explained earlier, comparison of the simulation results with the analytical solution of the Navier–Stokes equations is only a plausible practice against a few numbers of classic problems mainly with simplified boundary and initial conditions. Hence, verification and validation of a CFD simulations, which have normally a complex geometry with mixed-boundary conditions, are not a viable solution using the analytical solutions almost in all practical scenarios. One can add the impact of unsteadiness as well as the uncertainties arising from turbulence effect to the solutions. Nonetheless, we need a common mean to be able to trust the simulations even after a careful consideration of the longlist of required treatment in the triple stages of a CFD simulation process. Experimental benchmark studies are therefore the sole remedy to entrust a rigorous CFD simulation procedure.

Experimental studies are built upon numerous techniques invented from decades ago to observe a range of quantities such as velocity and temperature fields, mass flow, pressure, concentration, etc. It should be noted, however, not all the experimental observations are useful for a CFD validation. In this respect, especially for built environment applications, wind tunnel as well as in situ measurements are broadly applied in the CFD validation studies.

Amongst the popular velocimetry techniques, particle image velocimetry (PIV) and hot-wire probes velocimetry are practical approaches to observe velocity field as shown in Figure 11.13a,b, respectively. Hot-wire velocimetry, which includes a large spectrum of sensors, can provide velocities in multiple singular points while PIV can image a plane of the velocity field instead of multiple singular points. As shown in Figure 11.14, gas tracer technique is employed for mass flow measurement where a usually harmless gas is released in upstream of a flow through an opening, cavity, or air corridor, and its trace is then measured in the downstream. Thermocouples are capable of measuring air and surface temperatures while the infrared thermography is a well-known technique to monitor surface temperatures in a broader aspect as depicted in Figure 11.15. To measure pressure distribution on surfaces, multiple pressure transducers can be utilized to collect singular point pressures (see Figure 11.16) and then to interpolate the values between the measurement locations. In the proceeding examples, we demonstrate practical techniques in deployment of such measurements for the validation of CFD results.

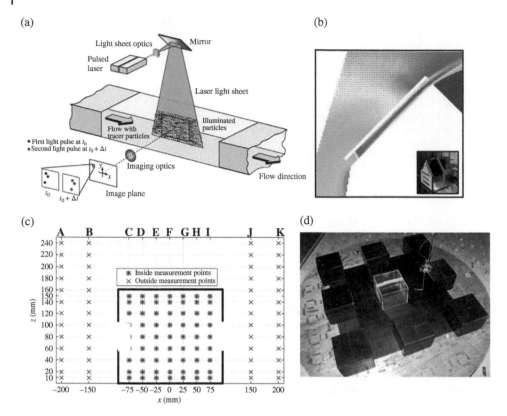

Figure 11.13 Velocimetry techniques used for CFD validation studies. (a) A schematic of particle image velocimetry (PIV). (b) A wind-tunnel PIV study of airflow in a building integrated photovoltaics. (c) Velocity measurement points, and (d) split fibre probe sensor with the horizontal support for the vertical velocity component measurement. *Source:* (a) Raffel et al. [8]/With permission of Springer Nature. (b) Based on Mirzaei and Carmeliet [9]. (c) Shirzadi et al. [10]/With permission of Elsevier.

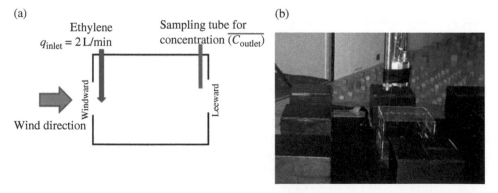

Figure 11.14 Mass flow measurement used for CFD validation studies. (a) Schematic of the airflow rate measurement with gas tracer, and (b) sampling tube inside the building model. *Source:* Shirzadi et al. [11]/With permission of Elsevier.

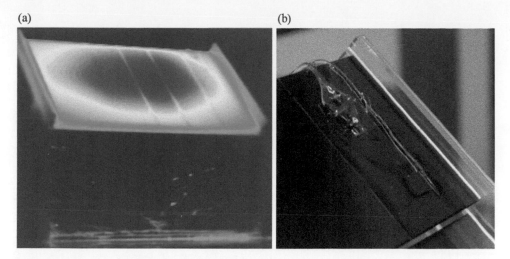

(a)   (b)

Figure 11.15 Temperature measurement techniques used for CFD validation studies. (a) thermography of surface temperature of a building integrated photovoltaics, and (b) measurement of surface temperature distribution over a building surface using thermocouples. *Source:* Mirzaei et al. [12]/With permission of Elsevier.

Example 11.16

A PIV study [14] is conducted to measure velocity at the cavity of the demonstrated building integrated photovoltaic in Figure 11.E16.1 with different upstream velocities. Discuss the validation of the study.

Solution

The CFD mesh details are discussed in [14]. The setup and boundary conditions are briefed below:

Boundary	Type	Treatment
Ground/ceiling/laterals walls/ building surfaces/radiator/PV holder/PV back and lateral surfaces	Wall	No-slip Not included in the radiation model
Front surface of PV	Wall	No-slip Emissivity = 0.9
Inflow	Velocity inlet	Constant Normal to the boundary Turbulent Intensity Hydraulic diameter = 1.54 m
Outflow	Pressure outlet	Gauge pressure = 0 Turbulent Intensity Hydraulic diameter = 1.54 m
Near-wall treatment	Enhanced wall function	
Pressure–velocity coupling	Simple	
Discretization scheme		
Pressure	Second order	
Momentum	Second-order Upwind	
Turbulent kinetic energy	First-order Upwind	
Turbulent dissipation rate	First-order Upwind	
Energy	Second-order Upwind	

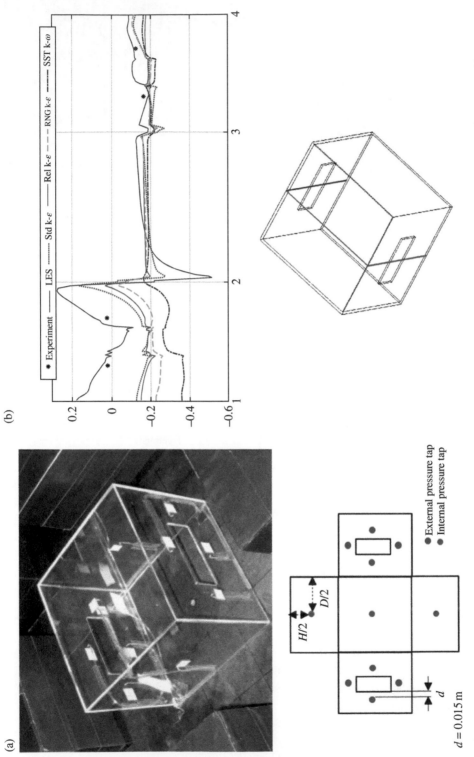

Figure 11.16 Pressure transducer technique used for CFD validation studies. (a) Pressure taps over a building model's surfaces, and (b) recorded and simulated pressure distribution. *Source:* Shirzadi et al. [13]/ with permission of Springer Nature.

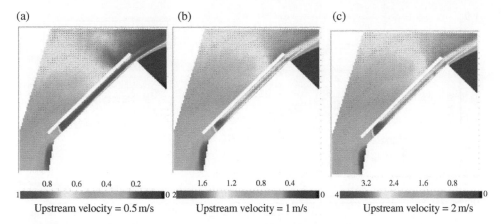

Figure 11.E16.1 shown with (a), (b), (c) panels:

(a) Upstream velocity = 0.5 m/s — scale: 0.8 0.6 0.4 0.2

(b) Upstream velocity = 1 m/s — scale: 1.6 1.2 0.8 0.4

(c) Upstream velocity = 2 m/s — scale: 3.2 2.4 1.6 0.8

Figure 11.E16.1 Velocity field around a building integrated photovoltaic against different upstream velocities. *Source:* Zhang et al. [14]/With Permission of Elsevier.

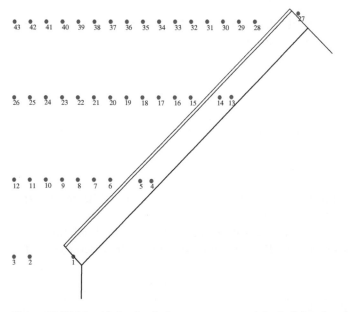

Figure 11.E16.2 Node-simulation sensors around the building integrated photovoltaic. *Source:* Zhang et al. [14]/With Permission of Elsevier.

Now, we can validate the CFD simulation against the experimental results using the defined node-simulation sensors as illustrated in Figure 11.E16.2.

The validation for the upstream velocity of 0.5 m/s results in the average and maximum differences between CFD and experiment inside the cavity to be obtained about 14.7% and 32.1%, respectively (see Figure 11.E16.3). The average deviation of the CFD model declines in the higher upstream velocities of 1 and 2 m/s as 10.1% and 9.9%, respectively. In general, the maximum error is almost halved (16.9%) in the upstream flow of 2 m/s while the average error of the velocity field is about 13.2% in the cavity, 7.2% in the upstream region and 8.0% in the whole domain.

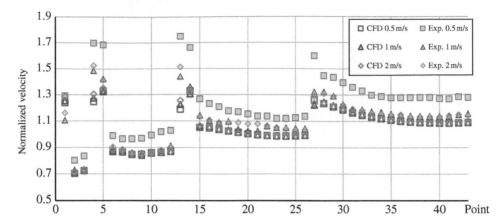

Figure 11.E16.3 Validation of velocity field results obtained from CFD simulations and particle image velocimetry (PIV) measurements using node-simulation sensors. *Source:* Zhang et al. [14]/With Permission of Elsevier.

The above findings state a successful validation of a CFD model using factual numbers calculated in node-simulation sensors in the critical points within and around the location of study, which is the cavity of the BIPV at this example.

Example 11.17

A split fibre probe is used to measure velocity at the demonstrated points in a cross-ventilation flow in an urban setting as shown in Figure 11.13c. Discuss the validation of the study.

Solution

For the meshing and CFD setup details, refer to [13]. When the CFD results are obtained, node-simulation sensors are created at similar points to the experiment. The vertical profiles of the stream-wise velocity obtained from CFD simulations are then compared as shown in Figure 11.E17. Here, advection terms are discretized with

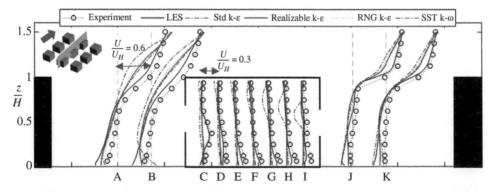

Figure 11.E17 Comparison of vertical profiles of the streamwise velocity obtained from CFD and measurement across multiple line-sensor simulations. *Source:* Shirzadi et al. [13]/ With permission of Springer Nature.

two methods denoted as the first-order Upwind differencing scheme (UP) and hybrid scheme (HR).

The validation is feasible with various defined metrics in the literature for the time-averaged velocity components. Here, over all 119 measurement points, we use four metrics, which are commonly used in built environment CFD problem (P_i are the experimental data and Q_i are the CFD simulation results):

Validation metric	Definition	Ideal value
Hit rate ($\boldsymbol{q}$)	$\dfrac{1}{N}\sum\limits_{i=1}^{N} n_i \quad \text{if} \quad \left\lvert\dfrac{P_i - Q_i}{P_i}\right\rvert \leq D_q \ \text{or}\ \lvert P_i - Q_i\rvert \leq W_q$ $n_i = 1 \quad \text{else}\ n_i = 0$	1
Fractional bias ($\boldsymbol{FB}$)	$\dfrac{[Q] - [P]}{0.5([Q] + [P])}$	0
Normalized mean square error ($\boldsymbol{NMSE}$)	$\dfrac{[(Q_i - P_i)^2]}{[Q][P]}$	0
Fraction of the predictions within a factor of 2 ($\boldsymbol{FAC2}$)	$\dfrac{1}{N}\sum\limits_{i=1}^{N} n_i \quad n_i = 1 \ \text{if}\ 0.5 \leq \dfrac{P_i}{Q_i} \leq 2 \ \text{else}\ n_i = 0$	1

Now, some of the calculated values for stream-wise velocity (U), vertical velocity (W), and turbulent kinetic energy (K) for the standard $k - \varepsilon$ turbulence model can be found as:

	FAC2$_U$	FAC2$_W$	q_U	q_W	q_k	FB$_k$	NMSE$_k$
Std $k - \varepsilon$	0.49	0.84	0.41	0.80	0.41	0.35	0.61

11.8.6 Best Practice

Guidelines and best practice for CFD applications are sets of recommendation provided by a series of successfully verified and validated simulations, mainly developed by different communities and associations, to help users to minimize their modelling uncertainty and numerical errors. These guidelines are stretched over all three CFD simulation stages, from geometry and mesh generation in pre-processing to streamline demonstration in post-processing stages. Choice of boundary conditions, mesh type, convergence and under-relaxation, solver selection, time step allocation, turbulence modelling, etc. are amongst parameters suggested in these best practices. Note that the generalization always deals with ambiguities and uncertainties, which seldom should be ignored in using guidelines. In other words, following the required steps as well as the associated best practices is the only perquisite to deliver a successful CFD simulation and, thus, user-experience and ongoing knowledge development in a deemed field of study should be always taken into considerations. We will provide some of the commonly used CFD best practices for built environment related problems as follows:

Name	Reference
AIJ guidelines for practical applications of CFD to pedestrian wind environment around buildings	[15]
Best Practice Guideline for the CFD Simulation of Flows in the Urban Environment	[16]

References

1 Anderson, D., Tannehill, J.C., and Pletcher, R.H. (2016). *Computational Fluid Mechanics and Heat Transfer*. Taylor & Francis.

2 Patankar, S. (1980). *Numerical Heat Transfer and Fluid Flow*. London, UK: Taylor & Francis.

3 Versteeg, H.K. (1995). *An Introduction to Computational Fluid Dynamics: The Finite Volume Method: Harlow, Essex*. England: Longman Scientific & Technical; New York: Wiley.

4 Yoshie, R., Jiang, G., Shirasawa, T., and Chung, J. (2011). CFD simulations of gas dispersion around high-rise building in non-isothermal boundary layer. *Journal of Wind Engineering and Industrial Aerodynamics.* 99 (4): 279–288.

5 Shirzadi, M., Mirzaei, P.A., and Naghashzadegan, M. (2017). Improvement of k-epsilon turbulence model for CFD simulation of atmospheric boundary layer around a high-rise building using stochastic optimization and Monte Carlo Sampling technique. *Journal of Wind Engineering and Industrial Aerodynamics.* 171: 366–379.

6 Chapman, J. (1991). *Data Accuracy and Model Reliability*. UK: BEPAC Canterbury.

7 Biljecki, F., Ledoux, H., and Stoter, J. (2016). An improved LOD specification for 3D building models. *Computers, Environment and Urban Systems* 59: 25–37.

8 Raffel, M., Willert, C.E., Scarano, F. et al. (2018). Introduction. In: *Particle Image Velocimetry: A Practical Guide* (ed. M. Raffel, C.E. Willert, F. Scarano, et al.), 1–32. Cham: Springer International Publishing.

9 Mirzaei, P.A. and Carmeliet, J. (2015). Influence of the underneath cavity on buoyant-forced cooling of the integrated photovoltaic panels in building roof: a thermography study. *Progress in Photovoltaics: Research and Applications.* 23 (1): 19–29.

10 Shirzadi, M., Tominaga, Y., and Mirzaei, P.A. (2019). Wind tunnel experiments on cross-ventilation flow of a generic sheltered building in urban areas. *Building and Environment.* 158: 60–72.

11 Shirzadi, M., Tominaga, Y., and Mirzaei, P.A. (2020). Experimental study on cross-ventilation of a generic building in highly-dense urban areas: Impact of planar area density and wind direction. *Journal of Wind Engineering and Industrial Aerodynamics.* 196: 104030.

12 Mirzaei, P.A., Paterna, E., and Carmeliet, J. (2014). Investigation of the role of cavity airflow on the performance of building-integrated photovoltaic panels. *Solar Energy.* 107: 510–522.

13 Shirzadi, M., Mirzaei, P.A., and Tominaga, Y. (2020). CFD analysis of cross-ventilation flow in a group of generic buildings: Comparison between steady RANS, LES and wind tunnel experiments. *Building Simulation.* 13 (6): 1353–1372.

14 Zhang, R., Mirzaei, P.A., and Carmeliet, J. (2017). Prediction of the surface temperature of building-integrated photovoltaics: development of a high accuracy correlation using computational fluid dynamics. *Solar Energy.* 147: 151–163.

15 Tominaga, Y., Mochida, A., Yoshie, R. et al. (2008). AIJ guidelines for practical applications of CFD to pedestrian wind environment around buildings. *Journal of Wind Engineering and Industrial Aerodynamics.* 96 (10): 1749–1761.

16 Franke, J., Hellsten, A., Schluenzen, H. et al. (2007). The COST 732 Best Practice Guideline for CFD simulation of flows in the urban environment: a summary. *International Journal of Environment and Pollution* 44 (1–4): 419–427.

12

Application of CFD in Buildings and Built Environment

12.1 CFD Models in Built Environment

As comprehensively explained in the previous chapters, unwanted errors can be triggered to negatively impact results of a computational fluid dynamics (CFD) simulation if a linear system of equations is not carefully resolved from a mathematical point of view. On the other hand, uncertainty related to incorrect modelling of a real-world problem can cause another major source of discrepancies in the results. Consequently, we are compelled to ensure a meaningful simplification of a certain problem with justified assumptions.

In this chapter, we attempt to clarify some of the common assumptions in the CFD modelling of buildings. We start by classifying such problems to different perspectives as explained in the preceding sections. Each aspect can dominantly impact on our geometry generation details, mesh generation approach, setup choices, computational resource considerations, validation techniques, input data acquisition, etc.

12.1.1 Spatial Scale

A CFD problem in the built environment can be defined as an individual room to a complete city. As it can be seen in Table 12.1, the spatial scale is the first decision in modelling of a real-world problem. For instance, when airflow is important on occupants' comfort, mechanical/natural ventilation design, indoor air quality, a single room or multiple ones in a building can suffice to employ CFD simulations. Alternatively, a community of buildings can be modelled, known as a microclimate, when pedestrian wind comfort, pollution dispersion, exterior surfaces' heat exchange is deemed to be investigated. Selection of a suitable scale is an essential factor in the geometry generation, number of cells as well as their types.

12.1.2 Temporal Variation

As it was explained in Chapter 10, the time-derivative of Navier–Stokes equations changes the type of the PDF and adds extra terms in their discretized equation form. Hence, the linearized system of equations shall be resolved in each time step (see Chapter 11), which can lead to intensive computational burdens, which sometime cannot be afforded. Therefore, it is essential to justify the necessity of a transient solution in a specific problem at a first instance.

Computational Fluid Dynamics and Energy Modelling in Buildings: Fundamentals and Applications, First Edition. Parham A. Mirzaei.
© 2023 John Wiley & Sons Ltd. Published 2023 by John Wiley & Sons Ltd.

Table 12.1 Spatial scale of CFD simulation problems in built environment.

	Single/multiple rooms	Multiple buildings
Example applications	• Ventilation design • Occupants' comfort • Indoor air quality	• Pedestrian comfort • Pollution dispersion • Surfaces' heat exchange
Domain scale (L)	• $L < 10\,\text{m}$	• $10\,\text{m} < L < 1000\,\text{m}$
Typical cell size (D)	• $D < 1\,\text{m}$	• $1\,\text{m} < D < 10\,\text{m}$
Typical mesh size (N)	• $N < 1\text{M}$	• $1\text{M} < N$
Typical mesh type	• Structured hex	• Hybrid hex/tet
Typical validation case measurement	• Wind tunnel • Test house	• Wind tunnel • In situ • Weather station
Typical input data	• 3D geometry • Inlet and outlet conditions	• 3D geometry • Weather condition • Surface vegetation and materials

Table 12.2 shows a range of steady state, quasi-steady, and transient problems in built environment. When isothermal solutions are expected from CFD simulation, a steady-state scenario in most of the cases can suffice. Examples are wind comfort in urban streets or wind load on buildings and infrastructures. One can use such information to calculate pressure coefficient on buildings surface for natural ventilation design purposes. Isothermal CFD simulations are widely being used in this context as demonstrated in Table 12.2. Some of the non-isothermal problems can be assumed to follow a steady-state behaviour for a shorter period of time, known as ***quasi-steady state*** problems. Examples are heat conduction through walls in a matter of few minutes/hour as it is broadly used in the energy simulation tools as explained in Chapters 8 and 9.

Table 12.2 Temporal scale of CFD simulation problems in built environment.

	Steady state	Quasi-steady	Transient
Example applications	• Ventilation design for rooms • Pedestrian comfort in streets	• Conduction through walls • Long-wave radiation in streets	• Pollution dispersion in streets • Aerosol in rooms
Time-step	• –	• –	• $\Delta t \ll 1$ s
Period of simulation	• –	• –	• Minutes/hours
Computational cost	• Low to intensive	• Low to intensive	• Heavy to unaffordable

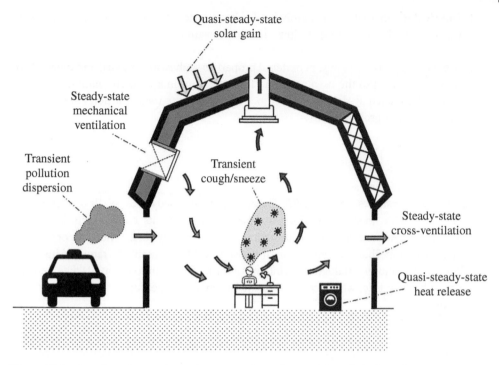

Figure 12.1   Steady-state, quasi-steady state, and transient problems in built environment.

Inversely, a range of problems related to the heat exchange within the systems of buildings, street canyons, and microclimates encounter non-isothermal conditions, which require the time variation to be considered in calculations (see Table 12.2). Climatic condition, occupancy of buildings, pedestrian activities in streets, long-wave radiation, etc. vary significantly on a typical day. Therefore, oversimplification of transient problems can provide incorrect outcomes on many occasions, which should be avoided, if possible (Figure 12.1). For example, pollution dispersion in cities is a transient phenomenon, which highly alters in respect to the source emission and environmental conditions (see Example 12.12).

12.2   Inputs to CFD Models

Three steps of preprocessing, solution, and post-processing to form a CFD solution were explained in the previous chapters. At the preprocessing stage, there are varieties of data necessary to form a CFD model. In a generic term, we can identify inputs for a built environment-related problem as:

1) **Geometrical data**: depending on the purpose of a simulation study, the details of an employed geometry significantly vary. Pre-/user-constructed CAD files, LiDAR data, GIS extracted data, etc. are some of the common methods to define CFD domains.

2) **Climatic data**: these data are mainly used in boundaries of a CFD domain, which can be extracted from local weather stations, in situ measurement, approximation/simplified models, etc.

3) **Material properties**: a range of material properties such as mechanical and thermal are necessary to assign to the used solid and fluid medium in a CFD simulation. Standard tables of the related associations, reference books and the related articles are useful to identify material properties.

12.2.1 Boundary Conditions

12.2.1.1 Enclosed Spaces

Figure 12.2 demonstrates a typical CFD domain for an enclosed space. Walls in such rooms are compliant with no-slip conditions. Inflow velocity with its components is directly obtained from measured velocity or indirectly from mass flow rate. When the opening is associated with a mechanical ventilation, it is also possible to obtain the required information from manufacturer catalogue. Turbulent intensity in the inflow boundary is another important characteristic, which can be obtained from direct measurement, manufacturer, or literature.

If the space does not include another opening to direct the airflow to exhaust, we can expect the room to be pressurized in a transient solution. Nonetheless, the steady state solution with air as the incompressible fluid requires a similar amount of mass to exist from the same inflow opening, which can cause divergence in the solution as the physics of the problem contradicts the conservation law. In such models, an extract opening should be always considered or a microclimate around the room should be included to avoid such misrepresentation of the realistic problem.

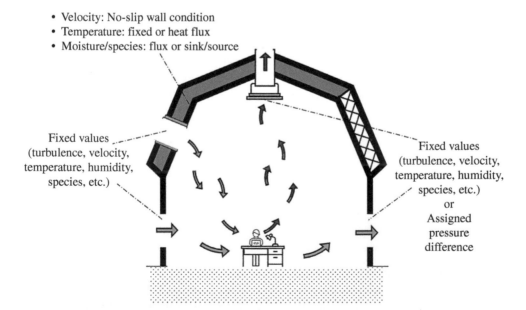

- Velocity: No-slip wall condition
- Temperature: fixed or heat flux
- Moisture/species: flux or sink/source

Fixed values (turbulence, velocity, temperature, humidity, species, etc.)

Fixed values (turbulence, velocity, temperature, humidity, species, etc.) or Assigned pressure difference

Figure 12.2 Boundary conditions in a typical enclosed space.

Incorrect/weak B.C:
- If $m_1 + m_2 \neq m_3 + m_4$ in a steady-state condition
- If p_3 or p_4 is set as outflow
- If air is set to be compressible
- If solar radiation is set as steady-state
- ...

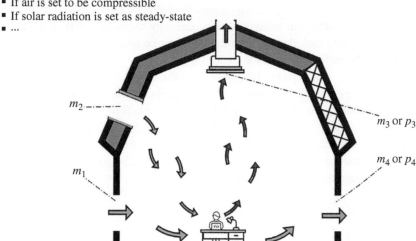

Figure 12.3 Incorrect/weak boundary conditions in enclosed spaces.

As shown in Figure 12.3, in the case of having multiple opening in an enclosed space, it is necessary to assign similar amount of air, entering from inflow boundaries, to at least one of the outlet openings although in an inverse direction, exhausting from the room to fulfil the conservation law. It should be noted that the turbulent intensity again can be either measured in an outflow boundary or it can be allocated in accordance with literature and recommended guidelines. If a fully developed boundary condition is considered for an outlet opening, kwon as outflow in many commercial packages, it should be ensured that the flow is not disturbed and the necessary condition for this assumption is met (see Chapter 2).

When non-isothermal scenarios are simulated, heat transfer from solid walls to the fluid requires to be imposed by a provided heat flux or temperature value identified on the walls as explained in detail in the previous chapters. The pair of inlet and outlet boundaries then can be set with temperature values. It is possible to set a temperature value at inlet when the outflow boundary condition is used for the outlet. A change in the density across the space should be expected by the state equation set to follow the ideal gas law or Boussinesq simplification.

12.2.1.2 Microclimates

As it was clarified earlier, very large microclimate domains result in a large number of cells, and mainly impossible to be simulated for transient applications such as pollution dispersion and heat island modelling. In the case of isothermal models, buildings', and infrastructures' walls as well as soil surfaces are modelled as no-slip boundary conditions with or without surface roughness. Non-isothermal modelling in a microclimate scale is an extremely complex procedure as the surface temperature of buildings should be defined as boundary conditions, which turns it into an impractical practice in many occasions. The soil temperature is mainly approximated with seasonal temperature variations or other simplified models.

For the purpose of practicality, the microclimate model only includes a part of the city while the other parts are only included in the shape of the wind profile. As demonstrated in Figure 12.4, wind at the inflow boundary is mainly approximated with log law or power law as described in Chapter 3. A certain distance around the target microclimate as illustrated in Figure 12.4 is expected in CFD domains to ensure that lateral and top boundary conditions are not distributing the airflow around the target microclimate. In this case, the outlet boundary can be assumed as outflow as the flow is largely fetched to be fully developed. For examples, the distances shown in Figure 12.4 are recommended by [2]. The vertical profiles of the turbulence kinetic energy (k_z) and turbulence dissipation rate (ε_z) also provided by the below suggested equations [2]:

$$k_z = \frac{2}{3}u'^2_z = \frac{2}{3}(I_z U_z)^2 \tag{12.1}$$

$$\varepsilon_z = C_\mu^{1/2} k_z \frac{U_{\text{met}}}{z_{\text{met}}} \alpha \left(\frac{z}{z_{\text{met}}}\right)^{\alpha-1} \tag{12.2}$$

where, I_z represents the turbulence intensity, which can be calculated by:

$$I_z = 0.1 \left(\frac{z}{\delta}\right)^{-\alpha-0.05} \tag{12.3}$$

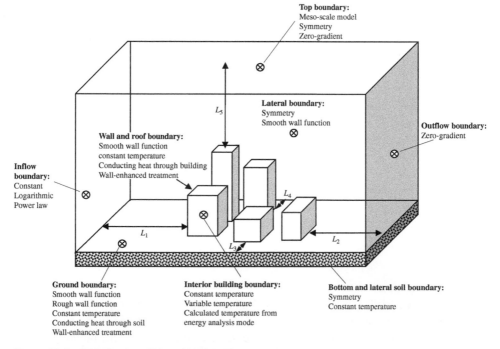

Figure 12.4 Boundary conditions in microclimates. *Source:* Mirzaei and Haghighat [1]/With permission of Elsevier.

12.3 Practical Examples

In this section, we practice about the key aspect of CFD modelling in built environment. This is based on the definitions and fundamental about preprocessing, simulation, and post-processing stages in CFD as explained in Chapters 10 and 11. While the following examples offer awareness about some of these capabilities, design and control of indoor environments, they focus on the necessary steps that should be carefully taken in a CFD simulation.

12.3.1 Conduction in Solid Materials

The first group of the examples are about conduction in solid materials where the velocity field is not resolved in the Navier–Stokes equations, resulting in these equations being reduced to the heat diffusion form. While a commercial tool is chosen to simulate many of the examples, the focus of this section is on understanding of mesh resolution and quality in addition to the validation of the thermal field mainly against analytical solutions for simplistic condition cases.

Example 12.1

Assume an isolated 1D wall without internal heat generation composed of concrete and 26cm in width as shown in Figure 12.E1.1 to constantly be exposed to indoor and outdoor temperatures of $T_{\text{room}} = 19\,^\circ\text{C}$ and $T_\infty = 31\,^\circ\text{C}$, respectively. The heat convection coefficients are $h_{\text{in}} = 3\ \text{W/m}^2\,\text{K}$ and $h_{\text{ext}} = 6\ \text{W/m}^2\,\text{K}$ for internal and external walls. Moreover, the radiation fluxes are neglected in this case study. Use a finite volume method (FVM) approach with $\Delta x = 0.01$ m to find the temperature distribution within the wall and validate your results.

Material properties:

	Wall
Material	Concrete
Roughness	Rough
Thickness (mm)	260
Conductivity (W/mK)	1.311
Density (kg/m³)	2240
Specific heat capacity (J/kg K)	836.8
Thermal absorptance	0.9
Solar absorptance	0.7
Visible absorptance	0.7

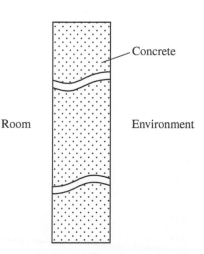

Figure 12.E1.1 One-layer isolated 1D wall without internal heat generation.

Solution

Using an analytical approach of 1D heat diffusion with Neumann boundary condition, the temperature distribution within the wall is linear as:

$$\frac{d}{dx}\left(k\frac{dT}{dx}\right) = 0$$

$$\Rightarrow \frac{dT}{dx} = c \Rightarrow T(x) = ax + b$$

Now, we apply boundary conditions:

$$@x = 0 \Rightarrow T(0) = T_{si} = b$$

$$@x = 0.26 \Rightarrow T(0.26) = T_{so} = 0.26a + b$$

$$@x = 0 \Rightarrow q''_{conv-indoor} = q''_{cond}$$

$$\Longrightarrow h_{in}(T_{room} - T_{si}) = -k\frac{dT}{dx}|_{x=0} = -ka \Longrightarrow 3(19 - b) = -1.311a \tag{$*$}$$

$$@x = 0.26 \Rightarrow q''_{conv-outdoor} = q''_{cond}$$

$$\Longrightarrow h_{ext}(T_{so} - T_{\infty}) = -k\frac{dT}{dx}|_{x=0.26} = -ka \Longrightarrow 6[(0.26a + b) - 31] = -1.311a \tag{$**$}$$

Now, from $(*)$ and $(**)$, we can find a and b, and thus obtain the exact solution as:

$$T(x) = -13.108x + 28.136$$

Regarding the FVM, this case is similar to Example 10.4 of Chapter 10. Thus, we only provide the results of a 1D mesh with 24 cells and two cell-faces. The results then can be compared and validated with the analytical solution to calculate the error as shown in the below table.

x	FVM temperature (°C)	Exact temperature (°C)	Error (%)
0	28.10	28.14	0.11
0.01	28.04	28.00	0.12
0.02	27.91	27.87	0.11
0.03	27.77	27.74	0.11
0.04	27.64	27.61	0.10
0.05	27.51	27.48	0.10
0.06	27.38	27.35	0.09
0.07	27.24	27.22	0.09
0.08	27.11	27.09	0.09
0.09	26.98	26.96	0.08
0.10	26.85	26.83	0.08
0.11	26.71	26.69	0.07
0.12	26.58	26.56	0.06
0.13	26.45	26.43	0.06
0.14	26.32	26.30	0.05
0.15	26.18	26.17	0.05

x	FVM temperature (°C)	Exact temperature (°C)	Error (%)
0.16	26.05	26.04	0.04
0.17	25.92	25.91	0.04
0.18	25.79	25.78	0.03
0.19	25.65	25.65	0.03
0.20	25.52	25.51	0.02
0.21	25.39	25.38	0.02
0.22	25.26	25.25	0.01
0.23	25.12	25.12	0.01
0.24	24.99	24.99	0.00
0.25	24.86	24.86	0.01
0.26	24.79	24.73	0.26
		Average error	0.07

Although the validation is successfully conducted, we can also show that the selected number of cells is a reasonable number. As it can be seen in Figure 12.E1.2, the average error is plotted against various number of cells. It can be seen that increasing the number of cells from 5 to 10 already results the error to be below 1% but increasing this number from 26 ($\Delta x = 0.01$ m) to 50 ($\Delta x = 0.005$ m) almost results in a very marginal change in the error. Thus, we can justify the employment of the mesh with 26 cells.

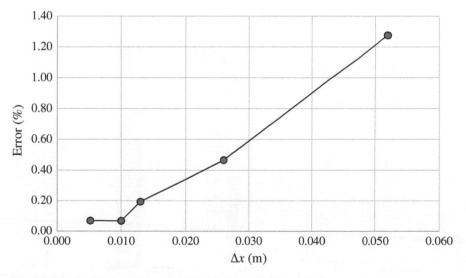

Figure 12.E1.2 Average error plot against various number of cells.

Example 12.2

Assume an isolated 2D wall similar to Example 12.1 while it is considered to be very long in the vertical direction ($L = 2.4$ m). First, generate a 2D mesh for the wall and examine the mesh quality. Then, validate the model with the analytical solution and compare the results with 1D model of Example 12.1. Furthermore, discuss the chosen temporal variation, assigned boundary conditions, and achieved convergence of your solution.

Solution

We assume the Neumann boundary on both side of the wall, so the 2D diffusion equation transforms to its steady-state form.

- **Employed Software:**

  ANSYS Fluent.

- **Mesh and quality:**

  As it can be seen in Figure 12.E2.1a, structured quadrilateral cells are used to mesh the geometry of the solid wall.

  The number of cells, length of the first layer cell, and mesh quality is also calculated as

Cell size/(first layer)		0.01 m × 0.04 m
Cell numbers		1560
Mesh	**Aspect ratio**	4.12
Quality	**Orthogonal quality**	1.00

(a) (b)

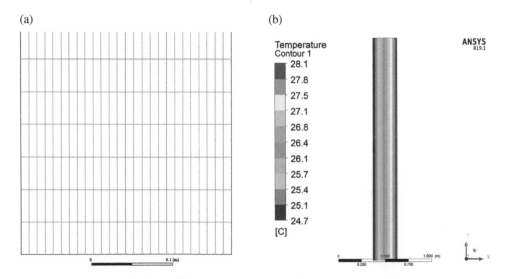

Figure 12.E2.1 Structured quadrilateral cells.

- **Setup:**
- **Boundaries**

We assign adiabatic (zero heat flux) on the top and bottom of the long wall. We also consider convection fluxes on both side of the wall with the known ambient temperatures and convective heat transfer coefficients as explained in Example 12.1.

- **Convergence**

Since the contributing material is only concrete (solid material), and velocities are zero, the only equation which is activated in the Navier–Stokes equation is the energy equation. This equation is the same as 2D heat diffusion equation. We also set 10^{-7} as the convergence criteria for the residuals as it is plotted in Figure 12.E2.2, which can be achieved in only five iterations. Note that with the current mesh resolution and setup, the residuals cannot further improve after reaching about 4×10^{-7}. So, if a better convergence for any reason in a specific problem is required, the mesh resolution and setup should be changed.

- **Mesh independency test:**

We skip this part in this example as it was observed in Example 12.1 that 1D heat diffusion equation can provide accurate results with only few cells though here we used 24 cells in the x-direction.

- **Validation:**

As it was stated in the previous chapters, heat conduction in common walls can be fairly assumed as 1D due to the nature of its heat transfer mechanism. Comparing 1D FVM results of Example 12.1 with a centre line of 2D FVM of Example 12.2 also acknowledges this point as shown in the below table in which the local and averaged errors are very small.

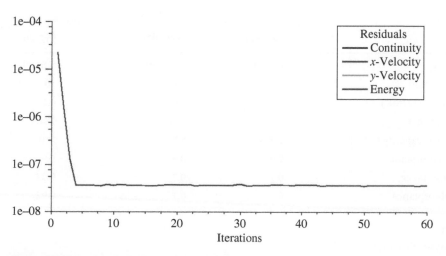

Figure 12.E2.2   Residual plot against iterations.

x	FVM 1D temperature (°C)	FVM 2D temperature (°C)	Difference FVM 1D– FVM2D (%)	x	FVM 1D temperature (°C)	FVM 2D temperature (°C)	Difference FVM 1D– FVM2D (%)
0	28.10	28.14	0.0000	0.14	26.32	26.30	0.0041
0.01	28.04	28.01	0.0003	0.15	26.18	26.17	0.0044
0.02	27.91	27.87	0.0006	0.16	26.05	26.04	0.0009
0.03	27.77	27.74	0.0009	0.17	25.92	25.91	0.0012
0.04	27.64	27.61	0.0011	0.18	25.79	25.78	0.0015
0.05	27.51	27.48	0.0014	0.19	25.65	25.65	0.0018
0.06	27.38	27.35	0.0017	0.20	25.52	25.52	0.0021
0.07	27.24	27.22	0.0020	0.21	25.39	25.38	0.0024
0.08	27.11	27.09	0.0023	0.22	25.26	25.25	0.0027
0.09	26.98	26.96	0.0026	0.23	25.12	25.12	0.0030
0.10	26.85	26.83	0.0029	0.24	24.99	24.99	0.0033
0.11	26.71	26.70	0.0032	0.25	24.86	24.86	0.0036
0.12	26.58	26.56	0.0035	0.26	24.79	24.73	0.0040
0.13	26.45	26.43	0.0038			Average	0.0022

Example 12.3

At this example, we would like to transform the wall of Example 12.2 to a more realistic one. Thus, we assume the wall to have three layers of gypsum (1 cm), insulation (5 cm) and concrete (20 cm). Now, repeat Example 12.2 for this wall.

Materials' properties:

	Wall		
Material	Concrete	Insulation	Gypsum
Roughness	Rough	Medium rough	Smooth
Thickness (mm)	200	50	10
Conductivity (W/mK)	1.311	0.049	0.16
Density (kg/m³)	2240	265	784.9
Specific heat capacity (J/kg K)	836.8	836.8	830
Thermal absorptance	0.9	0.9	0.9
Solar absorptance	0.7	0.7	0.92
Visible absorptance	0.7	0.7	0.92

Solution

- **Employed Software:**

  ANSYS Fluent.

Figure 12.E3.1 Three-layer isolated 1D wall without internal heat generation.

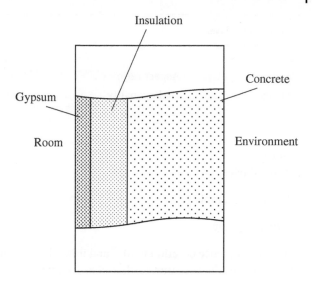

• **Mesh and quality:**

Again, we can demonstrate the generated mesh using structure quadrilateral cells in Figure 12.E3.2.

The number of cells, length of the first layer cell, and mesh quality is also calculated as:

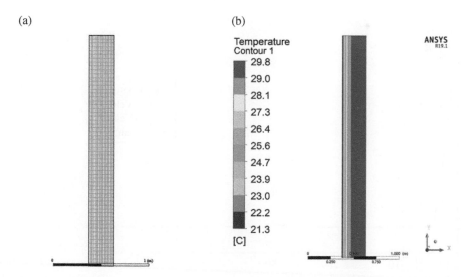

Figure 12.E3.2 Structured quadrilateral cells.

Cell size/(first layer)		0.01 m × 0.04 m (Gypsum 0.005 m × 0.04 m)
Cell numbers		1620
Mesh	**Aspect ratio**	8.10
Quality	**Orthogonality**	1.00

- **Setup:**
- **Boundaries**

Same as Example 12.2.

- **Convergence**

The convergence criterion is 10^{-7} and it is achieved quickly as plotted in Figure 12.E3.3. Again, the residuals cannot further improve after reaching about 3.45×10^{-7}.

- **Validation:**

If we assume linear temperature distribution in three layers and with a same procedure as before (see Example 7.5), we can obtain:

$T(x_1) = ax_1 + b$ for the concrete layer;

$T(x_2) = cx_2 + d$ for the insulation layer;

$T(x_3) = ex_3 + f$ for gypsum layer

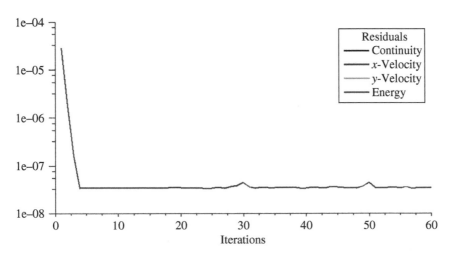

Figure 12.E3.3 Residual plot against iterations.

a	37.78
b	21.01
c	123.35
d	21.39
e	4.61
f	28.79

Thus, we can now compare the FVM results with the analytical solution and calculate the error as:

x (m)	Exact temperature ($^\circ$C)	FVM 2D temperature ($^\circ$C)	Error (%)	X (m)	Exact temperature ($^\circ$C)	FVM 2D temperature ($^\circ$C)	Error (%)
0	21.01	21.30	1.38	0.14	29.44	29.21	0.78
0.01	21.39	21.74	1.64	0.15	29.49	29.26	0.75
0.02	23.86	23.15	2.99	0.16	29.53	29.32	0.73
0.03	25.09	24.56	2.14	0.17	29.58	29.37	0.71
0.04	26.33	25.97	1.37	0.18	29.62	29.42	0.68
0.05	27.56	27.38	0.66	0.19	29.67	29.47	0.66
0.06	28.79	28.79	0.02	0.20	29.72	29.53	0.64
0.07	29.12	28.84	0.95	0.21	29.76	29.58	0.62
0.08	29.16	28.89	0.92	0.22	29.81	29.63	0.59
0.09	29.21	28.95	0.90	0.23	29.85	29.69	0.57
0.10	29.25	29.00	0.87	0.24	29.90	29.74	0.54
0.11	29.30	29.05	0.85	0.25	29.95	29.79	0.52
0.12	29.35	29.11	0.83	0.26	29.99	29.84	0.50
0.13	29.39	29.16	0.80			**Average error**	0.91

12.3.2 Wall Treatment of Boundary Layer

Wall effect on fluid flows was explained in Chapter 3. In CFD models, due to the extensive computational cost that resolving all the layers of a wall can cause, it is common to also bridge this layer with a logarithmic correlation, known as wall-function, as illustrated in Figure 12.5-left. Thus, if a wall-function is used, it is recommended that the first layer (wall-adjacent) cell's centroid is located within the log-law layer where $30 < y^+ < 300$. In the case that the wall effect is resolved with a fine mesh as shown in Figure 12.5-right, it is called enhanced wall treatment of the wall.

This section, thus, focuses mainly to include the fluid domain with a simplistic format around solid surfaces such as walls. Mesh independency, creation of boundary layer to

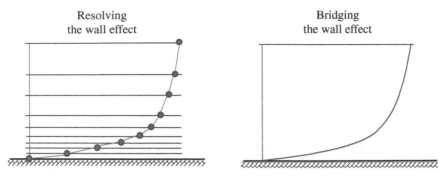

Figure 12.5 Resolving (enhanced wall treatment) and bridging (wall-function approximation) the wall effect on a surface.

control y^+, convergence of results, and validation of the flow fields are amongst the main goals of this section to be practiced.

Example 12.4

Assume the 2D wall of Example 12.2 to be in the isothermal condition. Consider an air medium around the wall with the dimension of $3\,m \times 2.4\,m$ as shown in Figure 12.E4.1, which is a weak representation of the microclimate around the buildings' walls. Resolve boundary layer over the wall surfaces using wall-function method and find the choice of the correct first layer size of the cells if the approaching airflow from the lateral sides has a velocity of 1.5 m/s and an angle of 45°. Assume that there is a lack of experimental study to validate your results, so perform a mesh independency test to minimize the grid sensitivity associated errors.

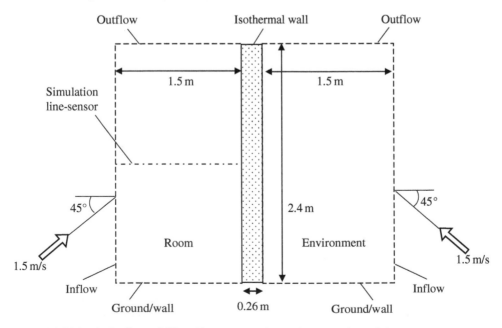

Figure 12.E4.1 An isothermal 2D wall exposed to the environmental condition.

Solution:

- **Employed Software:**

  ANSYS Fluent.

- **Mesh and quality:**

To ensure that y^+ is within the desired range, we generate six different meshes using structure quadrilateral cells. The fine and coarse meshes of M4.1, M4.6, and M4.7 are shown in Figure 12.E4.2.

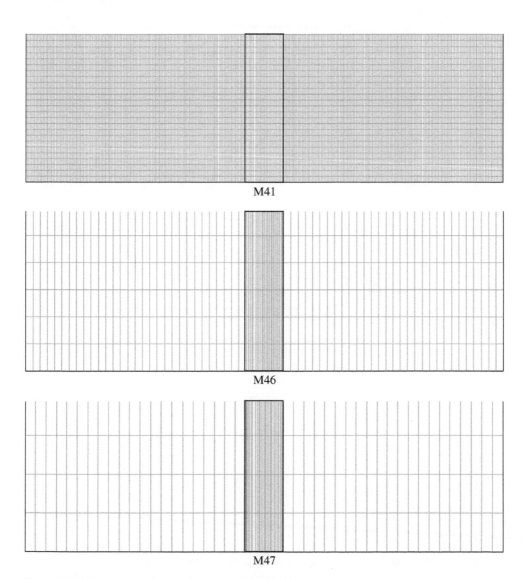

Figure 12.E4.2 Structured quadrilateral cells with various resolutions.

The first layer cell's centroid length and mesh quality are as below:

Mesh index		M4.1	M4.2	M4.3	M4.4	M4.5	M4.6	M4.7
Cell size/(first layer)		0.01 × 0.040	0.012 × 0.048	0.0144 × 0.058	0.022 × 0.086	0.033 × 0.129	0.050 × 0.019	0.072 × 0.279
Growth ratio		1.0	1.0	1.0	1.0	1.0	1.0	1.0
Cell numbers		19,620	13,850	9,954	4,620	2,261	1,131	621
Mesh	Aspect ratio	8.06	9.65	11.47	17.17	25.28	36.94	53.34
Quality	Orthogonal quality	1.00	1.00	1.00	1.00	1.00	1.00	1.00

As it can be seen, increase of the first layer size results in lower number of cells. The aspect ratio of meshes also increases when the first layer is increased. This can result in convergence and accuracy issues; however, for such simple case where the flow is parallel to the surface, it would reasonably perform as it is seen in Figure 12.E4.3. Moreover, as it can be

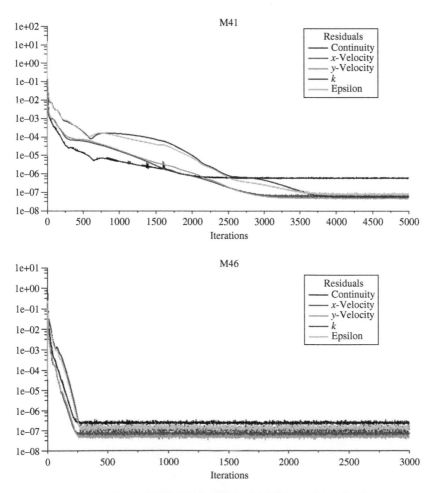

Figure 12.E4.3 Residual plot against iterations for M4.1 and M4.6 meshes.

seen in the below table, the average y^+ of the wall for M4.1 is below the required value. Also, the average y^+ of the ground for M4.6 is slightly above the threshold. M4.7 seems to be a bit high in the average while some of the local points are exceeding $y^+ = 500$. The rest of the meshes seems to comply within the required range.

Surface average y+	Wall	Ground
M4.1	27.49	52.00
M4.2	33.56	64.33
M4.3	40.64	78.83
M4.4	64.01	125.18
M4.5	99.68	198.56
M4.6	161.28	320.44
M4.7	256.97	485.08

- **Setup:**
- **Model**

The airflow is assumed to be steady state while the air properties are in a homogenous condition. The gravity effect on the fluid particles is also ignored. The energy equation is not contributing in the Navier–Stokes equations due to an isothermal condition. The selected turbulence model is Standard k-epsilon with the default coefficients.

- **Boundaries**

The velocity is set as inflow at the bottom of the mesh. The outflow is also assigned with the pressure outlet boundary condition with zero Pascal difference, which implies a zero-gauge pressure at that point. Walls are assumed to have no-slip condition.

- **Convergence**

The convergence criterion for continuity is selected to be obtained below 10^{-6} and for x-velocity, y-velocity, TKE, and epsilon to be below 10^{-7}. The residuals of M4.1 and M4.6 are plotted in Figure 12.E4.3. As it can be observed, all the residual criteria are met after 3,970 iterations (below 6.08×10^{-7}) for M4.1 while this number is only 310 (below 2.53×10^{-7}) for M4.6, which is mainly due to the lower number of cells in this mesh. Also from a flat residual, it can be seen that the potential of these meshes with these structure and setup is reached and lower convergence criteria cannot be achieved for these cases. Although values are fluctuating when a convergence is achieved, we can show them to be around the below numbers for each equation:

Mesh	M4.1	M4.2	M4.3	M4.4	M4.5	M4.6	M4.7
Continuity	6.08E−07	5.68E−07	4.27E−07	3.35E−07	2.65E−07	2.53E−07	2.04E−07
x-Velocity	6.21E−08	6.49E−08	6.43E−08	6.77E−08	5.96E−08	6.70E−08	6.14E−08

(Continued)

Mesh	M4.1	M4.2	M4.3	M4.4	M4.5	M4.6	M4.7
y-Velocity	4.96E−08	5.06E−08	4.86E−08	5.20E−08	4.60E−08	5.30E−08	5.05E−08
k	5.87E−08	6.20E−08	6.36E−08	7.80E−08	9.32E−08	1.21E−07	1.23E−07
Epsilon	7.75E−08	8.45E−08	8.30E−08	1.03E−07	1.13E−07	1.29E−07	1.20E−07

Now, we can show the velocity contour and vectors around the wall for the case of M4.1 as Figure 12.E4.4.

- **Mesh independency test:**

A simulation line-sensor as depicted in Figure 12.E4.1, at the middle of the fluid domain, is set to compare the normalized velocities (by the inflow velocity) obtained from different meshes. As the resolved boundary layer can be seen also in Figures 12.E4.4 and 12.E4.5 demonstrates these values across the line-sensor.

Also, the calculated average difference of the velocities between two consecutive meshes is depicted in Figure 12.E4.6. This implies that the results are significantly changing using a finer mesh until M4.2 and thus after refining the mesh further to M4.1, the results can be assumed to be independent of the grid size as the variation is not considerable anymore. Thus, M4.2 would be a suitable choice of the mesh in terms of quality and expected results.

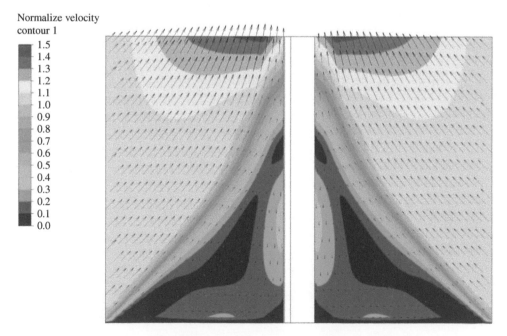

Figure 12.E4.4 Velocity contour and vectors around the wall for the case of M4.1.

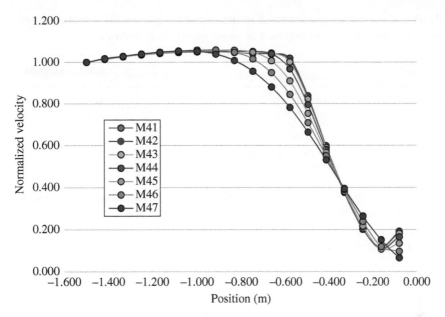

Figure 12.E4.5   Resolved boundary layer across the simulation line-sensor for various meshes.

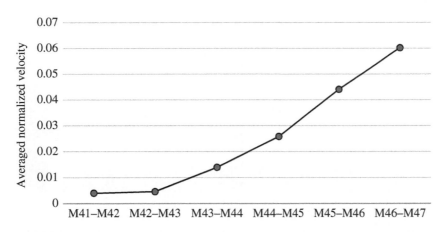

Figure 12.E4.6   Differences of the averaged normalized velocities for consecutive meshes.

Example 12.5

Resolve Example 12.4 with the enhanced wall treatment and compare the results for both solutions.

Solution

- **Employed Software:**

  ANSYS Fluent.

- **Mesh and quality:**

The mesh type is structured quadrilateral in the domain region while it has a denser arrangement in the boundary layer region. The mesh quality is as follows:

Cell size/(first layer)		0.00125 × 0.0005
Growth ratio		1.1
Cell numbers		52,569
Mesh	Aspect ratio	99.11
Quality	Orthogonality	1.00

To ensure that $y^+ < 1$ are resolved, we chose the first layer to be very small as depicted in Figure 12.E5.1. This caused the number of the cells to increase from about 1.1k in M46, as an acceptable mesh with wall-function approach, to about 52.5k.

The calculated y^+ is thus as below:

Averaged y+	Wall	Ground
	0.996	1.006

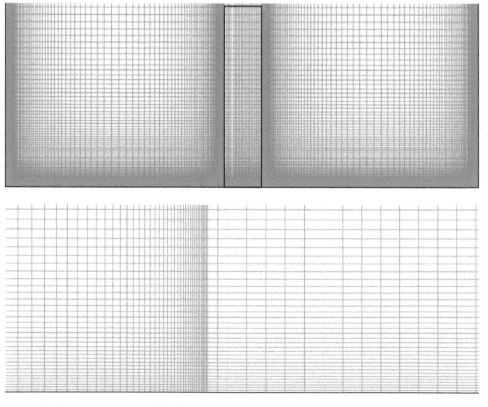

Figure 12.E5.1 Boundary layer mesh over the wall.

- **Setup:**
- **Model**

Similar as Example 12.4.

- **Boundaries**

Similar as Example 12.4.

- **Convergence**

Due to a considerably large number of cells, the convergence takes a higher number of iterations (2980) to reach the values below 10^{-7} as it is shown in Figure 12.E5.2. Also, the following table shows the residuals to fluctuate around them and to not be further improved.

Continuity	7.40E−07
x-Velocity	4.82E−08
y-Velocity	5.79E−08
k	5.59E−08
Epsilon	7.41E−08

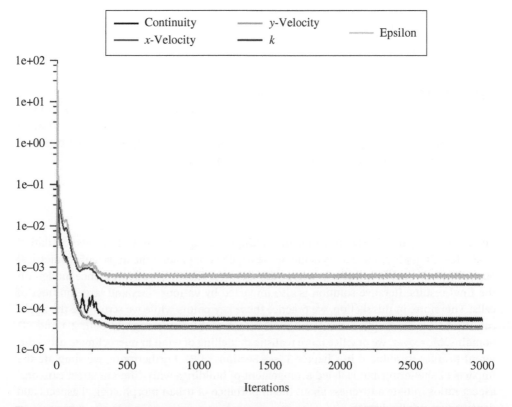

Figure 12.E5.2 Quadrilateral structured mesh with a denser arrangement in the boundary layer.

Normalize velocity
contour 1

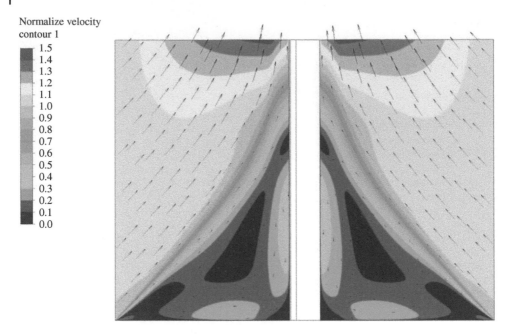

	1.5
	1.4
	1.3
	1.2
	1.1
	1.0
	0.9
	0.8
	0.7
	0.6
	0.5
	0.4
	0.3
	0.2
	0.1
	0.0

Figure 12.E5.3 Velocity contour and vectors around the wall.

- Results

After reaching the convergence, the velocity contours and vectors can be illustrated in Figure 12.E5.3.

Eventually, the resolved boundary layer is plotted in Figure 12.E5.4 and compared with the wall-function method used in M4.2, M4.6, and M4.7 of Example 12.4. While M4.2 can capture most of the velocity trend across the domain, it also follows the enhanced boundary layer with a fair approximation. Note that the number of cells is reduced from 52.5k to 13.8k using a wall-function method. Nonetheless, if y^+ is out of the suggested values and more towards the upper bound as shown in M4.6 and M4.7, the velocity is not well captured in both vicinity of the wall and across the entire domain.

12.3.3 Airflow in Microclimate

To be able to expand a CFD simulation in buildings to design and control various ventilation strategies, air quality, and energy demand, we need to understand the impact of geometrical domain, mesh quality and size in a large domain, and the choice of turbulence models on the final results. Iterative solution is also impacted by various sizes and large number of cells. Therefore, in this section, we practice these key points while we scrutinize the application of CFD in simple natural and mechanical ventilation design in both 2D and 3D domains. Moreover, we practice more realistic modelling of wind in microclimate using different profiles introduced in Chapter 3 (see Section 3.1.2). Furthermore, we simulate the impacts of sheltering and isolated arrangement of buildings with different street canyons' aspect ratios initiate awareness about the importance of urban morphological aspects and densities on urban climates.

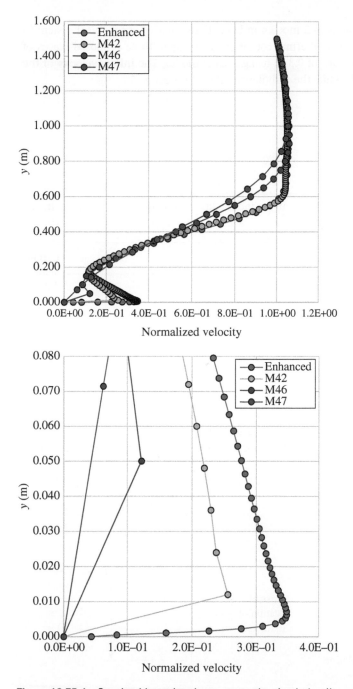

Figure 12.E5.4 Resolved boundary layer across the simulation line-sensor for various meshes.

Example 12.6

A 2D office with the dimensions of 3 m × 2.4 m is ventilated with a mechanical system as shown in Figure 12.E6.1 in an isothermal condition. Investigate three proposed places of the extractor with the dimensions of 0.3 m using a CFD model. The inflow velocity of the mechanical system is 0.5 m/s while the turbulence intensity is set as 5%.

Solution

- **Employed Software:**

  ANSYS Fluent.

- **Mesh and quality:**

  The mesh is structured quadrilateral in boundary layer and unstructured quad in the domain with the following quality:

Cell size/(first layer)		0.01 m
Growth ratio		1.2
Cell numbers		4221
Mesh	Aspect ratio	1.8
Quality	Orthogonality	0.97

- **Setup:**
- **Model**

  The model is a steady state isothermal flow in the room where the effect of gravity is neglected. Standard k-epsilon model is chosen as the turbulence model while the wall function is employed for the wall treatment purpose. Discretization methods are utilized

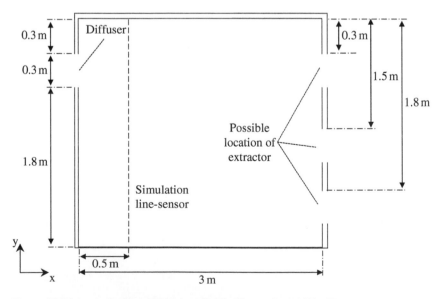

Figure 12.E6.1 An isothermal and mechanically ventilated 2D office.

Figure 12.E6.2 A mesh with structured quadrilateral cells in boundary layer and unstructured quadrilateral cells in the domain.

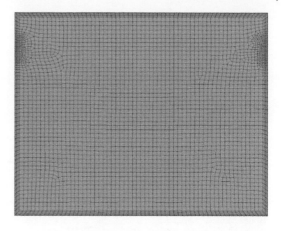

as second-order upwind and pressure-coupling scheme is selected as SIMPLE. The averaged $y+$ on the surface is about 15.

- **Boundaries**

The diffuser is set as inlet boundary condition where the velocity in a normal direction to the diffuser is assigned as 0.5 m/s with a turbulent intensity of 5% and a hydraulic diameter of 0.46m (where for the opening: $h_D = \frac{2ab}{a+b}$, $a = 0.3$ m, $b = 1$ m). The extractors in all three cases are set as the outlet boundary condition with a pressure difference of zero.

- **Convergence**

As it can be observed in Figure 12.E6.3, the convergence criteria for momentum, TKE, and epsilon is set as 10^{-7} while it is set to be 10^{-4} for the continuity equation. Also, after about 2000 iterations, there is no significant change in the residuals and thus an iterative convergence is achieved.

- **Mesh independency test:**

To ensure that the results are independent of the grid size, two more meshes are generated with coarser and finer resolutions as depicted in the below table.

Mesh resolution	Coarse	Medium	Fine
Face element size (m)	0.075	0.05	0.025
Edge element size (inlet/outlet)	0.015	0.015	0.015
Number of cell elements	2403	4221	8225

As it is demonstrated in Figure 12.E6.4, the discrepancy between three meshes are 9% and 4%, thus we can justify the utilization of the medium resolution mesh as with having almost double the number of cells in the fine mesh, the results are not significantly altered. Note that to reduce the number of simulations, only these three meshes are shown in this example though it might take further attempts as shown in the previous examples to obtain the mesh independency.

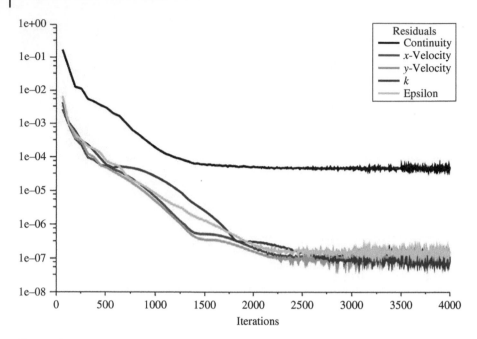

Figure 12.E6.3 Residual plot against iterations.

- **Results:**

Now, we can show the results as the contour and vectors of the velocity field in Figure 12.E6.5. The general flow demonstrates a well-mixed situation in all cases. Moreover, the averaged/velocity at the lower half of the room (occupied zone) is about 0.10,

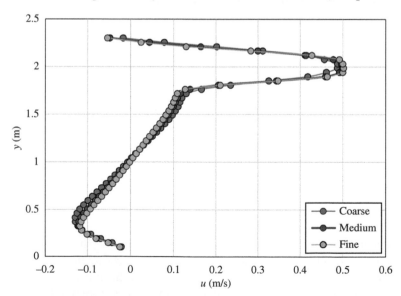

Figure 12.E6.4 Velocity profiles across the simulation line-sensor for various meshes.

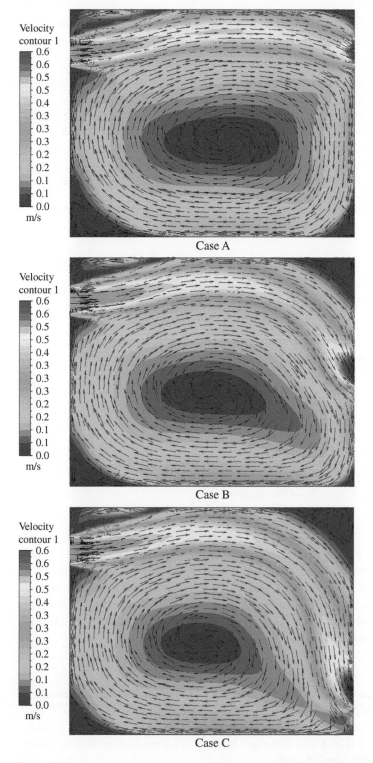

Figure 12.E6.5 Contours and vectors of the velocity field for different possible locations of the extractor.

0.12, and 0.15 m/s for Case A, Case B, and Case C, respectively. Thus, placing the extractor in the bottom of the room may help to slightly increase the velocity in the occupied zone.

Example 12.7

Assume a 2D isothermal cross-ventilation flow around the isolated buildings with a same dimension of 3 m × 2.4 m similar to Example 12.6 as shown in Figure 12.E7.1. The wind velocity is 1.5 m/s at a height of 10 m in all cases. Develop a CFD model and investigate the difference between utilization of suggested wind profile of log-law and power-law (see Chapter 3 – Section 3.1.2) and a uniform flow. Use AIJ guideline [2] for construction of the dimensions of the microclimate domain.

Solution

- **Employed Software:**

  ANSYS Fluent.

- **Mesh and quality:**

  The 2D mesh around the buildings is shown in Figure 12.E7.2. The mesh is structured quadrilateral in the domain and unstructured quadrilateral within and around the building. The mesh quality is also demonstrated as below:

Cell size/(first layer)		0.02 m
Growth ratio		1.2
Cell Numbers		23,022
Mesh	Aspect ratio	1.6
Quality	Orthogonality	0.98

- **Setup:**
- **Model**

  The CFD is simulated in a steady-state isothermal condition. Short- and long-wave radiations in addition to effect of gravity are neglected. Turbulence model is again set as Standard k-epsilon. Wall-function is used for the treatment of the wall effect. Discretization methods are utilized as second-order upwind and pressure-coupling scheme is selected as SIMPLE. The averaged $y+$ on the surface is about 20.

- **Boundaries**

  Boundary conditions are set as depicted in Figure 12.E7.1. While inlet will be changed according to different profiles, outlet is assigned as a constant zero-pressure. Default inlet profiles for turbulent intensity and epsilon are based on AIJ guideline [2] with $U_H = 1.5$ m/s, $H = 10$ m, $Z_G=350$, and $\alpha = 0.25$. Upper boundary is symmetry boundary condition, and all walls are assumed as no-slip smooth walls.

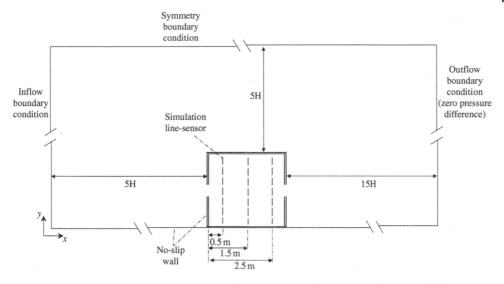

Figure 12.E7.1 Isothermal cross-ventilation flow around and through a 2D isolated building.

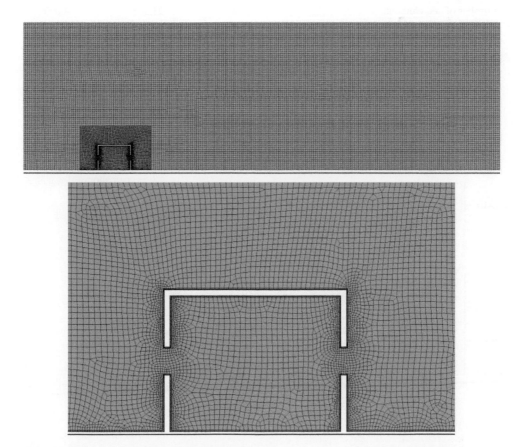

Figure 12.E7.2 A mesh with structured quadrilateral cells in the domain and unstructured quadrilateral cells within and around the building.

- **Convergence**

The convergence criteria for momentum, TKE, and epsilon is set as 10^{-5} while it is set to be 10^{-3} for the continuity equation. As it can be seen in Figure 12.E7.3, after about 1000 iterations, the simulation reaches to these values where most of the quantities will not change even after further iterations (using lower convergence criteria) as they have already attained their iterative convergence that can be seen as flat lines.

- **Mesh independency test:**

Three different meshes have been generated to ensure the least dependency of the results to the grid size. The characteristics of these meshes can be seen in the below table.

Mesh resolution	Coarse	Medium	Fine
Opening edge (number of elements)	3	3	3
Edge element (building wall)	33	40	48
Volume mesh size around the room (m)	0.12	0.1	0.08
Volume mesh size far from room (m)	0.24	0.2	0.16
Number of cell elements	13,942	23,022	34,000

Figure 12.E7.4 demonstrates the results of these meshes across the simulation line-sensors as shown in Figure 12.E7.1. The close values of these meshes justify the choice

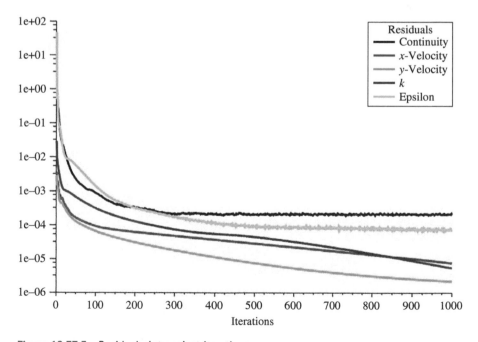

Figure 12.E7.3 Residual plot against iterations.

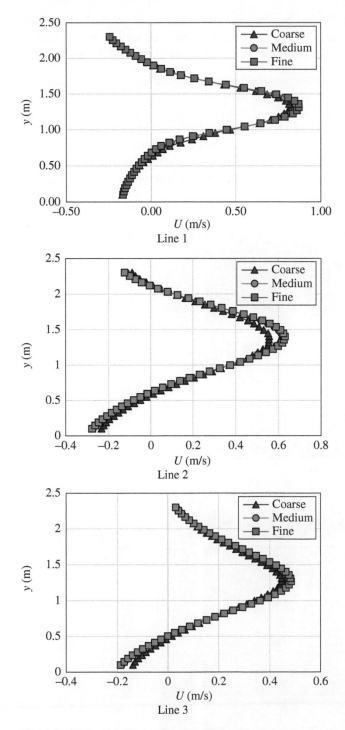

Figure 12.E7.4 Velocity profiles across the simulation line-sensors for various meshes.

of the medium mesh as the difference between the coarse and medium meshes is about 15.4%, which is reduced to about 2.2% when the medium and fine meshes are compared to each other. In specific, results in Line-2 show that the flow at the middle of the room is less accurate if the coarse mesh is used.

- **Results:**

Now, we can investigate the results using contours of velocity for three different employed profiles as shown in Figure 12.E7.5. It is evident that the uniform profile provides an exaggerated velocity, entering the room. This implies that the cross-ventilation can wrongly mislead the design of ventilation in rooms. On the other hand, both power-law and log-law

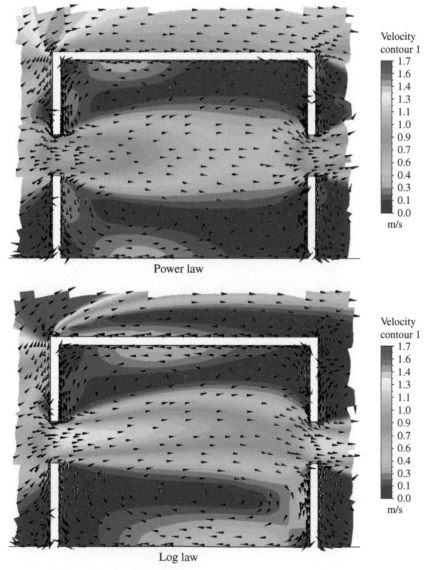

Figure 12.E7.5 Contours and vectors of the velocity field for different inflow boundary profiles.

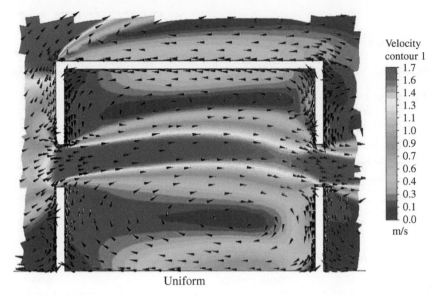

Uniform

Figure 12.E7.5 (Continued)

profiles result in a very close velocities inside the room. The mass flow rates against different wind profiles are also shown below:

Wind profile	Power law	Log law	Uniform
Mass flow (kg/s)	0.11	0.12	0.18

Example 12.8

Assume the cross-ventilation of Example 12.7. Now we add a building with a similar dimension in the upstream of the investigated building as seen in Figure 12.E8.1. Develop a CFD model and explore the mass flow rate of the sheltered building under different distances of $L = 2$ m, $L = 4$ m, and $L = 6$ m. Utilize the power-law as the wind profile.

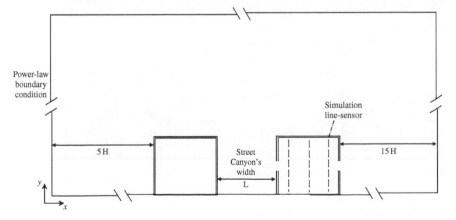

Figure 12.E8.1 Isothermal cross-ventilation flow around and through a 2D sheltered building.

Solution

- **Employed Software:**

ANSYS Fluent.

- **Mesh and quality:**

The microclimate domain of the study is meshed with structured quadrilateral cells though unstructured quadrilateral cells are used around the sheltered buildings as shown in Figure 12.E8.2. The mesh quality is presented as below:

Cell size/(first layer)		0.02 m
Growth ratio		1.2
Cell numbers		24,664
Mesh	Aspect ratio	1.7
Quality	Orthogonality	0.98

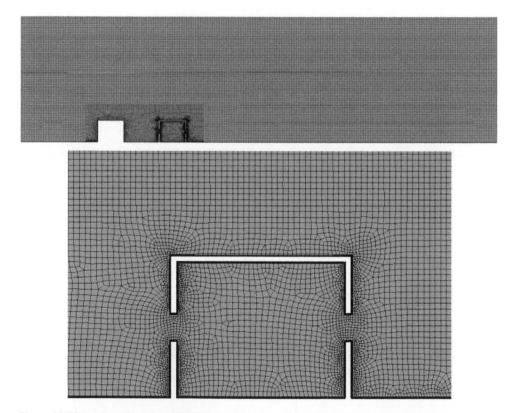

Figure 12.E8.2 A mesh with structured quadrilateral cells for microclimate domain and unstructured quadrilateral cells around the sheltered buildings.

- **Setup:**
- **Model**

Setups are similar as Example 12.7.

- **Boundaries**

Boundary conditions are similar as Example 12.7 except the inlet boundary, which is employed only with the power-law profile.

- **Convergence**

The convergence criteria are similar to Example 12.7 while it is achieved after about 1000 iterations as illustrated in Figure 12.E8.3.

- **Mesh independency test:**

Three meshes are again generated with very similar characteristics as those presented in Example 12.7 to compare their performance at the simulation line-sensors as depicted in Figure 12.E8.1. Here, the results are shown in Figure 12.E8.4. Similar to the previous example, the deference between coarse and medium meshes is about 26.8%, which is reduced to 18.6% when medium and fine meshes are compared. Hence, we can again justify the choice of the medium mesh in the CFD simulations.

- **Results:**

Velocity contours as depicted in Figure 12.E8.5 can be used to qualitatively compare the cross-ventilation through the room and assess it against the isolated building case of the

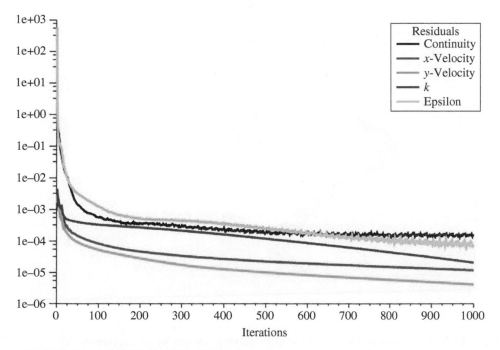

Figure 12.E8.3 Residual plot against iterations.

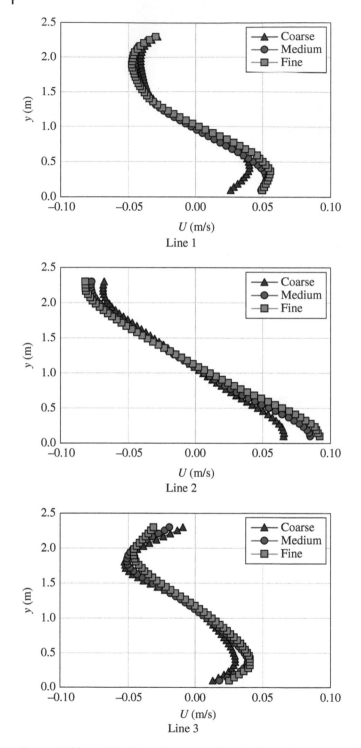

Figure 12.E8.4 Velocity profiles across the simulation line-sensors for various meshes.

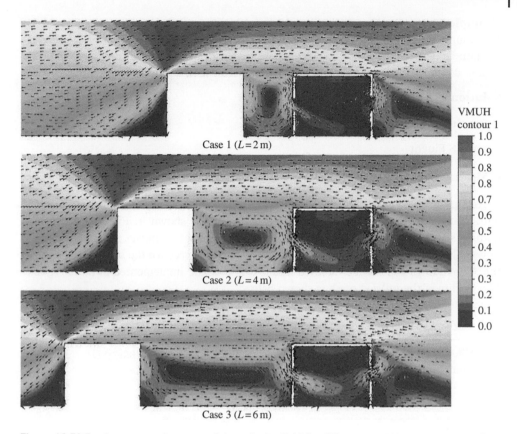

Figure 12.E8.5 Contours and vectors of the velocity field for different street canyon aspect ratios.

previous example. Additionally, we can investigate the impact of street canyon aspect ratio on the cross-ventilation. For this purpose, the mass flow rates are provided as below:

		L (m)	Mass flow rate (kg/s)
Sheltered	Case 1	2	0.008
	Case 2	4	0.019
	Case 3	6	0.036
Isolated	—	—	0.110

As it can be concluded, the isolated building case has an enormously higher mass flow rate while in reality, when the buildings are sheltered, the mass flow rate is significantly reduced. The decline in the mass flow rate is higher in the denser building arrangements. It should be noted that one of the main limitations of this example was associated to the 2D geometry of the buildings, which hinder the observation of the 3D airflow around them.

Example 12.9

We assume the cross-ventilation of Example 12.8 with $L = 4$ m to be conformed into a 3D domain as depicted in Figure 12.E9.1. The opening has a dimension of 1.38 m × 0.5 m. Again, develop a CFD model and explore the mass flow rate of the sheltered building.

Solution

- **Employed Software:**

  ANSYS Fluent

- **Mesh and quality:**

To improve the quality of flow simulation in a sheltered scenario, a 3D mesh is generated with structured hexahedral cells for the microclimate as shown in Figure 12.E9.2. The domain size is stretched $5H$ from the lateral sizes to avoid the impact of walls on the 3D flow within the microclimate. A cut-cell strategy is used to reduce the size of cells around the 3D buildings while still structured hexahedral cells are implemented in these regions. The quality of the utilized mesh is shown below:

Cell size/(first layer)		0.02 m
Growth ratio		1.2
Cell numbers		536,749
Mesh	**Aspect Ratio**	1.5
Quality	**Orthogonality**	0.97

- **Setup:**
- **Model**

Setups are similar as Example 12.8. Due to limitations in the size of cells, in particular, the first layer cell, y-plus is kept around $y^+ = 35$ to avoid ending up to a large mesh size.

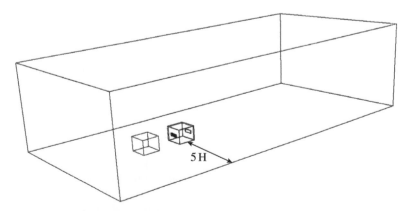

5 H

Figure 12.E9.1 Isothermal cross-ventilation flow around and through a 3D sheltered building.

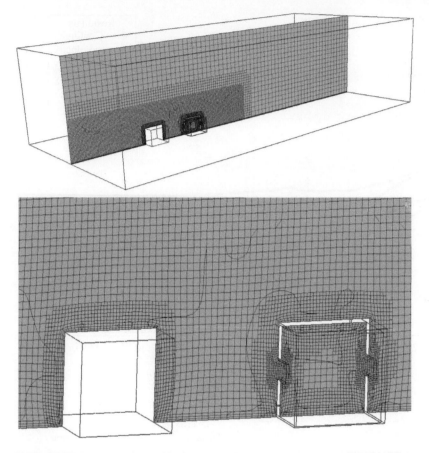

Figure 12.E9.2 A mesh with structured hexahedral cells for microclimate domain and unstructured hexahedral cells around the sheltered buildings.

- **Boundaries**

Boundary conditions are similar as Example 12.8. Lateral boundary conditions are set as symmetry.

- **Convergence**

The convergence criteria are set similar to Example 12.8 while these conditions are met after about 1200 iterations as it can be seen in Figure 12.E9.3.

- **Mesh independency test:**

Similar to the previous examples, three different meshes are generated as shown in the below table. The performance of these line simulation-sensors at the middle of the ventilated room are shown in Figure 12.E9.4. It is clear that the medium mesh has a close performance to the fine mesh though using the coarse mesh, specially at the middle of the room shown as Line-2, can result in a huge discrepancy in the final results as the mesh is still very much dependent of the cell sizes.

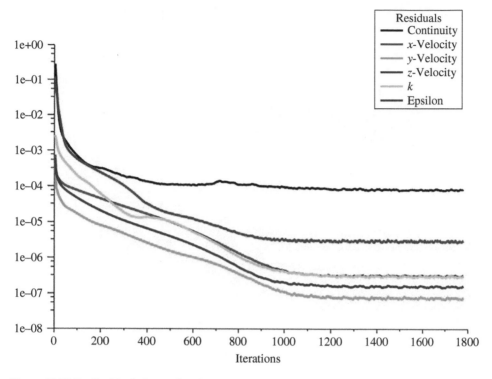

Figure 12.E9.3   Residual plot against iterations.

	Coarse	Medium	Fine
Layer numbers in BL	5 layers	5 layers	5 layers
y^+	$y^+ = 0.05$	$y^+ = 0.05$	$y^+ = 0.05$
Size of cell elements/size of the opening edges (m)	0.05	0.05	0.05
Mesh size around the room (m)	0.5	0.25	0.125
Mesh size far from the room (m)	1	0.8	0.16
Number of cell elements	397,000	536,749	796,652
Mass flow rate (kg/s)	0.138	0.132	0.133

- **Results:**

Now, we can expect the results obtained from the medium mesh to be relatively accurate. The velocity contours are depicted in the middle of the ventilated room as shown in Figure 12. E9.5. For the purpose of comparisons, the same results from the 2D model generated in Example 12.8 with $L = 4$ m is also illustrated in Figure 12.E9.5. Qualitatively, it can be clearly seen that the exaggerated 2D structure of the flow diminishes in the 3D model as the circulation becomes weaker. We can also calculate the mean velocity entering the opening in the cross-ventilation of Case 2 ($L = 4$ m) using an average streamwise velocity at the windward opening. The values are 0.15 m/s and 0.33 m/s in 2D and 3D models, respectively. The double time value of the 3D case emphasises on the importance of 3D structure of the flow in entering to the opening. Interested readers can refer to the following references for further details [3–6].

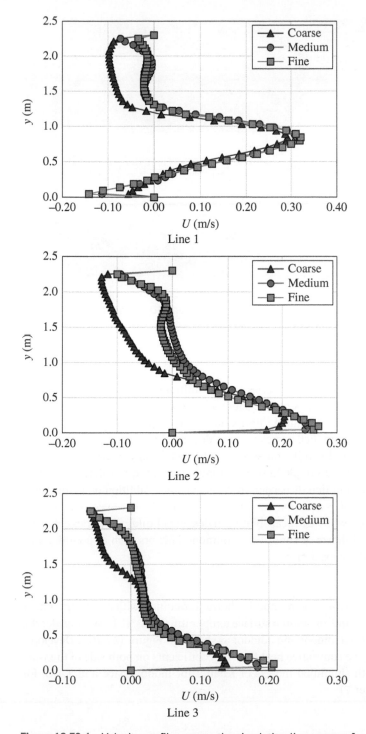

Figure 12.E9.4 Velocity profiles across the simulation line-sensors for various meshes.

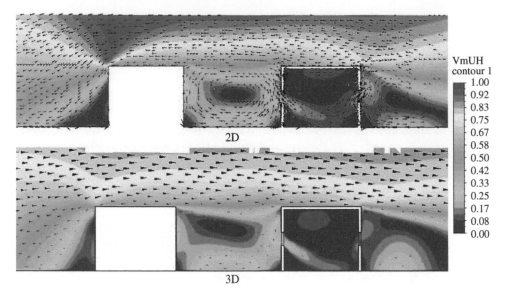

Figure 12.E9.5 Contours and vectors of the velocity field for 2D and 3D scenarios of the sheltered buildings.

12.3.4 Convective Heat Transfer Coefficient

In the following examples, we simulated a series of non-isothermal cases. The energy equation would be activated in the Navier–Stokes equations while air is assumed as an ideal gas. The benefits of CFD in providing detailed information for energy demand calculation of buildings rather than using simplified correlations as described in Chapters 7–10 are discussed. Modelling of the mixed ventilation systems when buoyancy effect is taken into the account is another practical example in this section. Moreover, we explore a transient solution in regard with the pollution dispersion in street canyons. Addition of quantities as new species to the Navier–Stokes equations and the transient iterative convergence are further exercised in this section while CFD capabilities in design of buildings' parameter such as opening conditions to minimize or control environmental factors such as pollution dispersion are presented in these examples.

Example 12.10

Assume the wall of Example 12.4 to be in a non-isothermal condition with a constant indoor surface temperature of 21°C and an outdoor surface temperature of 21°C. Now, consider the air velocity and temperature in indoor and outdoor as shown in Figure 12.E10.1. Develop a CFD model and calculate the convective heat transfer coefficients on both side of the wall. Compare your results with simplified models of Khalifa for indoor space and TARP for outdoor space (see Chapter 7).

Solution

- **Employed Software:**

  ANSYS Fluent

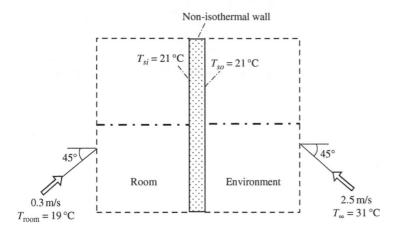

Figure 12.E10.1 A non-isothermal 2D wall exposed to the environmental condition.

- **Mesh and quality:**

Same as Example 12.4.

- **Setup:**
- **Model**

Same as Example 12.4, except the energy equation in Navier–Stokes equations, which is taken into the account (see Chapter 10). Buoyancy effect is not neglected, and the change of density is modelled assuming air as an ideal gas. Standard $k - \varepsilon$ turbulence model with wall-function treatment caused a divergence in the results, and an unsatisfactory convergence level with wall-enhanced treatment. Thus, SST K-Omega Turbulence model was employed in this problem, which demonstrated a quite better performance.

- **Boundaries**

Same as Example 12.4. Only constant temperatures of 21°C were assigned to the internal and external surfaces of the wall.

- **Convergence**

Nearly, 3500 iterations were conducted to achieve a convergence level of 10^{-7} for each equation as also shown in Figure 12.E10.2.

- **Mesh independency test:**

Same as Example 12.4.

- **Results:**

Here, we can see the velocity vectors embedded in the temperature contours of the air around the wall in Figure 12.E10.3.

For the external CHTC, first, we use TARP algorithm ($h_{\text{TARP}} = h_f + h_n$) (see Section 7.4.3). The concrete surface with the width of unity is exposed to the environment and thus it is in a

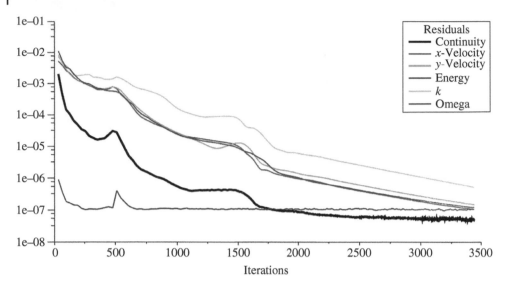

Figure 12.E10.2 Residual plot against iterations.

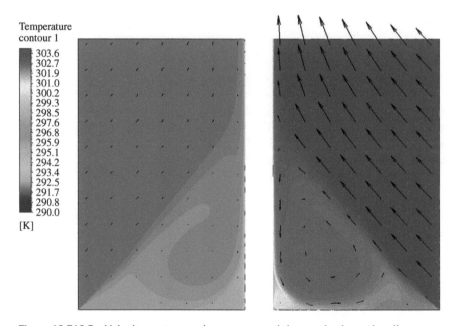

Figure 12.E10.3 Velocity contour and vectors around the non-isothermal wall.

windward position ($W_f = 1$). It is a medium rough surface ($R_f = 1.52$). Thus, the forced component of the equation can be found as:

$$h_f = 2.537 W_f R_f \left(\frac{PV_z}{A}\right)^{\frac{1}{2}}$$

$$\Longrightarrow h_f = 2.537 \times 1 \times 1.52 \left(\frac{2 \times (2.4 + 1) \times 2.5}{2.4 \times 1}\right)^{\frac{1}{2}} = 10.26 \, \text{W/m}^2\text{K}$$

And, for the natural component of the external vertical wall:

$$h_n = 1.31|\Delta T|^{\frac{1}{3}}$$

$$\Longrightarrow h_n = 2.82 \text{ W/m}^2\text{K}$$

And thus:

$$h_{\text{TARP}} = h_f + h_n$$

$$\Longrightarrow h_{\text{TARP}} = 13.09 \text{ W/m}^2\text{K}$$

For the internal wall, we use the below equation from Khalifa model (non-heated walls – see Section 7.5.1):

$$h_{\text{room}} = 2.30|\Delta \text{T}|^{0.24}$$

$$\Longrightarrow h_K = 2.72 \text{ W/m}^2\text{K}$$

Now, we can calculate the averaged h-values on both internal and external walls as demonstrated in Figure 12.E10.4 using CFD with the calculation of the total convective heat fluxes of 9.539 W and 243.70 W on the internal and external surfaces, respectively. Hence, for the internal wall, we can derive:

$$q_{\text{conv}-i} = h_{si}A\Delta T = 9.539 \ W$$

$$\Longrightarrow h_{si} = \frac{q_{\text{conv}-i}}{A\Delta T} = \frac{9.54}{2.4 \times (21 - 19)} = 1.99 \text{ W/m}^2 \text{ K}$$

And, for the external wall, we can find the convective heat flux as:

$$h_{so} = \frac{q_{\text{conv}-e}}{A \times \Delta T} = \frac{243.70}{2.4 \times (31 - 21)} = 10.15 \text{ W/m}^2 \text{ K}$$

Now, we can compare the average of both methods together and find a discrepancy of 36.7% and 28.9% for h-value of the indoor and outdoor spaces, respectively. We can, nevertheless, notice from Figure 12.E10.4 that CFD provides accurate local results due to a high-resolution solution of the airflow around the walls rather than using a bulk value as utilized in simplified correlations such as Khalifa and TARP. The local h-value is important in the design of natural and mechanical ventilations in addition to other applications in buildings.

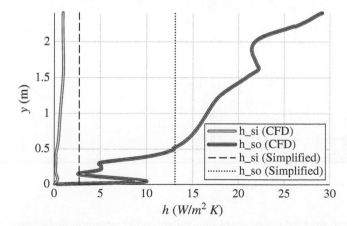

Figure 12.E10.4 Internal and external convective heat transfer coefficients using CFD and simplified models of Chapter 7.

Example 12.11

The office of Example 12.6 is heated with a mixed ventilation system during winter as shown in Figure 12.E11.1. The heating system consists of an electrical underfloor heating system, which can provide a constant heat flux of 55 W/m², and a mechanical ventilation system, which provides fresh air with the temperature of 17–19°C and velocity of 0.3–0.8 m/s. Calculate the required setting for the mechanical ventilation system to ensure the average room temperature of around 20–21°C while ensuring an average velocity of 0.2 m/s in the occupant zone as it is defined here as the lower half of the office where occupants are seated.

Solution

- **Employed Software:**

  ANSYS Fluent

- **Mesh and quality:**

  Same as Example 12.6.

- **Setup:**
- **Model**

  Same as Example 12.6, except the energy equation in the Navier–Stokes equations, which is solved due to a non-isothermal condition using an ideal gas assumption for air. The turbulence model was set as Realizable $k - \varepsilon$ adapted from the related literature of similar CFD simulations.

- **Boundaries (heated floor)**

  Same as Example 12.6. A line source term was set on the lower surface to mock the heater.

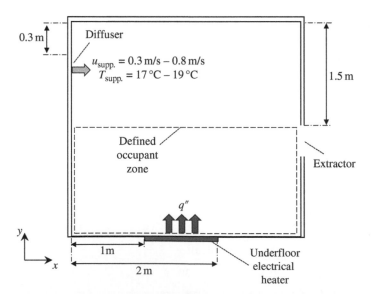

Figure 12.E11.1 An office with an electrical underfloor heating system and a mechanical ventilation system.

- **Convergence**

Convergence was achieved after nearly 10,000 iterations (see Figure 12.E11.2) to ensure that all the residual are below 10^{-6}. It should be noted that many quantities reached their iterative convergence way before this number.

- **Mesh independency test:**

Same as Example 12.6.

- **Results:**

Using the chosen mesh in the previous section, we run the simulation nine times by changing the inlet velocity as 0.3, 0.55, and 0.8 m/s and inlet temperature as 17, 18, and 19°C. These values are demonstrated in the below table. Note that these values are selected in accordance with the defined ranges for inlet flow velocity and temperature. Evidently, one can increase the accuracy of the results by adding more interval to the simulations while computational resource should always be assessed before starting multiple CFD simulations.

Case	Inlet velocity (m/s)	Inlet temperature (°C)	Occupied zone averaged temperature (°C)	Occupied zone averaged velocity (m/s)
1	0.3	17	21.0	0.09
2		18	22.0	0.09
3		19	23.0	0.09
4	0.55	17	19.0	0.21
5		18	20.0	0.21
6		19	21.0	0.21
7	0.8	17	18.3	0.32
8		18	19.3	0.32
9		19	20.3	0.32

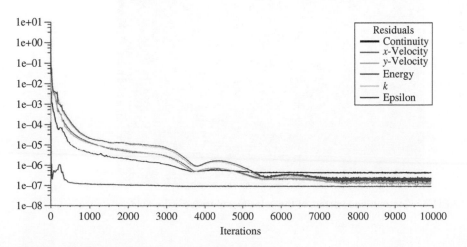

Figure 12.E11.2 Residual plot against iterations.

The surface average of temperature and velocity is calculated within the occupied zone defined as the lower half of the domain. The values are provided in the above table while the acceptable temperatures and velocities are highlighted as well, which are only Case 5 and Case 6. Furthermore, the velocity vectors contrasted with the temperature contours in addition to only the temperature contours for Case 5 are shown in Figure 12.E11.3. It can be concluded that the proposed design obviously results in the formation of a circulation in the room and occupied zone while the temperature remains in an acceptable range within most parts of the zone.

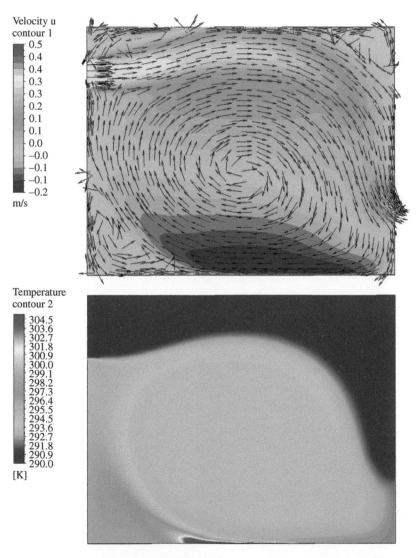

Figure 12.E11.3 Contours and vectors of the velocity and temperature fields.

Example 12.12

Assume the sheltered 2D cross-ventilation case of Example 12.8 with $L = 4$ m. As shown in Figure 12.E12.1, we add two occupants to the room with simplified geometries. Also, we simplify the pollution from cars with a constant source-point emission of NO with a half-sinusoidal flow rate of 0.024 kg/s (the release velocity is 0.1 m/s) and a constant temperature of 310 K. This is calculated to mock cars passing from the street with a 2-minute period. So, the unsteady release of pollutant can be expressed as $0.024 \sin\left(\frac{\pi}{60}t\right)$.

Now, develop a CFD model and evaluate the newly created mesh to ensure that it is still following almost a similar result as Example 12.8. Wind profile is approximated using the power-law with a constant temperature of $25°$C. If each occupant generates 80 W of heat, investigate the pollution concentration within the room using three different scenarios:

1) Cross-ventilation I (CV-I): when the wind is approaching from the left-hand side of the domain.

2) Cross-ventilation II (CV-II): when the wind is approaching from the right-hand side of the domain.

3) Single-sided ventilation (SV): when the wind is approaching from the right-hand side of the domain and the right-hand side opening in the target building is closed.

Solution

- **Employed Software:**

ANSYS Fluent

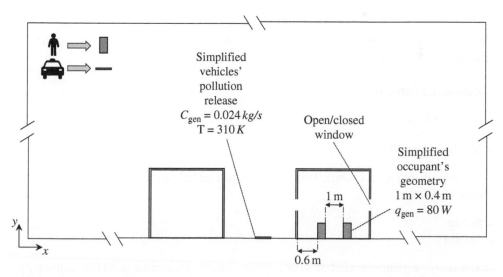

Figure 12.E12.1 2D pollution dispersion from a street canyon towards a sheltered building with occupants.

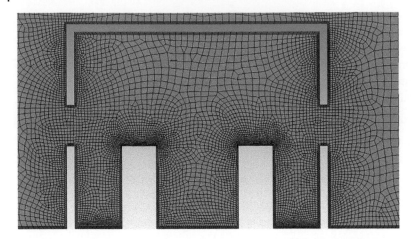

Figure 12.E12.2 A mesh with structured quadrilateral cells in boundary layer and unstructured quadrilateral cells in the domain.

- **Mesh and quality:**

As illustrated in Figure 12.E12.2, the mesh is generated structured quadrilateral in boundary layer and unstructured quadrilateral in the domain with the following quality:

Mesh parameters		Coarse	Medium	Fine
Cell size/(first layer)		0.01 m		
Growth ratio		1.08		
Cell numbers		33,394	48,414	78,248
Mesh	Aspect ratio	1.60	1.41	1.25
Quality	Orthogonality	0.97	0.98	0.99

- **Employed Software:**

ANSYS Fluent

- **Mesh and quality:**

Same as Example 12.8. Only, two occupants are added to the domain as depicted in Figure 12.E12.2.

- **Setup:**
- **Model**

The energy equation in addition to one species (quantity) equation as NO are activated to participate in the simulation. The air is assumed as an ideal gas. The turbulence model is selected as $RNG\ k - \varepsilon$ with standard wall-function for the wall treatment. The solution is transient for 10 minutes (five cycles) with a function of $0.024 \sin\left(\dfrac{\pi}{60}t\right)$. Each time step is set as 0.5 second with a maximum 20 iterations.

- **Boundaries (heated floor)**

The occupants' surfaces are set with a constant flux to replicate a $80W$ heat release from them. A similar boundary is set for the pollution source assigned to the line-source within the species (NO quantity) equation.

- **Convergence**

All residuals are obtained around 10^{-6} at each time step as the transient iterative convergence is shown in Figure 12.E12.3.

- **Mesh independency test:**

Similar to the previous examples, three meshes were generated with the following characteristics. The medium mesh again demonstrates a close match with the fine mesh as its mean discrepancy can be calculated as 3.7% (see Figure 12.E12.4). This is a valid point in almost all three lines across the studied ventilated room. It should be noted that in this case the difference between the medium and coarse meshes is also about 4.1%, which is a very acceptable number. Hence, we can conclude that both meshes can be utilized for further investigation of the proposed scenarios in this example.

Mesh resolution	Coarse	Medium	Fine
Opening edge (number of elements)	3	3	3
Edge element (building wall)	40	48	56
Volume mesh size around the room (m)	0.10	0.08	0.06
Volume mesh size far from room (m)	0.20	0.16	0.12
Number of cell elements	33,394	48,414	78,248

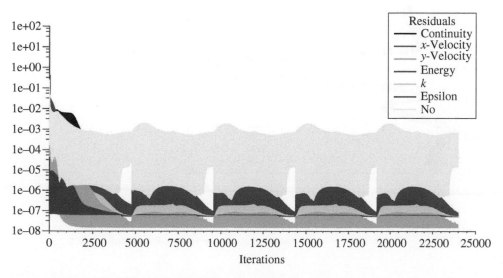

Figure 12.E12.3 Residual plot against iterations in different time steps.

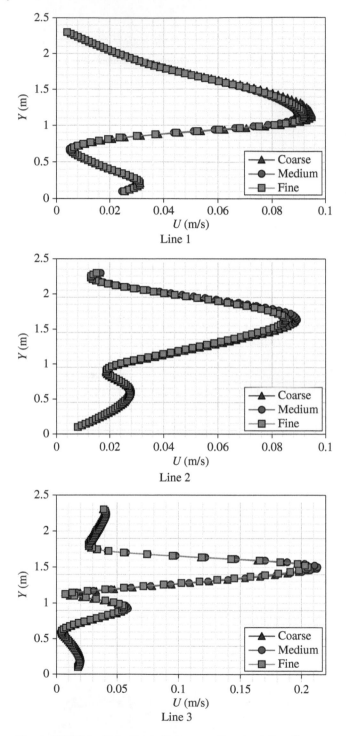

Figure 12.E12.4 Velocity profiles across the simulation line-sensors for various meshes.

- **Results:**

The velocity contours are demonstrated in Figure 12.E12.5 for all three scenarios. It is not difficult to determine that the cross-ventilation is more effective when the building is not sheltered against the upstream (CV-II). On the other hand, circulations in the size of the street canyon can be observed in other scenarios. Hence, it can be observed that the circulation in between two buildings in CV-I is mainly pushing pollution outside of the street canyon while in SV, the circulation directs the pollutants towards the room.

We can realize from Figure 12.E12.6 that the pollution concentration shown as iso-line distributions, as expected in the previous figure, is higher in scenario SV. While the flow direction is the essential parameter in the dispersion of the pollutant through the room, the design and place of openings such as windows and doors is a more important parameter in the quality of natural ventilation in buildings. As we could see in this example, CFD is powerful in the design and control of ventilation strategies in buildings.

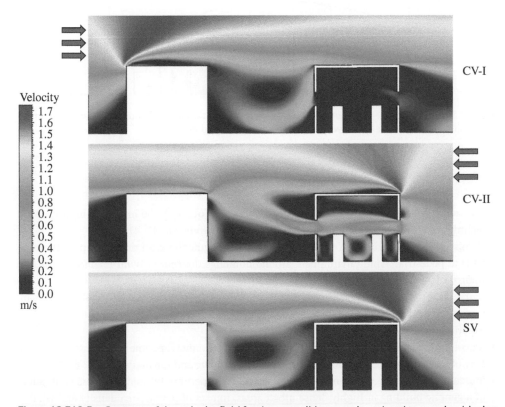

Figure 12.E12.5 Contours of the velocity field for three possible scenarios related to opening/closing of the windows and alteration of the wind direction.

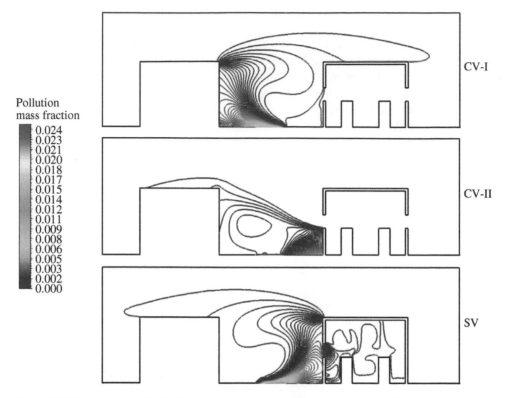

Pollution
mass fraction

0.024
0.023
0.021
0.020
0.018
0.017
0.015
0.014
0.012
0.011
0.009
0.008
0.006
0.005
0.003
0.002
0.000

Figure 12.E12.6 Iso-line distributions of pollution concentration for three possible scenarios related to opening/closing of the windows and alteration of the wind direction.

References

1 Mirzaei, P.A. and Haghighat, F. (2012). A procedure to quantify the impact of mitigation techniques on the urban ventilation. *Building and Environment.* 47: 410–420.

2 Tominaga, Y., Mochida, A., Yoshie, R. et al. (2008). AIJ guidelines for practical applications of CFD to pedestrian wind environment around buildings. *Journal of Wind Engineering and Industrial Aerodynamics.* 96 (10): 1749–1761.

3 Shirzadi, M., Mirzaei, P.A., and Tominaga, Y. (2020). CFD analysis of cross-ventilation flow in a group of generic buildings: Comparison between steady RANS, LES and wind tunnel experiments. *Building Simulation.* 13 (6): 1353–1372.

4 Shirzadi, M., Tominaga, Y., and Mirzaei, P.A. (2019). Wind tunnel experiments on cross-ventilation flow of a generic sheltered building in urban areas. *Building and Environment.* 158: 60–72.

5 Shirzadi, M., Tominaga, Y., and Mirzaei, P.A. (2020). Experimental study on cross-ventilation of a generic building in highly-dense urban areas: impact of planar area density and wind direction. *Journal of Wind Engineering and Industrial Aerodynamics.* 196: 104030.

6 Shirzadi, M., Tominaga, Y., and Mirzaei, P.A. (2020). Experimental and steady-RANS CFD modelling of cross-ventilation in moderately-dense urban areas. *Sustainable Cities and Society.* 52: 101849.

Index

Computational Fluid Dynamics and Energy Modelling in Buildings: Fundamentals and Applications,
First Edition. Parham A. Mirzaei.
© 2023 John Wiley & Sons Ltd. Published 2023 by John Wiley & Sons Ltd.